advanced

LEVEL

study aids

physics

Brian Arnold and Celia Bloor

JOHN MURRAY

Titles in this series:

Advanced Level Study Aids Biology	0 7195 7630 X
Advanced Level Study Aids Chemistry	0 7195 7631 8
Advanced Level Study Aids Physics	0 7195 7629 6

First published by John Murray Publishers Ltd, a member of the Hodder Headline Group
338 Euston Road
London NW1 3BH

Reprinted 2002, 2003, 2004

Layouts by Amanda Easter
Artwork by Oxford Illustrators & Designers
Cover design by John Townson/Creation

Typeset in 10/12 Garamond by Wearset, Boldon, Tyne and Wear
Printed and bound in Great Britain by St Edmundsbury Press Ltd, Bury St Edmunds

A catalogue entry for this title is available from the British Library

ISBN 0 7195 7629 6

Contents

Acknowledgements

I would like to thank all the students who tackled the draft versions of the questions and made helpful comments and the teachers who encouraged them. I acknowledge the influence of all the colleagues I have known at the schools and colleges where I have worked over the years, and appreciate the patience and understanding of those I left to get on with their jobs while I disappeared on sabbatical to produce the first drafts of this book, in which, without my husband's patient checking there would probably be many more errors. CB

I would like to thank all those who have helped me during the writing of this book: friends and colleagues at school who supported me when there didn't seem to be enough hours in the day to complete the task at hand. Finally I would like to thank my wife Jill, who throughout the writing of the book provided the encouragement I needed. BA

We would both like to acknowledge the support of Katie Mackenzie Stuart at John Murray Publishers for co-ordinating and overseeing the project, and Jane Roth for making many very valuable comments and suggestions.

We would also like to acknowledge our sincere gratitude to the author Don Mackean and Linda Ellis from the Schools Improvement and Advisory Service, Hertfordshire, who through the Hertfordshire Science Teaching Scholarship provided the time and opportunity to complete much of this work.

Thanks are due to the following for permission to reproduce copyright photographs: **p.95** Professor Henry A. Hill, University of Arizona.

The Publishers have made every effort to contact copyright holders. If any have been overlooked they will be pleased to make the necessary arrangements at the earliest opportunity.

AS and A level specification matrix

Board	AQA (A)					AQA (B)				Edexcel				OCR				WJEC			
Module	**1**	**2**	**3**	**4**	**5**	**1**	**2**	**4**	**5**	**1**	**2**	**4**	**5**	**2821**	**2822**	**2823**	**2824**	**PH1**	**PH2**	**PH4**	**PH5**
Topic 1	✓	✓	✓	✓	✓	✓	✓	✓	✓	✓	✓	✓	✓	✓	✓	✓	✓	✓	✓	✓	✓
2			✓			✓					✓				✓				✓		
3			✓			✓					✓				✓				✓		
4			✓			✓					✓				✓				✓		
5			✓			✓					✓				✓				✓		
6		✓				✓				✓				✓				✓			
7		✓				✓				✓				✓				✓			
8		✓				✓				✓				✓				✓			
9		✓				✓				✓				✓			✓	✓		✓	
10		✓				✓		✓		✓				✓			✓	✓			
11				✓			✓					✓				✓		✓			
12	✓															✓		✓			
13				✓			✓					✓				✓		✓			
14				✓			✓					✓				✓		✓			
15				✓			✓					✓				✓		✓			
16	✓							✓				✓			✓				✓		
17	✓						✓					✓			✓				✓		
18		✓						✓			✓						✓				
19		✓						✓			✓						✓				
20		✓						✓			✓						✓			✓	
21		✓						✓			✓						✓			✓	
22								✓			✓						✓			✓	
23				✓	✓		✓			✓							✓		✓		
24					✓		✓			✓							✓				
25				✓					✓												
26			✓			✓		✓						✓						✓	
27								✓				✓					✓			✓	
28				✓				✓				✓					✓			✓	
29				✓				✓					✓				✓			✓	
30				✓				✓	✓								✓			✓	
31				✓					✓				✓				✓				✓
32				✓					✓				✓				✓				✓
33				✓					✓				✓				✓				✓
34				✓					✓				✓				✓				✓
35				✓					✓				✓				✓				✓

*Modules in tinted columns make up the A2 part of the course

AS and A level specification matrix

Board	Salters Horners				Advancing Physics			
Module	**1**	**2**	**4**	**5**	**2860**	**2861**	**2863**	**2864**
Topic 1	✓	✓	✓	✓	✓	✓	✓	✓
2	✓		✓		✓			
3	✓				✓			
4	✓				✓			
5	✓	✓			✓			
6	✓					✓		
7	✓					✓		
8	✓					✓		
9	✓		✓			✓	✓	
10	✓		✓	✓		✓		
11	✓					✓		
12	✓	✓	✓	✓	✓			
13	✓	✓				✓		
14		✓				✓		
15	✓					✓		
16		✓				✓		
17	✓					✓		✓
18	✓							
19								
20				✓			✓	
21				✓			✓	
22							✓	
23			✓	✓				✓
24				✓			✓	
25			✓	✓				✓
26		✓		✓	✓			
27			✓	✓		✓		
28				✓			✓	
29			✓				✓	
30			✓				✓	
31				✓			✓	
32			✓					✓
33			✓				✓	✓
34			✓					✓
35			✓					✓

*Modules in tinted columns make up the A2 part of the course

Introduction

Students may readily grasp concepts and models in post-16 Physics courses but it is often the calculations and, to a lesser extent, the practical applications associated with these principles that cause difficulties. The aim of this book is to provide you the student with opportunities to practise and strengthen these skills, leading to higher levels of achievement and improved self-confidence.

The book is arranged by topic. Within each topic there is

- a short summary of the relevant theory
- a series of graded questions based upon the theory
- worked solutions with full explanations for all questions – not just the numerical ones.

Topics found at AS level are dealt with in the first part of the book (see the specification matrix on pages vii and viii). All the main topics included in the various A level specifications are covered and the final section includes a range of synoptic questions. You should revisit the AS and A level sections before you tackle the synoptic questions. Optional topics have not been included because they are dealt with in depth in separate textbooks.

One of the ideas behind the book is that you will test yourself by developing your own answers before you look at the worked solutions. Remember that there may be several different ways of tackling a question, and of getting the right answer. If you need to build up your confidence, start working on the easier questions.

Many of the questions test your ability to rearrange equations because this is a very important mathematical procedure. Units have been included throughout to aid your understanding of what the equations are doing. It is always a useful exercise to check that the units are correct as you progress through a calculation.

There are many ways in which this book might be used. It could be used as a course companion, allowing you to practise tackling questions on topics you are at present studying. The summary text and the questions can be used to catch up on lessons unavoidably missed. Alternatively it could be used as part of a revision programme, offering you the opportunity to improve your understanding of a topic of your choice, while at the same time sharpening your exam technique.

1 Units and orders of magnitude

Units

Scientists around the world use the same internationally agreed system of units. These are called SI (Système International) units. The system is built upon seven **base units**.

- The metre (m) is the base unit of length.
- The kilogram (kg) is the base unit of mass.
- The second (s) is the base unit of time.
- The ampere (A) is the base unit of electrical current.
- The kelvin (K) is the base unit of temperature.
- The mole (mol) is the base unit of the amount of a substance.
- The candela (cd) is the base unit of light intensity.

Quantities such as speed ($\mathrm{m\,s^{-1}}$) and density ($\mathrm{kg\,m^{-3}}$) which are not expressed in a single base unit are expressed in **derived units**. Table 1.1 shows some of the common derived units of the SI system.

Table 1.1 Derived units

Quantity	Symbol	Name of unit	Symbol for unit	Base units
speed or velocity	v		$\mathrm{m\,s^{-1}}$	$\mathrm{m\,s^{-1}}$
acceleration	a		$\mathrm{m\,s^{-2}}$	$\mathrm{m\,s^{-2}}$
force	F	newton	N	$\mathrm{kg\,m\,s^{-2}}$
energy	E	joule	J	$\mathrm{kg\,m^2\,s^{-2}}$
power	P	watt	W	$\mathrm{kg\,m^2\,s^{-3}}$
pressure	p	pascal	Pa	$\mathrm{kg\,m^{-1}\,s^{-2}}$
frequency	f	hertz	Hz	$\mathrm{s^{-1}}$
charge	Q	coulomb	C	A s
potential difference	V	volt	V	$\mathrm{A^{-1}\,kg\,m^2\,s^{-3}}$
resistance	R	ohm	Ω	$\mathrm{A^{-2}\,kg\,m^2\,s^{-3}}$
capacitance	C	farad	F	$\mathrm{A^2\,kg^{-1}\,m^{-2}\,s^4}$
magnetic flux	B	tesla	T	$\mathrm{A^{-1}\,kg\,s^{-2}}$

Quantities which have no units are described as being dimensionless – for example the refractive index of a material is a ratio of like quantities and therefore has no unit.

The base units of any quantity can be found by considering an equation that relates it to quantities whose units are known.

Example 1

Show that the base units of density are $\mathrm{kg\,m^{-3}}$.

$$\text{density} = \frac{\text{mass}}{\text{volume}}$$

so the units are $\dfrac{\mathrm{kg}}{\mathrm{m^3}}$ or $\mathrm{kg/m^3}$

This is more usually written as $\mathrm{kg\,m^{-3}}$ – division is indicated by a negative index.

Homogeneity of an equation

If an equation is written correctly it must be **homogeneous**; that is, the units of the quantities on the left hand side of the equation must be identical to those on the right hand side.

Example 2

The equation $F = mv^2/r$ describes the relationship between the force applied to an object of mass m so that it travels in a circle of radius r with speed v. Show that it is homogeneous.

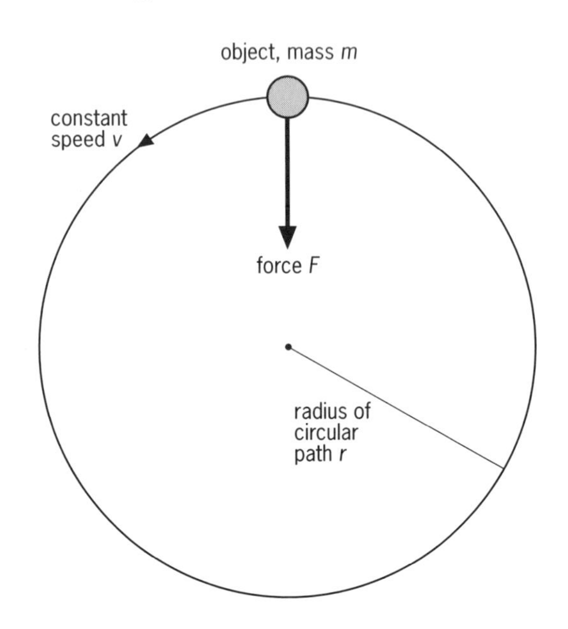

Figure 1.1

The units of the quantities are

F $\mathrm{N = kg\,m\,s^{-2}}$

m kg

v $\mathrm{m\,s^{-1}}$

r m

Therefore we can write

$$F = \frac{mv^2}{r}$$

$$\mathrm{kg\,m\,s^{-2}} = \frac{\mathrm{kg\,(m\,s^{-1})^2}}{\mathrm{m}}$$

$$= \frac{\mathrm{kg\,m^2\,s^{-2}}}{\mathrm{m}}$$

$$= \mathrm{kg\,m\,s^{-2}}$$

The equation is homogeneous. The powers of the units kg, m and s are the same on both sides of the equation.

Prefixes

Prefixes are used with units to change numerical values into a more convenient form. For example, the energy contained in a flash of lightning is approximately one thousand million joules or 1×10^9 J. Using a prefix this becomes 1 GJ (one gigajoule). Table 1.2 shows the most commonly used prefixes.

Table 1.2 Unit prefixes

Multiple	Prefix	Symbol	Example
10^{-12}	pico	p	pF picofarad
10^{-9}	nano	n	nm nanometre
10^{-6}	micro	μ	μA microamp
10^{-3}	milli	m	ms millisecond
10^{3}	kilo	k	kg kilogram
10^{6}	mega	M	MJ megajoule
10^{9}	giga	G	GW gigawatt

Note that $10^0 = 1$

Standard form

It is accepted practice to express numerical quantities as a number between 1 and 10 multiplied by the appropriate power of 10. This is known as standard form. Some examples are given in Table 1.3.

Table 1.3 Examples of standard form

Number	Number in standard form
1000	1.0×10^3
140 000 000	1.4×10^8
128 600	1.286×10^5
0.015	1.5×10^{-2}
0.003 86	3.86×10^{-3}

Significant figures

At first glance the values 2 cm, 2.0 cm and 2.00 cm may appear identical. There is, however, a very important difference between them. The first value is given to just one **significant figure**. This indicates that the true value of this length lies between 1.5 cm and 2.4 cm. The second value is given to two significant figures; the true value of this length lies between 1.95 cm and 2.04 cm. The third value is the most precise of the figures as it is given to three significant figures, indicating that its true value lies between 1.995 cm and 2.004 cm. The greater the number of significant figures given, the greater the implied precision of the measurement.

Note that the number of significant figures given and the number of decimal places are not necessarily the same. Writing the above values in m or km would alter the number of decimal places but not the number of significant figures or the implied precision of the values.

When calculating, the final answer should not be stated to more significant figures than the least precise of the given figures.

Estimating

Sometimes it is useful to have a rough idea of the likely answer to a problem before a detailed calculation or experiment is carried out. The secret to making a successful estimation is to spot the principles or equations that need to be applied to the problem and to estimate the size of the quantities involved. The skill of spotting the shortest and simplest route is one that grows with practice.

Example 3

Estimate the average power of a man who climbs from 2000 m to the top of Mount Everest in 3 days.

$$\text{power} = \frac{\text{work done}}{\text{time taken}}$$

$$= \frac{\text{change in potential energy}}{\text{time taken}}$$

$$= \frac{\Delta(mgh)}{t} \quad \text{or} \quad \frac{mg\,\Delta h}{t}$$

(see topic 10)

approximate mass of average man, $m = 80\,\text{kg}$

$g = 10\,\text{m}\,\text{s}^{-2}$

approximate height of Everest $= 10\,000$ m

$\therefore$ height climbed, $h = 10\,000 - 2000 = 8000$ m

time, $t = 3$ days $= 72$ hours $= 72 \times 60 \times 60$ s

$$\therefore \quad P \approx \frac{80 \times 10 \times 8000}{72 \times 60 \times 60}$$

$$\approx 25\,\text{W}$$

Often it is useful to estimate a quantity as an **order of magnitude**; that is, as a power of 10. Table 1.4 gives some examples.

Table 1.4 Orders of magnitude of some lengths

Quantity	Order of magnitude of length/m
distance to the boundary of the observable Universe	10^{28}
distance to the nearest star	10^{16}
distance to the Sun	10^{11}
wavelength of visible light	10^{-7}
diameter of an atom	10^{-10}

Level 1

1 These powers of 10 are used to show the order of magnitude of the listed physical quantities in SI units. However, the numbers and physical quantities have been muddled up. Match each power of 10 to a quantity.

10^{-31}	wavelength of visible light
10^{-27}	speed of light
10^{-19}	wavelength of radio waves
10^{-7}	mass of proton
10^{0} (= 1)	mass of electron
10^{8}	number of particles in a mole
10^{23}	charge on an electron

2 *Estimate* the mass of water in an average public swimming pool.

3 What are the base units of the volt?

4 An electric current, I, is proportional to $A \times v \times e$ where A is the cross-sectional area of a wire, v is the drift velocity of electrons in the wire and e is the charge on an electron.

a Find the base units of the constant of proportionality, k, in the equation

$$I = k \times A \times v \times e$$

b Explain the physical significance of the constant k.

5 Use a sequence of word equations to find the base units for the watt.

Level 2

6 Show that the equation

$$s = u \times t + \tfrac{1}{2} \times a \times t^2$$

is homogeneous.

7 Show that the equation

$$T = 2\pi \sqrt{\frac{l}{g}}$$

is homogeneous.

8 Which of these proportionalities is correct for the speed of sound?

Hint: Check the units on either side.

A $v \propto \dfrac{\text{pressure} \times \text{wavelength}}{\text{density}}$

B $v \propto \dfrac{\text{pressure}}{\text{density}}$

C $v \propto \sqrt{\dfrac{\text{pressure}}{\text{density}}}$

Level 3

9

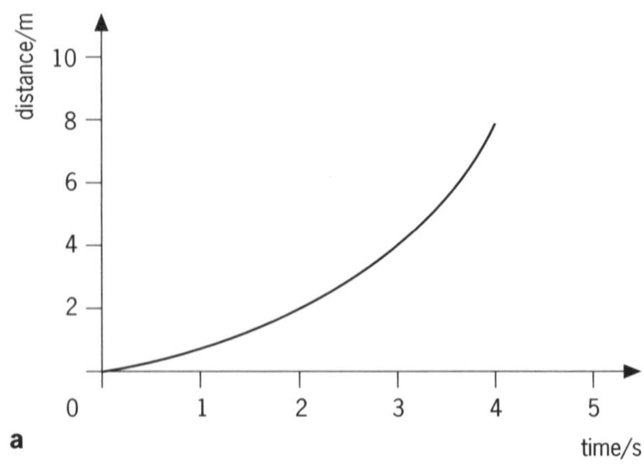

a

b

Figure 1.2

From the graphs in Figure 1.2 find

a the speed at 3.0 s (see topic 8)

b the resistance (see topic 4) when the current is 3.0 A.

Explain your method in each case.

10 The equation for the gravitational attraction between two masses, M_1 and M_2, a distance r apart, is

$$F = \frac{G \times M_1 \times M_2}{r^2}$$

What are the base units of the universal gravitational constant, G?

11 The power needed for a car to keep up a steady speed against air resistance may be calculated using the equation

$$P = 0.15 \times \rho \times v^3 \times A$$

where ρ is the air density, v is the velocity of the car and A is the frontal cross-section of the car. *Estimate* this power for a car on the motorway travelling at 70 mph, given that the density of air is about $1\,\text{kg}\,\text{m}^{-3}$ and that $10\,\text{mph} = 4.5\,\text{m}\,\text{s}^{-1}$.

12 Use the definition of electrical resistance (see topic 4) to work out its base units.

Level 1

1

10^{-31}	mass of electron (9.1×10^{-31} kg)
10^{-27}	mass of proton (1.6×10^{-27} kg)
10^{-19}	charge on an electron (1.6×10^{-19} C)
10^{-7}	wavelength of visible light ($4–7 \times 10^{-7}$ m)
10^{0}	wavelength of radio waves (e.g. 3 m)
10^{8}	speed of light (3×10^{8} m s^{-1})
10^{23}	number of particles in a mole (6.0×10^{23})

You should be familiar with these values by the end of your course.

2 If the pool is say $10\,\text{m} \times 25\,\text{m} \times 3\,\text{m}$, the volume, V, would be $750\,\text{m}^3$. If it is 33 m long, this goes up to about $1000\,\text{m}^3$ which is a nice round number. You should know that the density, ρ, of water is $1000\,\text{kg}\,\text{m}^{-3}$.

$$\text{density} = \frac{\text{mass}}{\text{volume}} \quad \text{or} \quad \rho = \frac{m}{V}$$

$$\therefore m = \rho V = 1000\,\text{kg}\,\text{m}^{-3} \times 1000\,\text{m}^3$$

So the mass is of the order of **10^6 kg** or **1000 tonnes** (1 tonne = 1000 kg).

3
$$\text{Volt} = \frac{\text{joule}}{\text{coulomb}} \quad \text{where joule} = \text{newton} \times \text{metre}$$
$$= \text{kg} \times \text{m}\,\text{s}^{-2} \times \text{m}$$
$$\text{and coulomb} = \text{ampere} \times \text{second}$$
$$= \text{A} \times \text{s}$$

$$\text{So volt} = \frac{\text{kg} \times \text{m} \times \text{s}^{-2} \times \text{m}}{\text{A} \times \text{s}} = \mathbf{A^{-1} \times kg \times m^2 \times s^{-3}}.$$

4 a The left hand side of the equation has the unit ampere (A) and the quantities given on the right hand side have units

A	v	e
m^2	$\text{m}\,\text{s}^{-1}$	C

but C = A × s, so the base units of the full equation are

$$I = k \times A \times v \times e$$
$$\text{A} = ? \times \text{m}^2 \times \text{m}\,\text{s}^{-1} \times \text{A}\,\text{s}$$
$$\text{A} = ? \times \text{m}^3 \times \text{A}$$

The constant k must have units of $\mathbf{m^{-3}}$ to make it homogeneous.

b The constant is the number of charge carriers per unit volume of the material (see topic 2) so m^{-3} is correct.

5 You can do this using either the equations involving physical quantities or the equations involving units. Both versions are given, starting from the knowledge that the watt is the unit of power.

Using quantities:

$$\text{power} = \frac{\text{work}}{\text{time}}$$

work = force × distance
force = mass × acceleration

$$\text{acceleration} = \frac{\text{change in speed}}{\text{time taken}}$$

$$\text{where speed} = \frac{\text{distance}}{\text{time}} = \text{m} \times \text{s}^{-1}$$

$$\text{so power} = \frac{\text{mass} \times \text{acceleration} \times \text{distance}}{\text{time}}$$
$$= \text{kg} \times \text{m} \times \text{s}^{-2} \times \text{m} \times \text{s}^{-1}$$

Using units:

$$\text{watt} = \frac{\text{joule}}{\text{second}}$$

joule = newton × metre
newton = kg m s^{-2}
joule = kg × m × s^{-2} × m
so watt = kg × m × s^{-2} × m × s^{-1}

So, either way, the units of power are kg × m^2 × s^{-3}.

Level 2

6 s is distance so the unit is m. We have to show that **each of the two terms** on the right hand side of the equation also has the unit of distance.

$(u \times t)$ is (speed × time) which has units $\text{m} \times \text{s}^{-1} \times \text{s} = \text{m}$

$(\frac{1}{2} \times a \times t^2)$ consists of a dimensionless constant, then $\text{m} \times \text{s}^{-2} \times \text{s}^2 = \text{m}$

We have shown that both terms on the right hand side of the equation have dimensions of length, or unit m, so the right hand side of the equation has the same unit as the left hand side.

7 T is a time and has the unit s. We have to show that the right hand side of the equation has the same unit.

2π is dimensionless
g is acceleration, which has units $\text{m} \times \text{s}^{-2}$
l is a length, which has unit m

$$\text{So } \sqrt{\frac{l}{g}} \text{ will have units } \sqrt{\frac{\text{m}}{\text{m} \times \text{s}^{-2}}} = \sqrt{\text{s}^2} = \text{s}$$

as required.

8 The expression must have the units of speed, which are $\text{m} \times \text{s}^{-1}$.

pressure has units $\text{Pa} = \text{N} \times \text{m}^{-2} = \text{kg} \times \text{m} \times \text{s}^{-2} \times \text{m}^{-2}$
$= \text{kg} \times \text{m}^{-1} \times \text{s}^{-2}$

density has units $\text{kg} \times \text{m}^{-3}$

A $\dfrac{\text{pressure} \times \text{wavelength}}{\text{density}}$ has units

$$\frac{\text{kg} \times \text{m}^{-1} \times \text{s}^{-2} \times \text{m}}{\text{kg} \times \text{m}^{-3}} = \text{m}^3 \times \text{s}^{-2}$$

which is **wrong**.

B $\dfrac{\text{pressure}}{\text{density}}$ has units $\dfrac{\text{kg} \times \text{m}^{-1} \times \text{s}^{-2}}{\text{kg} \times \text{m}^{-3}} = \text{m}^2 \times \text{s}^{-2}$

which is also **wrong**.

It should be clear that this latter expression would have the correct units, $\text{m} \times \text{s}^{-1}$, if the square root were taken.

C $\sqrt{\dfrac{\text{pressure}}{\text{density}}}$ is therefore the **correct** choice.

Level 3

9

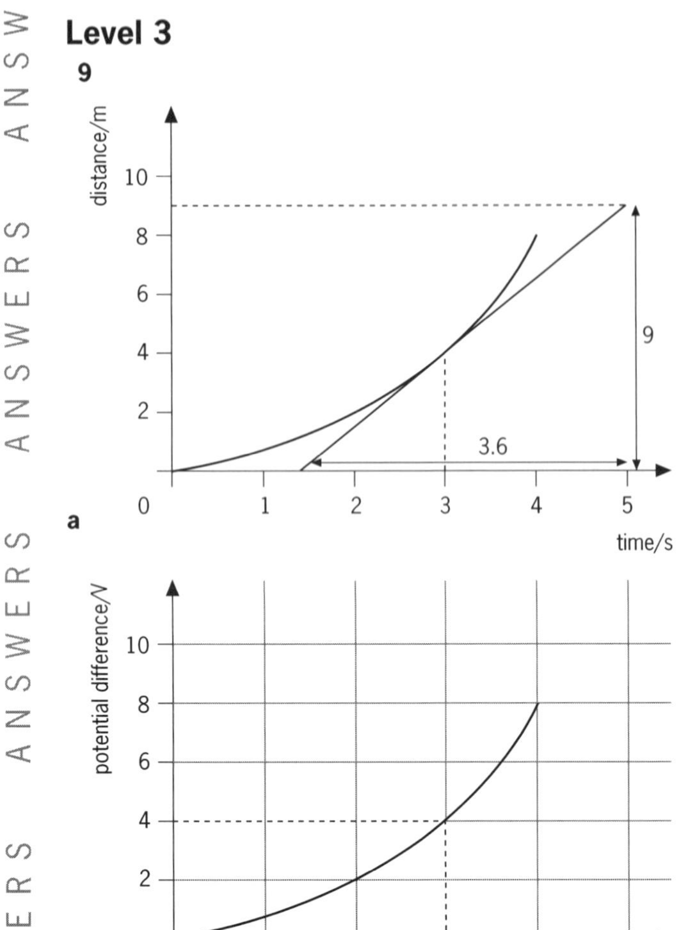

Figure 1.3

a Speed is the 'rate of change in distance with time',

$$\frac{\Delta s}{\Delta t}$$

To find the instantaneous speed you take the **gradient** of the curve at the point required (see topic 8) by drawing the tangent to the curve. The gradient at 3.0 s is **about 9/3.6 = 2.5 m s^{-1}**.

b Resistance is **not** the rate of change in potential difference with current; it is the **ratio** of potential difference to current (see topic 4) so you do **not** draw a tangent to the curve. The ratio is 4/3 = **1.3 Ω**.

10

$$F = \frac{G \times M_1 \times M_2}{r^2}$$

$$\therefore G = \frac{F \times r^2}{M_1 \times M_2}$$

The units are $\dfrac{\text{kg m s}^{-2} \times \text{m}^2}{\text{kg}^2}$

$$= \frac{\text{m}^3\,\text{s}^{-2}}{\text{kg}}$$

$$= \mathbf{kg^{-1}\,m^3\,s^{-2}}$$

11 The speed limit on motorways is 70 mph which is equivalent to about 30 m s^{-1}. (10 mph = 4.5 m s^{-1} so 70 mph = (70/10) × 4.5 = 31.5 m s^{-1}. As an estimate this is approximately 30 m s^{-1}.)

The front of a car is about 1.5 m wide by 1.0 m high, giving an area, A, of 1.5 m^2. So the power required,

$$P = 0.15 \times \rho \times v^3 \times A$$
$$= 0.15 \times 1 \times (30)^3 \times 1.5 \approx \mathbf{6\,kW}$$

12 By definition

$$R = \frac{V}{I} \text{ in units of } \frac{\text{joule/coulomb}}{\text{ampere}}$$

The joule has units of kg × m^2 × s^{-2} and the coulomb has units of A × s. Therefore R has units of

$$\frac{\text{kg} \times \text{m}^2 \times \text{s}^{-2} \times \text{A}^{-1} \times \text{s}^{-1}}{\text{A}}$$

$$= \mathbf{A^{-2} \times kg \times m^2 \times s^{-3}}$$

You could use your answer to question 3 to check.

2 Current and charge

An electric current is a flow of charge. In **metallic conductors** the charge is carried by electrons. The outer electrons of metal atoms are very weakly bound to the central, positive nucleus. Consequently some of them are able to move about between the atoms, creating a 'sea' of free electrons. Under normal circumstances the electrons in a piece of wire move haphazardly, so that the number of electrons moving in one direction equals the number moving in the opposite direction. There is therefore no net flow of charge.

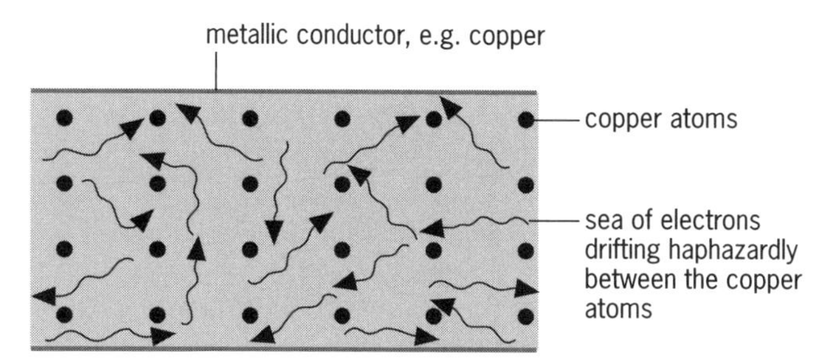

Figure 2.1 There is no net flow of charge and so no current through the conductor

If a potential difference is applied across the conductor more of the electrons move towards the positive terminal than in the opposite direction. There is therefore a net flow, or drift, of charge in one direction – there is a current.

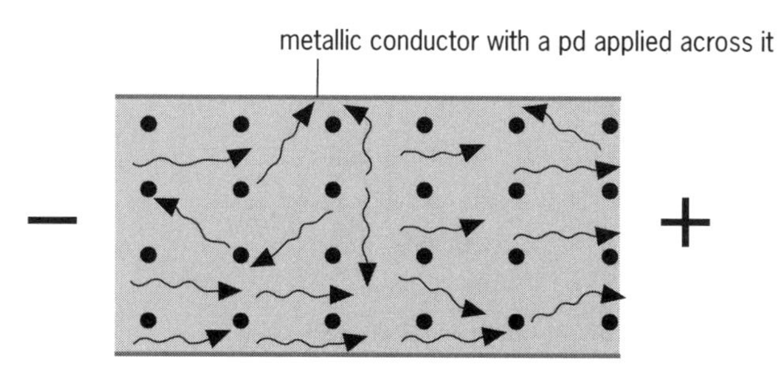

Figure 2.2 There is now an overall flow of charge in one direction – there is a current through the conductor

The size of an electric current, I, is a measure of the rate of flow of charge, Q, and is measured in the unit **ampere**. If the current in a wire is 1 ampere (A), then charge is flowing at the rate of 1 **coulomb** (C) per second. The charge carried by one electron is $e = -1.60 \times 10^{-19}$ C.

For steady currents

$$I = \frac{Q}{t}$$

where Q is the charge that flows past any point in a wire in time t seconds.

Small currents are often measured in mA (milliamps, 10^{-3} A) or μA (microamps, 10^{-6} A). Small charges are often measured in mC, μC or nC (nanocoulombs, 10^{-9} C).

If 5 C of charge flow along this wire in 1 s the current is 5 A

If 9 C of charge flow along this wire in 3 s the current is 3 A

Figure 2.3

A current of $1\,\text{A} = 1\,\text{C/s} = Ne$ where N is the number of electrons flowing per second. Therefore $N = 1\text{C}/e = 1/1.6 \times 10^{-19} = 6.24 \times 10^{18}$ electrons flowing past any point in 1 second. (In this calculation just the magnitude of e has been substituted.)

Example 1

A steady current of 0.5 A is used to recharge a cell. If the current flows for 1 hour calculate the amount of charge which flows through the cell.

$$\begin{aligned} \text{charge} &= \text{current} \times \text{time} \\ Q &= I \times t \\ &= 0.5\,\text{A} \times (1 \times 60 \times 60)\,\text{s} \\ &= 1800\,\text{C} \end{aligned}$$

Insulators Insulators, such as plastics, have no free charge carriers. Their electrons are bound tightly to the nuclei of their atoms and are therefore not able to carry charge from place to place.

Semiconductors Semiconducting materials contain some free charge carriers but their number is very small compared with the number present in metallic conductors. It is however often possible to change the number of charge carriers present in these materials by altering their temperature or exposing them to light. (See topic 5.)

Current and charge carriers

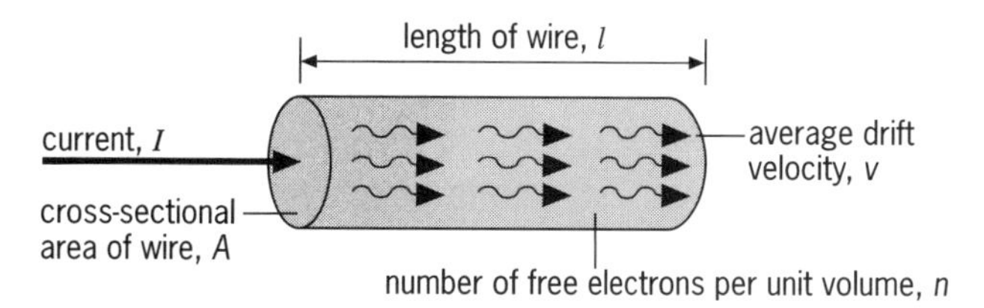

Figure 2.4

Consider a small section of a metallic conductor carrying a steady current, Figure 2.4. If the number of free electrons per unit volume is n then the number of free electrons in the wire is

$n \times \text{volume} = nAl$

The total charge, Q, carried by these free electrons is given by

$Q = nAle$

If the average **drift velocity** of the free electrons due to a potential difference is v, the time it takes for all the free electrons to travel through the section is distance/speed, that is $t = l/v$.

The current is given by

$$I = \frac{Q}{t}$$

Substituting from above,

$$I = \frac{nAle}{l/v} = nAle \times v/l$$

so

$$\boxed{I = nAve}$$

Typically the drift velocity in a metallic conductor is slightly less than 1 mm per second, that is approximately $10^{-3}\,\text{m s}^{-1}$.

Example 2

Calculate the current flowing through a piece of copper wire of cross-sectional area $1.25 \times 10^{-6}\,\text{m}^2$, if the number of charge carriers per m^3 is 10^{29}, the drift velocity of the electrons is $2.0 \times 10^{-4}\,\text{m s}^{-1}$ and the magnitude of the charge carried by each electron is $1.6 \times 10^{-19}\,\text{C}$.

$$\begin{aligned} I &= nAve \\ &= 10^{29}\,\text{m}^{-3} \times 1.25 \times 10^{-6}\,\text{m}^2 \times 2.0 \times 10^{-4}\,\text{m s}^{-1} \times 1.6 \times 10^{-19}\,\text{C} \\ &= 4.0\,\text{A} \end{aligned}$$

Flow of charge through liquids and gases

In liquids and gases the charge carriers are often ions. These are atoms which have lost or gained extra electrons. For example, molten lead bromide contains two ions, Pb^{2+} and Br^-. When a pd is applied across it the positive Pb^{2+} ions drift towards the negative electrode and the negative Br^- ions drift towards the positive electrode (Figure 2.5). This movement of the ions is an electric current through the liquid.

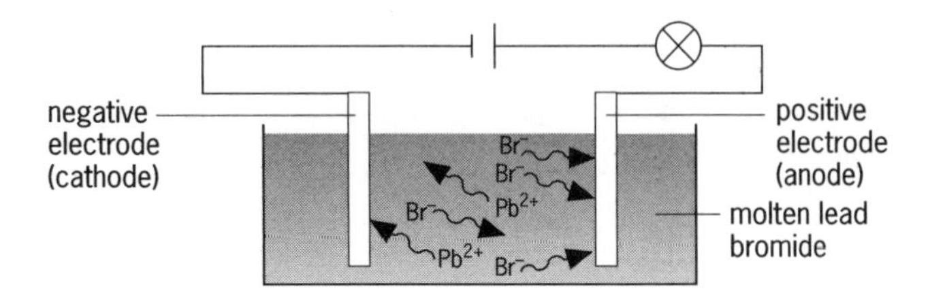

Fiigure 2.5 Charge is carried between the electrodes by the ions Pb^{2+} and Br^-

As a result of passing current through this liquid, lead forms on the negative electrode and bromine gas is released at the positive electrode. This separating of a compound using current is called electrolysis.

Current in series and parallel circuits

In a series circuit there are no branches. The rate of flow of charge, that is the current, is the same in all parts of the circuit.

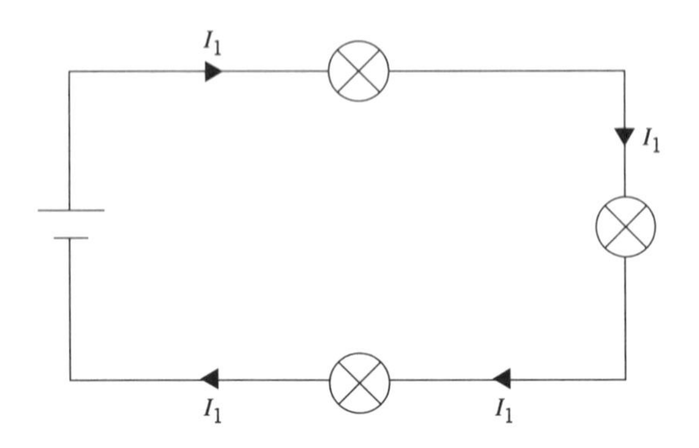

Figure 2.6 Current in a series circuit

In a parallel circuit there are branches; charges can therefore flow along different paths. However, the charge flowing into any junction in a certain time must equal the charge that flows out in the same time, Figure 2.7. This statement of the conservation of charge is known as **Kirchhoff's First Law** (see topic 5).

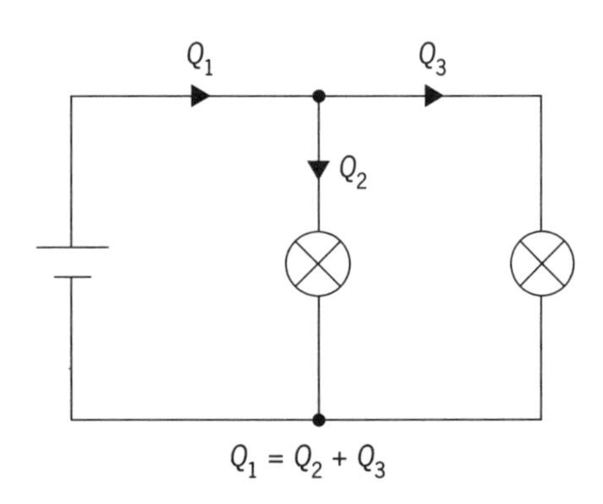

Figure 2.7 Flow of charge in a parallel circuit

This law is often expressed in terms of current: **the current flowing into a junction must equal the current flowing out of it.**

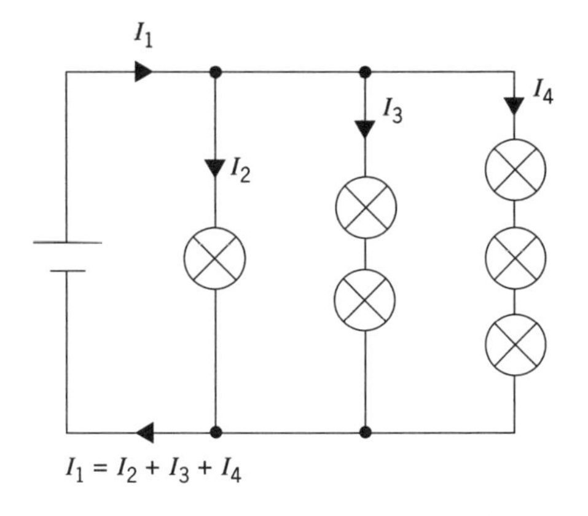

Figure 2.8 Current in a multi-branched parallel circuit

The charge on one electron is -1.6×10^{-19} C.

Level 1

1 What is the charge on the ions
a SO_4^{2-}
b PO_4^{3-}?

2 How many electrons are needed to make 1.0 coulomb of negative charge?

3 If a coulombmeter (a device which measures charge) reads -8 nC, how many surplus electrons are there?

4 The Avogadro constant, $N_A = 6.02 \times 10^{23}\,\text{mol}^{-1}$, defines the number of particles in **one mole** of any substance (see topic 20). What is the total charge of one mole of electrons?

5 **a** Water is pumped through a pipe at 0.50 litres per second. How many litres pass through the pipe in 14 s?
b A current of 0.50 A flows in a wire. How many coulombs pass through the wire in 14 s?
c 5000 cars per hour pass along the motorway. How many pass in 5 hours?
d 25 litres of water enter a bath in 10 s. What is the flow rate?
e Six cars enter a service station in 10 s. What is the flow rate?
f 240 nC flow from a dry cell through your body into a coulombmeter in 3 s. What is the current?

6 What are the missing readings on the ammeters **A**, **B** and **C** in Figure 2.9?

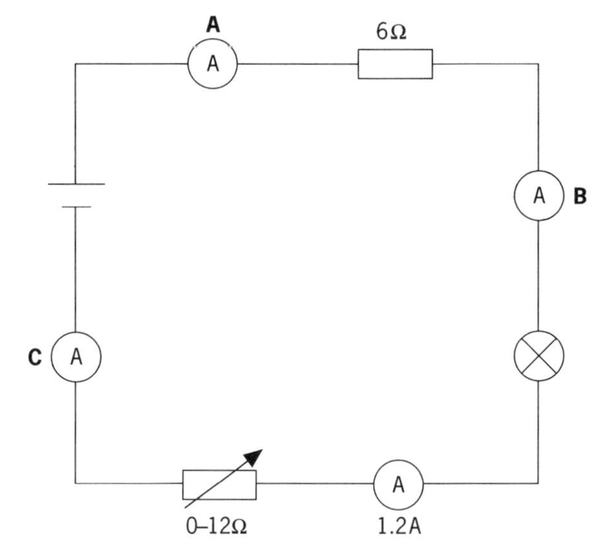

Figure 2.9

Level 2

7 Show that for a doubly ionised atom the current needed to deposit one mole of the atoms by electrolysis in 5.0 h is 10.7 A. (Take the Avogadro constant, N_A, as $6.0 \times 10^{23}\,\text{mol}^{-1}$.)

8 An electron beam in an evacuated tube is 0.30 m long. The electron speed is $6.0 \times 10^7\,\text{m s}^{-1}$. When the tube current is 1.0 mA, how many electrons are in the beam at any one time?

9 An electric current of 6.0 A is passed through a solution of silver nitrate for 500 s. Each silver ion carries a single electronic charge and its molar mass (see topic 20) is $108\,\text{g mol}^{-1}$. (Take the Avogadro constant as $6.0 \times 10^{23}\,\text{mol}^{-1}$.)
a What mass of silver is deposited?
b In a second run of the same duration, 8.4 g of silver was deposited. What was the size of the current this time?

10 A ping-pong ball shuttles to and fro between two metal plates (Figure 2.10) with a high potential difference between them. The period of oscillation T is 0.33 s. A microammeter in the circuit reads 6.0 μA. How much charge does the ball carry during each oscillation?

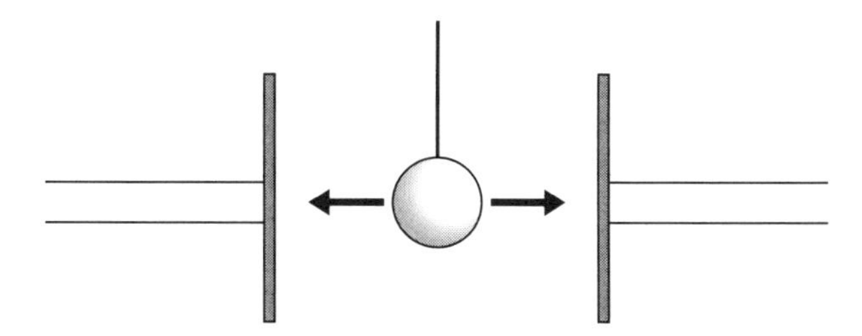

Figure 2.10

Level 3

11 **a** What is the difference between the drift velocities of the electrons in a good conductor and in a semiconductor of the same cross-sectional area when they are joined together in series with a potential difference applied to the extreme ends? Assume that a good conductor (for example copper) has 8×10^{28} charge carriers per m^3 and a semiconductor (for example germanium) has 6×10^{20} per m^3.
b What happens to the drift velocity of the electrons in the good conductor if its diameter is decreased?

12 The density of silver is $10\,500\,\text{kg m}^{-3}$ and its molar mass is $108\,\text{g mol}^{-1}$. Calculate the number of free electrons per m^3 if each atom contributes one electron.

13 When a nerve fibre is stimulated, singly charged ions travel through its outer membrane and set up an 'action potential'. If this lasts for 2.0 ms and the average current density is $1.2\,\text{A m}^{-2}$, find the number of singly charged ions per m^2 which cross the membrane.

Level 1

1 a In the chemical symbol SO_4^{2-}, the '2−' means that this ion has two extra electrons. So the total charge is twice the charge on one electron, *e*.

$$Q = 2 \times e = 2 \times -1.6 \times 10^{-19}\,\text{C}$$
$$= \mathbf{-3.2 \times 10^{-19}\,C}$$

If you are using a calculator, remember to use the EXP or EE button. It reduces the number of key presses. In this example:

[2] [×] [1.6] [+/−] [EXP] [+/−] [19] [=]

b PO_4^{3-} has three extra electrons. So the total charge is three times the charge on one electron, *e*.

$$Q = 3 \times e = 3 \times -1.6 \times 10^{-19}\,\text{C}$$
$$= \mathbf{-4.8 \times 10^{-19}\,C}$$

2 $$N = \frac{1.0\,\text{C}}{e}$$
$$= \frac{1.0\,\text{C}}{1.6 \times 10^{-19}\,\text{C}}$$
$$= \mathbf{6.3 \times 10^{18}\ electrons}$$

You can use the $\frac{1}{x}$ or x^{-1} button here. Remember that dividing by a very small number will give a very large one.

3 $$N = \frac{8 \times 10^{-9}\,\text{C}}{e}$$
$$= \frac{8 \times 10^{-9}\,\text{C}}{1.6 \times 10^{-19}\,\text{C}}$$
$$= \mathbf{5 \times 10^{10}\ electrons}$$

Mentally check that the power of 10 is correct. There is a 10^{-19} on the bottom which will become 10^{+19} when inverted. Then add the '−9' power to the '+19' power leaving a power of +10.

4 One mole always contains $N_A = 6.02 \times 10^{23}$ particles, so the charge on one mole of electrons, each with charge e, is

$$Q = N_A \times e = 6.02 \times 10^{23}\,\text{mol}^{-1} \times -1.6 \times 10^{-19}\,\text{C}$$
$$= \mathbf{-9.6 \times 10^4\,C\,mol^{-1}}$$

Mentally check that the power of 10 is correct: +23 minus 19 = 4.

This value of $96\,000\,\text{C}\,\text{mol}^{-1}$ is called the **Faraday constant**. You can use it to calculate the mass deposited when a known current has flowed for a known length of time. The mass, for any singly ionised substance, will be 1 molar mass for each 96 000 C.

5 a 0.5 litres flow each second for 14 seconds so

$$\text{total volume} = 0.5\,\text{litre}\,\text{s}^{-1} \times 14\,\text{s}$$
$$= \mathbf{7\ litres}$$

b 0.5 A is 0.5 C each second, so for 14 seconds

$$\text{total charge, } Q = It = 0.50\,\text{C}\,\text{s}^{-1} \times 14\,\text{s}$$
$$= \mathbf{7\,C}$$

c 5000 cars each hour for 5 hours, so

$$\text{total number of cars} = 5000\,\text{cars}\,\text{h}^{-1} \times 5\,\text{h}$$
$$= \mathbf{25\,000\ cars}$$

d A total of 25 litres flows in 10 seconds, so

$$\text{flow rate} = \frac{\text{volume}}{\text{time}}$$
$$= \frac{25\,\text{litres}}{10\,\text{s}} = \mathbf{2.5\ litre\,s^{-1}}$$

e A total of six cars enter a garage in 10 seconds, so

$$\text{number of cars each second} = \frac{6\ \text{cars}}{10\,\text{s}}$$
$$= \mathbf{0.6\ cars\,s^{-1}}$$

f 240 nC is 240×10^{-9} C. For a steady current,

$$\text{current} = \frac{\text{total charge}}{\text{total time}}$$
$$= \frac{240 \times 10^{-9}\,\text{C}}{3\,\text{s}}$$
$$= \mathbf{8.0 \times 10^{-8}\,A}\ \text{or}\ \mathbf{80\,nA}$$

6 Each of the meters will read 1.2 A because the current is the same everywhere in a series circuit. The electrical energy may be converted into other forms as the current flows but the number of charge carriers leaving and entering the cell per unit time is always the same.

Level 2

7 Current is the rate of flow of charge; first we have to find out how much charge is required to deposit one mole. One mole has $N_A = 6.0 \times 10^{23}$ particles (see question 4), so the charge needed for one mole of *doubly* ionised atoms is *twice* 'N_A times the charge on one electron, *e*':

$$Q = 2 \times N_A \times e$$
$$= 2 \times 6.0 \times 10^{23} \times 1.6 \times 10^{-19}\,\text{C}$$
$$= 192\,000\,\text{C}$$

$$\text{current} = \frac{\text{charge}}{\text{time}}$$ *but remember that the time must be in seconds*

$$5\ \text{hours} = 5 \times 60 \times 60\,\text{s} = 18\,000\,\text{s}$$

$$\text{current} = \frac{\text{charge}}{\text{time}}$$
$$= \frac{192\,000\,\text{C}}{18\,000\,\text{s}} = \mathbf{10.7\,A}$$

8 A current of 1.0 mA is equivalent to a transfer of 1.0 mC per second. We saw in question 2 that 6.25×10^{18} electrons have a total charge of 1.0 C. One thousandth of this number, 6.25×10^{15}, will have a total charge of 1.0 mC. This number must pass along the beam in one whole second.

But each electron is only in the beam for the time it takes to travel 0.30 m, at a speed of $6.0 \times 10^7\,\text{m}\,\text{s}^{-1}$.

$$\text{speed} = \frac{\text{distance}}{\text{time}}$$

can be rearranged to give

$$\text{time} = \frac{\text{distance}}{\text{speed}}$$

So for these electrons,

$$\text{time spent in the beam} = \frac{0.30\,\text{m}}{6.0\times10^7\,\text{m s}^{-1}}$$

$$= 5.0\times10^{-9}\,\text{s or } 5.0\,\text{ns}$$

To work out what to do next, imagine that the beam was longer and that the electrons were in the beam for 2 seconds. To find the number in the beam at any moment you would just multiply the number travelling down in 1 second by 2. You would multiply the number passing along in one second by the time in the beam.

So the number of electrons in a beam which is only '5 nanoseconds long' is

$$N = 6.25\times10^{15}\,\text{s}^{-1}\times5.0\times10^{-9}\,\text{s}$$
$$= \mathbf{3.1\times10^7\ electrons}$$

Compare this with the answer to question 5c. The method of working is the same.

9 a Total charge passed = current × time
= 6.0 A × 500 s
= 3000 C

$$\therefore\ \text{number of ions deposited} = \frac{3000\,\text{C}}{e}$$

$$= \frac{3000\,\text{C}}{1.6\times10^{-19}\,\text{C}}$$

$$= 1.9\times10^{22}$$

When a number of silver ions equal to the Avogadro constant is deposited, the mass is 108 g (one mole).

$$\therefore \frac{\text{mass deposited}}{108\,\text{g}} = \frac{1.9\times10^{22}\,\text{ions}}{6.0\times10^{23}\,\text{ions}}$$

mass deposited = 0.0317 × 108 g = **3.4 g**

b Number of ions deposited to give a mass of 8.4 g

$$= \frac{6.0\times10^{23}\times8.4\,\text{g}}{108\,\text{g}} = 4.7\times10^{22}$$

$\therefore$ charge passed $= 4.7\times10^{22}\times1.6\times10^{-19}\,\text{C}$
= 7500 C to 2 significant figures

For this charge to flow in 500 seconds the current must be

$$\frac{\text{charge}}{\text{time}} = \frac{7500\,\text{C}}{500\,\text{s}}$$

$$= \mathbf{15\,A}$$

10 With each oscillation of the ball, charge is carried across in only one direction. Electrons move on to the ball from the negative plate and off at the positive. There will be three complete trips per second.

To get this answer, divide 1.0 s by time for one complete trip $T = 0.33$ s, or recognise that the frequency $f = 1/T = 1/0.33\,\text{s} = 3\,\text{Hz}$.

The current is 6.0 μA, so the charge transferred is 6.0 μC every second. To carry 6.0 μC in three trips, the ball must carry **2.0 μC** across on each trip.

Level 3

11 With the conductor and semiconductor in series the current, I, is the same throughout. But $I = n\times A\times v\times e$, where n is the number of charge carriers (electrons) per unit volume, A is the cross-sectional area, v is the average drift velocity of the charge carriers and e is the charge on an electron. So $n\times A\times v\times e$ will be constant throughout the circuit.

a $n_1\times A_1\times v_1\times e = n_2\times A_2\times v_2\times e$ where 1 refers to copper and 2 to germanium

If the cross-sectional areas are same:

$$n_1\times v_1 = n_2\times v_2$$

$$\frac{n_1}{n_2} = \frac{v_2}{v_1} = \frac{8\times10^{28}\,\text{m}^{-3}}{6\times10^{20}\,\text{m}^{-3}} = 1.3\times10^8$$

$$\therefore\quad \mathbf{v_2 = 1.3\times10^8\,v_1}$$

The drift velocity in the semiconductor is very much higher than in the metal conductor.

b $n_1\times A_1\times v_1\times e = n_2\times A_2\times v_2\times e$ where 2 now refers to the situation in the copper of reduced diameter. The cross-sectional area A is reduced but the material is the same so the number of charge carriers per unit volume remains constant. Then this reduces to:

$$A_1\times v_1 = A_2\times v_2$$

$$\frac{v_2}{v_1} = \frac{A_1}{A_2}$$

So **v will increase** in inverse proportion to the decrease in A. (If A is halved, v is doubled.)

This relationship also applies to liquids flowing in pipes. If the pipe narrows, the velocity of the liquid must increase in order to maintain the volume flow rate.

12 Find the number of moles in 1 m^3 of material by dividing the mass of 1 m^3 by the mass of one mole. The mass of 1 m^3 of material is its density.

So, for silver, 1 m^3 contains $\dfrac{10\,500\,\text{kg}}{0.108\,\text{kg mol}^{-1}} = 97\,200\,\text{mol}$

Each mole of silver contains N_A electrons (the Avogadro constant, $N_A = 6.0\times10^{23}\,\text{mol}^{-1}$).
So, the total number of electrons will be
$97\,200\,\text{mol}\times6.0\times10^{23}\,\text{mol}^{-1} = \mathbf{5.8\times10^{28}}$.

13 Consider 1 m^2 of nerve membrane.

total charge passing across membrane = current × time
$= 1.2\,\text{A}\times2.0\times10^{-3}\,\text{s}$
= 2.4 mC

The total charge must equal the number of charge carriers (N) × charge per carrier. If the charge carriers are singly charged ions then:

$Q = N\times e$ where e is the charge of an electron (1.6×10^{-19} C)

$$\therefore N = \frac{Q}{e} = \frac{2.4\times10^{-3}\,\text{C}}{1.6\times10^{-19}\,\text{C}}$$

$= \mathbf{1.5\times10^{16}}$ singly charged ions.

3 Electromotive force and potential difference

Electromotive force

When charges pass through a power supply, such as a cell or battery, they receive energy. The energy given to each coulomb of charge is determined by the **electromotive force** or **emf** of the supply, measured in volts (V).

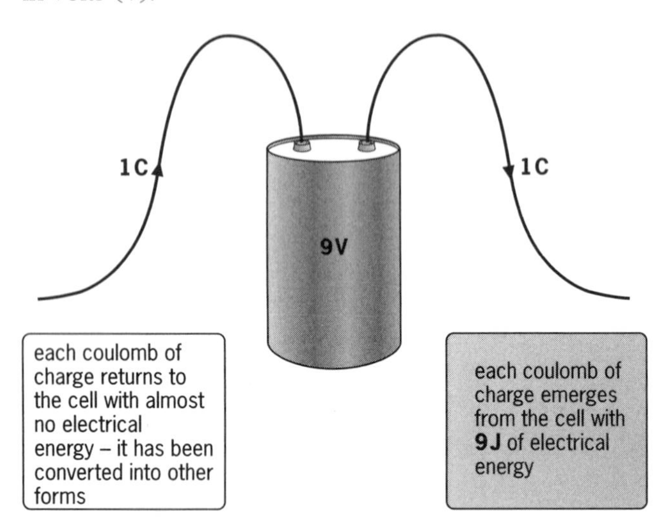

Figure 3.1 A cell of emf 9V gives 9J of energy to each coulomb of charge that passes through it

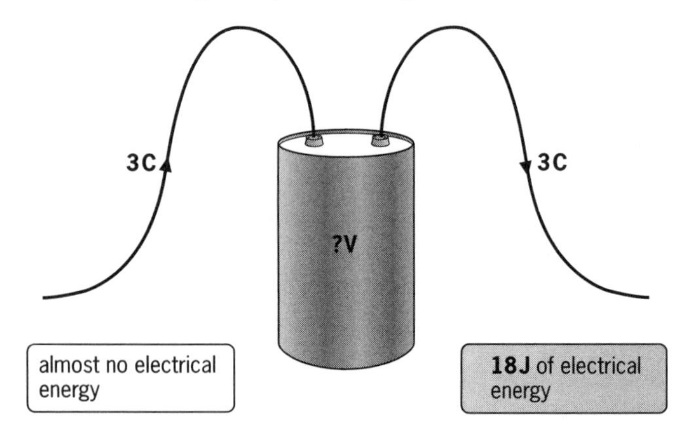

Figure 3.2 If 3C of charge receives a total of 18J of energy, the emf of the cell is 18J/3C = 6V

$$\text{emf of source} = \frac{\text{energy given}}{\text{charge}} \quad \text{or} \quad V = \frac{W}{Q}$$

1 volt = 1 joule per coulomb
1 joule = 1 volt × 1 coulomb

Example 1

Calculate the total energy, W, given to 50C of charge as it passes through a battery of emf 12V.

$$W = Q \times V$$
$$= 50\,\text{C} \times 12\,\text{V}$$
$$= 600\,\text{J}$$

Potential difference

As charge moves around a circuit its energy is dissipated – it is changed into other forms.

If a voltmeter is connected in parallel with any component in the external part of the circuit it will measure the **potential difference** or **pd** across it. This value, measured in volts, indicates how much energy is dissipated when 1C of charge flows through this component.

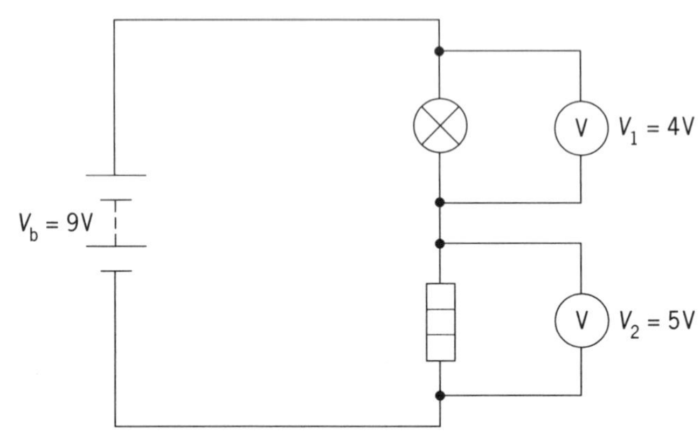

Figure 3.3 All the energy received by the charge from the battery is changed into other forms by the bulb and the heater

The battery in the circuit of Figure 3.3 gives 9J of energy to each coulomb of charge that passes through it. The bulb converts 4J of this electrical energy into 4J of heat and light energy each time 1C of charge passes through it. The heater converts the remaining 5J of electrical energy into 5J of heat energy each time 1C of charge passes through it. It follows therefore that

$$V_b = V_1 + V_2$$

The relationship between the energy dissipated by a component, the charge that passes through it and the pd across is described by the equation

$$\text{energy dissipated} = \text{charge} \times \text{pd} \quad \text{or} \quad W = Q \times V$$

Example 2

Calculate the pd across a bulb if 12J of electrical energy are converted into heat and light energy when 3C of charge pass through it.

$$W = Q \times V$$
$$12\,\text{J} = 3\,\text{C} \times V$$
$$V = 4\,\text{V}$$

Energy is dissipated in an electrical circuit because work is done in moving charge through the various components present. If 1J of work is done in moving 1C of charge between two points, the pd between these two points is 1V.

Electrical power and energy

The pd (V) between two points in a circuit is defined as being the energy dissipated (W) when 1 C of charge flows between them. Current (I) is defined as the number of coulombs that flow past a point in the circuit each second. Since we have

$$V = \frac{W}{Q} \text{ and } I = \frac{Q}{t}$$

it follows that the total energy dissipated each second between these two points is $V \times I$.

$$V \times I = \frac{W}{Q} \times \frac{Q}{t} = \frac{W}{t}$$

The rate at which energy is dissipated by a component in a circuit is its **power**, P, and is measured in J s^{-1} or more usually watts (W).

power = pd × current

$$P = V \times I$$

Note: W is the symbol for the physical quantity energy or work but W is the symbol for the unit of power, the watt. In handwritten material and some printed material these symbols appear the same.

Assuming that the values of V and I remain constant, the total energy dissipated after a time t is $V \times I \times t$.

$$W = V \times I \times t \quad \text{or} \quad W = P \times t$$

Example 3

Calculate the power rating of an electric kettle which requires a 240 V supply and a current of 5 A. If the kettle is turned on for 5 minutes how much electrical energy has been dissipated by the kettle's heating element?

power = pd × current

$$P = V \times I$$
$$= 240\,\text{V} \times 5\,\text{A}$$
$$= 1200\,\text{W or } 1.2\,\text{kW}$$

energy = power × time

$$W = P \times t$$
$$= 1200\,\text{W} \times 300\,\text{s}$$
$$= 360\,000\,\text{J or } 360\,\text{kJ}$$

Level 1

1 The potential difference across a light bulb is 6.0 V. *(This means that every coulomb passing through the bulb changes 6 J of electrical energy into heat and light.)* The current is 2.0 A.

a How many coulombs per second pass through the bulb?

b How many joules per second are converted into heat and light?

c At what power is the bulb operating?

2 A heater is working at 100 W. *(This means it converts 100 J of electrical energy to heat every second.)* The current is 1.0 A, so 1.0 C passes through the heater every second. This charge has passed through a potential difference. How many volts is this potential difference?

3 The potential difference across a lamp is 3.0 V.

a If 6.0 C pass through it, how much energy is converted?

b If the 6.0 C flow steadily taking 2.0 s, what is the power, in joules per second?

c Check that this is equal to $V \times I$.

4 A 12 V car battery delivers 30 A for 10 s (Figure 3.4).

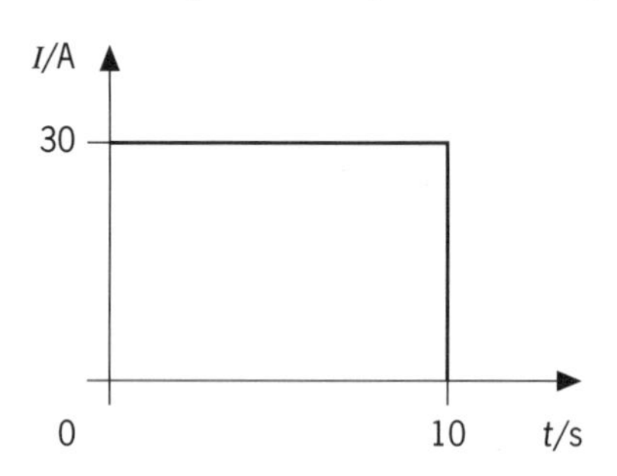

Figure 3.4

a How many coulombs of charge flow out of the battery?

b How much energy does each coulomb carry?

c What is the total energy converted?

d What is the power of operation of the battery?

5 An 8.0 MW power station generates electricity at 40 kV. What is the current in the transmission cables?

Level 2

6 One rechargeable NiCad battery is marked 1.2 V, 1.6 A h. A second battery is marked 8.4 V, 120 mA h.

a Which battery stores more energy?

b How many joules of energy does this battery store?

7 Your car battery is labelled 12 V, 24 A h, and is fully charged. You leave your sidelights (4 × 6.0 W bulbs) switched on for 4 hours. Should you be concerned?

8 A 2.3 V torch bulb and a 230 V mains bulb both pass a current of 0.25 A.

a Is the rate of flow of charge the same in each? Explain your answer.

b Is the rate of energy transfer the same in each? Explain your answer.

Level 3

9 An X-ray tube operates at 90 kV with a tube current of 100 mA. Only 1% of the power supplied produces X-rays; the rest is wasted as heat. Calculate:

a the power drawn from the supply

b the rate at which heat is produced.

10 Electrons in a cathode ray tube pass through a potential difference of 5.0 kV.

a What is the increase in energy of each electron, in joules?

b If each electron is stationary at first, find its final velocity.
(mass of an electron $m_e = 9.1 \times 10^{-31}$ kg)

11 A resistor rated at 1.0 W is carrying a current of 0.20 A at 2.3 V.

a Calculate the power at which it is working.

b Your friends say it will be all right to double the current, as you are working at only half the maximum power. Are they correct? Explain your answer.

12 Explain what you would see if a 6 V, 12 W bulb and a 230 V, 100 W bulb were connected in series across a 230 V supply.

Hint: Find the resistance of each bulb when lit.

Level 1

1 a A current of 2.0A is equal to a rate of flow of charge of

2.0 coulombs per second ($I = Q/t$).

b When 1 coulomb of charge moves through a potential difference of 1 volt there is an energy change of 1 joule. So, for a potential difference of 6.0 volts there will be an energy change of 6 joules per coulomb.

In this case the rate of flow charge, I, is 2.0 coulombs per second, so the rate of change of energy

$$P = V \times I$$
$$= 6.0\,\mathrm{J\,C^{-1}} \times 2.0\,\mathrm{C\,s^{-1}}$$
$$= \mathbf{12\,J\,s^{-1}}$$

c Power in watts = rate of change of energy in joules per second.

So $P = 12$ joules per second = **12W**

2 The heater is working at a rate of 100W which means that the 1.0 coulomb per second is generating 100 joules per second. When 1 coulomb per second moves through a potential difference of 1 volt there is an energy change of 1 joule per second. So the charge in this case must be moving through a potential difference of **100 volts**.

3 a The potential difference across the lamp is 3.0V so each coulomb passing through the lamp will generate 3.0J of heat and light.

$$W = Q \times V$$

so 6.0C will generate $6.0 \times 3.0\,\mathrm{J} = \mathbf{18\,J}$

b 18J is converted in 2.0s, so the rate of change of energy or power, P, is

$P = W/t = 18\,\mathrm{J}/2.0\,\mathrm{s} = \mathbf{9.0\,J\,s^{-1}}$ or **9.0W**

c 6.0C flows steadily in 2.0s, so

$$\text{rate of flow of charge, } I = \frac{6.0\,\mathrm{C}}{2.0\,\mathrm{s}}$$
$$= 3.0\,\mathrm{C\,s^{-1}}$$
$$= 3.0\,\mathrm{A}$$

potential difference, $V = 3.0\,\mathrm{V}$

$P = V \times I = 3.0\,\mathrm{V} \times 3.0\,\mathrm{A} = \mathbf{9.0\,W}$

4 a 30A flows steadily for 10 seconds.

$$Q = I \times t$$

so total charge, $Q = 30\,\mathrm{A} \times 10\,\mathrm{s} = \mathbf{300\,C}$

b When 1 coulomb of charge moves through a potential difference of 1 volt there is an energy change of 1 joule. So, for a potential difference of 12V, for each coulomb, there is an energy change of **12J**.

c The total charge that passes is 300C and for each coulomb there is an energy change of 12J, so

total energy, $W = Q \times V = 300\,\mathrm{C} \times 12\,\mathrm{J\,C^{-1}} = \mathbf{3600\,J}$

d This change in energy takes place in 10s.

$$\text{power, } P = \frac{W}{t} = \frac{3600\,\mathrm{J}}{10\,\mathrm{s}} = \mathbf{360\,J\,s^{-1}} \text{ or } \mathbf{360\,W}$$

5 Power = voltage × current

$$P = V \times I$$

which can be rearranged to give $I = \dfrac{P}{V}$

$$\text{So the current in the cables} = \frac{8.0\,\mathrm{MW}}{40\,\mathrm{kV}}$$
$$= \frac{8.0 \times 10^6\,\mathrm{W}}{40 \times 10^3\,\mathrm{V}} = \mathbf{200\,A}$$

Level 2

6 a When a current flows in a circuit,

total energy (joules) = potential difference (volts) × current (ampere) × time

$$W = V \times Q = V \times I \times t$$

If the time is in hours the energy may be measured in watt-hours (Wh).

For the first battery:

total energy available = $1.2\,\mathrm{V} \times 1.6\,\mathrm{Ah} = 1.9\,\mathrm{Wh}$

For the second battery:

total energy available $= 8.4\,\mathrm{V} \times 120\,\mathrm{mAh}$
$= 8.4\,\mathrm{V} \times 0.120\,\mathrm{Ah} = 1.0\,\mathrm{Wh}$

The **first battery** stores more energy.

b To calculate the energy available in joules, the time must be in seconds, not hours.

energy stored $= 1.9\,\mathrm{Wh} = 1.9\,\mathrm{Wh} \times 3600\,\mathrm{s\,h^{-1}}$
$= 6.8 \times 10^3\,\mathrm{W\,s} = \mathbf{6.8\,kJ}$

7 First find the current through a bulb using

power = voltage × current

$P = V \times I$ which can be rearranged to give $I = \dfrac{P}{V}$

$$\text{So the current} = \frac{P}{V} = \frac{6\,\mathrm{W}}{12\,\mathrm{V}} = 0.5\,\mathrm{A}.$$

Car lights are always wired in parallel, so that if one fails it does not make all the lights go off. The currents through the bulbs must be added together to get the total current supplied by the battery.

The four bulbs will therefore draw a total of 2.0A from the battery. In 4 hours the energy drawn from the battery will be:

$$12\,\mathrm{V} \times 2.0\,\mathrm{A} \times 4\,\mathrm{h} = 12\,\mathrm{V} \times 8.0\,\mathrm{Ah}$$

This is only one third of the energy available (if the battery was fully charged) so you should still have 67% of your battery energy left, but don't make a habit of it!

8 a **Yes**. The rates of flow of charge are the same because the currents are the same.

b **No**. In the 230 V bulb each coulomb of charge will convert 230 joules of electrical energy to heat and light, whereas in the 2.3 V bulb each coulomb will only convert 2.3 joules of electrical energy to heat and light.

(The powers of the bulbs are given by

power = voltage × current

For the 230 V bulb, power = 230 V × 0.25 A = 57.5 W.
For the 2.3 V bulb, power = 2.3 V × 0.25 A = 0.575 W.)

Level 3

9 a Power input to the X-ray tube = tube voltage × tube current

$$P = V \times I$$
$$= 90 \times 10^3\,\text{V} \times 100 \times 10^{-3}\,\text{A}$$
$$= \mathbf{9.0\,kW}$$

b Only 1% of the power input is used usefully to produce X-rays.

useful power = $9.0 \times 10^3\,\text{W} \times 0.01 = 90\,\text{W}$

$\therefore$ 9000 − 90 = **8910 W** is 'wasted' as heat

10 a Energy acquired = potential difference × charge on an electron

$$W = V \times e$$
$$= 5.0 \times 10^3\,\text{V} \times 1.6 \times 10^{-19}\,\text{C}$$
$$= \mathbf{8.0 \times 10^{-16}\,J}$$

b If this energy is converted into kinetic energy (of the electron), then

$$\tfrac{1}{2} \times m_e \times v^2 = 8.0 \times 10^{-16}\,\text{J}$$

Rearranging this equation gives

$$v^2 = \frac{8.0 \times 10^{-16}}{\tfrac{1}{2} m_e}\,\text{J}$$
$$= 2 \times \frac{8.0 \times 10^{-16}\,\text{J}}{9.1 \times 10^{-31}\,\text{kg}} = 1.8 \times 10^{15}\,\text{m}^2\,\text{s}^{-2}$$

$\therefore v = \mathbf{4.2 \times 10^7\,m\,s^{-1}}$

Such a large speed is possible for electrons.

11 a Power = voltage × current
= 2.3 V × 2.0 A = **0.46 W**

b If you want to double the current you would have to double the voltage. This would mean that the power would go up by a factor of 4:

$$\text{power} = (2.3 \times 2)\,\text{A} \times (0.20 \times 2)\,\text{V}$$
$$= 4.6 \times 0.40 = 1.84\,\text{W}$$

This is almost twice the rated power of the resistor. The resistor would get very hot, very quickly, and, although it might not be destroyed, could well suffer irreversible damage. You should not exceed the rating of any component.

12 To find out what would be seen, you need to find out how much power would be dissipated in each of the two bulbs. This will depend on the potential difference across the bulb.

First find the resistance of the bulb, under normal operating conditions, based on the quoted power and voltage.

$$P = V \times I \quad \text{and} \quad R = \frac{V}{I} \quad \text{(see topic 4)}$$

$$\text{so} \quad R = \frac{V}{P/V} = \frac{V^2}{P}$$

For the low voltage bulb

$$R = \frac{(6\,\text{V})^2}{12\,\text{W}} = 3\,\Omega$$

For the high voltage bulb

$$R = \frac{(230\,\text{V})^2}{100\,\text{W}} = 529\,\Omega$$

The total resistance of the two bulbs in series is 532 Ω; the potential difference across the bulbs will be shared in proportion to their individual resistances.

The low voltage bulb will get 3/532 of 230 V = 1.3 V, so it **would not be lit.**

The high voltage bulb will get 529/532 of 230 V = 229 V, so it **would be fully lit**.

It is worth noting that the resistance of the elements in traditional light bulbs depends very strongly on temperature. The resistance when the bulb is on may be more than ten times greater than the resistance when it is off.

Resistance

Resistance is the opposition to the flow of charge. The following experiment investigates the opposition to flow of charge through a metallic conductor.

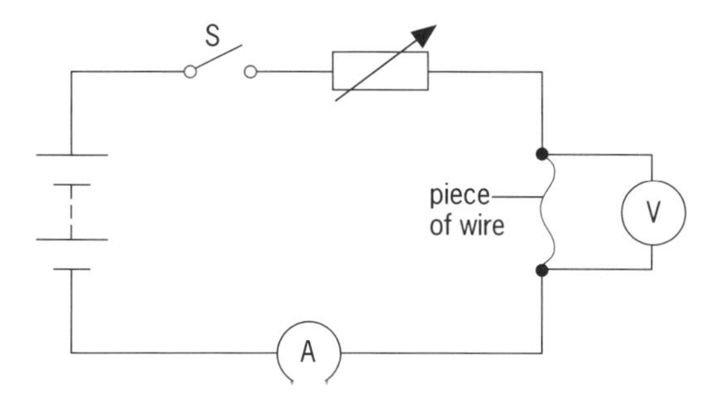

a

Current in wire, I/mA	pd across ends of wire, V/V
0	0.0
25	2.0
50	4.0
75	6.0
100	8.0
125	10.0

b

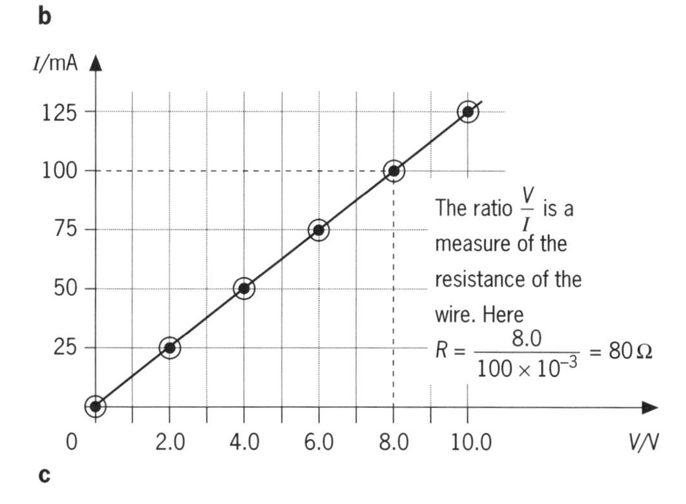

c

Figure 4.1 Investigating the resistance of a piece of wire

The switch S in the circuit of Figure 4.1a is closed and the current flowing through the wire, I, and the pd across it, V, noted. The variable resistor is altered and the new values of I and V noted. This procedure is repeated several times and a table of results constructed, Figure 4.1b. From this table a graph of I against V is drawn, Figure 4.1c. Note that the temperature of the wire must remain constant.

The graph shows that for a metallic conductor an increase in the pd across its ends results in a proportional increase in the current flowing through it, for example if the pd is doubled the current doubles. Conductors such as these are called **ohmic** conductors as they obey **Ohm's Law**. This states that **the current flowing through a metallic conductor is directly proportional to the potential difference across its ends provided the temperature remains constant**.

The resistance of a conductor is defined by the equation

$$\text{resistance} = \frac{\text{potential difference}}{\text{current}}$$

or

$$\boxed{R = \frac{V}{I}}$$

where R is measured in **ohms** (Ω) when V is in volts and I in amps. The equation may be rearranged to give

$$V = IR \quad \text{and} \quad I = \frac{V}{R}$$

Example 1

Calculate the resistance of a piece of wire if a current of 0.50 A flows when a pd of 12 V is applied across its ends.

$$R = \frac{V}{I}$$

$$= \frac{12\,\text{V}}{0.50\,\text{A}}$$

$$= 24\,\Omega$$

Conductors for which the graph of I against V is not a straight line through the origin do not obey Ohm's Law; they are called **non-ohmic** conductors (Figure 4.2).

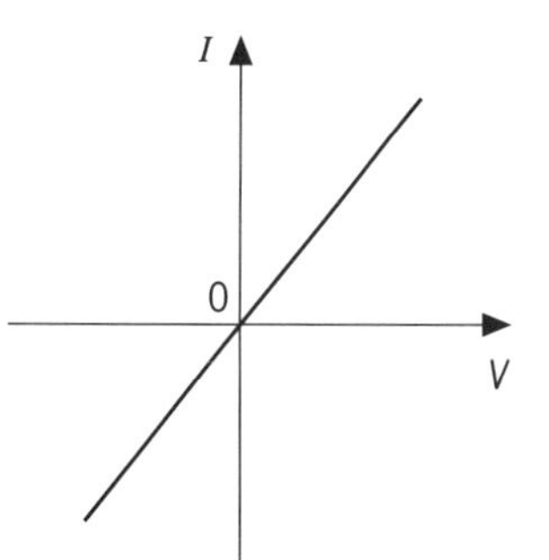

ohmic conductor, e.g. a metal at constant temperature

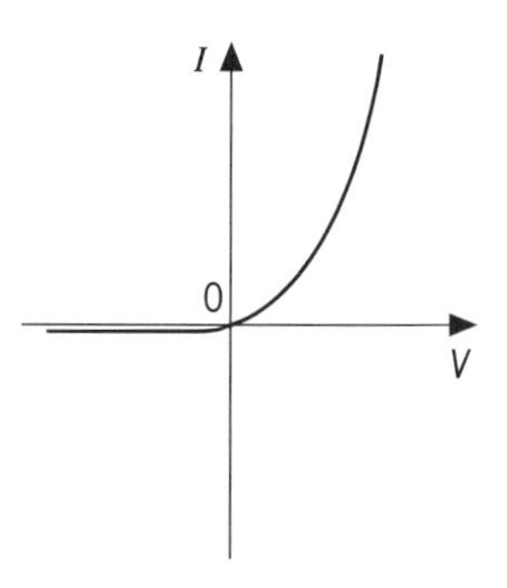

non-ohmic conductor: a diode

Figure 4.2 I against V for ohmic and non-ohmic conductors

Power and resistance

As we have already seen, the power dissipated by a component is given by

power = potential difference × current
$P = V \times I$

Using the Ohm's Law equation ($V = IR$) we can write the above equation as

$$P = I^2 R \quad \text{or} \quad P = \frac{V^2}{R}$$

Example 2

Calculate the power dissipated by a 100 Ω resistor when a current of 100 mA passes through it.

$P = I^2 R$

100 mA = 0.10 A

$P = (0.10\,\text{A})^2 \times 100\,\Omega$
$= 1.0\,\text{W}$

Resistors in series

Consider several resistors connected in series (Figure 4.3).

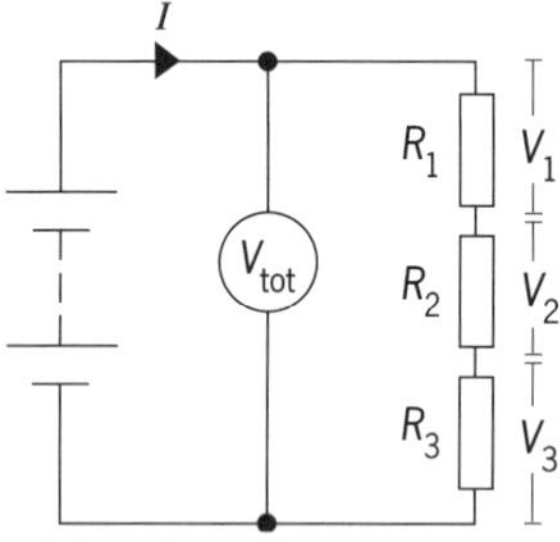

Figure 4.3 Resistors in series

1. The current, I, flowing through each of the resistors will be the same.
2. If the connecting wires have negligible resistance, the sum of the pds across each resistor is equal to the pd across the network of resistors:

   $V_{tot} = V_1 + V_2 + V_3$ (see Kirchhoff's Second Law, page 24)

3. But from the equation $V = IR$, substituting for V,

   $IR_{tot} = IR_1 + IR_2 + IR_3$
   $= I(R_1 + R_2 + R_3)$

$$R_{tot} = R_1 + R_2 + R_3$$

Example 3

Calculate the total effective resistance when a 2 Ω, a 3 Ω and a 4 Ω resistor are connected in series.

$R = R_1 + R_2 + R_3$
$= 2\,\Omega + 3\,\Omega + 4\,\Omega$
$= 9\,\Omega$

Resistors in parallel

Consider several resistors connected in parallel (Figure 4.4).

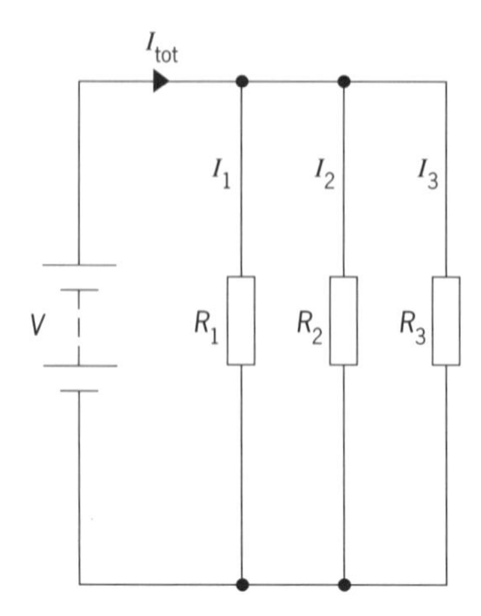

Figure 4.4 Resistors in parallel

1. The pd across each resistor, V, will be the same.
2. Using Kirchhoff's First Law (see page 24)

   $I_{tot} = I_1 + I_2 + I_3$

3. But $I = V/R$. So, substituting for I,

$$\frac{V}{R_{tot}} = \frac{V}{R_1} + \frac{V}{R_2} + \frac{V}{R_3}$$

$$= V\left(\frac{1}{R_1} + \frac{1}{R_2} + \frac{1}{R_3}\right)$$

$$\therefore \quad \frac{1}{R_{tot}} = \frac{1}{R_1} + \frac{1}{R_2} + \frac{1}{R_3}$$

Example 4

Calculate the total effective resistance when a 2 Ω, a 3 Ω and a 4 Ω resistor are connected in parallel.

$$\frac{1}{R} = \frac{1}{R_1} + \frac{1}{R_2} + \frac{1}{R_3}$$

$$= \frac{1}{2} + \frac{1}{3} + \frac{1}{4}$$

$$= \frac{6}{12} + \frac{4}{12} + \frac{3}{12} = \frac{13}{12}\,\Omega^{-1}$$

This is inverted to give

$$R = \frac{12}{13}\,\Omega \text{ or } 0.92\,\Omega$$

Resistivity

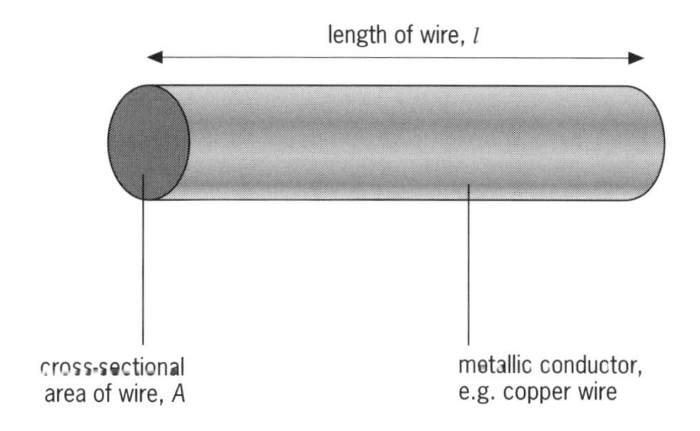

Figure 4.5

The resistance of a metallic conductor, such as a piece of wire (Figure 4.5), at a fixed temperature, depends upon

- the length of the wire, l – the longer the wire, the greater the resistance;
- the cross-sectional area of the wire, A – the thicker the wire, the smaller the resistance;
- the material from which the wire is made.

These factors are related by the equation

$$R = \rho \frac{l}{A}$$

where ρ is called the **resistivity** of the material and is defined as being the resistance of a piece of wire of that material, 1 m long and with unit cross-sectional area (i.e. 1 m^2) at a particular temperature. ρ is measured in Ω m.

Example 5

Calculate the resistivity of copper if a piece of copper wire 1.0 m long, of cross-sectional area 1.0×10^{-6} m^2, has a resistance of 0.02 Ω.

Rearrange

$$R = \frac{\rho l}{A}$$

to give

$$\rho = \frac{RA}{l}$$

$$= \frac{0.02\,\Omega \times 1.0 \times 10^{-6}\,\text{m}^2}{1.0\,\text{m}}$$

$$= 2 \times 10^{-8}\,\Omega\,\text{m}$$

Resistivity and temperature

Metallic conductors

At any temperature above zero kelvin the lattice structure of a metallic conductor vibrates. At low temperatures the vibrations are small and there is little opposition to the flow of electrons. As the temperature increases the vibrations become more violent and the flow of charge through the metal becomes more difficult. The resistivity, and hence resistance, of metallic conductors thus increases as their temperature increases.

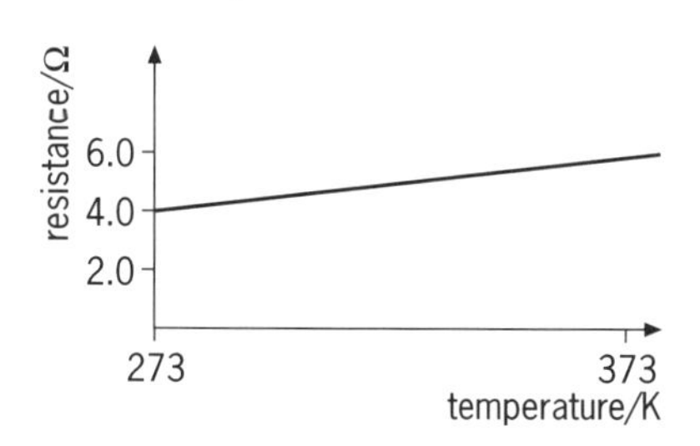

Figure 4.6 Increase in resistance of a metallic conductor with temperature

Some metals, such as tin, zinc and lead, exhibit a phenomenon known as **superconductivity** as their temperature approaches zero kelvin. Their resistance to the flow of charge through their lattice structure becomes zero. Some compounds show this effect at higher temperatures, such as 80 K.

Non-metallic conductors

The rate of flow of charge through non-metallic conductors, for example semiconductors, is determined by the number of charge carriers available to move charge through the structure. In most of these materials an increase in temperature frees more charge carriers, resulting in the resistivity of the material decreasing.

The change in the resistance of a semiconductor can be quite large even for small changes in temperature. This property is extremely useful in temperature-sensing devices such as fire alarms or thermostats. Resistors that are sensitive to changes in temperature are called **thermistors**.

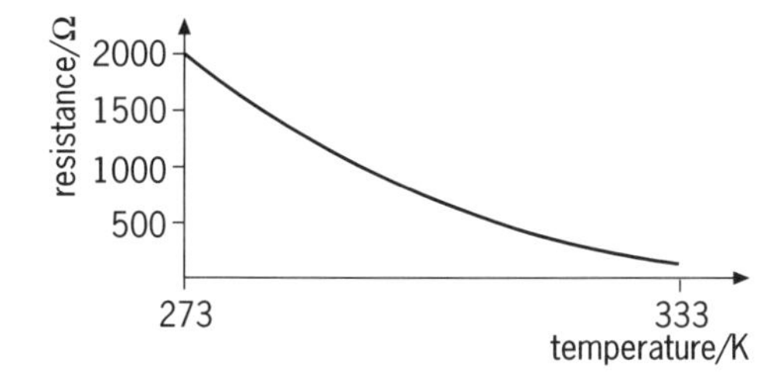

Figure 4.7 Decrease in resistance of a thermistor with temperature

With some materials it is possible to provide the energy necessary to release extra charge carriers using light energy – so the resistance decreases in brighter light. Resistors that are sensitive to light are called **light dependent resistors** (LDRs).

Level 1

1 When the potential difference between the ends of a length of wire is 3.4 V a current of 0.62 A flows.
  a What is the resistance of the wire?
  b When the potential difference is doubled to 6.8 V, the wire gets hot and the resistance increases to 7.4 Ω. What is the new current?

2 The resistance of a light emitting diode (LED) is 200 Ω. What potential difference will produce a current of 10 mA?

3 When the current in the LED of question 2 reaches 30 mA, the resistance has fallen to 100 Ω. What is the new potential difference across it?

4 The current flowing through a fully lit 100 W mains light bulb is 0.43 A. What is the resistance of the bulb?

Hint: Use $P = I^2R$.

5 What is the total resistance of the elements of a 2.0 kW mains electric fire if the current through it is 8.7 A when all the bars are on?

Hint: Use $P = I^2R$.

6 What is the resistance of a 60 W, 230 V mains light bulb when fully lit?

Hint: Use $P = V^2/R$.

7 Which of the circuits **A** to **F** in Figure 4.8 have resistors in series?

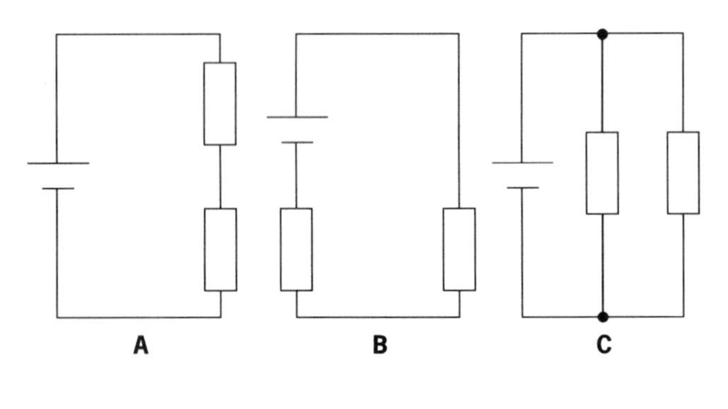

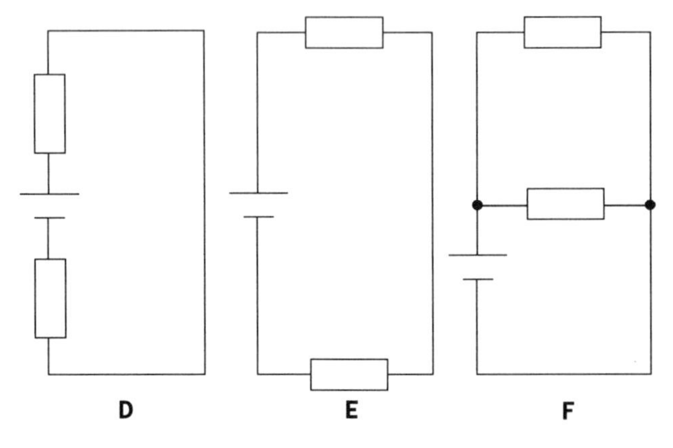

Figure 4.8

8 Which of the circuits **A** to **F** in Figure 4.8 have resistors in parallel?

Level 2

9 A hand-held neon mains-tester allows a current no larger than 1.0 mA to flow through your body. Roughly what resistance should the manufacturers have included in the circuit to make it safe for the person holding it?

Hint: What is the approximate value of mains voltage?

10 The curve in Figure 4.9 shows the potential difference vs current for a car sidelight bulb.

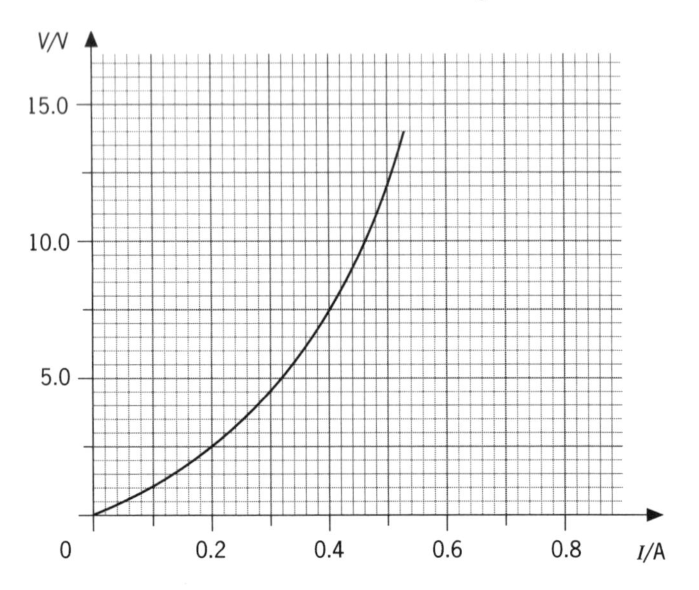

Figure 4.9

What is the resistance of the bulb when it has a potential difference of 9.0 V across it?

11 What are the missing readings on ammeters B and C in Figure 4.10? Assume the meters and connecting wires have negligible resistance.

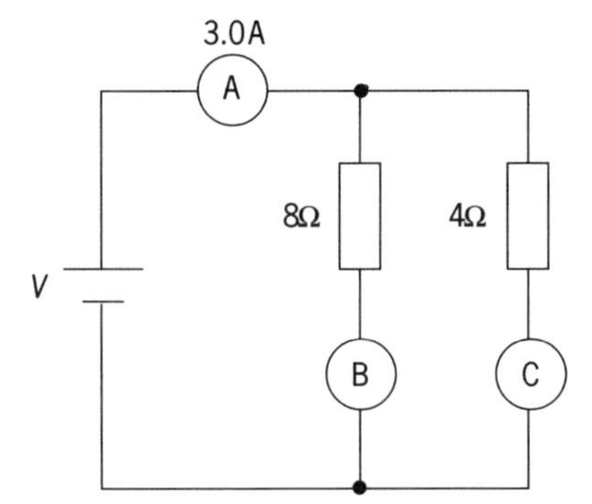

Figure 4.10

12 A constantan wire of resistivity $\rho = 47 \times 10^{-8}\,\Omega\,m$ has an area of cross-section of 0.060 mm² (or $0.060 \times 10^{-6}\,m^2$).
  a What is the resistance of a 3.0 m length of this wire?
  b Using your answer to part a, write down the resistance of:
    i a 6.0 m length
    ii a 1.0 m length
    iii 3.0 m of wire of half the cross-sectional area
    iv 3.0 m of wire of half the diameter.

13 Explain:

a the increase in resistivity with temperature of metals

b the decrease in resistivity with temperature of some types of semiconductor, used for thermistors.

Level 3

14 An electrode 0.010 m by 0.010 m is placed on the skin where there is a layer of fat 0.014 m thick. The resistivity of fat $\rho = 25\,\Omega\,\text{m}$. What is the resistance of this layer of fat to current passing through to the electrodes?

15 Rods X and Y are the same length and made from the same material. The diameter of X is three times that of Y. If the resistance of X, R_X, is $3.0\,\Omega$, what is the resistance of Y, R_Y?

16 If the resistivity ρ of the 'lead' in a pencil is $9.5 \times 10^{-5}\,\Omega\,\text{m}$, find the thickness (depth) of a pencil mark 2.0 mm wide by 20 mm long on a piece of card (Figure 4.11) if there is a resistance of $6.2\,\text{k}\Omega$ between the ends of the mark.

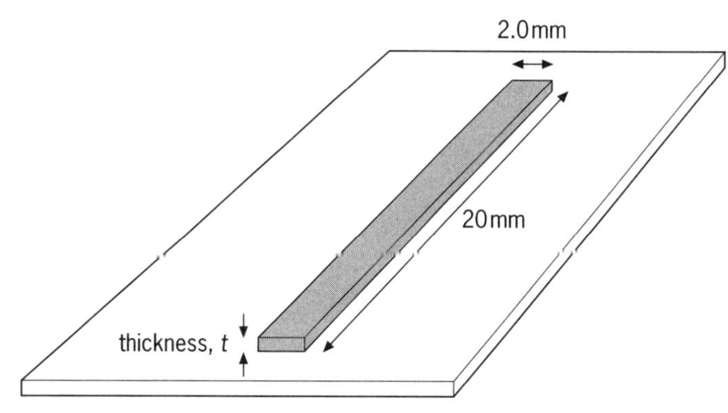

Figure 4.11

Level 1

1 a Resistance $= \dfrac{\text{potential difference}}{\text{current}}$

$$R = \frac{V}{I} = \frac{3.4\,\text{V}}{0.62\,\text{A}} = \mathbf{5.5\,\Omega}$$

b Current $= \dfrac{\text{potential difference}}{\text{resistance}}$

$$I = \frac{V}{R} = \frac{6.8\,\text{V}}{7.4\,\Omega} = \mathbf{0.92\,A}$$

2 Voltage = current × resistance

$V = I \times R = 10 \times 10^{-3}\,\text{A} \times 200\,\Omega = \mathbf{2.0\,V}$

3 Voltage = current × resistance

$V = I \times R = 30 \times 10^{-3}\,\text{A} \times 100\,\Omega = \mathbf{3.0\,V}$

4 When you are told the size of the current it is simplest to use $P = I^2R$, which can be rearranged to give

$$R = \frac{P}{I^2} = \frac{100\,\text{W}}{(0.43\,\text{A})^2} = \mathbf{540\,\Omega}$$

5 $P = I^2R$

$$R = \frac{P}{I^2} = \frac{2000\,\text{W}}{(8.7\,\text{A})^2} = \mathbf{26\,\Omega}$$

6 This time the potential difference, V, is supplied so it is simplest to use $P = V^2/R$, which can be rearranged to give

$$R = \frac{V^2}{P} = \frac{(230\,\text{V})^2}{60\,\text{W}} = \mathbf{880\,\Omega}$$

If you can only remember $P = VI$, find the quantity not supplied and then find R using $R = V/I$.

7 **A**, **B**, **D** and **E** are in series.

Hint: Look to see if the same current has to go through both resistors. If it does then they are in series.

8 **C** and **F** show a parallel arrangement.

Hint: If there is a junction so that some current goes through one of the resistors while some goes through another, before joining up again, then the resistors are in parallel.

Level 2

9 To be safe, at say, 250 V, you would need

$$R = \frac{V}{I} = \frac{250\,\text{V}}{1 \times 10^{-3}\,\text{A}} = \mathbf{0.25\,M\Omega} \text{ or larger}$$

10 $R = \dfrac{V}{I} = \dfrac{9.0\,\text{V}}{0.44\,\text{A}}$ (from the graph)

$\therefore R = \mathbf{20.5\,\Omega}$

Notice that at 12.0 V the ratio of V to I is 24.0 Ω. The resistance increases. Also note that the resistance is ***not*** *equal to the slope of the curve.*

11 **B reads 1.0 A** and **C reads 2.0 A**, because the branch with meter B has twice the resistance (and the same pd) and so only half of C's current will go that way. The currents in the two branches must add up to 3.0 A, by Kirchhoff's First Law.

12 a $R = \dfrac{\rho \times l}{A}$

$$= \frac{47 \times 10^{-8}\,\Omega\,\text{m} \times 3.0\,\text{m}}{0.060 \times 10^{-6}\,\text{m}^2} = \mathbf{23.5\,\Omega}$$

b **i** The resistance of a wire is directly proportional to its length, so with twice the length the resistance will double, to $2 \times 23.5\,\Omega = \mathbf{47\,\Omega}$.

ii If the length is reduced to one-third (of 3.0 m) the resistance is also reduced to one-third, to $\dfrac{23.5\,\Omega}{3} = \mathbf{7.8\,\Omega}$.

iii Resistance is inversely proportional to the area of cross-section, so if the cross-sectional area of a wire is halved the resistance is doubled, to $2 \times 23.5\,\Omega = \mathbf{47\,\Omega}$.

iv If the diameter is halved the cross-sectional area is divided by 4, so the resistance goes up by a factor of 4, to $4 \times 23.5\,\Omega = \mathbf{94\,\Omega}$.

13 a As the temperature rises the energy of the lattice atoms increases and they vibrate more. This makes it harder for the electrons to flow so there is a higher resistivity.

b The extra thermal energy in this case frees more charge carriers. This increases the current flowing so it decreases the resistivity of the semiconductor.

Level 3

14 Cross-sectional area $= (0.010\,\text{m})^2 = 1.0 \times 10^{-4}\,\text{m}^2$

$$R = \frac{\rho l}{A} = \frac{25\,\Omega\,\text{m} \times 0.014\,\text{m}}{1.0 \times 10^{-4}\,\text{m}^2}$$

$= \mathbf{3500\,\Omega}$

15 The diameter of X is three times the diameter of Y, so, because cross-sectional area is proportional to the square of the diameter, cross-sectional area of X = 3^2 × cross-sectional area of Y.

area of X = 9 × area of Y

But resistance is inversely proportional to cross-sectional area, so

resistance of X $= \dfrac{\text{resistance of Y}}{9}$

and resistance of Y = 9 × resistance of X

$\therefore \quad R_Y = 9 \times R_X = 9 \times 3.0\,\Omega = \mathbf{27\,\Omega}$

16 Rearrange the equation $R = \dfrac{\rho l}{A}$ to give

$$A = \frac{\rho l}{R}$$

The mark is 2.0 mm (2.0×10^{-3} m) wide, so the cross-sectional area $A = 2.0 \times 10^{-3} \times t\,\text{m}^2$, where t is the thickness in metres, and l = 20 mm = 0.020 m.

So $\quad 2.0 \times 10^{-3}\,\text{m} \times t\,\text{m} = \dfrac{9.5 \times 10^{-5}\,\Omega\,\text{m} \times 0.020\,\text{m}}{6.2 \times 10^{3}\,\Omega}$

and $\quad$ thickness, $t = \dfrac{9.5 \times 10^{-5}\,\Omega\,\text{m} \times 0.020\,\text{m}}{6.2 \times 10^{3}\,\Omega \times 2.0 \times 10^{-3}\,\text{m}}$

$= \mathbf{1.5 \times 10^{-7}\,m}$

5 Circuit behaviour

Internal cell resistance

In real circuits, not all of the energy given to the charge flowing through a cell (see topic 3) is dissipated in the external part of the circuit. Some energy is dissipated within the cell itself, which has **internal resistance**.

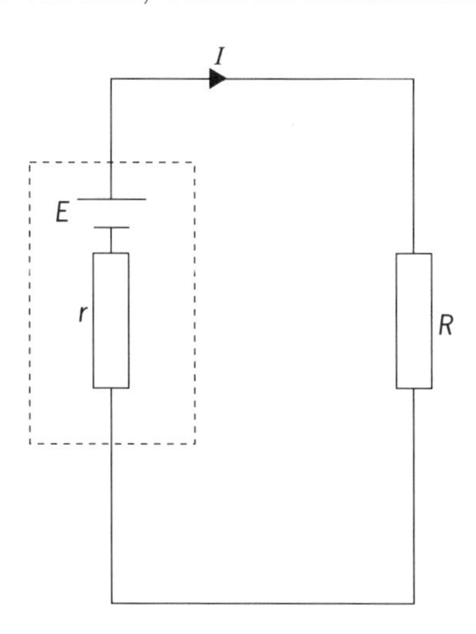

E = emf of cell
r = internal resistance of cell
R = external resistance of circuit
I = current flowing around circuit

Figure 5.1 Internal resistance of a cell

From the conservation of energy:

energy supplied by the cell per second = energy dissipated in the external part of the circuit per second + energy dissipated in the cell per second

that is, $EI = VI + vI$

where E is the emf of the cell, V is called the terminal pd of the cell and v is known as the 'lost volts'. As I is the same all round a series circuit, this will cancel, giving

$$E = V + v$$

or (using $V = IR$) $E = IR + Ir$

$$\boxed{E = I(R + r)}$$

where R is the resistance of the external part of the circuit and r is the internal resistance of the cell (Figure 5.1).

Example 1

A current of 0.2 A flows when a cell is connected across a 5.0 Ω resistor. Calculate the emf of the cell if its internal resistance is 1.0 Ω.

$$E = I(R + r)$$
$$= 0.2\,\text{A}\,(5.0\,\Omega + 1.0\,\Omega)$$
$$= 1.2\,\text{V}$$

Note: If a voltmeter is connected across a cell that is not producing current, it will measure the emf of the cell. (Figure 5.2a). If a voltmeter is connected across a cell which is producing current, it will measure the terminal pd of the cell (Figure 5.2b).

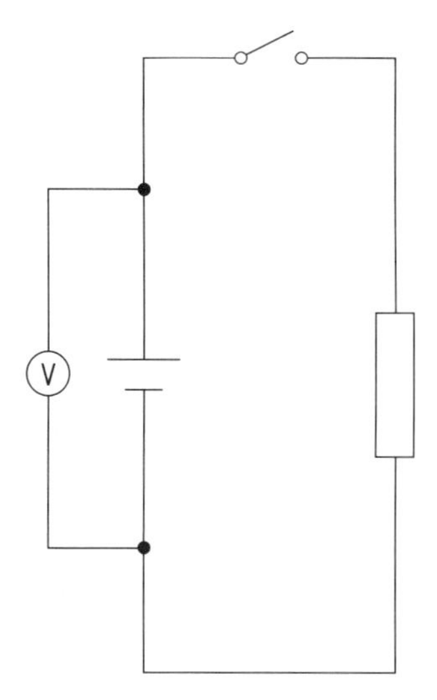

a With the switch open the voltmeter measures the emf of the cell

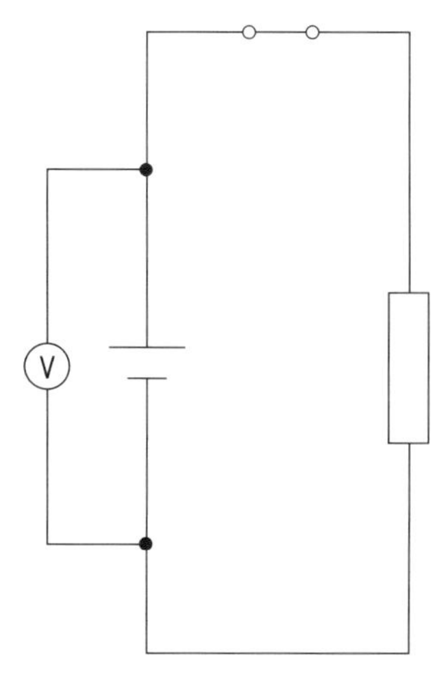

b With the switch closed the voltmeter measures the terminal pd of the cell

Figure 5.2

Kirchhoff's Laws

These are laws that describe the behaviour of currents and voltages in circuits.

Kirchhoff's First Law

The sum of the currents flowing into a circuit junction must be equal to the sum of the currents flowing out of the junction.

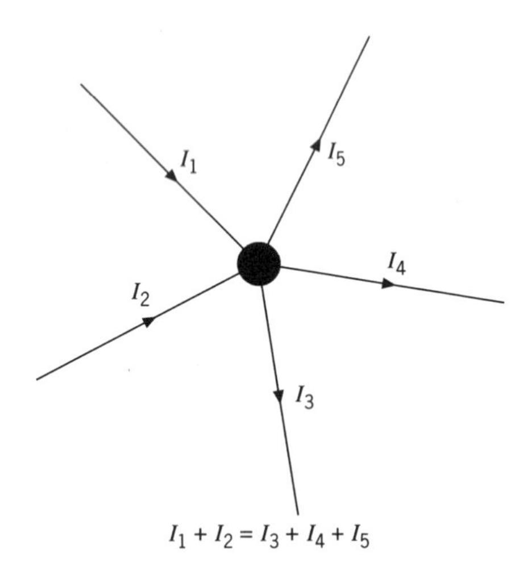

Figure 5.3 Kirchhoff's First Law at a circuit junction

Example 2

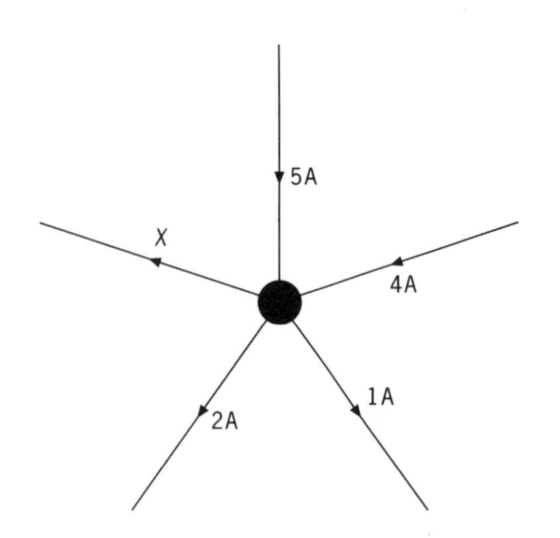

Figure 5.4

Calculate the current X leaving the junction.

Using Kirchhoff's First Law:

current into junction = current out of junction

$$5\text{A} + 4\text{A} = 1\text{A} + 2\text{A} + X$$

$$\therefore \quad X = 6\text{A}$$

Kirchhoff's Second Law

In any closed loop the algebraic sum of the emfs is equal to the algebraic sum of the pds.

$$\sum E = \sum IR$$

'Algebraic' means that emfs and pds causing current in one direction (say clockwise) are taken as positive and those causing current in the opposite (anticlockwise) direction as negative.

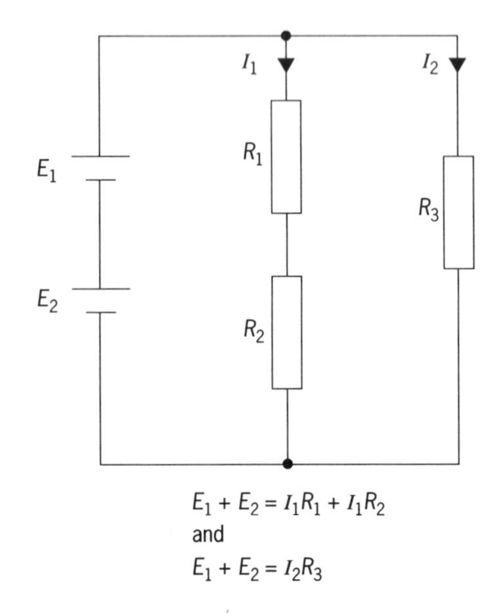

Figure 5.5 Kirchhoff's Second Law in a parallel circuit

Example 3

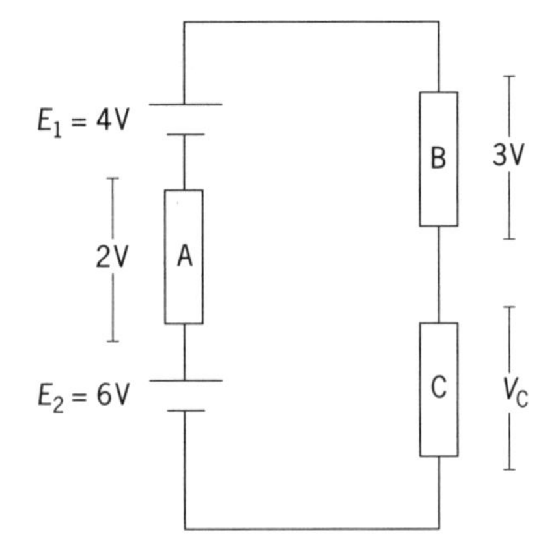

Figure 5.6

Calculate the pd across the resistor C if the internal resistances of cells E_1 and E_2 are negligible and the pds across resistors A and B are 2V and 3V respectively.

Using Kirchhoff's Second Law:

$$\sum E = \sum IR$$

Both cells 'face the same way' so their emfs add.

$$4\text{V} + 6\text{V} = 2\text{V} + 3\text{V} + V_C$$

$$\therefore \quad V_C = 5\text{V}$$

The potential divider

If two or more resistors are connected in series, as shown in Figure 5.7, they can form a **potential divider**. The pd applied across them is divided between AB and BC.

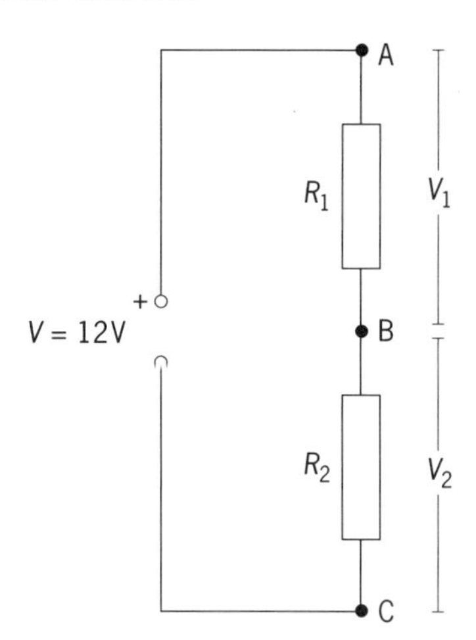

Figure 5.7 A simple potential divider

The magnitude of the pd across each resistor depends upon the relative resistances of the two resistors. If both resistors have the same value they will each have the same pd, here 6V. If R_1 has a value which is twice R_2, the pd across R_1 will be twice that across R_2.

In general, the pd across R_1 is given by

$$V_1 = \left(\frac{R_1}{R_1 + R_2}\right)V$$

and that across R_2 is given by

$$V_2 = \left(\frac{R_2}{R_1 + R_2}\right)V$$

By varying the values of R_1 and R_2 we can adjust the values of the pd across each resistor.

Example 4

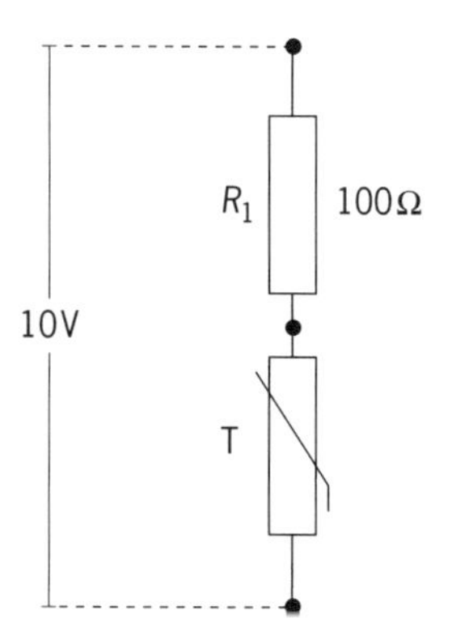

Figure 5.8

The resistance of thermistor T decreases as its temperature increases. As the resistance decreases the pd across T falls and the pd across R_1 increases. Calculate the pd across R_1 when the resistance of T falls to 120 Ω.

$$V_1 = \left(\frac{R_1}{R_1 + R_2}\right)V$$

where R_2 here is the resistance of T.

$$V_1 = \left(\frac{100\,\Omega}{100\,\Omega + 120\,\Omega}\right)10\,\text{V}$$
$$= 4.5\,\text{V}$$

Sometimes a potential divider is in the form of a length of wire of uniform resistance with a sliding contact (Figure 5.9). As the contact is moved along the wire the pd between A and C and that between B and C vary continuously, but they always add up to the pd between A and B.

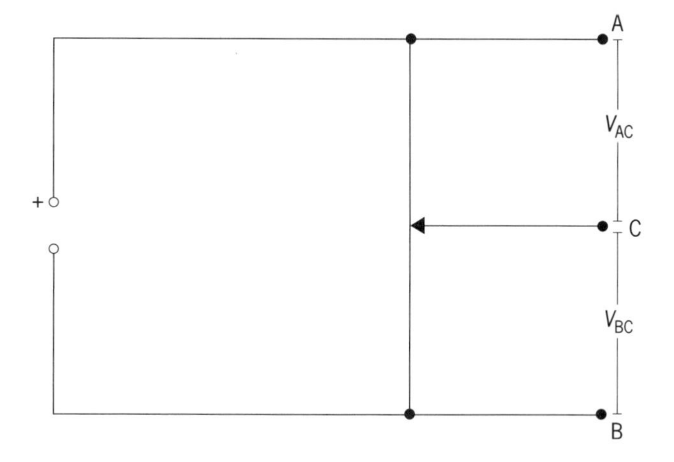

Figure 5.9 A continuously variable potential divider

Level 1

1 A cell has internal resistance, r, of 0.80 Ω and is connected to a circuit consisting of a 10.0 Ω resistor, R. If a current of 0.14 A flows, what is the emf of the cell?

2 When a current of 2.0 A flows the pd across a battery's terminals falls from 12.0 V to 9.0 V.

a What is the value of the 'lost volts'?

b What is the internal resistance of the battery?

3 **a** Calculate the current marked I in Figure 5.10.

Hint: Use Kirchhoff's First Law.

b Calculate the 'lost volts'.

c Calculate the pd across the parallel resistors.

Hint: Use Kirchhoff's Second Law.

d Use this result to verify that the current you calculated in part a is correct.

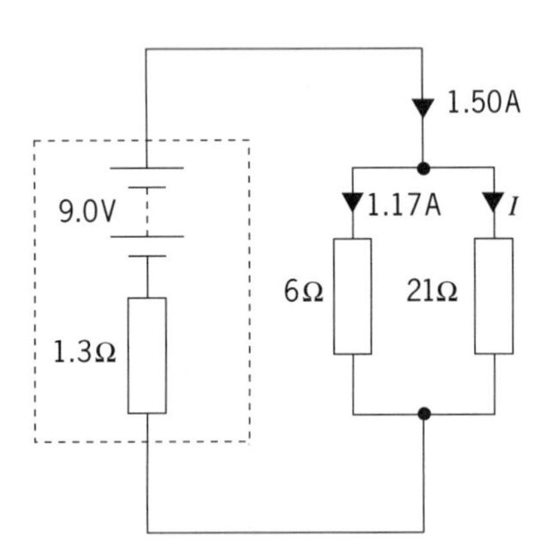

Figure 5.10

4 Figure 5.11 shows a potential divider.

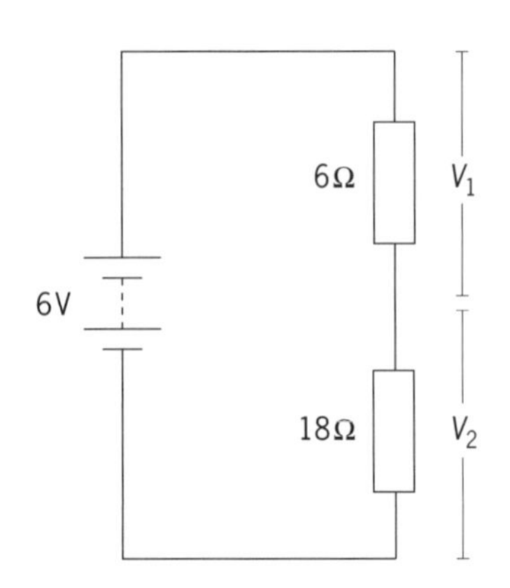

Figure 5.11

a What is the value of V_1?

b What is the value of V_2?

5 Calculate the voltmeter readings you would expect to see on a meter connected across each marked position in circuits A, B and C in Figure 5.12. Justify the value of the total effective resistance of the circuit in each case.

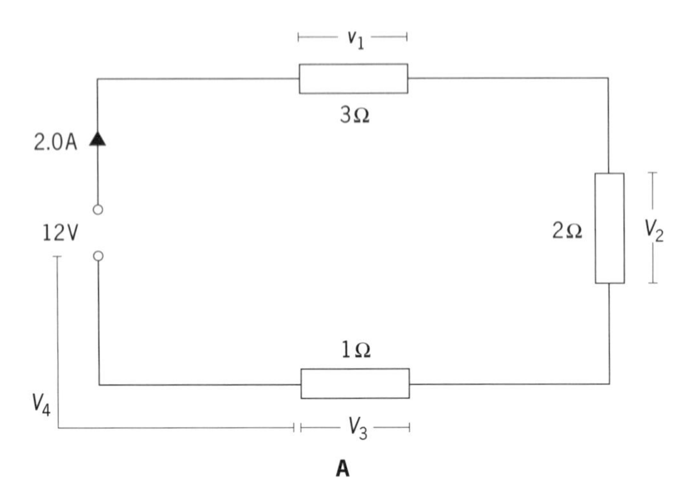

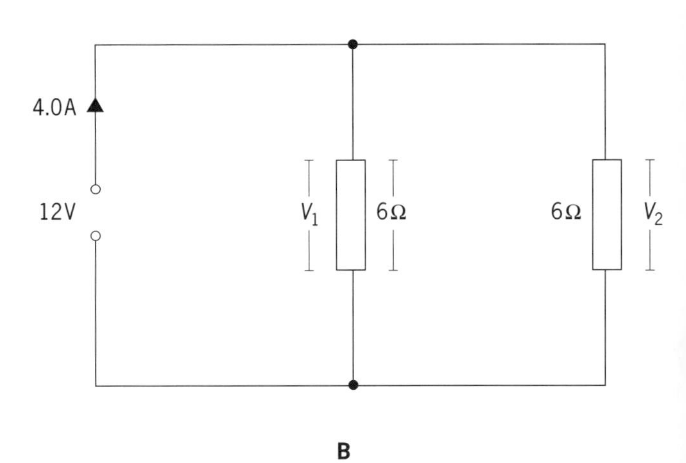

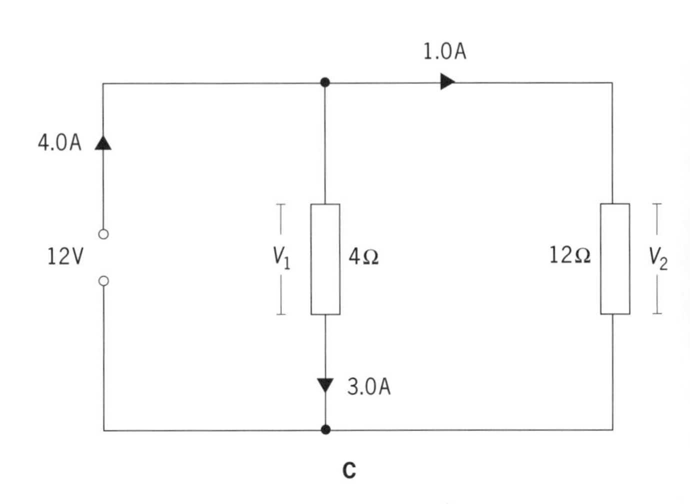

Figure 5.12

Level 2

6

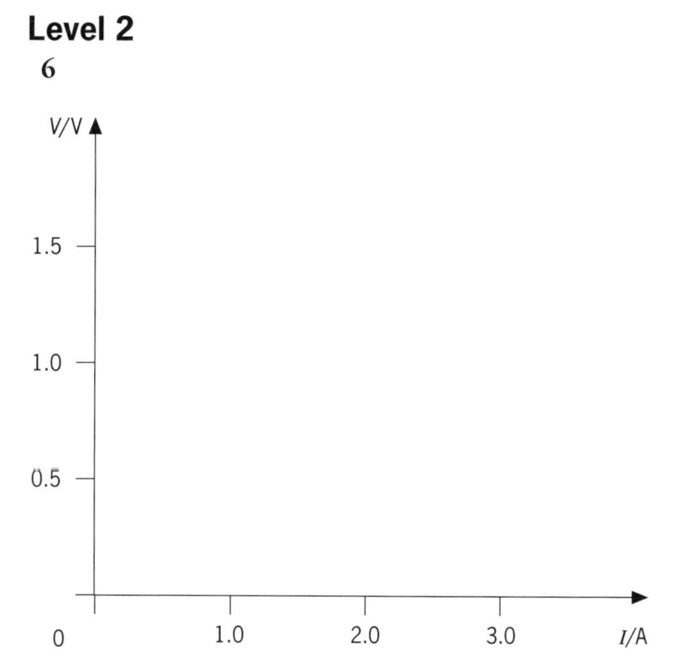

Figure 5.13

Copy the axes in Figure 5.13 and sketch a graph to show the fall in pd when currents of

a 1.0 A
b 2.0 A
c 3.0 A

are drawn from a cell. The internal resistance, r, is $0.50\,\Omega$ and the emf of the cell is 1.5 V.

7 a Calculate the current through the battery in circuits A and B, shown in Figure 5.14.

Hint: You need to calculate the total effective resistance. Combine the parallel resistors first, then add the series resistors.

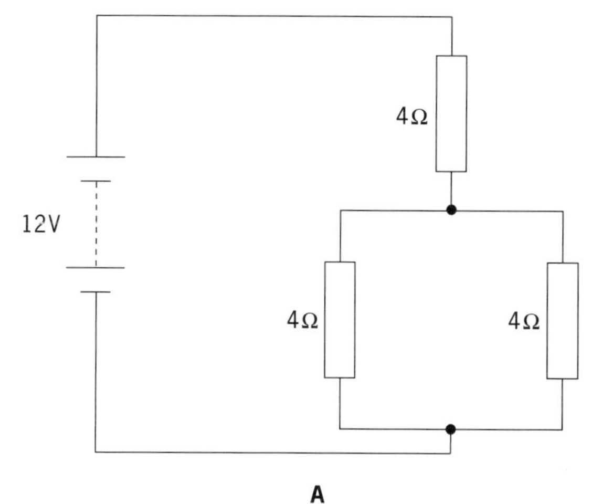

Figure 5.14

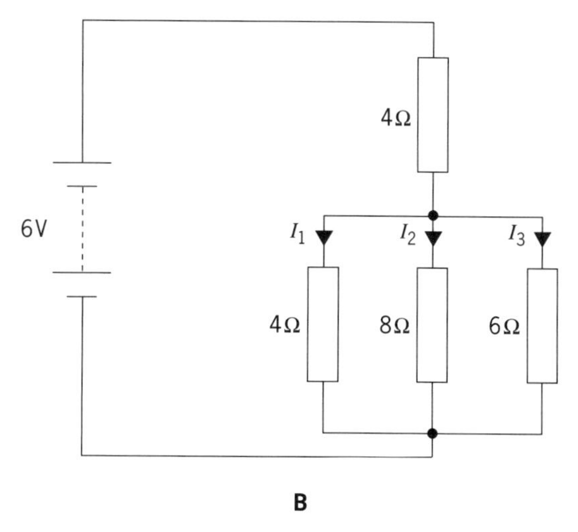

Figure 5.14 continued

b Calculate the current through each of the parallel resistors and the potential difference across each of the resistors in circuits A and B in Figure 5.14.

8 The label on my 12 volt car battery claims that it will deliver 335 A when short-circuited for 30 s. What is the maximum possible value of its internal resistance?

9 The wire XY in the circuit of Figure 5.15 is made of a material of high resistivity and is of uniform cross-sectional area so that its resistance, R, is proportional to its length, l. Its resistivity does not depend on the current flowing through it (it obeys Ohm's Law) so the potential difference, V, between one end and any other point, for a given value of current, is also proportional to l.

What is the potential difference between X and the 65 cm mark in the circuit? (Assume all connecting wires have zero resistance.)

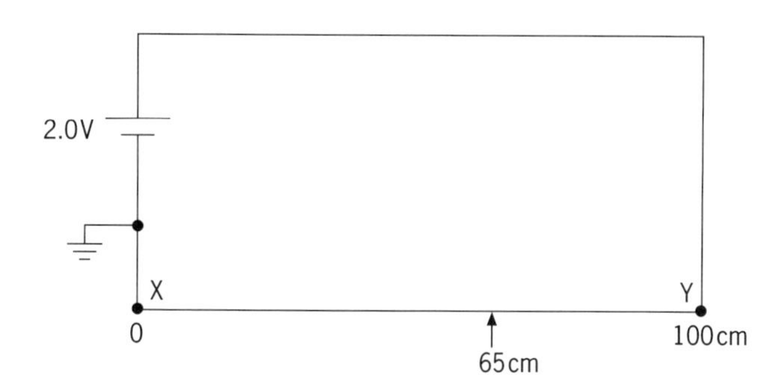

Figure 5.15

Level 3

The internal resistance of a power supply can be treated exactly like a resistor in series; there is no need for different equations. Compare the two parts of question 10.

10 a In circuit A, Figure 5.16, calculate the current flowing round the circuit and the potential difference across each of the resistors, assuming that the battery has zero internal resistance.

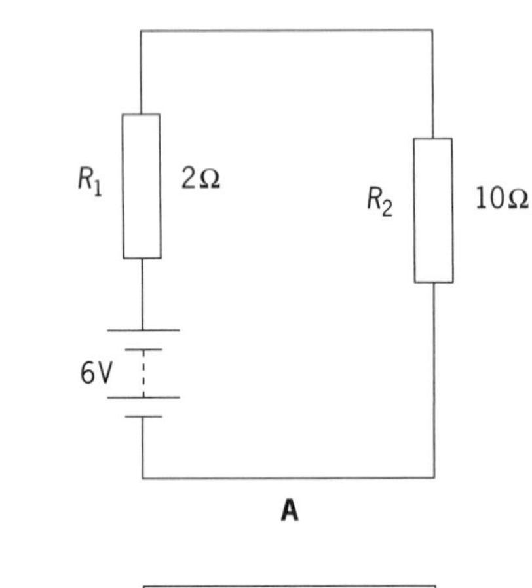

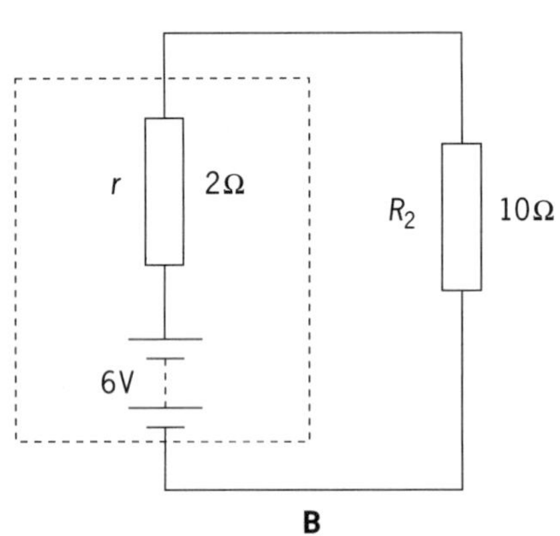

Figure 5.16

b Now repeat the calculation, taking the resistance R_1 to be the internal resistance of the battery, r, as shown in circuit B.

11 Three cells of emf 1.5 V and internal resistance $r = 0.30\,\Omega$ are joined

a in series
b in parallel
c two in parallel, in series with the third.

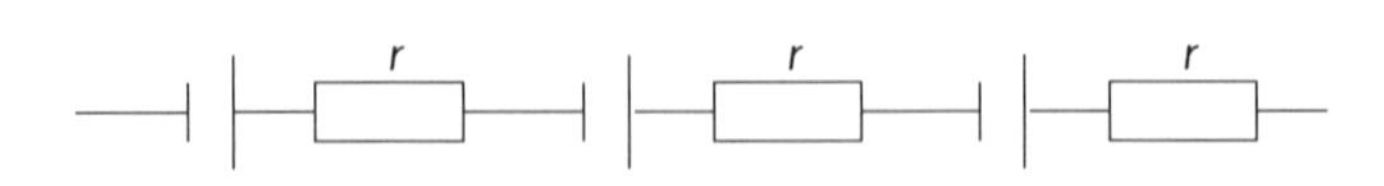

a

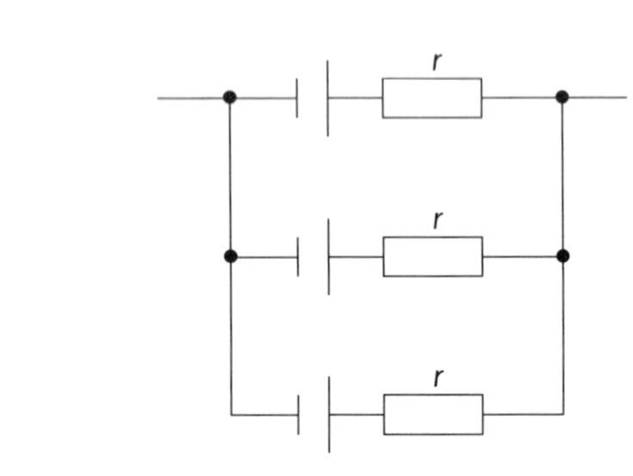

b

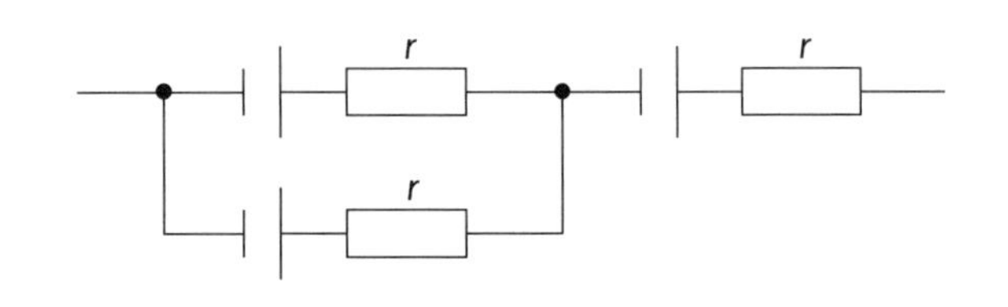

c

Figure 5.17

Calculate the effective emf and internal resistance in each case.

Level 1

1 The emf, E, is equal to the sum of the potential differences around the circuit.

$$E = IR + Ir = I(R + r)$$
$$= 0.14\,\text{A}(10.0\,\Omega + 0.80\,\Omega)$$
$$= 0.14 \times 10.8\,\text{V}$$
$$= \mathbf{1.5\,V}$$

2 a The 'lost volts' is the difference between the emf of 12.0 V and the terminal pd of 9.0 V, that is 12.0 − 9.0 = **3.0 V**, which are not available to the outside circuit and so are 'lost'.

This fall in voltage occurs because energy is used as the current flows through the cell itself.

b The 'lost volts' $v = Ir$ where r is the internal resistance; this can be rearranged to give

$$r = \frac{v}{I} = \frac{3.0\,\text{V}}{2.0\,\text{A}} = \mathbf{1.5\,\Omega}$$

3 a The current flowing into the junction equals the sum of the currents flowing out (Kirchhoff's First Law).

$$1.50\,\text{A} = I + 1.17\,\text{A}$$
$$\therefore \quad I = 1.50\,\text{A} - 1.17\,\text{A}$$
$$= \mathbf{0.33\,A}$$

b The 'lost volts' $= Ir = 1.50\,\text{A} \times 1.3\,\Omega$
$= \mathbf{2.0\,V}$

c The emf = the sum of the potential differences around the circuit (Kirchhoff's Second Law).

emf = 'lost volts' + pd across resistors, V

$$9.0\,\text{V} = 2.0\,\text{V} + V$$
$$\therefore \quad V = \mathbf{7.0\,V}$$

d pd across the 21 Ω resistor $= IR = 7.0\,\text{V}$
so $I = 7.0\,\text{V}/21\,\Omega$
$= \mathbf{0.33\,A}$
which confirms the answer to part a.

4 The two resistors are in the ratio 3:1 so the pds will be in the same ratio.

$V_2{:}V_1$ will be **4.5 V:1.5 V**

Or the pds can be calculated from:

$$V_1 = \frac{R_1}{R_1 + R_2} \times V_{tot} = \frac{6}{24} \times 6\,\text{V} = \mathbf{1.5\,V}$$

$$V_2 = \frac{R_2}{R_1 + R_2} \times V_{tot} = \frac{18}{24} \times 6\,\text{V} = \mathbf{4.5\,V}$$

5 The potential difference across a resistor equals the current through it, I, times the size of the resistor, R.

$$V = I \times R$$

In circuit A

$$V_1 = 2.0\,\text{A} \times 3\,\Omega = \mathbf{6\,V}$$
$$V_2 = 2.0\,\text{A} \times 2\,\Omega = \mathbf{4\,V}$$
$$V_3 = 2.0\,\text{A} \times 1\,\Omega = \mathbf{2\,V}$$

The wire connecting the resistor to the negative terminal of the cell will have a resistance that is negligible (probably less than 0.005 Ω), so

$$V_4 = 2.0\,\text{A} \times 0\,\Omega = \mathbf{0\,V}$$

Check: The sum of the voltage drops = 6 + 4 + 2 + 0 = 12 V = the emf of the battery. This is Kirchhoff's Second Law.

The total effective resistance is given by

$$R_{tot} = \frac{V}{I} = \frac{12\,\text{V}}{2.0\,\text{A}} = \mathbf{6\,\Omega}$$

which is equal to the sum of the three significant resistances: 3 + 2 + 1 = 6 Ω.

In circuit B

The current will split into two equal parts at the junction if there are two equal resistors in parallel. So the current in each resistor is 2.0 A. This is Kirchhoff's First Law.

Using $V = I \times R$,

$$V_1 = 2.0\,\text{A} \times 6\,\Omega = \mathbf{12\,V}$$
$$V_2 = 2.0\,\text{A} \times 6\,\Omega = \mathbf{12\,V}$$

Note that the voltage drops are always equal when the resistors are in parallel, whether or not the resistors are equal.

$R = \frac{V}{I}$ so the effective total resistance of the circuit is given by

$$R_{tot} = \frac{12\,\text{V}}{4.0\,\text{A}} = \mathbf{3\,\Omega}$$

which is **not** equal to the sum of the separate resistors. The expression for resistors in parallel is

$$\frac{1}{R_{tot}} = \frac{1}{R_1} + \frac{1}{R_2}$$

In this case

$$\frac{1}{6\Omega} + \frac{1}{6\Omega} = \frac{2}{6\Omega} = \frac{1}{3\Omega}$$

then take the reciprocal to get the total resistance, 3 Ω.

Note that two identical resistors in parallel halve the effective resistance. Check what happens with three, four, and so on. Use a simple number in your calculations. Can you see the pattern? Use this as a quick way of calculating the resistance of any number of identical resistors and amaze your friends!

In circuit C

You are told the currents through each resistor so it is easy to find the potential differences using $V = I \times R$.

$$V_1 = 3.0\,\text{A} \times 4\,\Omega = \mathbf{12\,V}$$
$$V_2 = 1.0\,\text{A} \times 12\,\Omega = \mathbf{12\,V}$$

The potential differences are equal because the resistors are in parallel.

Effective resistance, $R_{tot} = \frac{12\,\text{V}}{4.0\,\text{A}} = \mathbf{3\,\Omega}$

Check: $\frac{1}{R_{tot}} = \frac{1}{4\Omega} + \frac{1}{12\Omega}$

$$= \frac{3}{12\Omega} + \frac{1}{12\Omega} = \frac{4}{12\Omega}$$

so $R_{tot} = \frac{12\Omega}{4} = 3\Omega$

Level 2

6 First calculate the 'lost volts' $= Ir$.

a $1.0\,A \times 0.50\,\Omega = 0.5\,V$

b $2.0\,A \times 0.50\,\Omega = 1.0\,V$

c $3.0\,A \times 0.50\,\Omega = 1.5\,V$

Subtract the 'lost volts' from the emf.

a $1.5\,V - 0.5\,V = 1.0\,V$

b $1.5\,V - 1.0\,V = 0.5\,V$

c $1.5\,V - 1.5\,V = 0\,V$

So the graph is as shown in Figure 5.18.

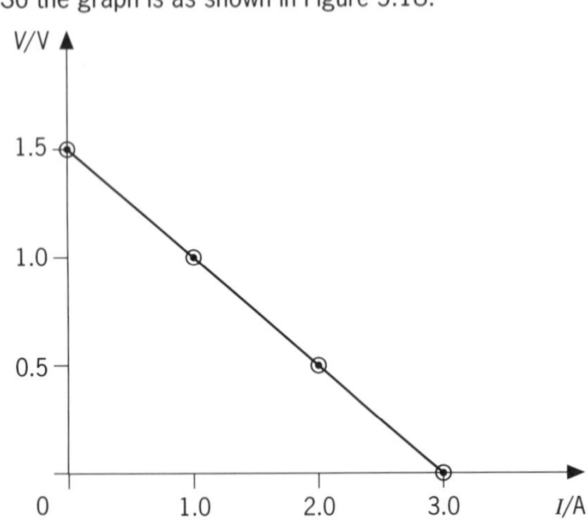

Figure 5.18

7 a *In circuit A*

Combine the two $4\,\Omega$ resistors in parallel using

$$\frac{1}{R_{tot}} = \frac{1}{R_1} + \frac{1}{R_2} = \frac{1}{4} + \frac{1}{4} = \frac{2}{4} = \frac{1}{2}\,\Omega^{-1}$$

$$R_{tot} = \frac{2}{1} = 2\,\Omega$$

This $2\,\Omega$ is in series with another $4\,\Omega$ resistor, so the total effective resistance of the circuit is $2\,\Omega + 4\,\Omega = 6\,\Omega$. This effective resistance of the whole circuit determines the current flowing from the battery.

$$I = \frac{V}{R} = \frac{12\,V}{6\,\Omega} = \mathbf{2.0\,A}$$

In circuit B

Work out the total resistance of the combination of three resistors in parallel first using

$$\frac{1}{R_{tot}} = \frac{1}{R_1} + \frac{1}{R_2} + \frac{1}{R_3} = \frac{1}{4} + \frac{1}{8} + \frac{1}{6} = 0.54\,\Omega^{-1}$$

Try different ways of doing this calculation. You could use the $\frac{1}{x}$ or x^{-1} button, or use fractions with 24 as the common denominator.

So for the three resistors in parallel,

$$R_{tot} = \frac{1}{0.54\,\Omega^{-1}} = 1.85\,\Omega$$

Then we add this to the $4\,\Omega$ which is in series with the three resistors in parallel. So the total effective resistance of the circuit is $4\,\Omega + 1.85\,\Omega = 5.85\,\Omega$. Calculate the current using

$$I = \frac{V}{R} = \frac{6\,V}{5.85\,\Omega} = \mathbf{1.03\,A}$$

b *In circuit A*

The current through the resistor in series with the battery is 2.0 A. This current must split equally so that the two resistors in parallel each pass a current of **1.0 A**.

You can find the potential difference across the series resistor by using

$$V = I \times R = 2.0\,A \times 4\,\Omega = \mathbf{8.0\,V}$$

You can find the potential difference across each of the two parallel resistors – which is the same for both resistors – by again using

$$V = I \times R = 1.0\,A \times 4\,\Omega = \mathbf{4.0\,V}$$

Note that these two potential differences add up to the emf of the battery, $8 + 4 = 12\,V$, as required by Kirchhoff's Second Law.

In circuit B

The $4\,\Omega$ resistor in series with the battery has all the current through it so the potential difference across it is

$$V = I \times R = 1.03\,A \times 4\,\Omega = \mathbf{4.1\,V}$$

This means the potential difference across each of the parallel resistors must be, by Kirchhoff's Second Law,

$$6\,V - 4.1\,V = \mathbf{1.9\,V}$$

The currents through the parallel resistors will be

$$I_1 = \frac{V_1}{R_1} = \frac{1.9\,V}{4\,\Omega} = \mathbf{0.47\,A}$$

$$I_2 = \frac{V_2}{R_2} = \frac{1.9\,V}{8\,\Omega} = \mathbf{0.24\,A}$$

$$I_3 = \frac{V_3}{R_3} = \frac{1.9\,V}{6\,\Omega} = \mathbf{0.32\,A}$$

Note that these currents should add up to the current through the series resistor, by Kirchhoff's First Law:

$$0.47\,A + 0.24\,A + 0.32\,A = 1.03\,A$$

8 If the battery is short-circuited with a cable that has a very low resistance then the only significant resistance in the circuit will be the internal resistance of the battery. You can calculate the internal resistance using $E = I(R + r)$. As the resistance R is insignificant this becomes $E = Ir$ so

$$r = \frac{E}{I} = \frac{12\,V}{335\,A} = \mathbf{0.036\,\Omega}$$

9 There is a potential difference of 2.0 V across 100 cm of uniform wire. This means there is a potential gradient of

$$\frac{2.0\,V}{100\,cm} = 0.02\,V\,cm^{-1}$$

So at 65 cm from X, the low potential end, the potential difference must be

$$65\,cm \times 0.02\,V\,cm^{-1} = \mathbf{1.3\,V}$$

The potential difference between Y and this point would be $35\,cm \times 0.02\,V\,cm^{-1} = 0.7\,V$, which demonstrates Kirchhoff's Second Law.

Level 3

10 a In circuit A the resistors are in series, so the total resistance of circuit is given by

$$R_{tot} = 2\,\Omega + 10\,\Omega = 12\,\Omega$$

so the current flowing in the circuit is

$$I = \frac{V}{R_{tot}} = \frac{6\,\text{V}}{12\,\Omega} = \mathbf{0.5\,A}$$

You can calculate the potential difference across the $2\,\Omega$ resistor using

$$V = I \times R$$
$$= 0.5\,\text{A} \times 2\,\Omega = \mathbf{1\,V}$$

Similarly the potential difference across the $10\,\Omega$ resistor is given by

$$V = 0.5\,\text{A} \times 10\,\Omega = \mathbf{5\,V}$$

These potential differences add up to 6 V, the emf of the battery. This is Kirchhoff's Second Law.

b The calculation is identical in circuit B but because the $2\,\Omega$ resistance is 'internal' we now say that the 1 V is the 'lost volts' used to push current through the battery. The terminal potential difference is only 5 V. As before the potential differences add up to the emf (Kirchhoff's Second Law).

11 a The emfs and the internal resistances are in series and are additive.

total emf $= 1.5\,\text{V} + 1.5\,\text{V} + 1.5\,\text{V} = \mathbf{4.5\,V}$
total internal resistance $= 0.30\,\Omega + 0.30\,\Omega + 0.30\,\Omega = \mathbf{0.90\,\Omega}$

b The emfs are in parallel and the total emf is equal to the emf of one cell.

total emf $= \mathbf{1.5\,V}$

The internal resistances have to be combined in the usual way for parallel resistances using

$$\frac{1}{r_{tot}} = \frac{1}{r_1} + \frac{1}{r_2} + \frac{1}{r_3}$$

$$\frac{1}{r_{tot}} = \frac{1}{0.30} + \frac{1}{0.30} + \frac{1}{0.30} = 10.0\,\Omega^{-1}$$

$$r_{tot} = \frac{1}{10.0\,\Omega^{-1}} = \mathbf{0.10\,\Omega}$$

c The single series cell will contribute 1.5 V to the total emf and the two cells in parallel will together contribute a further 1.5 V, giving

total emf $= 1.5\,\text{V} + 1.5\,\text{V} = \mathbf{3.0\,V}$

The single series cell will contribute $0.30\,\Omega$ to the total internal resistance, while the internal resistances of the two cells in parallel must be combined in the usual way using

$$\frac{1}{r_{tot}} = \frac{1}{r_1} + \frac{1}{r_2}$$

$$\frac{1}{r_{tot}} = \frac{1}{0.30} + \frac{1}{0.30} = 6.67\,\Omega^{-1}$$

$$r_{tot} = \frac{1}{6.67\,\Omega^{-1}} = 0.15\,\Omega$$

giving a final value for the effective internal resistance of $0.30\,\Omega + 0.15\,\Omega = \mathbf{0.45\,\Omega}$

6 Scalars and vectors

A **scalar** is a quantity that has only size, or magnitude. Distance, mass, density, energy, power and speed are all examples of scalar quantities. The overall value of several measurements of a scalar quantity is found by adding the individual values together. For example, a piece of chocolate has a mass of 20 g and contains 50 kJ of energy. Four identical pieces of this chocolate will therefore have a total mass of 80 g and contain 200 kJ of energy. Similarly, a hiker who walks 4.0 km in the morning and a further 3.0 km in the afternoon will have travelled a total distance equal to 4.0 km + 3.0 km = 7.0 km.

A **vector** quantity has size *and* direction. Displacement, velocity, acceleration and force are all examples of vector quantities. It is not possible to find the overall value of several measurements of a vector quantity by simply adding together their numerical values. Their directions must also be taken into account. In the above example of the hiker, the total distance and direction from his starting point (his displacement) depends upon the directions in which he was walking. The following two simple cases illustrate this.

If the hiker walks 4.0 km North in the morning and a further 3.0 km North in the afternoon his total displacement is equal to 4.0 km North + 3.0 km North = 7.0 km North (Figure 6.1a).

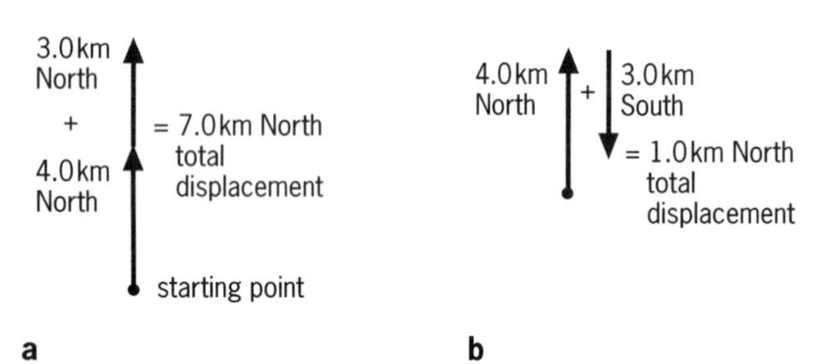

Figure 6.1 Adding displacements

However, if the hiker walks 4.0 km North in the morning and 3.0 km South in the afternoon his total displacement will be equal to 4.0 km North + 3.0 km South = 1.0 km North (Figure 6.1b).

Similarly, because forces are vectors their sizes *and* directions must be taken into account when determining their total effect (Figure 6.2).

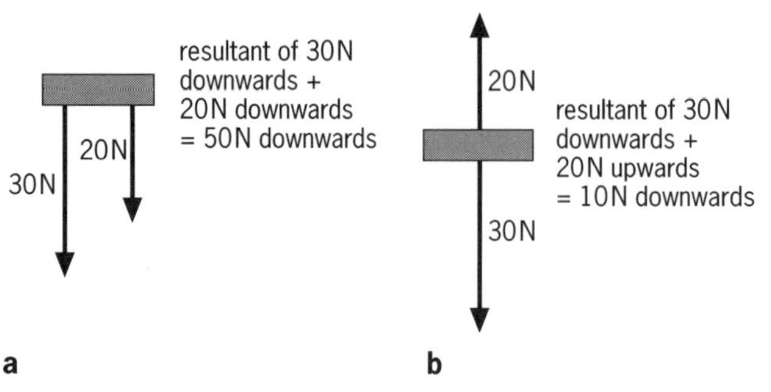

Figure 6.2 Determining the resultant of forces

Vector diagrams

If we have two vectors which do not lie along the same line their resultant can be found by drawing a vector scale diagram.

Example 1

A girl walks 3.0 m due East from point A, stops and then walks 4.0 m due North. How far is the girl from her starting point?

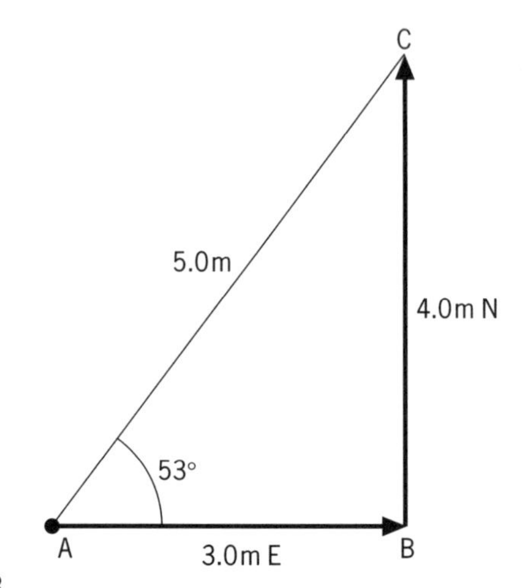

Figure 6.3

From this scale diagram we can see that the girl will be 5.0 m from her starting point in a direction 53° North of East. Her displacement is 5.0 m in the direction 53° North of East.

Alternatively, as the diagram is a right-angled triangle, the final displacement can be found by calculation. Using Pythagoras to find the magnitude:

$$\begin{aligned} AC^2 &= AB^2 + BC^2 \\ &= 3.0^2\,m^2 + 4.0^2\,m^2 \\ &= 25\,m^2 \\ AC &= 5.0\,m \end{aligned}$$

To find the direction:

$$\begin{aligned} \tan\theta &= 4/3 \\ \theta &= 53° \end{aligned}$$

This method can be used for any vectors that are at right angles to each other.

Triangle of vectors

When the two vectors to be added are not at right angles, the more general method of the **triangle of vectors** can be used to find the resultant.

Example 2

What single force could replace a horizontal force of 8.0 N and a force of 6.0 N acting at 45° to the horizontal at some point A, as shown in Figure 6.4?

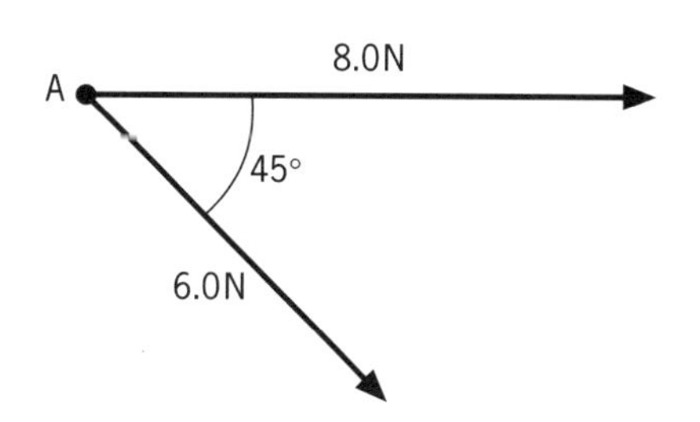

Figure 6.4

The above diagram is first redrawn to scale, so that the applied vectors are head to tail. Then the triangle of vectors is completed by drawing in the resultant vector AC, Figure 6.5.

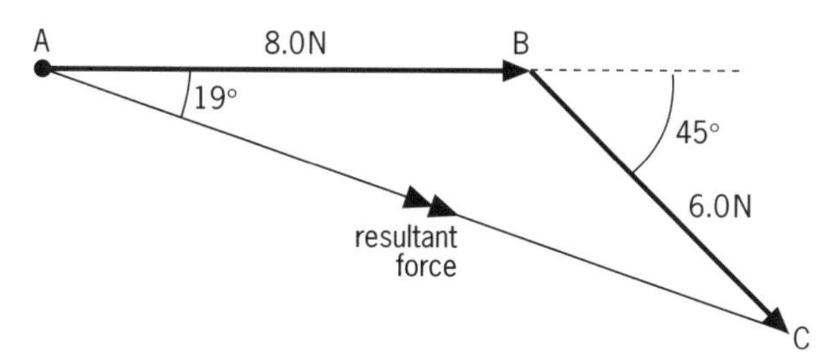

Figure 6.5 Finding a resultant by the triangle of vectors

From the scale diagram it can be seen that the two forces can be replaced by a single force of 13.0 N in a direction of 19° to the horizontal.

Parallelogram of vectors

A second method of adding together vectors that are not at right angles is to use the **parallelogram of vectors**. For example, a passenger walks across a ship at 5.0 m s^{-1} at an angle of 60° to the direction in which the ship is sailing. If the ship has a velocity of 10 m s^{-1} East, how can we calculate the true velocity of the passenger?

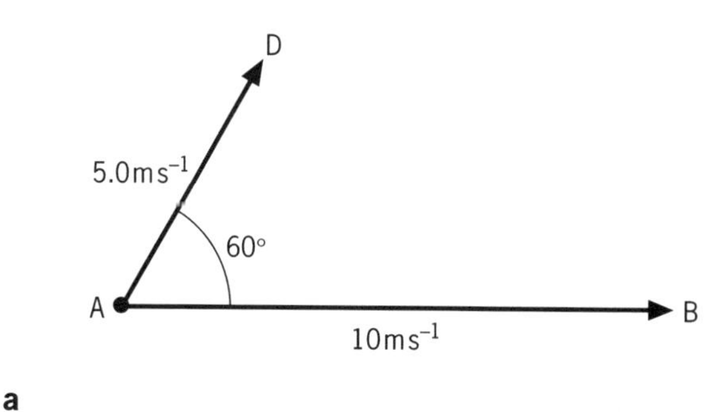

a

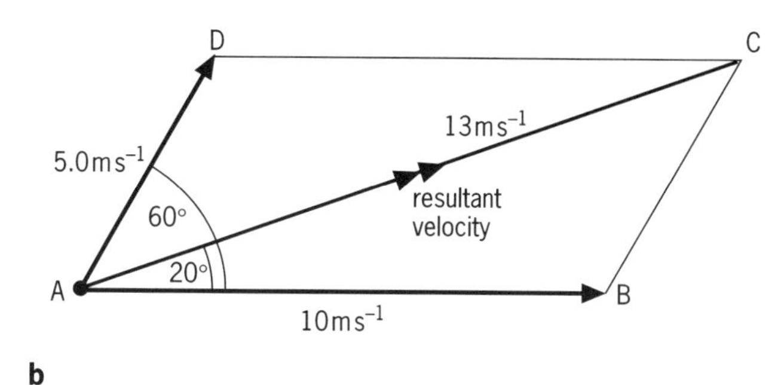

b

Figure 6.6 Finding a resultant by the parallelogram of vectors

1. The vectors are drawn to scale and in the correct directions (AB and AD)
2. The remaining two sides of the parallelogram are drawn (BC and DC)
3. The diagonal of the parallelogram AC represents the magnitude and direction of the resultant.

Balanced forces

Consider the three forces F_1, F_2 and F_3, acting at a point as shown in Figure 6.7a. If these forces are balanced so that the system is in equilibrium then the resultant of any two of the forces, for example F_1 and F_2, must be equal to the third force F_3 but in the opposite direction to it.

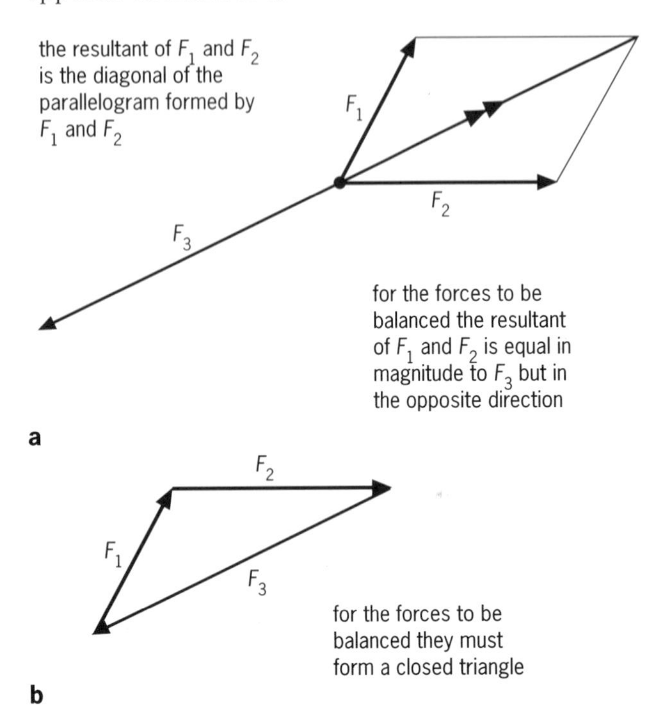

Figure 6.7 Balanced forces

If the three forces are redrawn head to tail, Figure 6.7b, they form a closed triangle if the system is in equilibrium.

Resolving vectors

Whilst adding vectors to find their resultant is useful there are also occasions when we want to carry out the reverse process, that is take a single vector and resolve it into two perpendicular components.

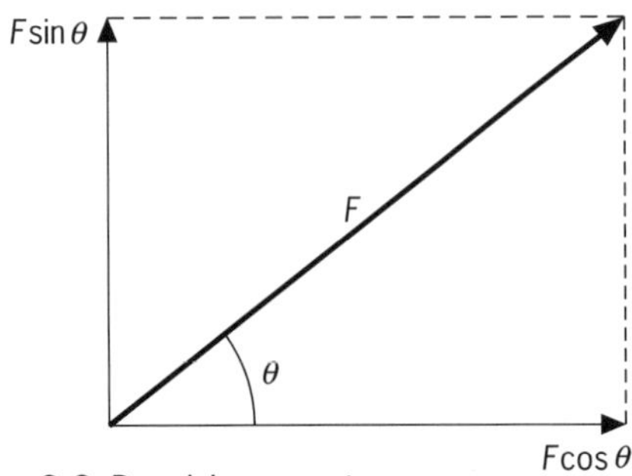

Figure 6.8 Resolving a vector

The force F in Figure 6.8 has been resolved into two components, a horizontal component equal to $F\cos\theta$ and a vertical component equal to $F\sin\theta$. (The component adjacent to the angle used is always the cosine component.)

Example 3

A bullet travels at a constant speed of $400\,\text{m s}^{-1}$ at an angle of 60° to the horizontal. How far will the bullet have travelled **a** horizontally and **b** vertically after 5.0 s? (Ignore any gravitational effects.)

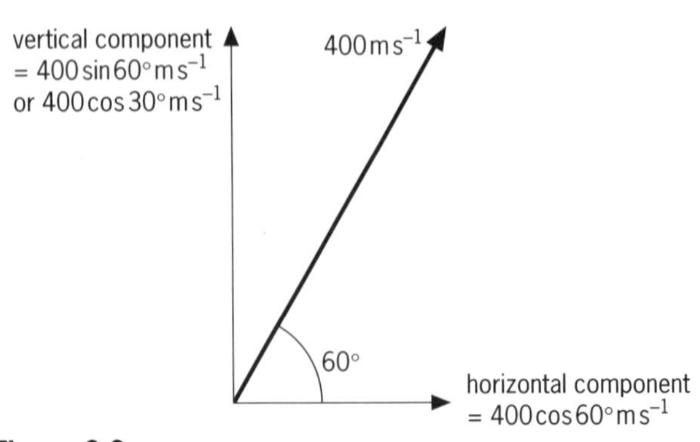

Figure 6.9

a Resolving horizontally,

$$\begin{aligned} v_h &= v\cos 60^\circ \\ &= 400\,\text{m s}^{-1} \times \cos 60^\circ \\ &= 200\,\text{m s}^{-1} \end{aligned}$$

$$\begin{aligned} \text{horizontal distance travelled} &= v_h \times t \\ &= 200\,\text{m s}^{-1} \times 5.0\,\text{s} \\ &= 1000\,\text{m} \end{aligned}$$

b Resolving vertically,

$$\begin{aligned} v_v &= v\cos 30^\circ \\ &= 400\,\text{m s}^{-1} \times \cos 30^\circ \\ &= 346\,\text{m s}^{-1} \end{aligned}$$

$$\begin{aligned} \text{vertical distance travelled} &= v_v \times t \\ &= 346\,\text{m s}^{-1} \times 5.0\,\text{s} \\ &= 1732\,\text{m} \end{aligned}$$

Example 4

Figure 6.10 shows a weight hanging from the centre of a piece of string. Calculate the tension in the string.

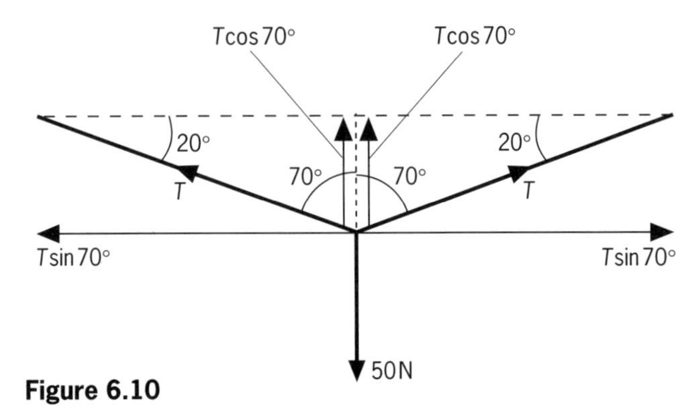

Figure 6.10

The system is in equilibrium so the components of the forces in any one direction must be balanced. Therefore, resolving vertically,

$$\begin{aligned} T\cos 70^\circ + T\cos 70^\circ &= 50\,\text{N} \\ T \times 0.342 + T \times 0.342 &= 50\,\text{N} \\ T &= 73\,\text{N} \end{aligned}$$

(The two horizontal components of the tension are in opposite directions and cancel.)

Level 1

You will need to draw a diagram to explain your answers to most of these questions.

1 Write out the following equations and underline the vector quantities.

$F = ma$ $\quad W = mg$

$\text{speed} = \dfrac{\text{distance}}{\text{time}}$ $\quad \text{velocity} = \dfrac{\text{displacement}}{\text{time}}$

$F = Bvq$ $\quad V = \frac{4}{3}\pi r^3$ $\quad Q = mc\Delta T$

2 Figure 6.11 shows three pairs of vectors. Draw the resultant vector R obtained when each pair is added.

Remember that it is the length and the direction of a vector that matters – the position can be changed.

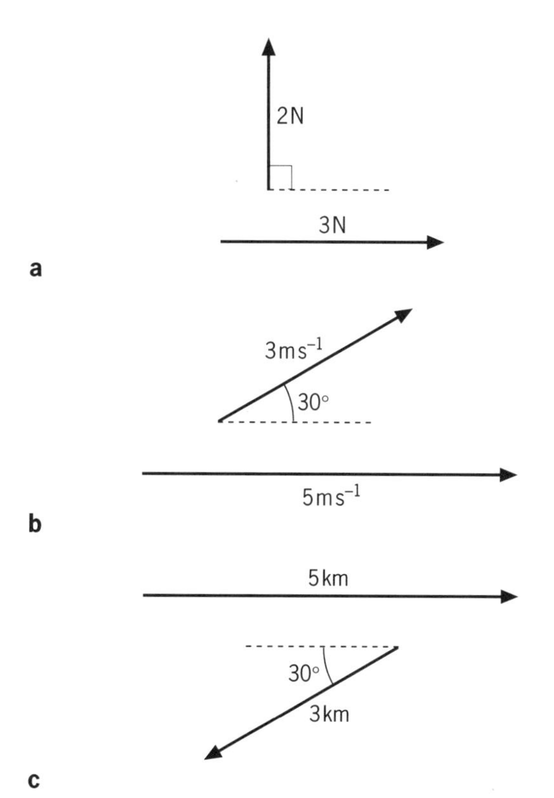

Figure 6.11

3 What is the tension, T, in the section of string in Figure 6.12?

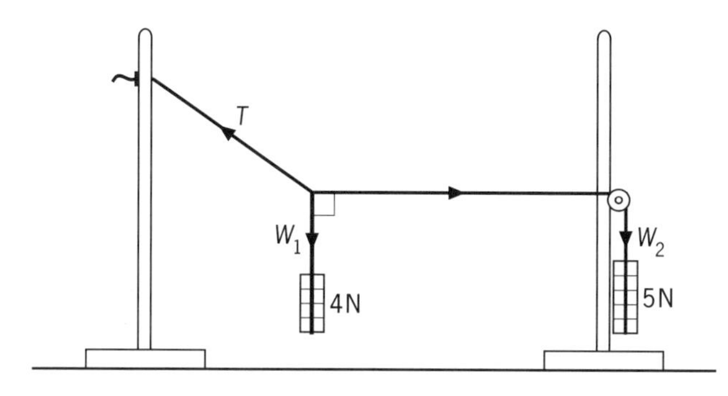

Figure 6.12

4 Trailing around the Shopping Mall you go 120 m East, 200 m North, 80 m West and 40 m South. Calculate

a the total distance you walked

b your final displacement relative to the starting point.

Don't forget a diagram! You can describe the direction using a compass bearing.

5 Three forces, each of 6 N, can be applied together at any angles.

a Sketch the arrangement that gives the largest resultant force and say how large it is.

b Sketch the arrangement that gives the smallest resultant force and say how large it is.

c How could they be arranged to give a resultant of 6 N?

6 a Draw a diagram illustrating the addition of any two vectors A and B. Label the resultant vector R.

b Then repeat this showing the resultant vector $R = A - B$.

Hint: Minus B is in the opposite direction to B.

7 A car of mass 700 kg is on a slope of 20° to the horizontal, Figure 6.13. What is the size of the component of the car's weight acting down the slope?

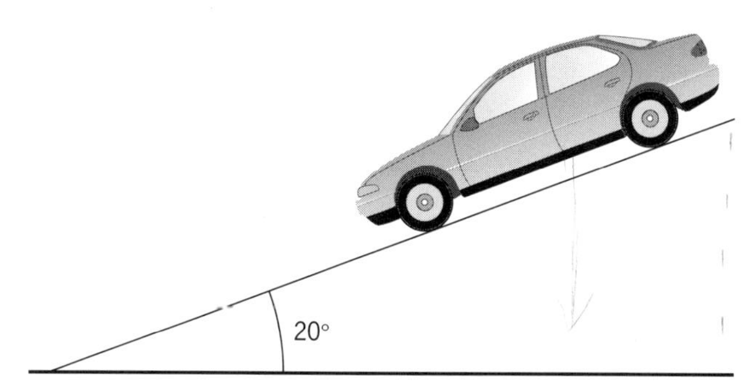

Figure 6.13

8 Your mass is 65 kg. You stand on bathroom scales, which read in newtons. At first they are horizontal so your full weight is recorded. But they are fixed to a board that can be raised at one end. The scales will read only the component of your weight that is at 90° to them.

What would they read as the angle with the horizontal was increased to 20°, 40°, 60°, and 80°?

(Obviously you would need excellent grip between your feet and the scales!)

Level 2

9 A boat moves forwards at $10.0\,\mathrm{m\,s^{-1}}$. You walk across the deck at $1.5\,\mathrm{m\,s^{-1}}$ at an angle of 30° as shown in Figure 6.14.

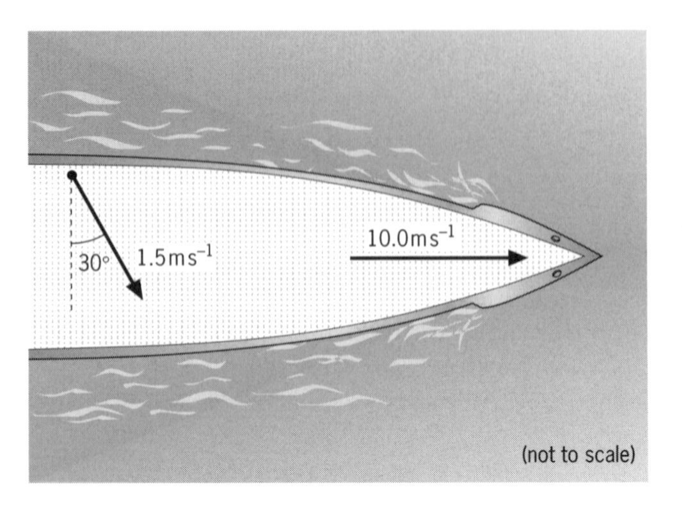

Figure 6.14

a What is the forward component of your velocity relative to the boat?
b What is your total forward velocity?

10 A baby buggy plus baby has a mass of 25 kg. On soft ground it is easier to pull it rather than to push it. You pull with a force of 100 N at an angle of 40° to the horizontal, Figure 6.15.
a What is the force moving the buggy along horizontally?
b What is the force on the ground?

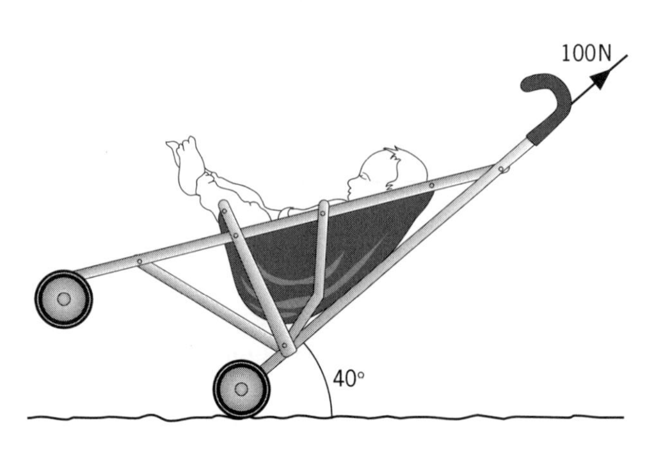

Figure 6.15

c If you **pushed** with the same force at the same angle what would the answers to **a** and **b** be?

11 A horse pulls a canal-boat with a rope at an angle of 15° to the bank, Figure 6.16. If the tension in the rope is 1000 N what is the component of the tension acting along the bank?
Why is it better to use a long rope rather than a short one?

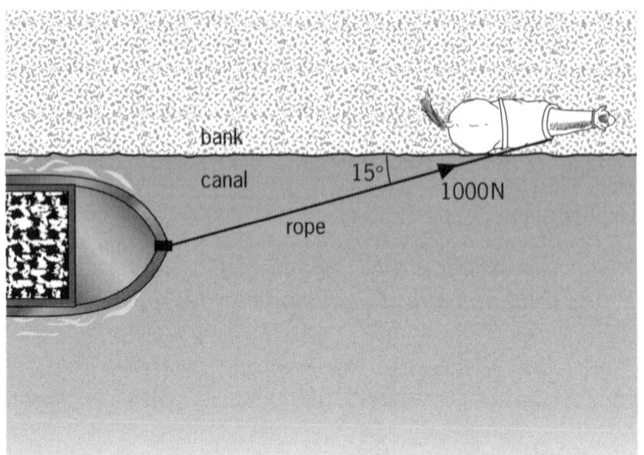

Figure 6.16

Level 3

12 The system shown in Figure 6.17 is in equilibrium. Explain with diagrams how this can be used to check the rule of vector addition.

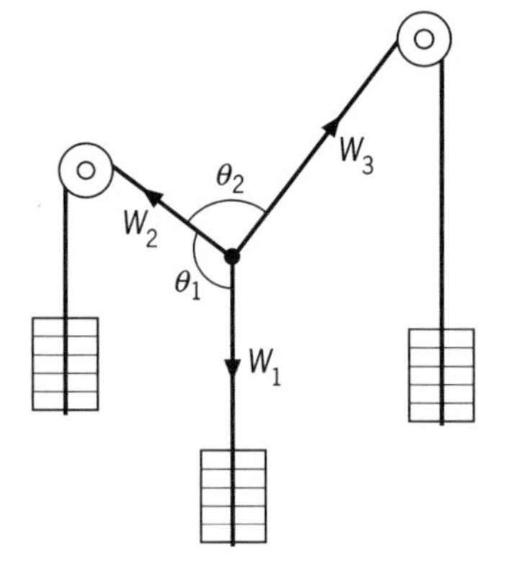

Figure 6.17

13 A boat heads North at $2.4\,\mathrm{m\,s^{-1}}$ across a river that flows East at $1.6\,\mathrm{m\,s^{-1}}$.
a Calculate the resultant velocity of the boat.
b The river is 30 m wide. What is the displacement from the starting point when the boat reaches the far bank?

Level 1

1 The vector quantities are
F, a, W, g, velocity, displacement, F, B, v.

2 a You can use Pythagoras' theorem to calculate the size of the resultant and the tangent gives you the angle.

$$R^2 = 2^2 + 3^2 = 4 + 9 = 13\,\text{N}^2$$
$$\therefore R = \sqrt{13} = 3.6\,\text{N}$$

$$\tan\theta = \frac{\text{opposite}}{\text{adjacent}} = \frac{2}{3} = 0.677$$

$$\therefore \quad \theta = \tan^{-1} 0.667 = 34° \quad \text{(see Figure 6.18a)}$$

$\tan^{-1}$ means 'the angle with a tangent of…'. Most calculators have a $\tan^{-1}$ button.

b and **c** are best solved by drawing to scale, finding the resultant by the triangle of vectors or the parallelogram of vectors and measuring the angle. See Figures 6.18b and c and the following note.

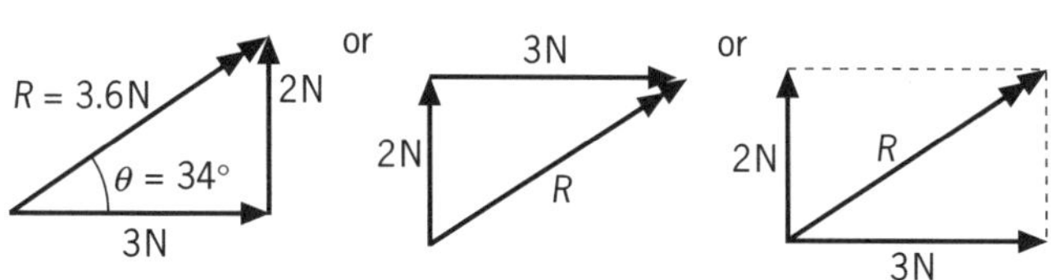

R = **3.6N at 34°**

a

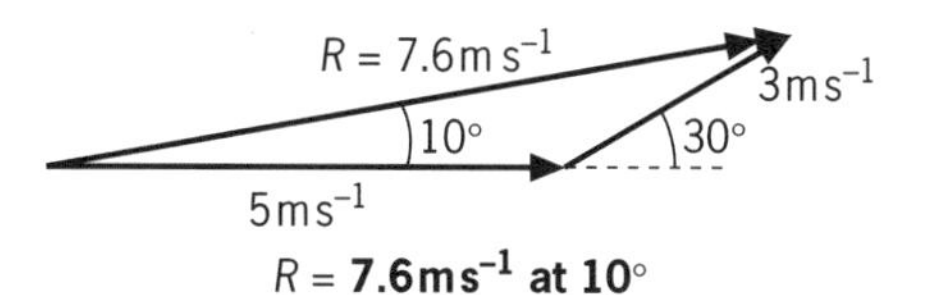

R = **7.6ms⁻¹ at 10°**

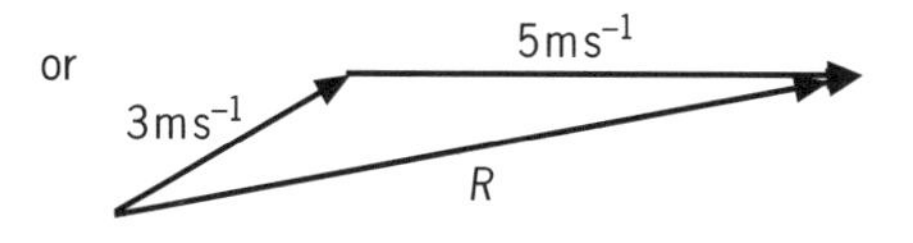

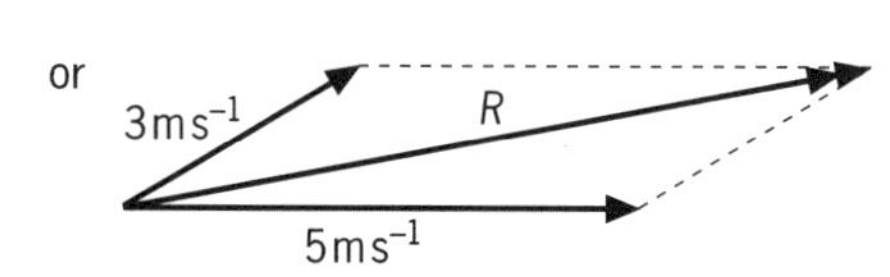

b

5km

34° 30°

R = 2.9km 3km

R = **2.9km at 34°**

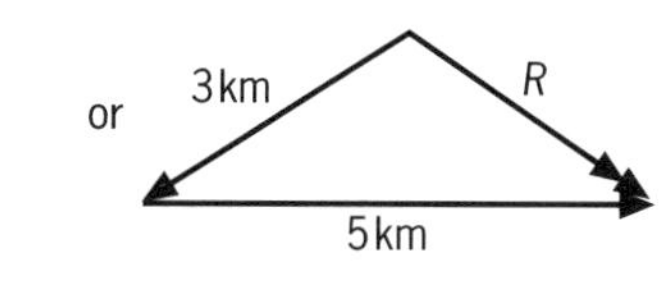

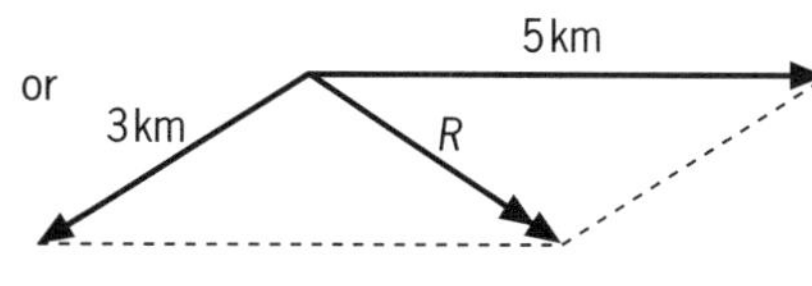

c

Figure 6.18

Note:

- *There are two ways of arranging the vectors in a triangle.*
- *If you use the parallelogram rule be sure to pick the correct diagonal.*
- *The resultant should be the same whichever method is used.*

3

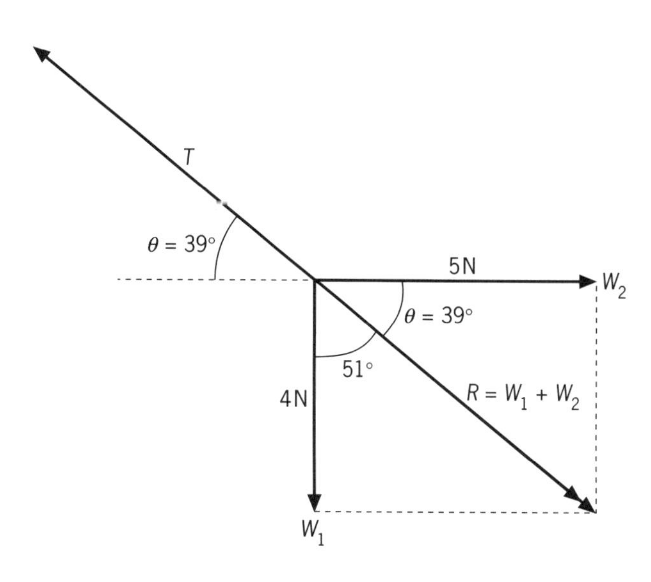

Figure 6.19

T is equal and opposite to the resultant of $W_1 + W_2$. See Figure 6.19. The resultant is calculated using Pythagoras' theorem:

$$T^2 = 5^2 + 4^2 = 25 + 16 = 41\,\text{N}^2$$
$$\therefore T = \sqrt{41} = \mathbf{6.4\,N}$$

$$\tan\theta = \frac{\text{opposite}}{\text{adjacent}} = \frac{4}{5}$$

$\therefore$ angle $\theta = \tan^{-1} 0.8 = 39°$ downwards, clockwise from horizontal (or 51° from vertical)

The tension is the same size but in the opposite direction, that is **39° upwards**, clockwise from horizontal (or 51° from vertical).

4 a The total distance travelled is
$120\text{m} + 200\text{m} + 80\text{m} + 40\text{m} = \mathbf{440\,m}$.

b The displacement is the vector R, from the start to the finish, in Figure 6.20 overleaf. Using Pythagoras' theorem on $\triangle$OXY,

$$R^2 = 40^2 + 160^2 = 1600 + 25\,600 = 27\,200\,\text{m}^2$$
$$R = \sqrt{27200} = 165\,\text{m}$$

$$\tan\theta = \frac{\text{opposite}}{\text{adjacent}} = \frac{160}{40} = 4.0$$

$\therefore$ angle $\theta = \tan^{-1} 4.0 = 76°$ anticlockwise from East

This can be expressed as a bearing of 014° East of North.
So the displacement is **165 m** on bearing **014° East of North**.

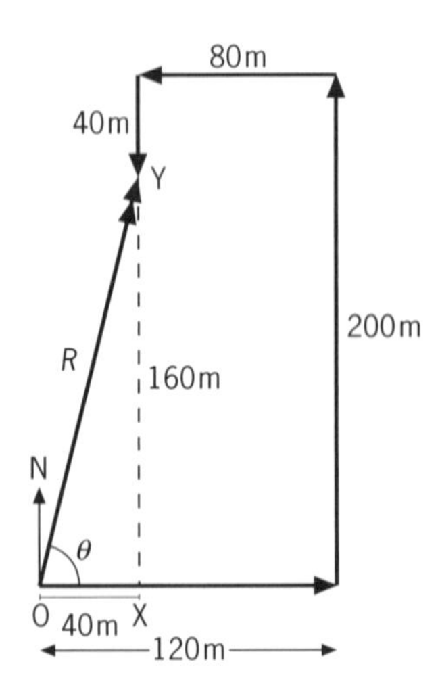

Figure 6.20

5 **a** The maximum resultant force is produced when all three forces act in the same direction (Figure 6.21a) and the magnitude is $6\,N + 6\,N + 6\,N =$ **18 N**.

b When the forces are arranged to form a closed equilateral triangle (Figure 6.21b) the resultant force is **zero**.

c The resultant force is 6 N when the forces are aligned to form three sides of a square (Figure 6.21c).

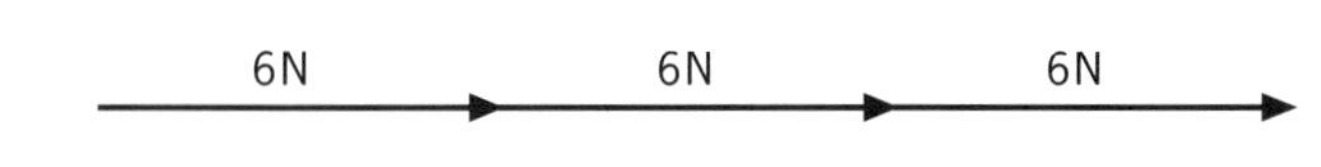

a

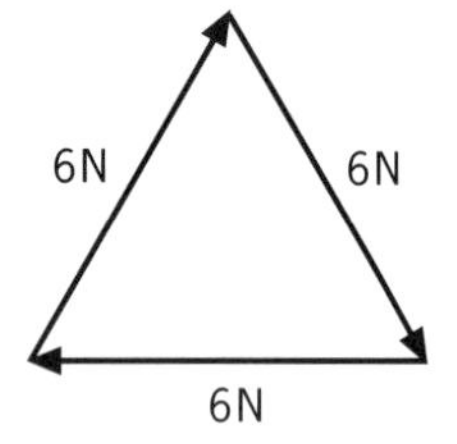

b

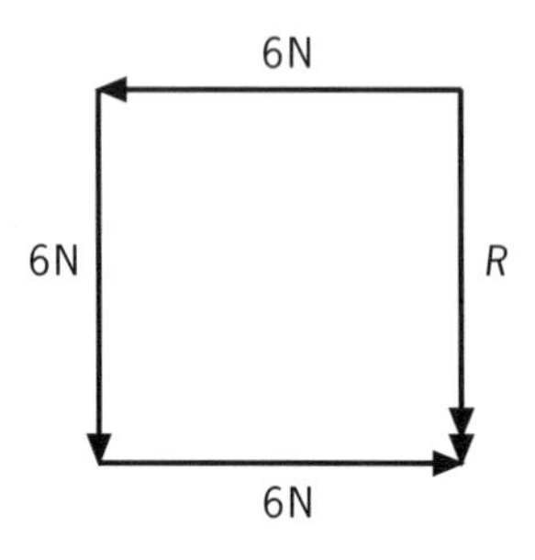

c

Figure 6.21

6 **a** See Figure 6.22a.

b Reverse vector B, to become $-B$, and add this to vector A to give $R = A + (-B)$. See Figure 6.22b.

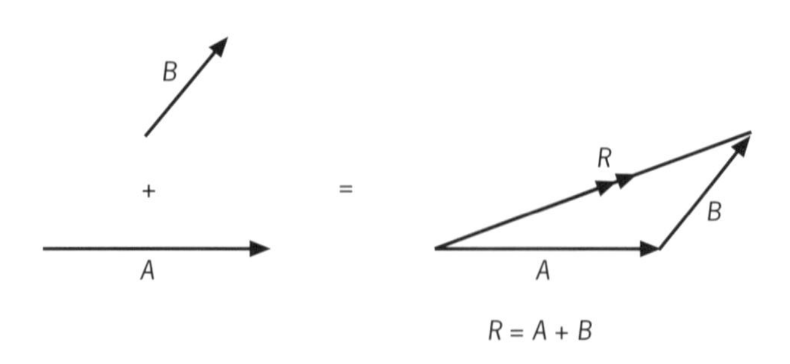

a

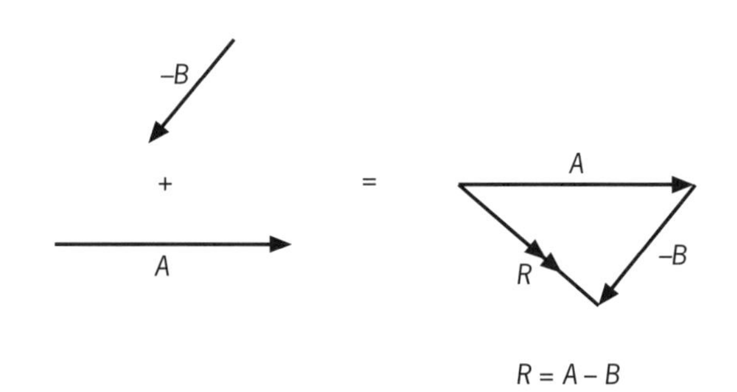

b

Figure 6.22

7 The car's weight is

$W = m \times g = 700\,\text{kg} \times 9.8\,\text{m s}^{-2} = 6860\,\text{N}$

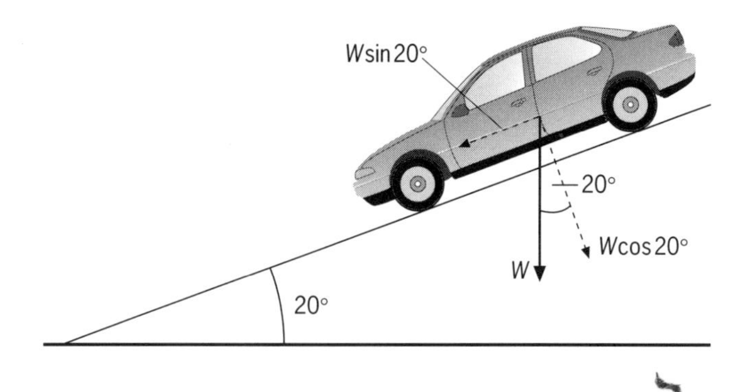

Figure 6.23

The component of the car's weight acting down the slope (Figure 6.23)

$= W \times \sin 20° = 6860 \times 0.342 =$ **2350 N**

Remember to make the components look different to the actual force by colour or dashes. They are not extra forces. The component alongside the marked angle is always the cosine component.

8

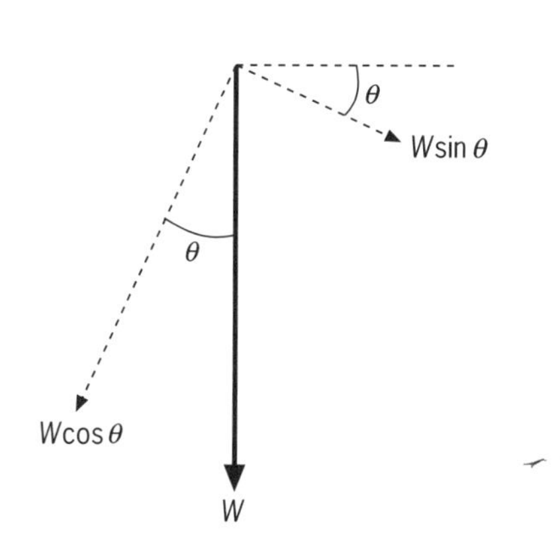

Figure 6.24

The component of your weight, W, acting along the line in which the scales are supposed to be used is $W \times \cos\theta$ where

$W = m \times g = 65\,\text{kg} \times 9.8\,\text{m}\,\text{s}^{-2} = 637\,\text{N}$

and θ is the angle of the scales to the horizontal (Figure 6.24).

At 20° the scales read

	$W \times \cos 20° = 637\,\text{N} \times 0.940 = \mathbf{599\,N}$
40°	$W \times \cos 40° = 637\,\text{N} \times 0.766 = \mathbf{488\,N}$
60°	$W \times \cos 60° = 637\,\text{N} \times 0.50 = \mathbf{318\,N}$
80°	$W \times \cos 80° = 637\,\text{N} \times 0.174 = \mathbf{111\,N}$

If the scales were vertical they would read zero. (You would have fallen off anyway!)

Level 2

9 a The forward component of your velocity relative to the boat is (Figure 6.25a)

$$v \times \sin 30° = 1.5\,\text{m}\,\text{s}^{-1} \times \sin 30° = \mathbf{0.75\,m\,s^{-1}}$$

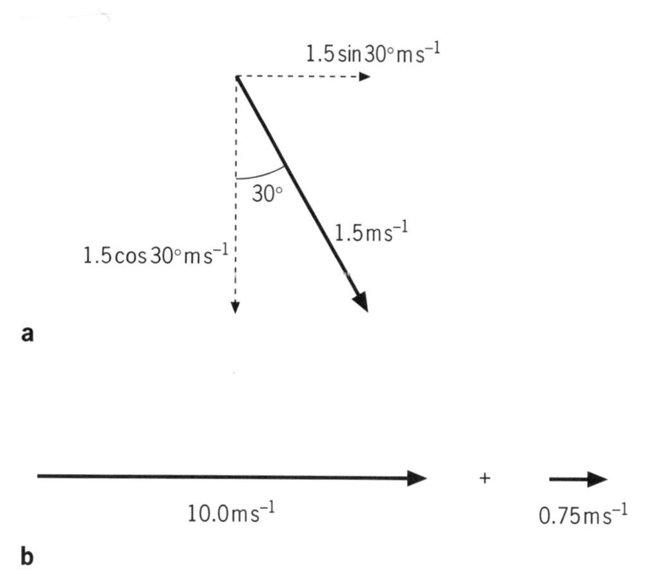

Figure 6.25

b This component must be added to the boat's forward velocity of $10.0\,\text{m}\,\text{s}^{-1}$. Your total forward velocity (Figure 6.25b) will be
$10.0\,\text{m}\,\text{s}^{-1} + 0.75\,\text{m}\,\text{s}^{-1} = \mathbf{10.75\,m\,s^{-1}}$

10 The weight of the buggy plus baby $= m \times g$
$= 25\,\text{kg} \times 9.8\,\text{m}\,\text{s}^{-2} = 245\,\text{N}$.

a The components of the pulling force are (Figure 6.26)

vertical component $= 100\,\text{N} \times \sin 40°$
$= 64\,\text{N}$ upwards

horizontal component $= 100\,\text{N} \times \cos 40°$
$= \mathbf{77\,N}$

b The resultant vertical force on the ground

$= 245\,\text{N}$ (down) $- 64\,\text{N}$ (up) $=$ **181 N down**

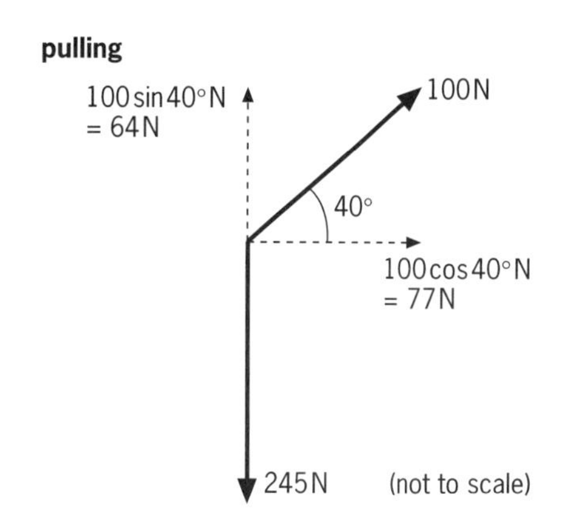

Figure 6.26

c When the buggy is being pushed with a force of 100 N, the horizontal component remains the same at **77 N**. The vertical component, 64 N, is now downwards (see Figure 6.27). So the total vertical force on the ground

= 245 N (down) + 64 N (down) = **309 N (down)**

This will tend to make the buggy sink into the ground and so it will be harder to push.

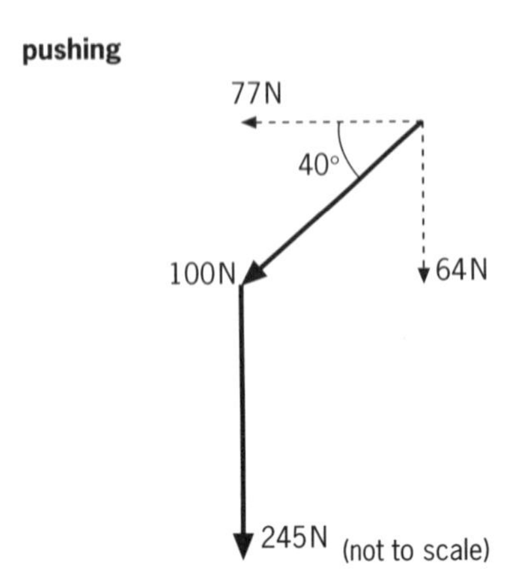

Figure 6.27

11

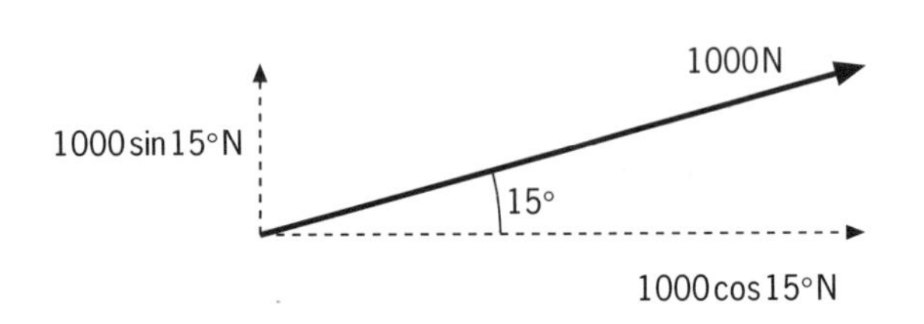

Figure 6.28

The component of the force parallel to the bank is (Figure 6.28)

$1000\,\text{N} \times \cos 15° =$ **966 N**

A long rope is better because you can keep the angle small and so keep the forward component large. If the angle increased to 30° the useful force would go down to $1000\,\text{N} \times \cos 30° = 866\,\text{N}$.

Level 3

12 W_1 must be equal in magnitude and opposite in direction to $W_2 + W_3$. If you draw to scale starting with W_1 vertically down and add the other two, they should form a closed triangle (Figure 6.29a). Or you could find the resultant of $W_2 + W_3$ and show that it is equal but opposite to W_1 (Figures 6.29b, c).

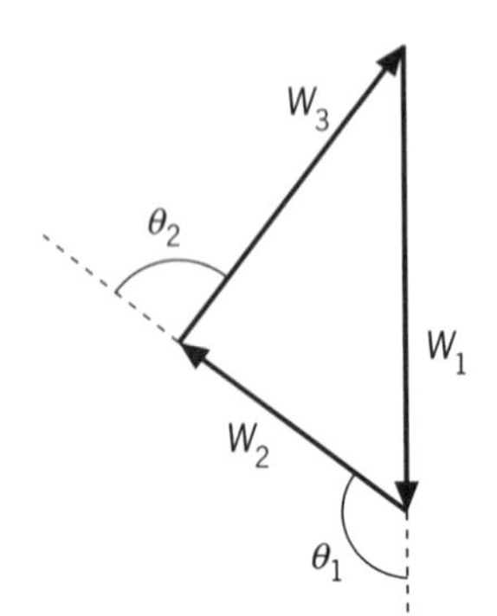

a

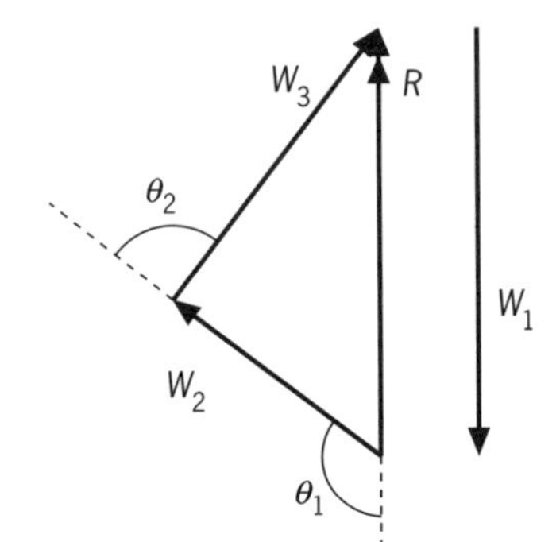

b

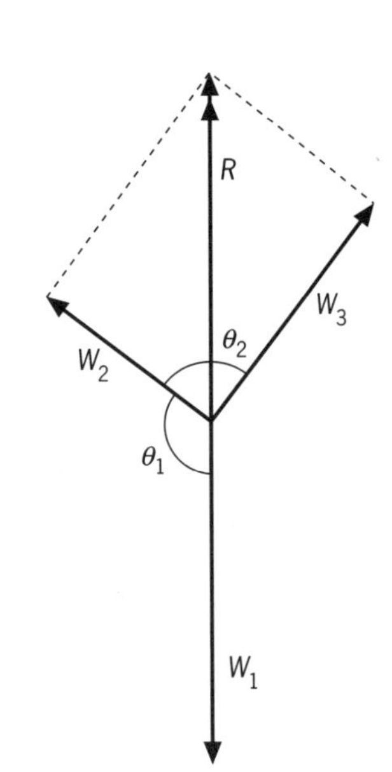

c

Figure 6.29

13 a You need to add the velocity vectors of the boat and the river.

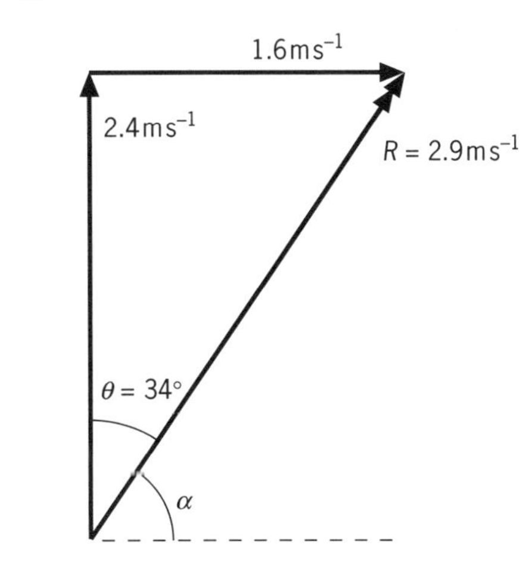

Figure 6.30

The velocities are at right angles (Figure 6.30) so you can calculate the resultant, R, by Pythagoras' theorem:

$R^2 = 2.4^2 + 1.6^2 = 5.76 + 2.56 = 8.32\,\mathrm{m\,s^{-1}}$
$\therefore R = \sqrt{8.32} = \mathbf{2.9\,m\,s^{-1}}$

The resultant direction is given by

$$\tan\theta = \frac{1.6}{2.4} \quad \text{which gives } \theta = 34°$$

so the direction is **34°** from North.
(Or you could find the angle α with the river, given by $\tan\alpha = \frac{2.4}{1.6}$, so $\alpha = 56°$.)

b Figure 6.31 shows a triangle of displacements.

Remember never to mix velocities and displacements in one triangle.

You can draw it because you know two angles and the length of the side between them.

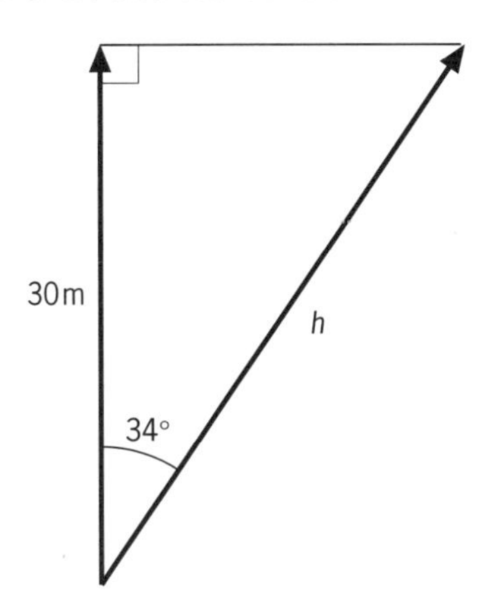

Figure 6.31

The displacement required is the hypotenuse, h, of the triangle. This is given by

$$\cos 34° = \frac{30\,\mathrm{m}}{h}$$

$$\therefore \quad h = \frac{30\,\mathrm{m}}{0.83} = \mathbf{36\,m}$$

Or you could use similar triangles, as the triangle of displacements is geometrically similar to the triangle of velocities. This means that the ratios of corresponding sides are the same, that is

$$\frac{(\text{displacement})_1}{(\text{velocity})_1} = \frac{(\text{displacement})_2}{(\text{velocity})_2}$$

so $$\frac{30\,\mathrm{m}}{2.4\,\mathrm{m\,s^{-1}}} = \frac{h}{2.9\,\mathrm{m\,s^{-1}}}$$

$$\therefore \quad h = \frac{2.9\,\mathrm{m\,s^{-1}} \times 30\,\mathrm{m}}{2.4\,\mathrm{m\,s^{-1}}}$$

$$= 36\,\mathrm{m}$$

The displacement is **36 m** at an angle **34°** from North.

7 Moments, torques and equilibrium

Moment of a force

Sometimes when a force is applied to an object it causes the object to turn or rotate. This turning effect of a force is called a **moment**. The size of a moment can be calculated using the equation

$$T = F \times d$$

where T is the moment in Nm
F is the magnitude of the force in N
d is the *perpendicular distance* in m between the axis of rotation and the line of action of the force.

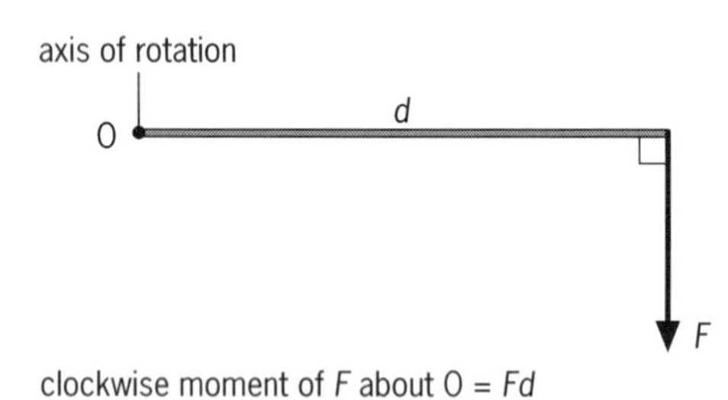

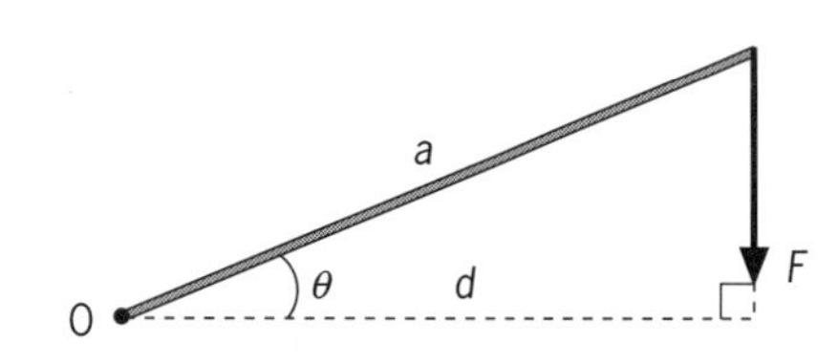

Figure 7.1 The moment of a force about an axis

Example 1

Calculate the moment created around A by the force shown in Figure 7.2.

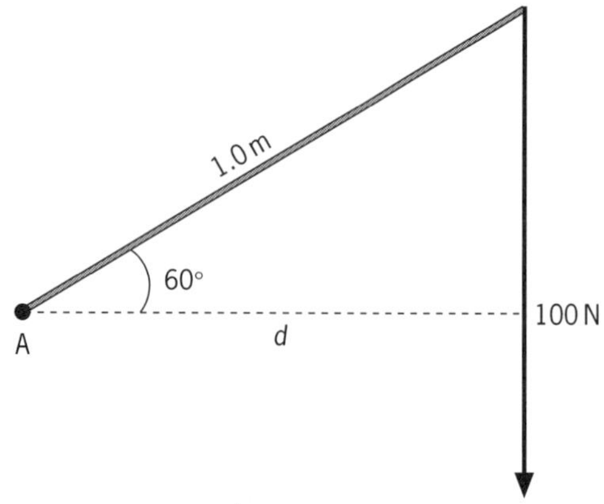

Figure 7.2

$$T = F \times d$$
$$= 100\,\text{N} \times 1.0\,\text{m} \times \cos 60°$$
$$= 50\,\text{Nm clockwise}$$

Principle of moments

The bar in Figure 7.3 is experiencing the effects of several moments created by the forces F_1, F_2 and F_3. If the bar does not rotate, that is, it is in equilibrium, the moments trying to turn the bar clockwise must be equal to the moments trying to turn the bar anticlockwise. This is the principle of moments.

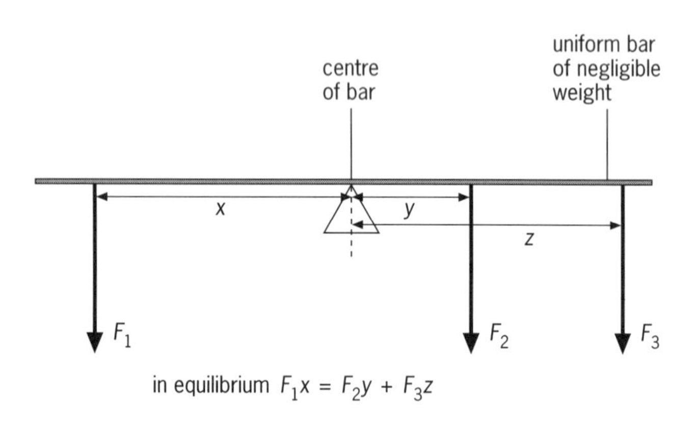

Figure 7.3 The principle of moments

Example 2

Calculate the magnitude of the force F needed to keep the bar shown in Figure 7.4 in equilibrium.

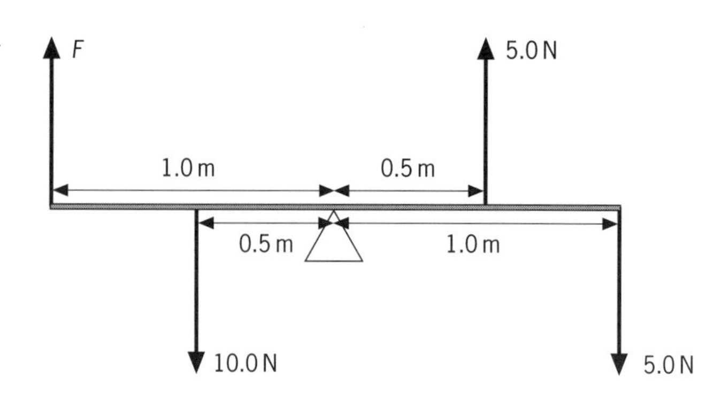

Figure 7.4

For the bar to be balanced,

$$\text{clockwise moments} = \text{anticlockwise moments}$$
$$F \times 1.0\,\text{m} + 5.0\,\text{N} \times 1.0\,\text{m} = 10.0\,\text{N} \times 0.5\,\text{m} + 5.0\,\text{N} \times 0.5\,\text{m}$$
$$F \times 1.0\,\text{m} + 5.0\,\text{Nm} = 5.0\,\text{Nm} + 2.5\,\text{Nm}$$
$$F = 2.5\,\text{N upwards}$$

Centre of mass (or centre of gravity)

It is often convenient to imagine all the mass of an object as being concentrated at one point rather than spread throughout the object. The weight of the object can then be considered to act at that point. This point is known as the **centre of mass** (or centre of gravity) of the object. For uniform, regularly shaped objects this point is at the geometric centre of the object.

If an object is supported directly below (or above) its centre of mass it will balance, as the weight of the object creates no moment about the support. If an object is supported at any other point the weight creates a turning moment (Figure 7.5).

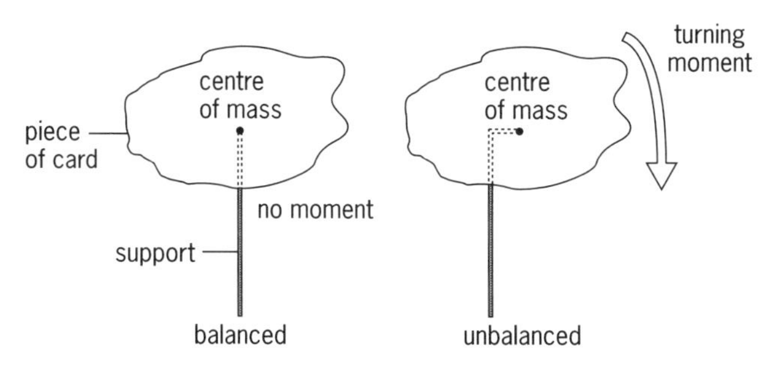

Figure 7.5 Whether or not an object balances depends on the position of its centre of mass

The stability of a three-dimensional object depends on the position of its centre of mass relative to its point of support. See Figure 7.6.

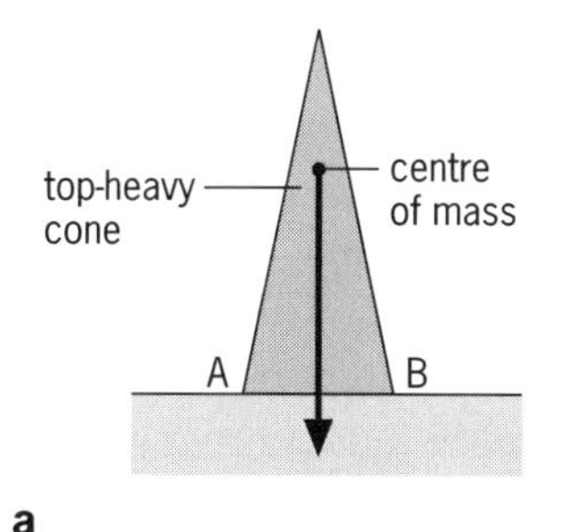

the centre of mass is directly above the base AB and so the weight is creating no moment

a

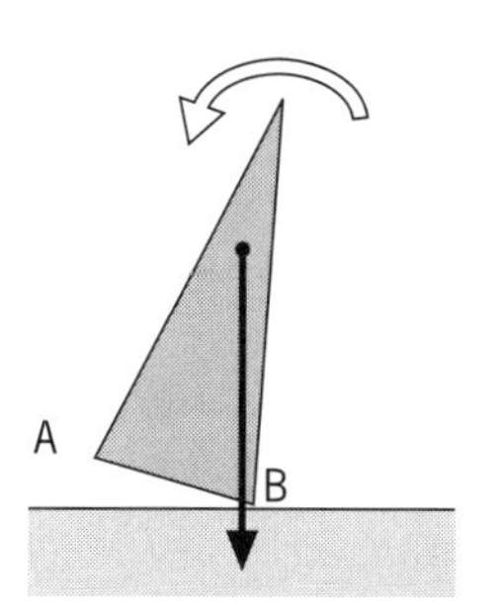

the weight is acting slightly to the left of B and therefore creates an anticlockwise, restoring moment

b

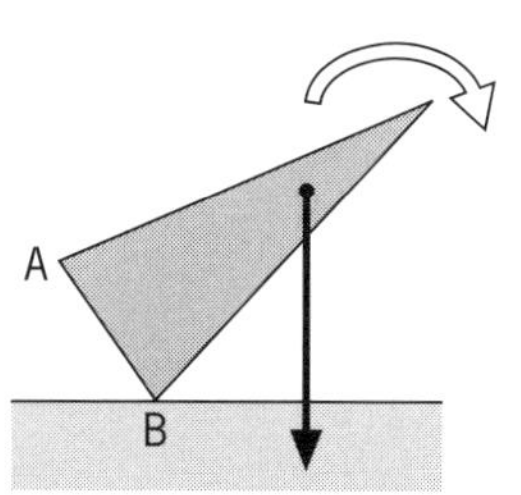

the weight is now acting to the right of B and therefore creates a clockwise moment causing the object to topple

c

Figure 7.6 Whether or not an object topples depends on the position of its centre of mass

Translational and rotational motion

There are two types of motion an object may experience:

- translational motion – an object has pure translational motion if every part of it is moving in the same direction with the same speed at a particular instant;
- rotational motion – an object has pure rotational motion if every part of it moves in a circle about some axis of rotation.

Many objects when subjected to external forces have both translational and rotational motion.

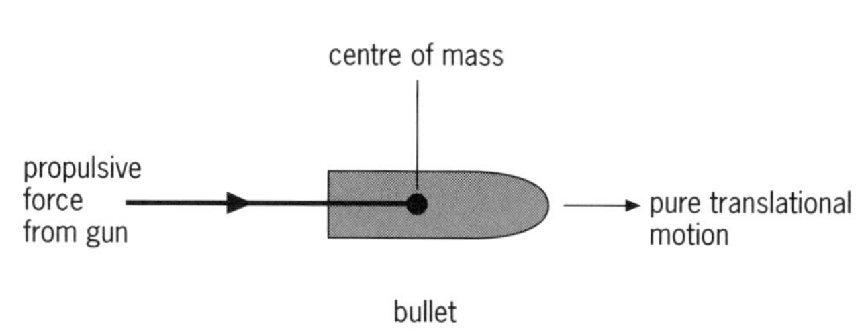

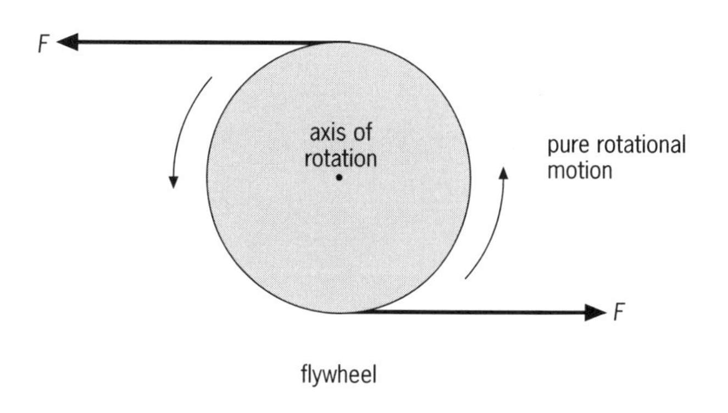

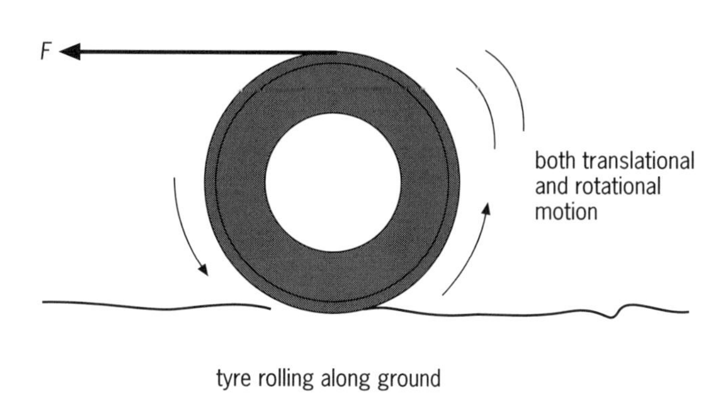

Figure 7.7 Types of motion

If a single force or a resultant force acts through the centre of mass of an object then translational motion is produced.

If two equal, opposite forces not acting through the same point are applied to an object then pure rotational motion is produced. A pair of equal but opposite forces which produce just a turning effect is called a **couple**. The total turning effect or moment of a couple is called a **torque**. The moment of a couple, or the torque, can be calculated using the equation:

moment of a couple (or torque) = magnitude of one of the forces × perpendicular distance between the forces

In Figure 7.8 the couple creates a torque of

$$20\,\text{N} \times 2.0\,\text{m} = 40\,\text{N}\,\text{m}$$

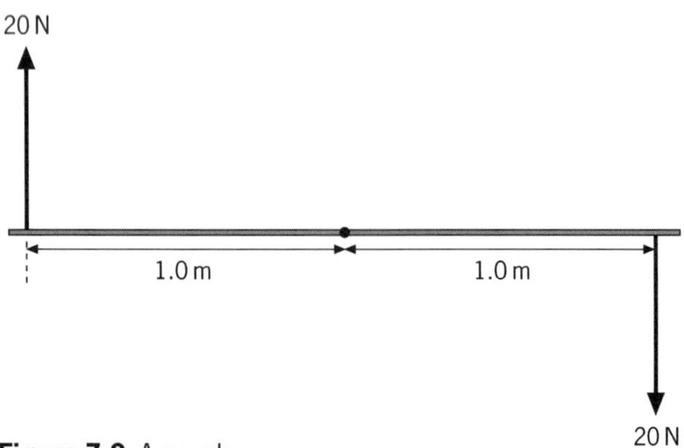

Figure 7.8 A couple

Equilibrium

If an object is described as being in equilibrium then two conditions must apply:

- the resultant of all the applied forces in any direction must be zero, otherwise there would be translational motion, *and*
- the moments taken around any axis must balance, otherwise there would be rotational motion.

Simple systems where the forces all act in the same plane are called **coplanar** systems.

Example 3

A lorry of weight 30 000 N is stopped one third of the way across a small, uniform beam bridge of 30 m in length. If the total weight of the bridge is 40 000 N, calculate the load on each of the supporting pillars.

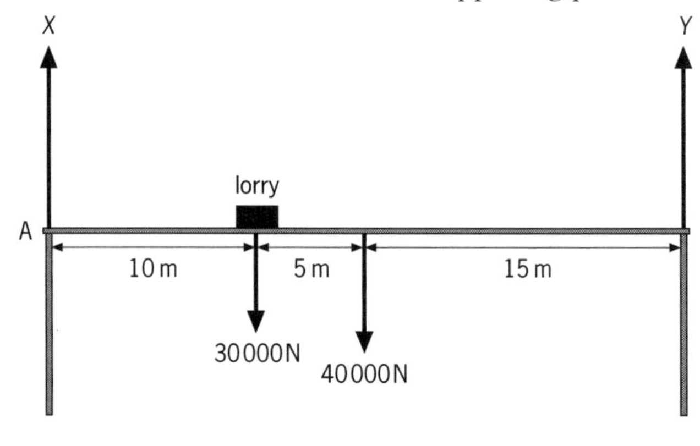

Figure 7.9

Figure 7.9 shows the forces on the beam. X and Y are the reaction forces at the supporting pillars and are equal in magnitude to the loads supported by each pillar.

If the system is in equilibrium there is no rotational motion, so clockwise moments about any point of the system must equal anticlockwise moments about that point. Taking moments about A (this eliminates X):

$$30\,000\,\text{N} \times 10\,\text{m} + 40\,000\,\text{N} \times 15\,\text{m} = Y \times 30\,\text{m}$$

$$\therefore \quad Y = 30\,000\,\text{N}$$

If the system is in equilibrium there is no translational motion, so the resultant vertical force must be zero.

$$X + Y = 30\,000\,\text{N} + 40\,000\,\text{N}$$
$$X + 30\,000\,\text{N} = 30\,000\,\text{N} + 40\,000\,\text{N}$$
$$\therefore \quad X = 40\,000\,\text{N}$$

Level 1

1 Draw free body diagrams for the forces on each of the objects named first in the list below. The objects are in equilibrium.

A free body diagram shows one object only and the forces upon it. So in ***a****, for example, you should draw the mass but not the spring.*

Label each force and say which forces are equal in magnitude.

a a mass hanging from a spring
b a plane flying straight and level
c a raft floating and supporting a passenger
d a climber chimneying (see Figure 7.10)
e a parachutist travelling at terminal velocity

Figure 7.10 A climber chimneying

2 A uniform rod 1.0 m long, of mass 0.50 kg, is suspended at X, 0.40 m from one end. A lamp of mass 0.20 kg hangs from the far end (Figure 7.11).

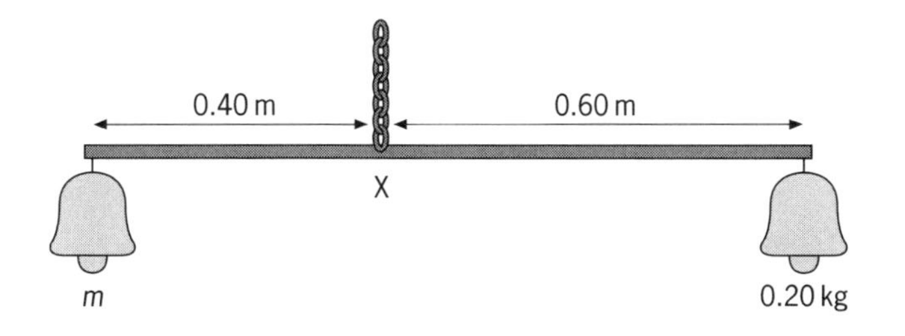

Figure 7.11

What mass would a lamp placed at the other end need to have in order to balance the rod? Take the value of g to be $10\,\mathrm{m\,s^{-2}}$.

3 a What moment is needed to start lifting the boulder in Figure 7.12?
b What force F, applied vertically to the other end of the crowbar, will provide this moment?

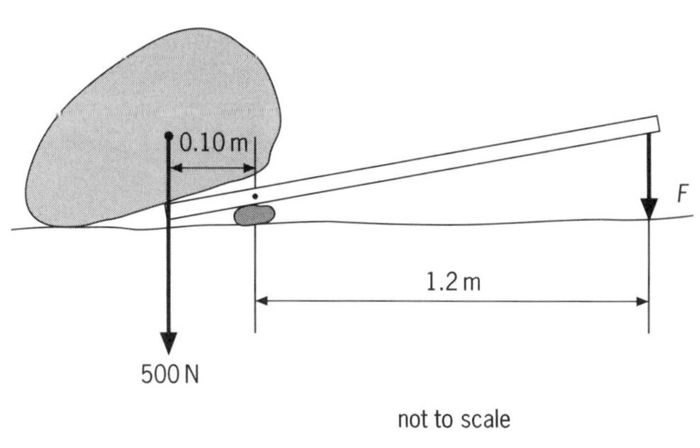

Figure 7.12

4 The forces on each side of a current coil in a moving coil meter are 0.15 N, and they are separated by a distance of 1.0×10^{-2} m, as shown in Figure 7.13. The coil turns until this torque is balanced by the restoring torque of a coiled spring of diameter 3.0×10^{-3} m.

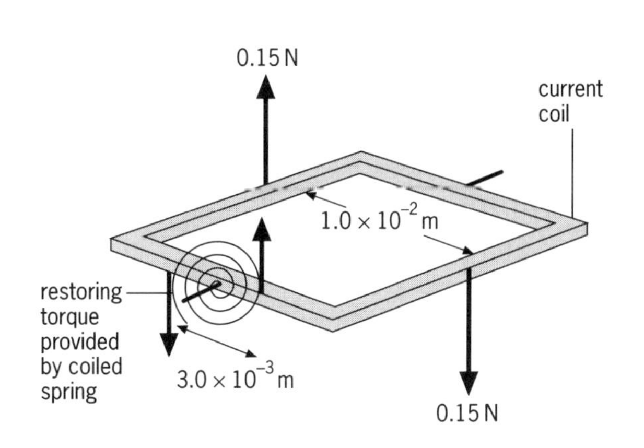

Figure 7.13

Calculate

a the torque on the coil
b the force exerted by each side of the spring when the coil is stationary.

Level 2

5 When considering a mass stationary on a slope at an angle θ to the horizontal, the opposing forces can be analysed in many ways. Two obvious ways are:

a parallel to and at 90° to the sloping surface
b vertically and horizontally.

Write the equations that would be obtained by each method. Use W for the weight, R for the normal contact force (reaction) and F for the friction contact force, as shown in Figure 7.14.

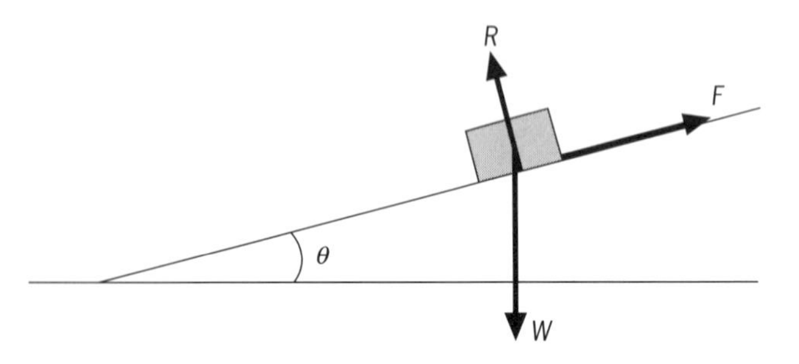

Figure 7.14

6 A *uniform* plank is 3.0 m long and weighs 200 N. Two people, A and B, are carrying it. A is 0.1 m from one end and B is 1.0 m from the other end.

a Draw a force diagram of the plank.
b Who takes the most weight? Calculate the load carried by each person and check your answer by taking moments about the end nearest to A.

7 A wine glass has a circular base 70 mm in diameter. When empty the centre of mass of the glass is 130 mm above the table. At what angle to the table would the base be when it tips over?

8 Two tugs are pulling a ship at constant velocity as shown in Figure 7.15. Find the tension T in the cables.

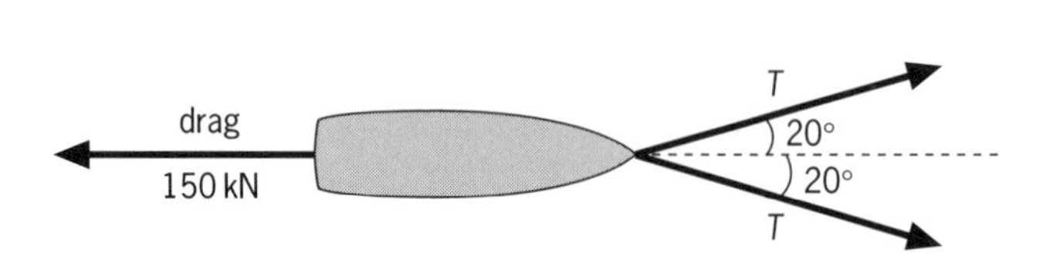

Figure 7.15

9 The wind on the sail of a yacht produces a moment which is counterbalanced by a second moment created by the weight of the sailor. How far from the mast will the sailor's centre of mass have to be in the example shown in Figure 7.16?

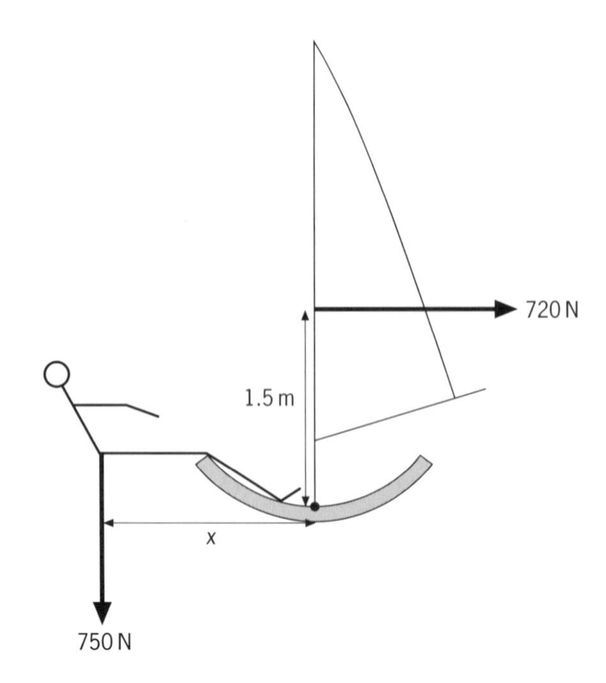

Figure 7.16

Level 3

10 A 5.0 m long uniform ladder of mass 15 kg has its foot 3.0 m from a smooth wall as shown in Figure 7.17. The ground is rough.

Draw a labelled diagram showing the forces on the ladder.

Hint: There are four forces in all. Smooth means frictionless; rough means friction is acting.

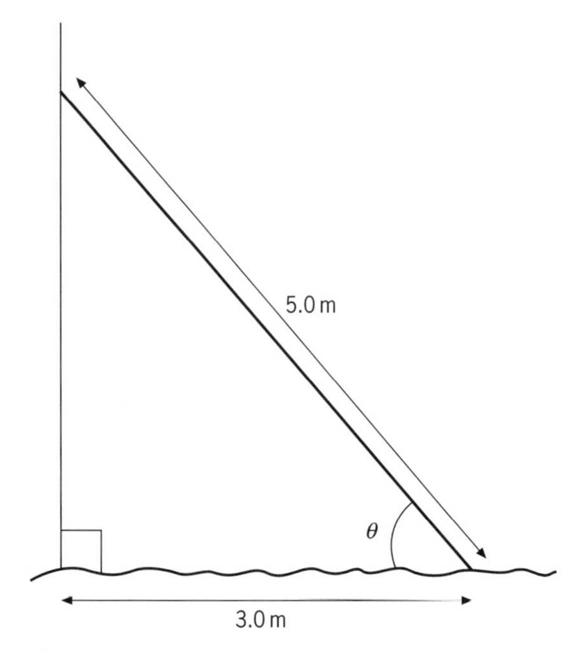

Figure 7.17

Find

a the angle between the ladder and the ground
b the angle at which the ground exerts a force on the foot of the ladder.

Take the value of g to be $9.8\,\text{m}\,\text{s}^{-2}$.

11 A cantilevered balcony juts out as shown in Figure 7.18.

a If, without a load at D, it is just about to tip, what will be the normal contact force from the lower wall at D?

b Calculate the load that has to be added at D to prevent tipping.

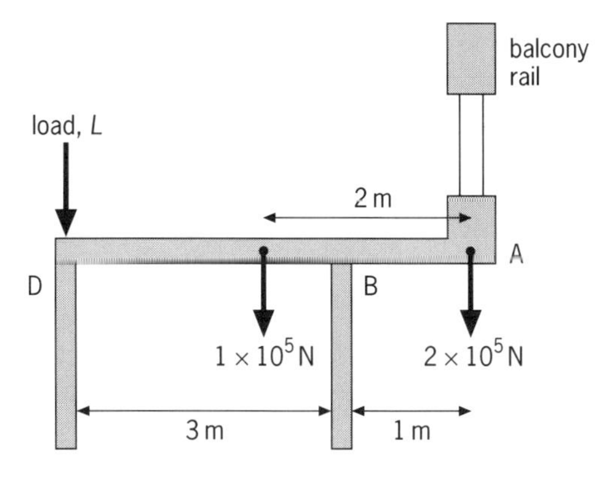

Figure 7.18

12 Bending and lifting incorrectly often cause back injury.

a How big is the vertical component, T_v, of the tension in the back muscles in the situation shown in Figure 7.19?

b Calculate the total force *T*.

c Sketch a triangle of the forces including the reaction *R* along the spine needed for equilibrium.

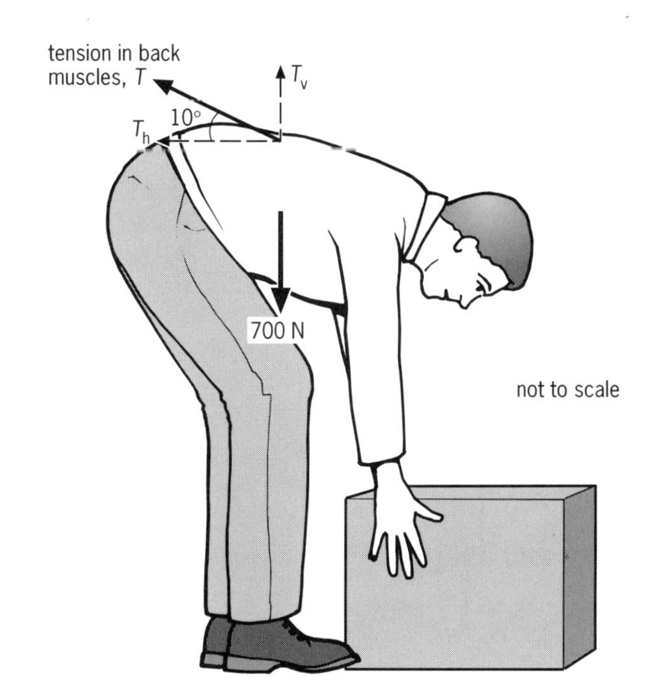

Figure 7.19

Level 1

1

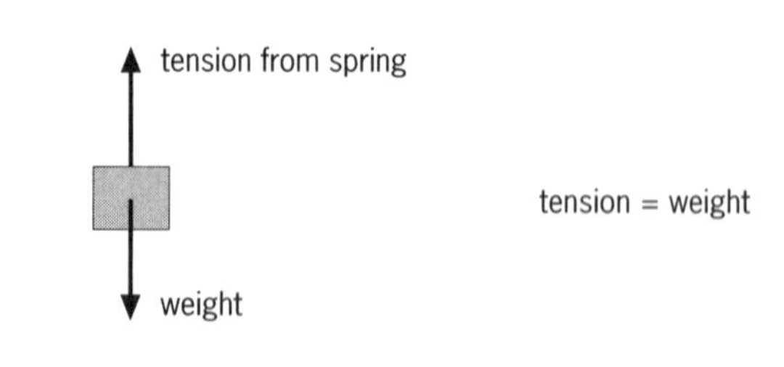

a

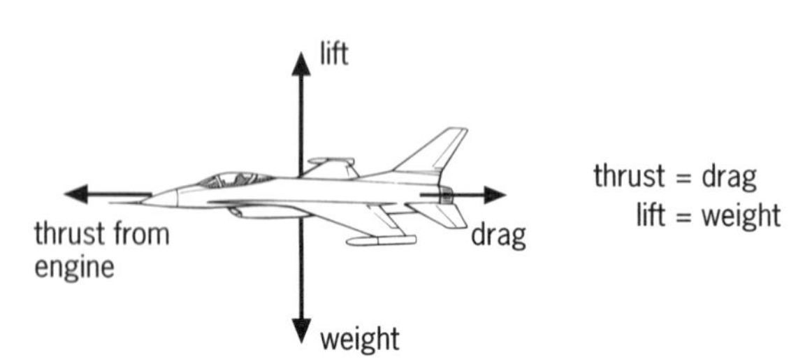

b

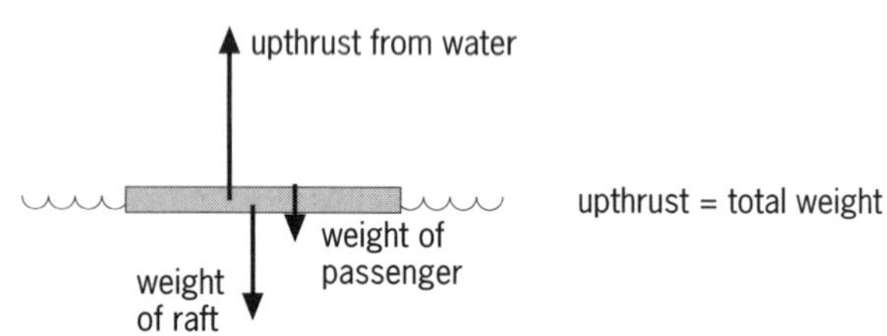

c

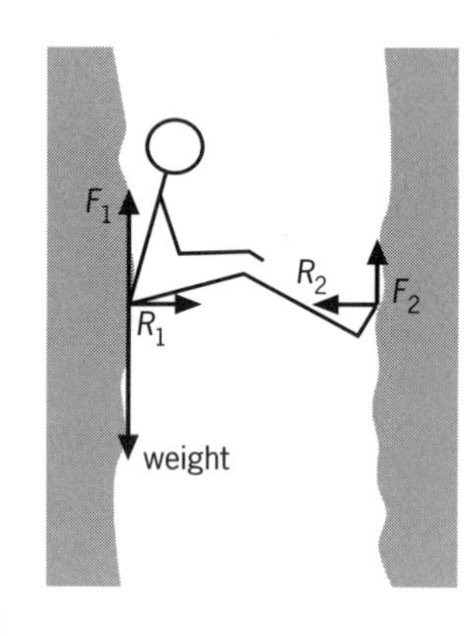

F_1 is friction of rock surface at back
F_2 is friction of rock surface at feet
R_1 is push of rock on back
R_2 is push of rock on feet
$R_1 = R_2$
total friction $F_1 + F_2 =$ weight

d

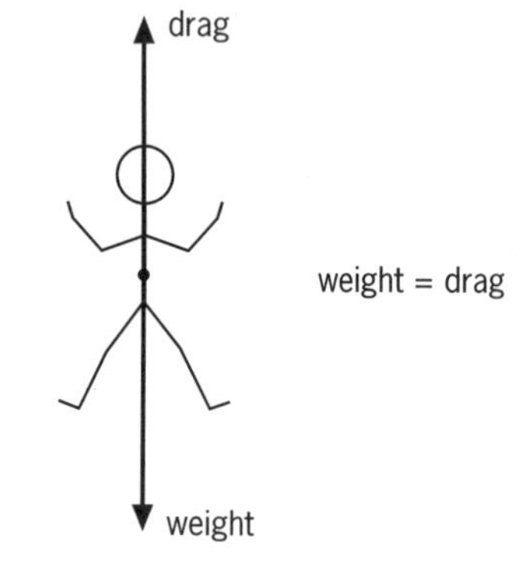

e

Figure 7.20

2 As the rod is uniform, the centre of gravity is at the centre: 0.50 m from either end.

The weight of the rod is mass $\times g = 0.5\,\text{kg} \times 10\,\text{m s}^{-2} = 5\,\text{N}$.

Similarly, the weight of the first lamp is mass $\times g = 0.2\,\text{kg} \times 10\,\text{m s}^{-2} = 2\,\text{N}$.

Let the weight of the second lamp be $W = m \times g$.

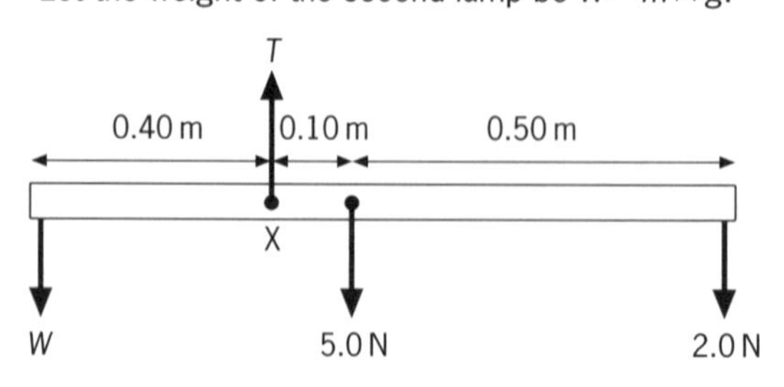

Figure 7.21

Take moments about the point of suspension X because the unknown force T there has no turning moment about X (because the distance from the pivot is zero), so it doesn't matter that we don't know it.

For equilibrium,

anticlockwise moments = clockwise moments

$$W \times 0.40\,\text{m} = 5.0\,\text{N} \times 0.10\,\text{m} + 2.0\,\text{N} \times 0.60\,\text{m}$$

$$0.40\,\text{m} \times W = 0.5\,\text{Nm} + 1.2\,\text{Nm} = 1.7\,\text{Nm}$$

$$\therefore \quad W = \frac{1.7\,\text{Nm}}{0.40\,\text{m}} = 4.25\,\text{N}$$

$$W = m \times g$$

$$\therefore \quad m = \frac{W}{g}$$

$$= \frac{4.25\,\text{N}}{10\,\text{m s}^{-2}} = \mathbf{0.425\,kg}$$

3 Figure 7.22 is a simplified diagram showing the lines of action of the forces.

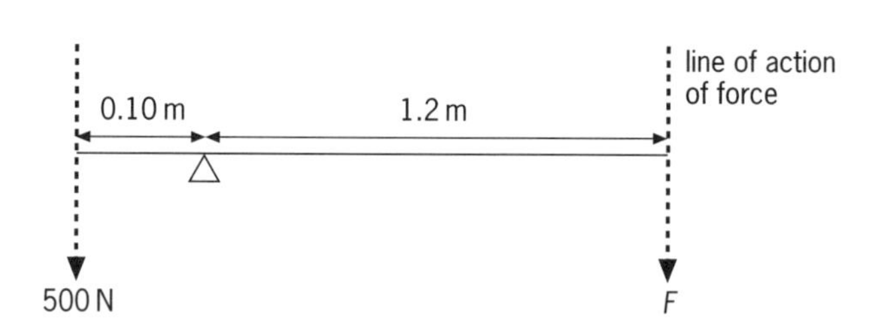

Figure 7.22

The moment is the force $\times$ the perpendicular distance from the line of action of the force to the turning point. These forces and distances are shown in the diagram.

a Moment of weight of boulder $= 500\,\text{N} \times 0.10\,\text{m} = 50\,\text{Nm}$ anticlockwise.
An equal and opposite moment, **50 Nm clockwise**, will be needed to lift the boulder.

b To provide this moment, $F \times 1.2\,\text{m}$ must equal 50 Nm.

$$F = \frac{50\,\text{Nm}}{1.2\,\text{m}} = \mathbf{42\,N}$$

This result could have been obtained by using the fact that the ratio of the distances is 12 : 1, so the forces must be in the ratio 1 : 12, or 42 : 500. This is because when the boulder just moves

$$F_1 \times d_1 = F_2 \times d_2$$

$$\text{so} \quad \frac{F_1}{F_2} = \frac{d_2}{d_1}$$

4 The torque, T, is equal to one of the forces multiplied by the perpendicular distance between them. So, on the coil,

$$T = 0.15\,\text{N} \times 1.0 \times 10^{-2}\,\text{m} = \mathbf{1.5 \times 10^{-3}\,N\,m}$$

b For equilibrium, this torque must be balanced by an equal and opposite torque. So, for the spring, if f is one of the pair of forces produced by the twist, then

$$f \times \text{diameter of spring} = 1.5 \times 10^{-3}\,\text{N m}$$

$$\therefore \quad f = \frac{1.5 \times 10^{-3}\,\text{N m}}{3.0 \times 10^{-3}\,\text{m}} = \mathbf{0.50\,N}$$

Level 2

5 a See Figure 7.23a. The component of the weight W acting along the slope is $W \sin\theta$ and must be opposed by the frictional force F.

$$W \sin\theta = F$$

The component of the weight W acting at 90° to the slope is $W \cos\theta$ and must be opposed by the normal contact force (reaction) R.

$$W \cos\theta = R$$

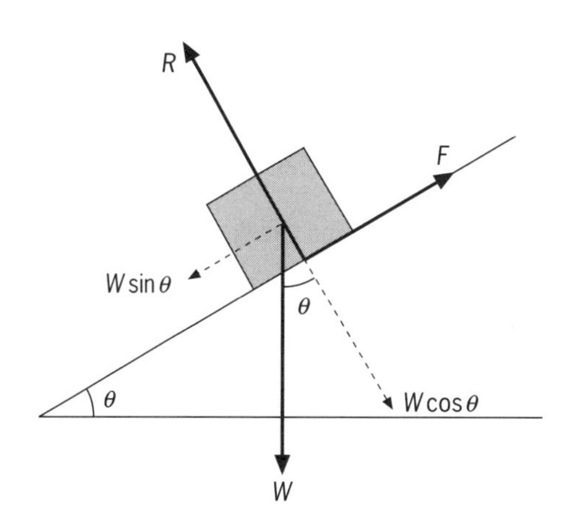

a

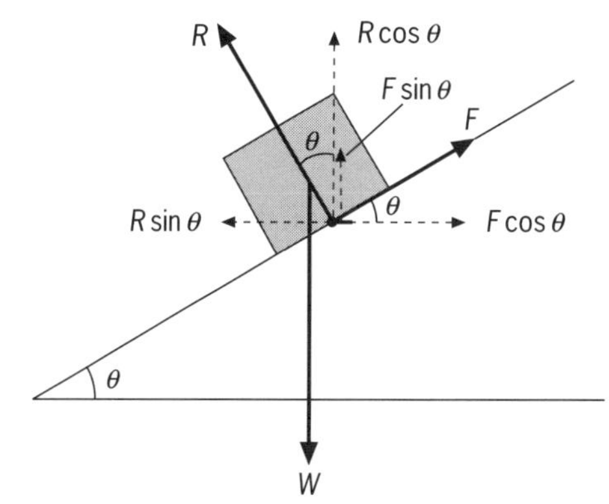

b

Figure 7.23

b We must resolve the normal contact force R and the friction contact force F into their vertical and horizontal components. See Figure 7.23b. The sum of their vertical components must be equal in magnitude but opposite in direction to the weight.

$$R \cos\theta + F \sin\theta = W$$

The sum of their horizontal components must be equal in magnitude but opposite in direction, because there is no horizontal component of the weight.

$$F \cos\theta = R \sin\theta$$

6 a For a uniform plank the centre of gravity is at the centre, so it is 1.5 m from either end. The weight W of the plank is considered to act here. There are two upward forces R_A and R_B from the people. See Figure 7.24.

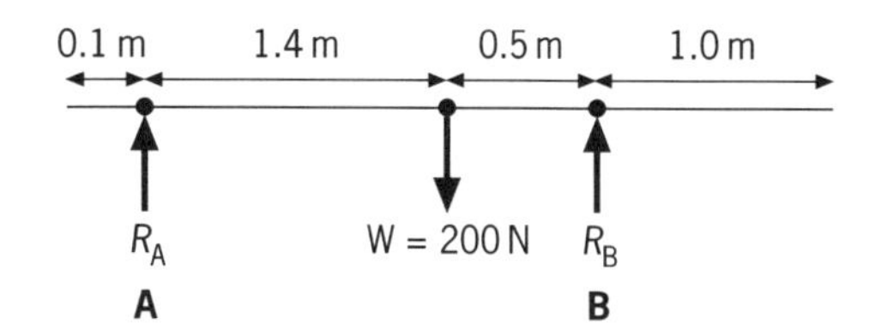

Figure 7.24

b Because the plank is in equilibrium, the upward forces must be equal in magnitude but opposite in direction to the downward forces.

$$R_A + R_B = W = 200\,\text{N}$$

$$\text{So} \quad R_B = 200\,\text{N} - R_A$$

Also, the sum of the clockwise moments must equal the sum of the anticlockwise moments taken about any point.

Taking moments about point B (*you could use point A instead*):

$$\text{clockwise moments} = \text{anticlockwise moments}$$

$$R_A \times 1.9\,\text{m} = 200\,\text{N} \times 0.5\,\text{m}$$

$$R_A = \frac{200\,\text{N} \times 0.5\,\text{m}}{1.9\,\text{m}}$$

$$\therefore \quad R_A = \mathbf{53\,N}$$

and

$$R_B = 200\,\text{N} - R_A = 200 - 53 = \mathbf{147\,N}$$

So **person B**, the person further from an end of the plank, takes most weight.

Check by taking moments about the end of the plank nearest to A. The anticlockwise moments come from A and B:

$$R_A \times 0.1\,\text{m} + R_B \times 2.0\,\text{m}$$
$$= 52.6\,\text{N} \times 0.1\,\text{m} + 147.4\,\text{N} \times 2.0\,\text{m}$$
$$= 300\,\text{N m anticlockwise}$$

The clockwise moments come from the weight of the plank:

$$200\,\text{N} \times 1.5\,\text{m} = 300\,\text{N m clockwise}$$

So the anticlockwise moments balance the clockwise moments.

7 The glass will tip if its centre of gravity is beyond the point of support, which will happen if

$$\tan\theta \text{ is greater than } \frac{35}{130} = 0.269$$

where θ is the angle between the base of the glass and the table. See Figure 7.25.

So the glass will begin to tip when

$$\theta = \tan^{-1} 0.269 = \mathbf{15^\circ}$$

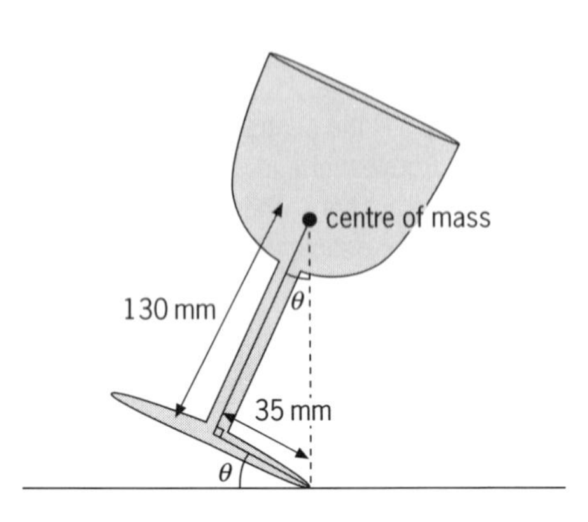

Figure 7.25

8 If the ship is going at constant velocity then it is in equilibrium (see topic 9) and the sum of the two forward components of the tensions in the tugs' cables must be equal in magnitude but opposite in direction to the backwards (drag) force on the ship.

Each cable has a component in the direction of motion of $T \times \cos 20^\circ$. So

$$2 \times T \times \cos 20^\circ = 150\,\text{kN}$$
$$2 \times T \times 0.94 = 150\,\text{kN}$$
$$T = \frac{150\,\text{kN}}{2 \times 0.94} = \mathbf{80\,kN}$$

9 Figure 7.26 is a simplified diagram of the forces and distances involved. The clockwise moment of the force of the wind about the base of the mast is

$$720\,\text{N} \times 1.5\,\text{m} = 1080\,\text{N\,m}$$

To give the same size moment the sailor's centre of mass must be at a distance x from the base of the mast, such that

$$750\,\text{N} \times x = 1080\,\text{N\,m}$$
$$\therefore \quad x = \frac{1080\,\text{N\,m}}{750\,\text{N}} = \mathbf{1.4\,m}$$

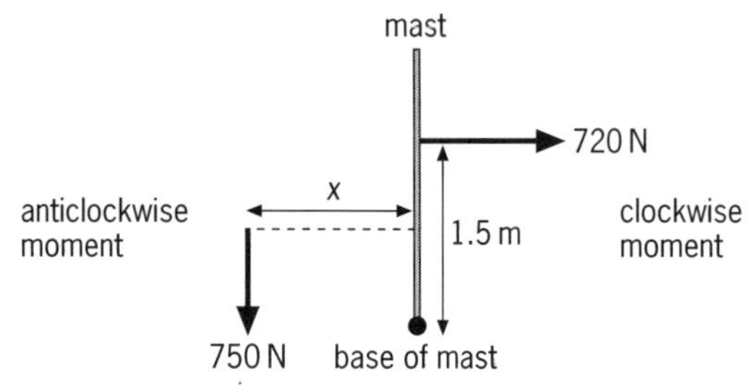

Figure 7.26

Level 3

10

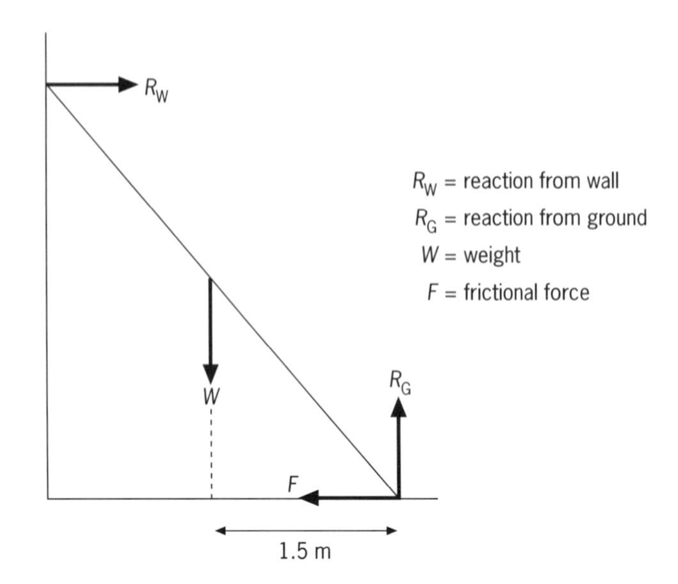

Figure 7.27

The total weight of a uniform ladder acts at its centre. The fact that the wall is smooth means that there is no friction where the ladder touches the wall. Friction would generate forces that would act along the surface of the wall, but in the absence of friction, the only force acting at the top of the ladder is the normal contact force (reaction) of the wall on the ladder.

The fact that the ground is rough means that there is friction where the ladder touches the ground. There is also the normal contact force (reaction) of the ground on the ladder, so that the combined force at the foot will be at an angle less than 90° to the ground.

a The angle between the ladder and the ground, θ, is given by

$$\cos\theta = \frac{\text{adjacent}}{\text{hypotenuse}} = \frac{3.0\,\text{m}}{5.0\,\text{m}} = 0.60$$
$$\therefore \quad \theta = \cos^{-1} 0.60 = 53^\circ$$

You may have spotted the 3, 4, 5 right-angled triangle formed by the ladder, the ground and the wall which means that the ladder touches the wall at a height of 4.0 m above the ground. You can check this using Pythagoras' theorem:

$$3^2 + 4^2 = 9 + 16 = 25 = 5^2$$

b The force of the ground on the base of the ladder is the vector sum of F and R_G, so we need to evaluate these forces to find the direction of their resultant.

Equating the magnitudes of the horizontal forces,

$$R_W = F$$

and equating the magnitude of the vertical forces,

$$R_G = W = m \times g = 15\,\text{kg} \times 9.8\,\text{m\,s}^{-2} = 147\,\text{N}$$

Taking moments about the foot of the ladder because the forces acting there are the most complicated and will be eliminated,

$$\text{anticlockwise moments} = \text{clockwise moments}$$

$$147\,\text{N} \times 1.5\,\text{m} = R_W \times 4.0\,\text{m}$$

(see Figure 7.28a)

$$\therefore \quad R_W = \frac{147\,\text{N} \times 1.5\,\text{m}}{4.0\,\text{m}} = \mathbf{55\,N}$$

and since $F = R_W$,

$F = \mathbf{55\,N}$ (but acts in the opposite direction)

The angle between the ground and the force of the ground on the foot of the ladder, α (see Figure 7.28b), is given by

$$\tan\alpha = \frac{\text{opposite}}{\text{adjacent}} = \frac{147\,\text{N}}{55\,\text{N}} = 2.673$$

$$\therefore \quad \alpha = \tan^{-1} 2.673 = \mathbf{69°}$$

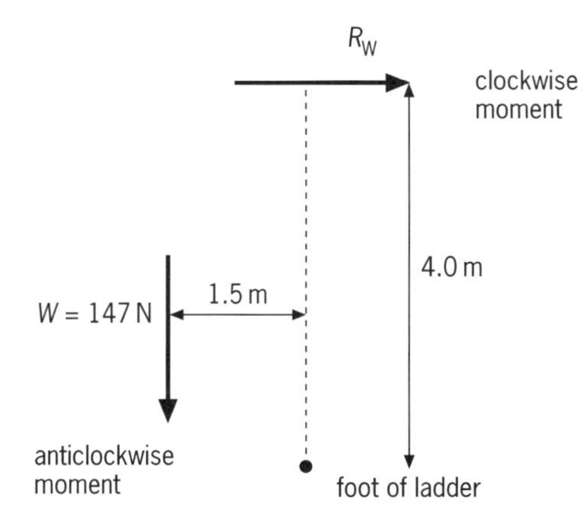

a

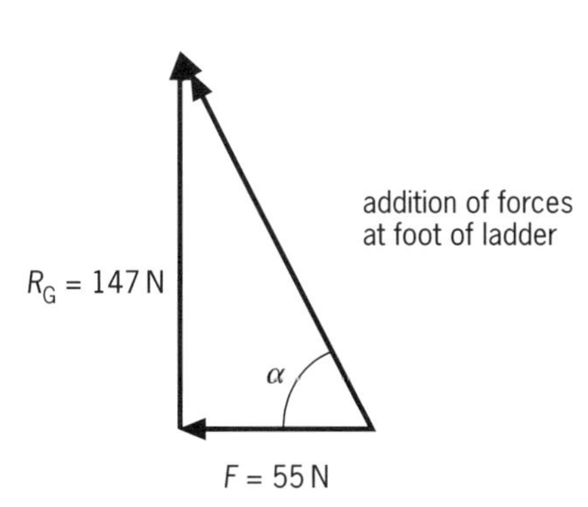

b

Figure 7.28

11 a When the balcony is just about to tip the vertical force down at D will become zero as contact is lost. The normal contact force (reaction) is, therefore, zero.

b To prevent tipping the load at D must balance the floor, plus balcony rail, supported at B. Taking moments about B, for equilibrium:

$$\text{clockwise moments} = \text{anticlockwise moments}$$

$$2 \times 10^5\,\text{N} \times 1\,\text{m} = 1 \times 10^5\,\text{N} \times 1\,\text{m} + L \times 3\,\text{m}$$

$$L = \frac{1 \times 10^5\,\text{Nm}}{3\,\text{m}}$$

$$= \mathbf{0.3 \times 10^5\,N\ downwards}$$

The loading will need to be at least this large.

12 a The vertical component of T, T_V, must equal the weight, so

$$T_V = \mathbf{700\,N}$$

b $T_V = T \times \sin 10° = 700\,\text{N}$

$$\therefore T = \frac{700\,\text{N}}{0.174} = \mathbf{4030\,N}$$

This is nearly six times the body weight of 700 N.

c

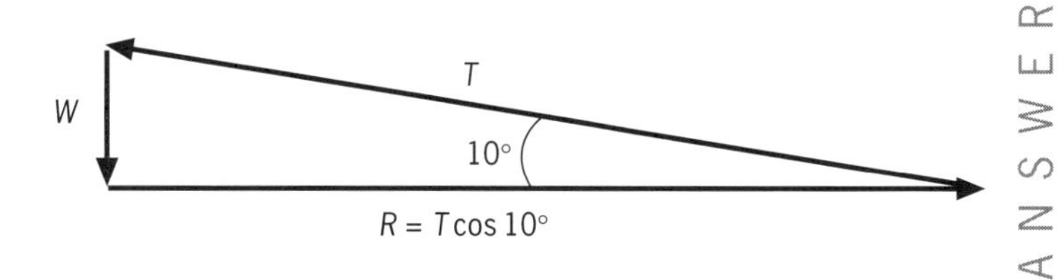

Figure 7.29

The reaction at the base of the spine is only a little less than T, as it is equal to the horizontal component of T, T_h.

$$R = T_h = T \times \cos 10°$$

$$= 4030\,\text{N} \times 0.985$$

$$= \mathbf{3970\,N}$$

8 Uniform motion

Equations of motion

If an object is moving with **uniform acceleration** its velocity is changing at a constant rate, such as $5\,m\,s^{-1}$ each second, which is written as $5\,m\,s^{-2}$. Uniform motion in a straight line can be described by three equations:

$$s = ut + \tfrac{1}{2}at^2$$

$$v = u + at$$

$$v^2 = u^2 + 2as$$

where s is the distance travelled
u is the initial velocity
v is the final velocity
t is the time
a is the acceleration

Example 1

A car starting from rest accelerates at $4.0\,m\,s^{-2}$ for 10 s. Calculate the distance travelled by the car during this time.

$s = ?$
$u = 0\,m\,s^{-1}$
$v =$ not given
$t = 10\,s$
$a = 4.0\,m\,s^{-2}$

Using $s = ut + \tfrac{1}{2}at^2$

$$s = 0 \times 10\,s + \tfrac{1}{2} \times 4.0\,m\,s^{-2} \times (10\,s)^2$$
$$= 200\,m$$

Example 2

A ball is thrown vertically downwards with an initial velocity of $20\,m\,s^{-1}$. What is the final velocity of the ball just before it hits the ground, 5.0 s later? The acceleration due to gravity is $10\,m\,s^{-2}$.

$s =$ not given
$u = 20\,m\,s^{-1}$
$v = ?$
$t = 5.0\,s$
$a = 10\,m\,s^{-2}$

Using $v = u + at$

$$v = 20\,m\,s^{-1} + 10\,m\,s^{-2} \times 5.0\,s$$
$$= 70\,m\,s^{-1}$$

Example 3

Calculate the deceleration of a bullet initially travelling at $400\,m\,s^{-1}$, if it is brought to rest after travelling 10 cm through a block of wood.

$s = 10\,cm = 0.10\,m$
$u = 400\,m\,s^{-1}$
$v = 0\,m\,s^{-1}$
$t =$ not given
$a = ?$

Using $v^2 = u^2 + 2as$ which can be rearranged to give $v^2 - u^2 = 2as$ and hence

$$a = \frac{v^2 - u^2}{2s}$$

$$a = \frac{0^2 - (400\,m\,s^{-1})^2}{2 \times 0.10\,m}$$

$$= \frac{-160\,000\,m^2\,s^{-2}}{0.20\,m}$$

$$= -800\,000\,m\,s^{-2}$$

The acceleration of the bullet is $-800\,000\,m\,s^{-2}$, that is the deceleration of the bullet is $800\,000\,m\,s^{-2}$.

Graphs of motion

Note the following definitions of important terms.

Displacement	The distance moved in a particular direction, e.g. 5 m North
Velocity	The rate of change of displacement, e.g. $5\,m\,s^{-1}$ North
Acceleration	The rate of change of velocity, e.g. $5\,m\,s^{-2}$ North
Uniform motion	Motion that has a constant acceleration

It is often convenient to describe the motion of an object in the form of a graph. See Figures 8.1 and 8.2.

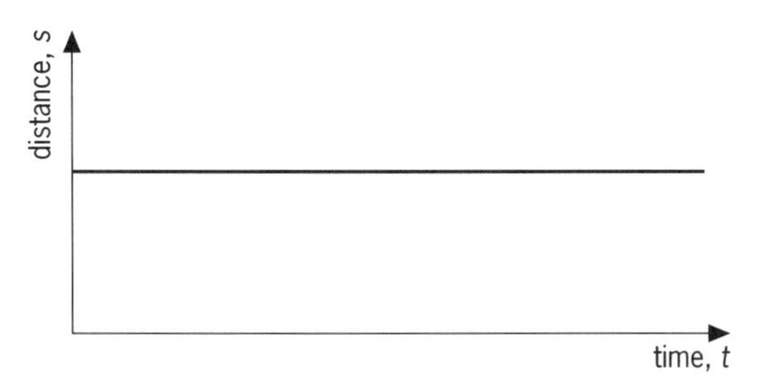

The distance travelled by this object is not changing with time. The object is stationary

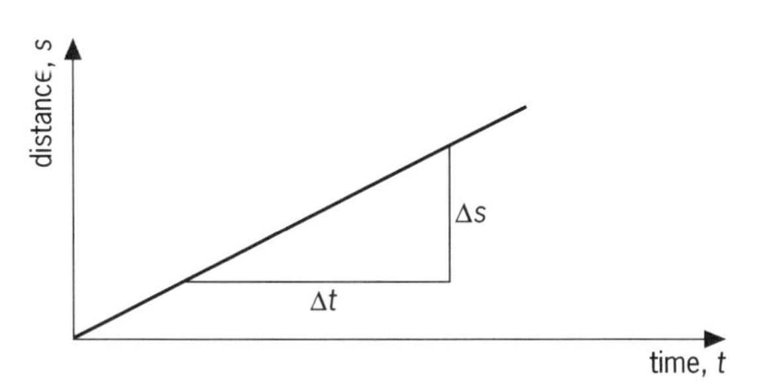

The distance travelled by this object is changing uniformly with time. is moving with a constant speed. The magnitude of the speed is equa to the gradient of the graph

$= \frac{\Delta s}{\Delta t}$

Figure 8.1 Distance–time graphs

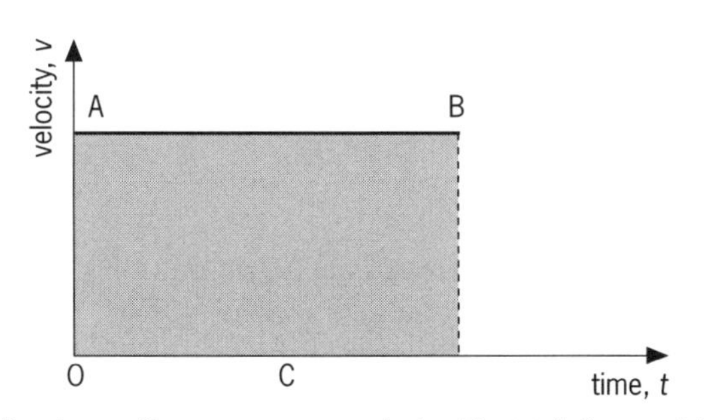

This object is travelling at a constant velocity. The total distance it has travelled is equal to the area under the graph (OABC)

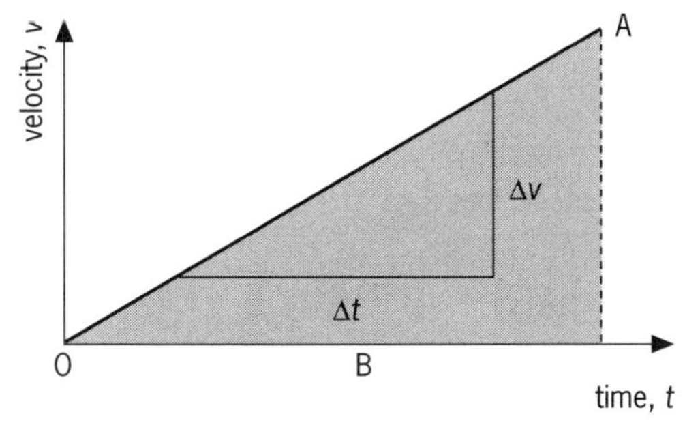

The velocity of this object is increasing regularly with time. The object is accelerating uniformly. The magnitude of the acceleration is equal tc the gradient of the graph

$= \frac{\Delta v}{\Delta t}$

The total distance travelled is again equal to the area under the graph (OAB)

Figure 8.2 Velocity–time graphs

Acceleration due to gravity

One of the most common accelerations we meet is that due to gravity. If we ignore all resistive and frictional forces, the acceleration due to gravity on the surface of the Earth (g) is the same for all objects. It has a value of $9.81\,\mathrm{m\,s^{-2}}$ (often taken as $10\,\mathrm{m\,s^{-2}}$).

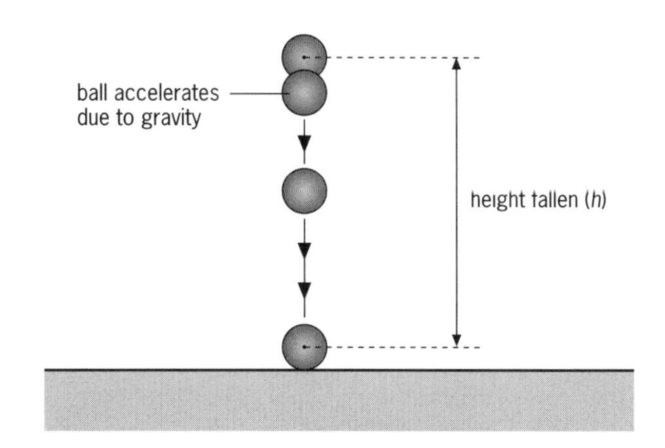

Figure 8.3 A falling ball

If a ball is dropped from a height h, Figure 8.3, the equation of motion

$$s = ut + \tfrac{1}{2}at^2$$

can be modified to

$$h = \tfrac{1}{2}gt^2 \qquad (u = 0,\ a = g,\ s = h)$$

$$\therefore\ g = \frac{2h}{t^2}$$

If the height from which the ball is dropped and the time it takes to fall are known, a value of the acceleration due to gravity can be calculated.

Example 4

a Calculate the acceleration due to gravity if a ball dropped from a height of 45 m strikes the ground after 3.0 s.

b Calculate the velocity of the ball just before it strikes the ground.

a Using $g = \frac{2h}{t^2}$

$$g = \frac{2 \times 45\,\mathrm{m}}{(3.0\,\mathrm{s})^2}$$

$$= 10\,\mathrm{m\,s^{-2}}$$

b

$v = ?$

$u = 0\,\mathrm{m\,s^{-1}}$

$a = 10\,\mathrm{m\,s^{-2}}$

$t = 3.0\,\mathrm{s}$

Using $v = u + at$

$$v = 0 + 10\,\mathrm{m\,s^{-2}} \times 3.0\,\mathrm{s}$$

$$= 30\,\mathrm{m\,s^{-1}}$$

Projectiles

If a ball is thrown upwards at an angle θ to the horizon it will follow a path similar to that shown in Figure 8.4. As the diagram shows, there is movement in both the horizontal and the vertical directions. Projectile motion like this is easier to analyse if the horizontal and the vertical motions are dealt with separately.

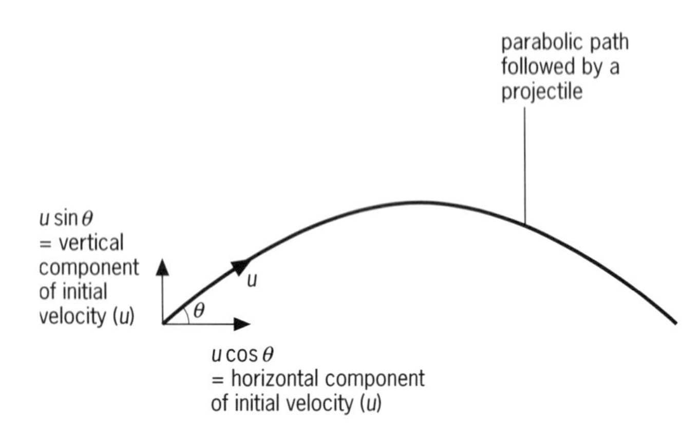

Figure 8.4 The path of a projectile is a curve called a parabola

The ball travels horizontally with a constant velocity. In this case the horizontal component of the velocity is $u \cos \theta$.

The initial vertical component of the velocity is $u \sin \theta$. But the vertical velocity of the ball is continually changing due to the effects of gravity – the ball experiences a constant downwards acceleration g. We can use any of the equations of motion to describe the vertical motion.

Example 5

A ball is thrown horizontally from the top of a tower 45 m high. The ball has a horizontal velocity of $30\,\mathrm{m\,s^{-1}}$.

a After how long will the ball hit the ground?
b How far from the base of the tower will the ball hit the ground?

a Considering the vertical motion:

$$s = ut + \tfrac{1}{2}at^2$$
$$s = 45\,\mathrm{m}$$
$$u = 0\,\mathrm{m\,s^{-1}}$$
$$t = ?$$
$$a = 10\,\mathrm{m\,s^{-2}}$$

Since $u = 0$ this becomes

$$s = \tfrac{1}{2}at^2$$

so $$t^2 = \frac{2s}{a} = \frac{2 \times 45\,\mathrm{m}}{10\,\mathrm{m\,s^{-2}}} = 9\,\mathrm{s}^2$$

$$\therefore \quad t = 3\,\mathrm{s}$$

The horizontal motion does not affect the time taken to hit the ground.

b Considering the horizontal motion:

$$\text{distance travelled} = \text{speed} \times \text{time} = 30\,\mathrm{m\,s^{-1}} \times 3\,\mathrm{s} = 90\,\mathrm{m}$$

The vertical motion does not affect the horizontal motion; they are independent.

Example 6

An arrow is fired upwards at an angle of 30° to the vertical and with an initial speed of $40\,\mathrm{m\,s^{-1}}$. Calculate:

a the initial vertical speed of the arrow
b the initial horizontal speed of the arrow.

a $u_\mathrm{v} = u \cos \theta = 40\,\mathrm{m\,s^{-1}} \cos 30° = 35\,\mathrm{m\,s^{-1}}$

b $u_\mathrm{h} = u \sin \theta = 40\,\mathrm{m\,s^{-1}} \sin 30° = 20\,\mathrm{m\,s^{-1}}$

Level 1

1

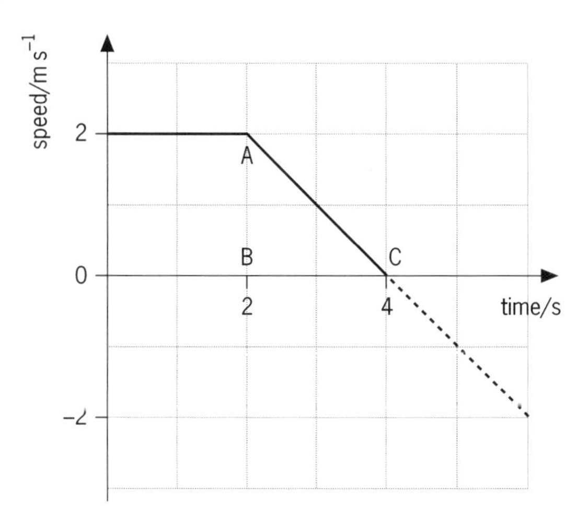

Figure 8.5

a Describe the motion shown in the graph, Figure 8.5.
b What distance is travelled in the first 2 seconds?
c What distance is travelled altogether?
d What is the value of the deceleration (or negative acceleration) from 2 s to 4 s?
e What would have been happening if the line had continued below the axis (dashed line)?

2 How many *g* are experienced if you accelerate from 0 to 260 m s^{-1} in 5 s? (A rocket-powered car might achieve this!)

3 A train going at 53 m s^{-1} slows down steadily, taking 1600 m to stop. What is its acceleration?

Hint: You can either use an appropriate equation or consider the average speed and find the time taken to stop. Try both ways!

4 What is the change in velocity when an initial velocity of 4 m s^{-1} due South changes to 12 m s^{-1} due North?

Level 2

5 A bullet is fired horizontally at 200 m s^{-1}, at a target 40 m away. Neglecting air resistance:
a how far below the point of aim will it hit the target? (that is, how far does it drop?)
b at what angle will it enter the target?

6 A glider on an air track travels to the end and bounces off a stretched elastic band without loss of energy. Sketch graphs of
a the displacement
b the velocity
c the acceleration
against time, for three traverses of the track.

7 An aeroplane accelerates at 3.1 m s^{-2} to reach its take-off speed of 100 m s^{-1}.
a How long must the runway be?
b How long does it take before the plane takes off?

Level 3

8 A golf ball leaves the club at 46 m s^{-1}, at an angle of 50° upwards from the horizontal, across level ground. Neglect air resistance.
a What is the vertical component of the velocity?
b What is the horizontal component of the velocity?
c How long does it take to reach the top of its flight?
d How high does it go?
e After how long does it hit the ground?
f What is the horizontal distance travelled?

9 The graph in Figure 8.6 shows the velocity of a sky diver.

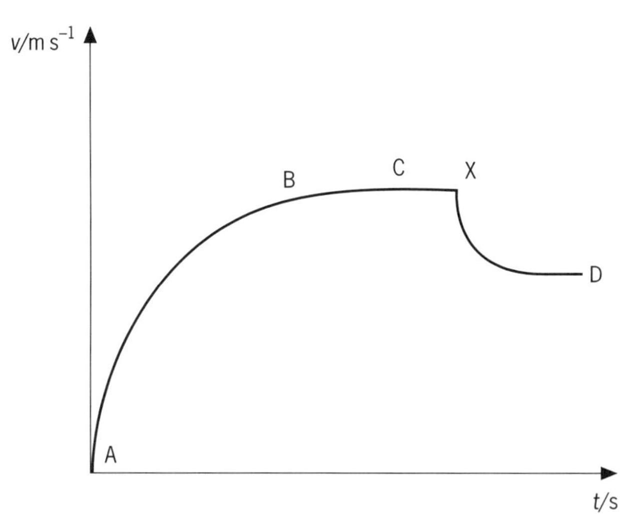

Figure 8.6

a What would you expect her acceleration to be at A?
b How would you measure it from the graph?
c Explain the curve from A to B.
d What is happening at C?
e What do you think happens at X?
f Why is the velocity at D lower than at C?

10 A car is travelling along the motorway at 35 m s^{-1}.

Note: 70 mph = 31 m s^{-1}

The driver does not notice a stationary police car, which then gives chase. The police car accelerates at 2.5 m s^{-2} until it overtakes the speeding car.
a How long does it take for the police car to draw level with the speeding car?
b How far does the police car travel before this happens?
c Sketch a graph of distance against time, showing the positions of both cars.

11 a Describe the motion represented by the velocity–time graph in Figure 8.7. Comment on each of the lettered sections or points.

b What distance is travelled in the first 2 s?

c Sketch a graph showing the first 6 s of a similar motion on the Moon, where $g = 1.6\,\text{m}\,\text{s}^{-2}$.

d What distance will be travelled before the velocity becomes zero on the Moon?

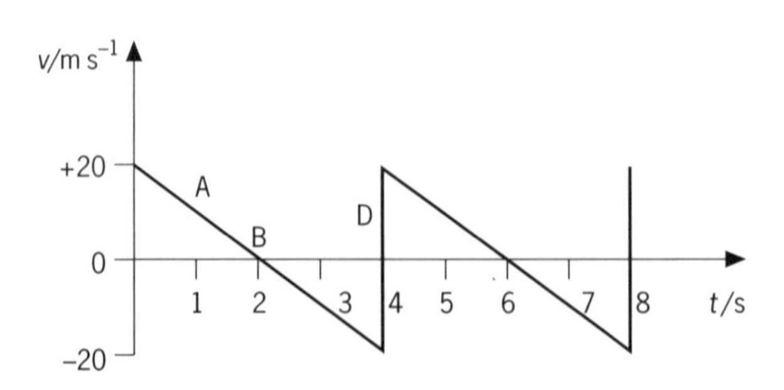

Figure 8.7

Level 1

1 a The object has a constant speed of $2\,\mathrm{m\,s^{-1}}$ for 2 s and then slows down steadily until it stops at 4 s.

b Speed $= \dfrac{\text{distance}}{\text{time}}$ or $v = \dfrac{s}{t}$

so $s = v \times t = 2\,\mathrm{m\,s^{-1}} \times 2\,\mathrm{s} = \mathbf{4\,m}$

c The distance is the area under the graph which is the 2×2 square plus the triangle ABC. The area of the triangle is $\frac{1}{2} \times$ base $\times$ height $= \frac{1}{2} \times 2\,\mathrm{s} \times 2\,\mathrm{m\,s^{-1}} = 2\,\mathrm{m}$

Total distance $= 4\,\mathrm{m} + 2\,\mathrm{m} = \mathbf{6\,m}$

d Acceleration $= \dfrac{\text{change of speed}}{\text{time taken}}$ or $a = \dfrac{v-u}{t}$

so $a = \dfrac{0 - 2\,\mathrm{m\,s^{-1}}}{2\,\mathrm{s}} = -1\,\mathrm{m\,s^{-2}}$.

This is a deceleration of $\mathbf{1\,m\,s^{-2}}$.

e The object had zero speed (it had stopped) at C and it would then have started to travel in the opposite direction with steadily increasing speed.

2 Acceleration $a = \dfrac{\text{change in speed}}{\text{time taken}}$

so $a = \dfrac{260\,\mathrm{m\,s^{-1}}}{5\,\mathrm{s}} = 52\,\mathrm{m\,s^{-2}}$

This is about $\mathbf{5 \times g}$.

3 You need an equation that involves velocity, acceleration and distance. So $v^2 = u^2 + 2as$ looks useful.

Final velocity, $v = 0\,\mathrm{m\,s^{-1}}$
Initial velocity, $u = 53\,\mathrm{m\,s^{-1}}$
Distance, $s = 1600\,\mathrm{m}$

You need to rearrange the equation to give a as the subject:

$$2as = v^2 - u^2$$

$$\therefore \quad a = \frac{v^2 - u^2}{2s} = \frac{0 - (53\,\mathrm{m\,s^{-1}})^2}{2 \times 1600\,\mathrm{m}}$$

$$= -\frac{2809\,\mathrm{m\,s^{-2}}}{3200} = \mathbf{-0.88\,m\,s^{-2}}$$

An alternative way is to say that the average speed must be half the original speed, because it drops steadily from $53\,\mathrm{m\,s^{-1}}$ to $0\,\mathrm{m\,s^{-1}}$.

$$\text{average speed} = \frac{53\,\mathrm{m\,s^{-1}} + 0\,\mathrm{m\,s^{-1}}}{2} = 26.5\,\mathrm{m\,s^{-1}}.$$

Time taken to go 1600 m at that average speed =

$$\frac{1600\,\mathrm{m}}{26.5\,\mathrm{m\,s^{-1}}} = 60.4\,\mathrm{s}$$

Acceleration $a = \dfrac{\text{change in speed}}{\text{time taken}}$

so $a = \dfrac{0\,\mathrm{m\,s^{-1}} - 53\,\mathrm{m\,s^{-1}}}{60.4\,\mathrm{s}} = \mathbf{-0.88\,m\,s^{-2}}$

4 As these velocities are at 180° to each other you do not need a velocity triangle.
$4\,\mathrm{m\,s^{-1}}$ due South is reduced to zero by a change of $4\,\mathrm{m\,s^{-1}}$ due North.
Then the velocity has to increase to $12\,\mathrm{m\,s^{-1}}$ due North.
So the total change is $\mathbf{16\,m\,s^{-1}}$ **to the N.**

A diagram always helps. See Figure 8.8.

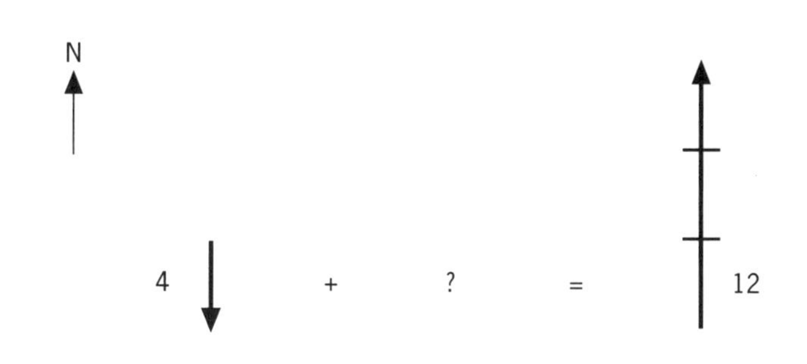

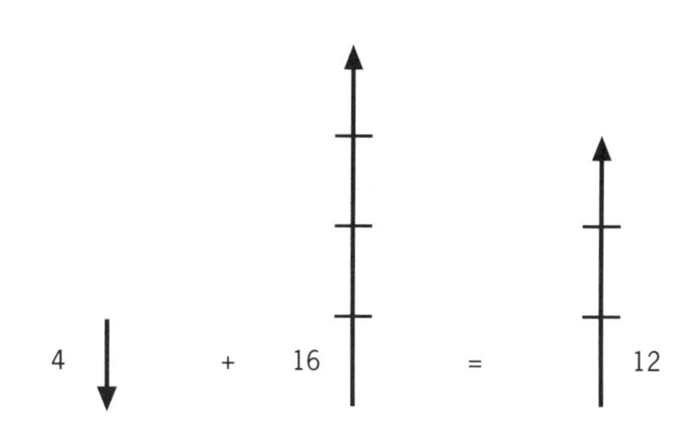

Figure 8.8

Level 2

5 a At a speed of $200\,\mathrm{m\,s^{-1}}$ the time taken for the bullet to reach the target, which is 40 m away, is given by

$$\text{time, } t = \frac{\text{distance}}{\text{speed}} = \frac{40\,\mathrm{m}}{200\,\mathrm{m\,s^{-1}}} = 0.20\,\mathrm{s}$$

During this time the bullet is being accelerated downwards by gravity. Its vertical velocity after time t will be given by the equation

$$v = u + at$$

where the initial downward velocity, $u = 0\,\mathrm{m\,s^{-1}}$, and the downward acceleration, a, is the acceleration due to gravity $= g = 9.8\,\mathrm{m\,s^{-2}}$.

The initial vertical component of velocity is zero because the bullet was travelling horizontally when it left the barrel of the gun.

So, when the bullet hits the target after 0.20 seconds, its vertical (downwards) velocity is

$$v = 0\,\mathrm{m\,s^{-1}} + 9.8\,\mathrm{m\,s^{-2}} \times 0.20\,\mathrm{s} = 1.96\,\mathrm{m\,s^{-1}}$$

This vertical component of velocity is independent of the horizontal velocity.

To calculate how far the bullet falls before it hits the target you can use the equation

$$s = ut + \tfrac{1}{2}at^2$$

for the vertical motion, where the initial vertical velocity $u = 0\,\mathrm{m\,s^{-1}}$.

So $s = 0 + \tfrac{1}{2} \times 9.8\,\mathrm{m\,s^{-1}} \times (0.20\,\mathrm{s})^2 = \mathbf{0.196\,m}$

Alternatively, you can think of the bullet travelling downwards at its average speed of

$$\frac{u+v}{2} = \frac{(0 + 1.96\,\mathrm{m\,s^{-1}})}{2} = 0.98\,\mathrm{m\,s^{-1}}$$

for 0.20 s, so the distance it travels = average speed × time

$$= 0.98\,\mathrm{m\,s^{-1}} \times 0.20\,\mathrm{s} = \mathbf{0.196\,m}$$

b To find θ, the angle to the horizontal at which the bullet enters the target, use a vector diagram to add the horizontal and vertical components of velocity, Figure 8.9.

$$\tan\theta = \frac{1.96\,\mathrm{m\,s^{-1}}}{200\,\mathrm{m\,s^{-1}}} \approx 0.0098$$

so $\theta = \tan^{-1} 0.0098 = \mathbf{0.56°}$

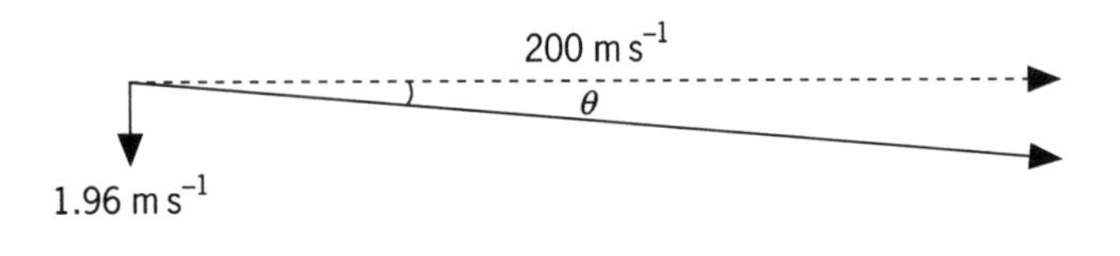

Figure 8.9

6 On an air track, friction is almost eliminated so you can assume that speeds remain constant.

a Displacement is measured from the end of the track at which the glider starts. Assume this is the left end. The displacement increases and then decreases steadily because, while the glider is moving, the speed is constant. See Figure 8.10a.

b Remember that the velocity reverses when the glider is at each end of the track, so these are the only times when there is an acceleration. See Figure 8.10b.

c When the glider hits the elastic band at the right end of the track, the acceleration is to the left to bring the glider to a halt and get it going towards the left end instead. The reverse happens when the glider hits the left end. See Figure 8.10c.

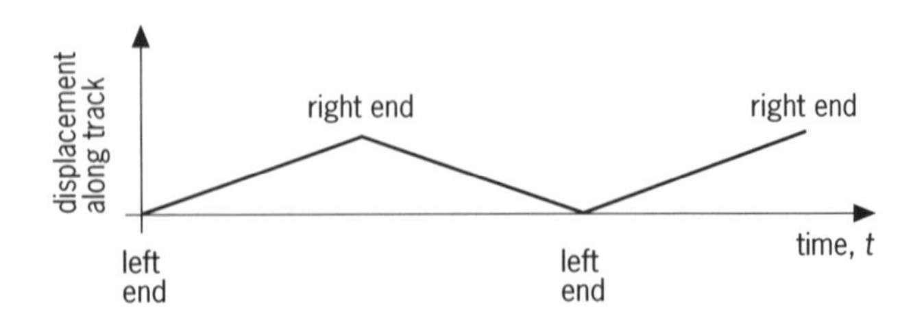

a

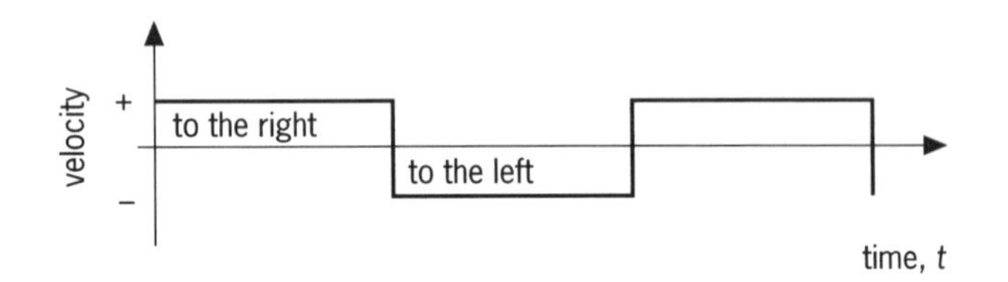

b

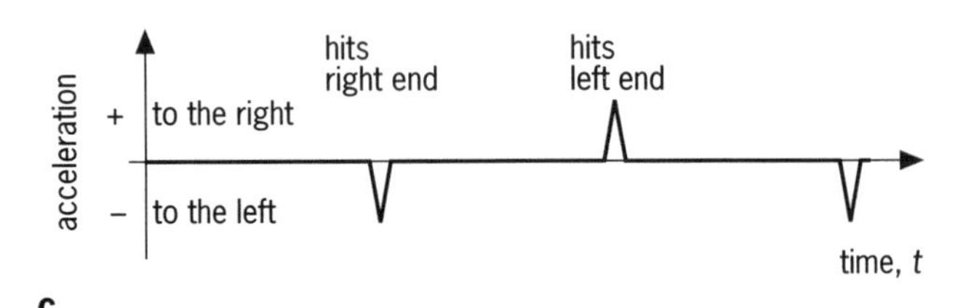

c

Figure 8.10

7 a To find the length of the runway we need to use the equation

$$v^2 = u^2 + 2as$$

In this case, the initial velocity of the aeroplane, $u = 0\,\mathrm{m\,s^{-1}}$.
Rearrange the equation to get s as the subject:

$$s = \frac{v^2 - u^2}{2a} = \frac{(100\,\mathrm{m\,s^{-1}})^2 - 0^2}{2 \times 3.1\,\mathrm{m\,s^{-2}}}$$

$$= \frac{10\,000\,\mathrm{m^2\,s^{-2}}}{6.2\,\mathrm{m\,s^{-2}}} = \mathbf{1613\,m} \text{ or } \mathbf{1610\,m} \text{ to 3 significant figures}$$

b To find the time taken to reach the velocity at take-off, use the equation

$$v = u + at$$

where the initial velocity, $u = 0\,\mathrm{m\,s^{-1}}$. This can be rearranged to give

$$t = \frac{v}{a} = \frac{100\,\mathrm{m\,s^{-1}}}{3.1\,\mathrm{m\,s^{-2}}} = \mathbf{32\,s}$$

Level 3

8 a The vertical component of the initial velocity is (Figure 8.11)

$$46\,\text{m}\,\text{s}^{-1} \times \sin 50° = \mathbf{35.2\,m\,s^{-1}}$$

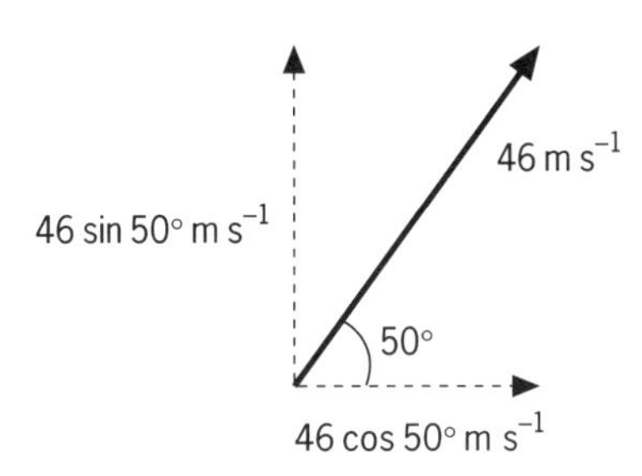

Figure 8.11

b The horizontal component is

$$46\,\text{m}\,\text{s}^{-1} \times \cos 50° = \mathbf{29.6\,m\,s^{-1}}$$

c The vertical component of the velocity will be zero at the top of the ball's flight. The vertical acceleration is $-g$ so the time t taken to reach the top of the ball's flight is given by

$$v = u + at$$

with $v = 0$ and $a = -g$. So

$$t = \frac{-u}{-g} = \frac{-35.2\,\text{m}\,\text{s}^{-1}}{-9.8\,\text{m}\,\text{s}^{-2}} = \mathbf{3.6\,s}$$

d To find the maximum height, use the vertical component of velocity from part a, remembering that the acceleration is negative, in the equation

$$s = ut + \tfrac{1}{2}at^2$$
$$\therefore\ s = 35.2\,\text{m}\,\text{s}^{-1} \times 3.6\,\text{s} - \tfrac{1}{2} \times 9.8\,\text{m}\,\text{s}^{-2} \times (3.6\,\text{s})^2$$
$$= 126.6 - 63.3 = \mathbf{63.3\,m}$$

Alternatively you could calculate the average vertical velocity,

$$v_{av} = \frac{35.2\,\text{m}\,\text{s}^{-1}}{2} = 17.6\,\text{m}\,\text{s}^{-1}$$

then simply multiply by the time the ball takes to reach the maximum height:

$$s = v_{av} \times t = 17.6\,\text{m}\,\text{s}^{-1} \times 3.6\,\text{s} = \mathbf{63.3\,m}$$

Or you can use this method to check your answer.

e If air resistance is ignored the horizontal velocity is constant and so the flight of this idealised golf ball is symmetrical about the point at which it reaches its maximum height. The time between it being hit and landing is simply twice the time to reach the top:

$$t_{tot} = 2 \times 3.6\,\text{s} = \mathbf{7.2\,s}$$

f To calculate how far it travels before it hits the ground, use the horizontal component of the velocity, which stays constant, and multiply by the time the ball is in the air, t_{tot}:

$$\text{distance} = \text{speed} \times \text{time}$$
$$= 29.6\,\text{m}\,\text{s}^{-1} \times 7.2\,\text{s} = \mathbf{210\,m} \text{ to 2 significant figures}$$

9 a Right at the start, at A, when the diver has zero vertical velocity, there is no vertical drag so you would expect her to fall with acceleration $= g$.

b You would measure the slope of the graph at the origin by drawing a tangent to the curve.

c As the vertical velocity increases the drag increases so the **unbalanced force** downwards on the diver gradually **reduces** from a value equal to her weight at point A, to zero at C. At C the drag is equal to her weight, but acts in the opposite direction, so there is no resultant accelerating force on her.

d At C her velocity becomes constant and is called her terminal velocity.

e At X the drag suddenly increases, probably because she opens her parachute (or she could have changed position from vertical to horizontal). Anything that presents a greater area to the air will mean that the drag becomes greater than the weight. This means that there is now a resultant force upwards which reduces the downward velocity. There is a negative acceleration for a while (opposite to g).

f The new terminal velocity is less because the required drag for equilibrium is achieved at a lower velocity due to the extra area.

10 a The distance, s, from the point where the police car was parked to where it catches up with the speeding car is the same for both vehicles, and the time taken to cover this distance is also the same for both, say t.

The distance covered by accelerating police car

$$s = ut \times \tfrac{1}{2}at^2 = \tfrac{1}{2}at^2 \quad \text{as } u = 0$$

The distance covered by car at steady velocity

$$s = v \times t$$

So $\quad \tfrac{1}{2}a \times t^2 = v \times t$

Dividing both sides of the equation by t gives

$$\tfrac{1}{2}at = v \quad \text{or} \quad t = 2\frac{v}{a}$$
$$\therefore\ t = \frac{2 \times 35\,\text{m}\,\text{s}^{-1}}{2.5\,\text{m}\,\text{s}^{-2}}$$
$$= \mathbf{28\,s}$$

b Distance $s = v \times t = 35\,\text{m}\,\text{s}^{-1} \times 28\,\text{s} = \mathbf{980\,m}$

or $\quad s = \tfrac{1}{2}a \times t^2 = \tfrac{1}{2} \times 2.5\,\text{m}\,\text{s}^{-2} \times (28\,\text{s})^2 = \mathbf{980\,m}$

c

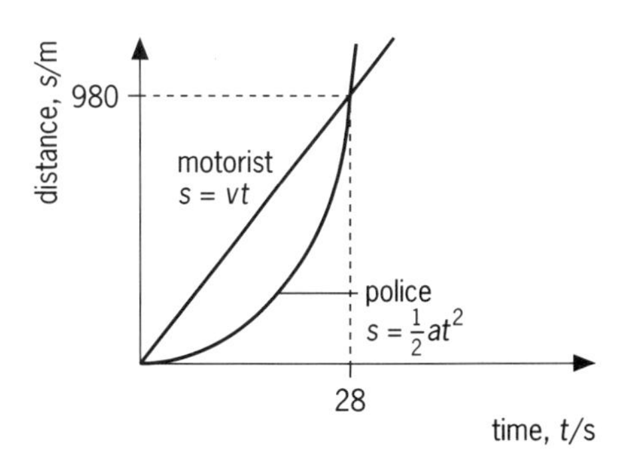

Figure 8.12

11 a In section A the velocity decreases steadily from $+20\,\mathrm{m\,s^{-1}}$ to zero at B and then increases steadily in the other direction in section C to $-20\,\mathrm{m\,s^{-1}}$. The acceleration is given by the slope of the graph, which is $-10\,\mathrm{m\,s^{-1}}$ every second. This could be the acceleration due to gravity and it acts in the negative direction.

This looks like the motion of an object thrown upwards at $20\,\mathrm{m\,s^{-1}}$, gradually slowing down until it stops (at B) and then accelerating back down again. When it reaches the same value of velocity as it had at the start it is given a sudden reversal of velocity in section D. This would happen if it hit the ground and bounced back up again. If little energy is lost this will carry on for more bounces.

b The area under the graph gives the distance travelled. In this case the area is a triangle so

$$\text{area} = \tfrac{1}{2}\,\text{base} \times \text{height} = \tfrac{1}{2} \times 2\,\mathrm{s} \times 20\,\mathrm{m\,s^{-1}} = \mathbf{20\,m}$$

This is how high the object goes.

c

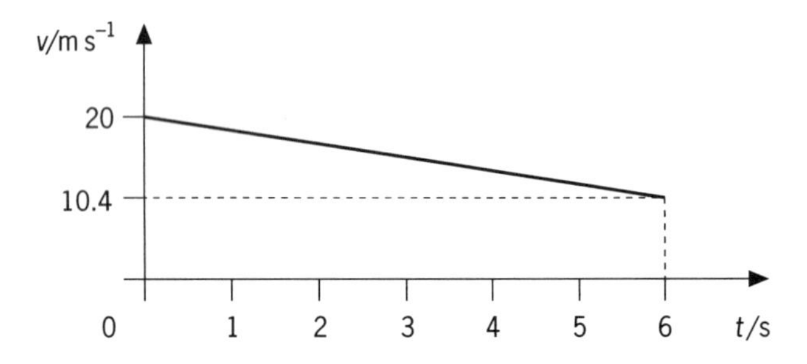

Figure 8.13

The slope of the new graph, Figure 8.13, will be $-1.6\,\mathrm{m\,s^{-2}}$ so after 6 s the velocity will have decreased by only $6 \times 1.6 = 9.6\,\mathrm{m\,s^{-1}}$ and the object is still moving upwards.

d To find the time taken to stop, use the equation

$$v = u + at$$

where $v = 0\,\mathrm{m\,s^{-1}}$, $u = 20\,\mathrm{m\,s^{-1}}$ and $a = -1.6\,\mathrm{m\,s^{-2}}$. So

$$t = \frac{v-u}{a} = \frac{0 - 20\,\mathrm{m\,s^{-1}}}{-1.6\,\mathrm{m\,s^{-2}}} = 12.5\,\mathrm{s}$$

If it is 12.5 s before it stops it will have gone a lot higher; the distance will be given by the area under the new line. This is the area of a triangle of base 12.5 s and height $20\,\mathrm{m\,s^{-1}}$, that is

$$\tfrac{1}{2}\,\text{base} \times \text{height} = \tfrac{1}{2} \times 12.5\,\mathrm{s} \times 20\,\mathrm{m\,s^{-1}} = \mathbf{125\,m}$$

9 Newton's laws of motion and momentum

Inertia

When we travel in a car which suddenly slows down we feel as if we are being thrown forwards. Similarly, as the car turns a corner we feel that we are being pushed to the outside of the bend. In reality neither of these feelings are correct descriptions of what is happening. As the velocity of the car changes we experience our body's reluctance to change its velocity. It tries to continue to move at the same speed and in the same direction. This reluctance of an object to accelerate (or decelerate) is called its **inertia**. Mass is a measure of the inertia of an object; it is a scalar quantity and is measured in kg. To overcome the inertia of a body a force must be applied to it.

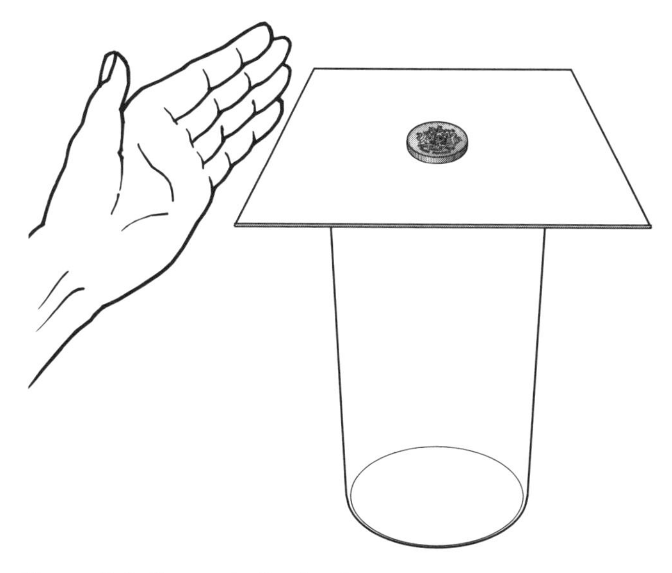

Figure 9.1 Demonstrating inertia

In the 'trick' shown in Figure 9.1, the coin's inertia causes it to remain almost stationary as the card is accelerated sideways by a sharp tap, and it then falls into the tumbler.

Newton's First Law of Motion

Newton, in his First Law of Motion, stated formally that a force is needed to change the velocity of an object. He said that **an object will remain stationary, or continue to move at a constant velocity, unless an unbalanced force is applied to it**. See Figure 9.2.

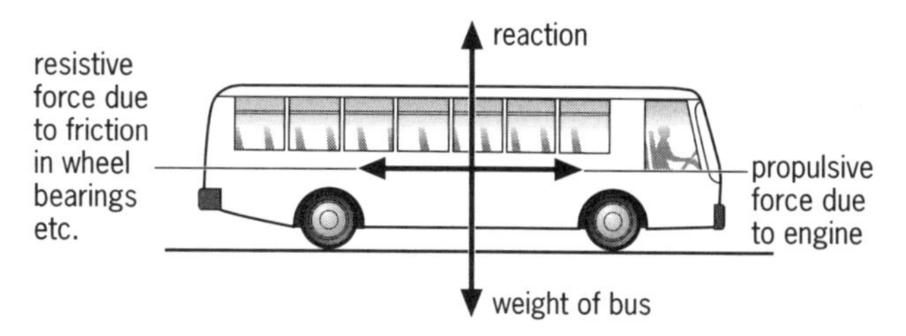

The bus is stationary. When equal and opposite forces are applied to it, it remains stationary

a

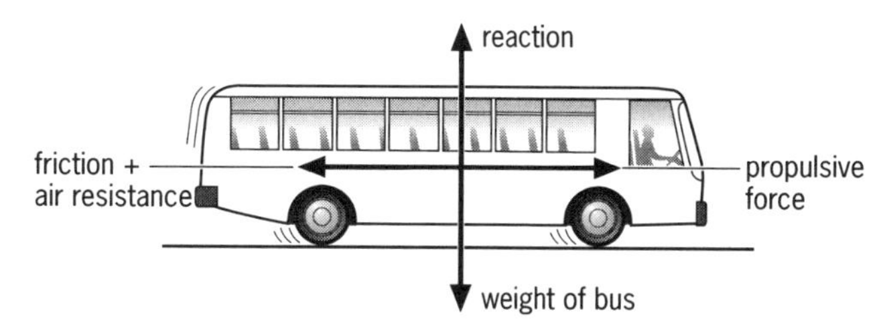

The bus is moving with constant velocity. When equal and opposite forces are applied to it, it continues to move with constant velocity

b

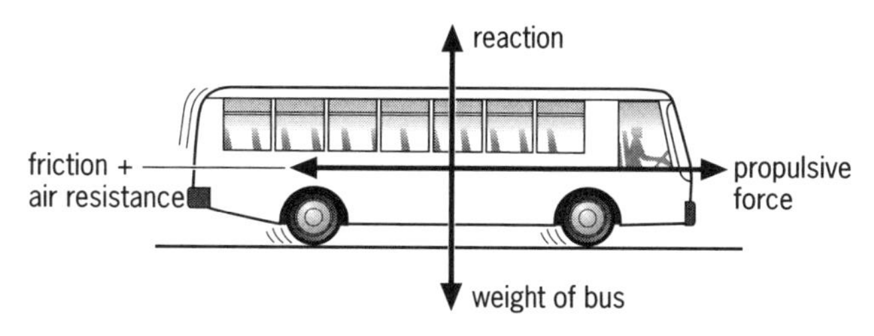

The propulsive force is larger than the resistive force, so the bus will accelerate

c

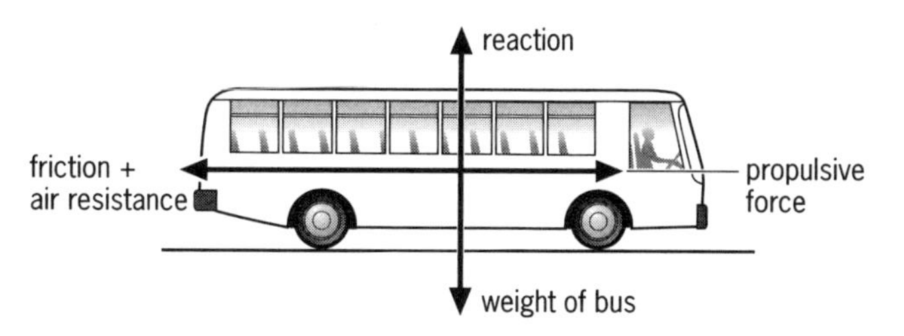

The resistive force is larger than the propulsive force, so the bus will decelerate

d

Figure 9.2 A resultant force is needed to change the state of an object's motion (points of action of the forces have been simplified in order to illustrate the principle)

Measuring forces

The magnitude of a force in newtons can be measured using a newtonmeter, Figure 9.3.

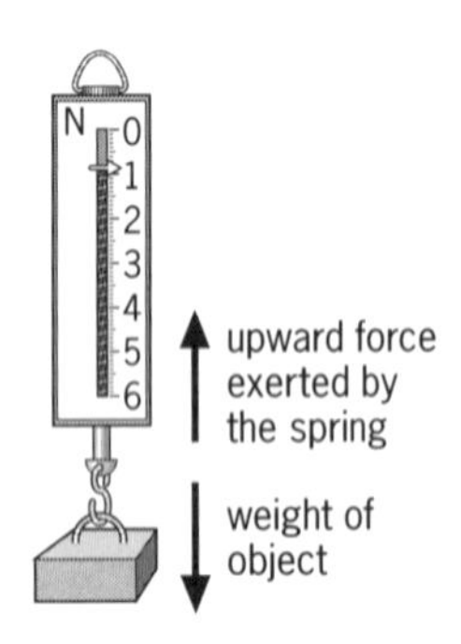

Figure 9.3 The newtonmeter

The mass hung on the newtonmeter shown is stationary. It is experiencing two equal but opposite forces

- the gravitational pull downwards of the Earth – we call this force the **weight** of the object; and
- the upward force exerted on the mass by the extended spring.

Momentum

If an object is moving it has **momentum**. Momentum is a vector quantity equal to the product of an object's mass and its velocity.

$$\text{momentum} = \text{mass} \times \text{velocity}$$

$$p = m \times v$$

The unit of momentum is thus kg m s^{-1}.

Example 1

Calculate the momentum of

a an object of mass 5 kg moving with a velocity of $4\,\text{m s}^{-1}$

b an object of mass 100 g moving with a velocity of $20\,\text{m s}^{-1}$.

a momentum, $p = m \times v$
$= 5\,\text{kg} \times 4\,\text{m s}^{-1}$
$= 20\,\text{kg m s}^{-1}$

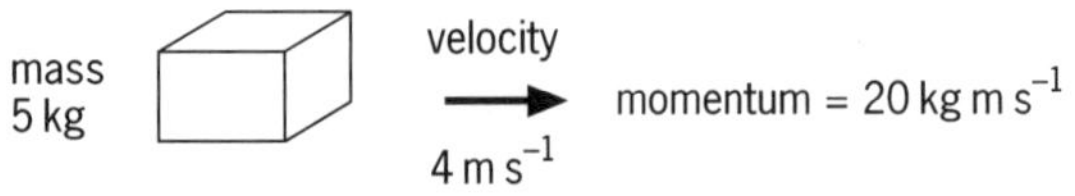

mass 5 kg — velocity $4\,\text{m s}^{-1}$ → momentum = $20\,\text{kg m s}^{-1}$

Figure 9.4a

b momentum, $p = m \times v$
$= 0.1\,\text{kg} \times 20\,\text{m s}^{-1}$
$= 2\,\text{kg m s}^{-1}$

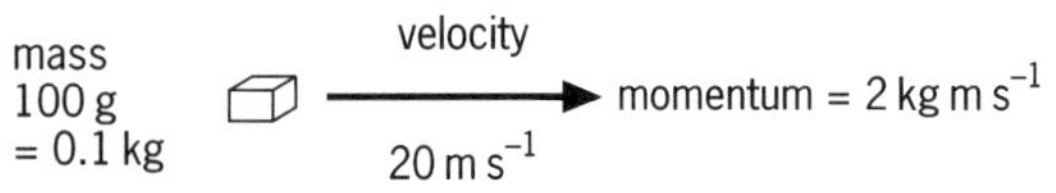

mass 100 g = 0.1 kg — velocity $20\,\text{m s}^{-1}$ → momentum = $2\,\text{kg m s}^{-1}$

Figure 9.4b

Newton's Second Law of Motion

If an unbalanced force is applied to an object its velocity and therefore its momentum changes. Newton's Second Law of Motion states how the size of the applied force is related to the rate at which the momentum of an object changes.

Figure 9.5 The larger the force you apply, the greater the rate of change of momentum of the trolley

Newton stated that **the rate of change of momentum of an object is proportional to the resultant force applied to it and is in the direction of this force**.

For a steady rate of change of momentum we can say

$$F \propto \frac{\text{final momentum} - \text{initial momentum}}{\text{time taken to change}}$$

that is $$F \propto \frac{(mv - mu)}{t}$$

If F is in N, m in kg, v and u in m s^{-1} and t in s this proportionality becomes an equality, and we have the equation:

$$\text{force} = \text{rate of change of momentum}$$

$$F = \frac{(mv - mu)}{t} \quad \text{or} \quad F = \frac{\Delta p}{t}$$

Example 2

Calculate the force needed to change the momentum of an object from $50\,\text{kg m s}^{-1}$ to $250\,\text{kg m s}^{-1}$ in 10 s.

$$F = \frac{(mv - mu)}{t}$$

$$= \frac{(250\,\text{kg m s}^{-1} - 50\,\text{kg m s}^{-1})}{10\,\text{s}} = \frac{200\,\text{kg m s}^{-1}}{10\,\text{s}}$$

$$= 20\,\text{N}\ (\text{kg m s}^{-2})$$

Newton originally stated his second law in terms of momentum but if the mass of the object and its acceleration are constant it can be written in a different form.

$$F = \frac{(mv - mu)}{t}$$

$$F = \frac{m(v - u)}{t}$$

But $(v-u)/t$ is rate of change of velocity (if this is constant), that is the acceleration, a. Therefore

$$F = ma$$

Note that the unit of this is $N = kg\,m\,s^{-2}$.

This equation can be considered as the definition of the inertial mass of an object – the force needed to give the object unit acceleration.

Example 3

Calculate the acceleration of an object of mass 40 kg when a force of 5.0 N is applied to it.

$$F = ma$$

which can be rearranged to give

$$a = \frac{F}{m}$$

$$\text{so} \quad a = \frac{5.0\,N}{40\,kg} = 0.125\,m\,s^{-2}$$

Impulse

The quantity $F \times t$ is known as **impulse** and has the unit N s. From Newton's Second Law of Motion

$$F = \frac{(mv - mu)}{t}$$

$$\therefore \quad Ft = mv - mu = \Delta p$$

that is impulse = change in momentum

Example 4

Calculate the change in momentum (impulse) of a cricket ball if a force of 50 N is applied to it for 0.05 s.

$$\text{change in momentum} = F \times t$$
$$= 50\,N \times 0.05\,s$$
$$= 2.5\,N\,s$$

If a graph of F against t is drawn, the area under the graph is the impulse (Figure 9.6).

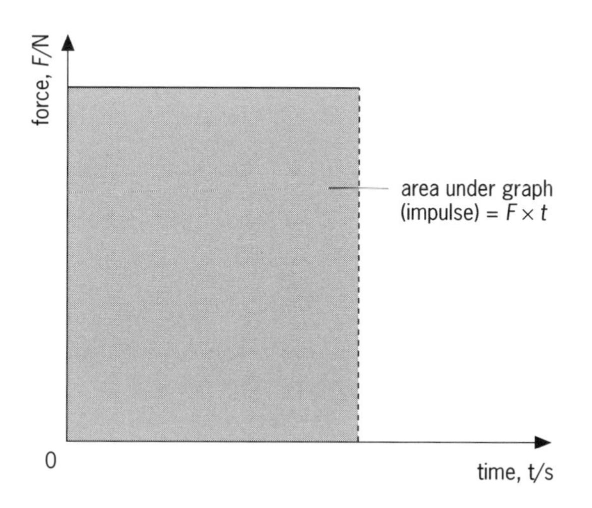

Figure 9.6 Impulse is the area under a force–time graph

Newton's Third Law of Motion

Forces always exist in pairs. They are often known as the action force and the reaction. As the rower in Figure 9.7 steps off her boat one force moves her to the right and the reaction moves the boat to the left. The force of the rower on the boat is equal and opposite to the force of the boat on the rower.

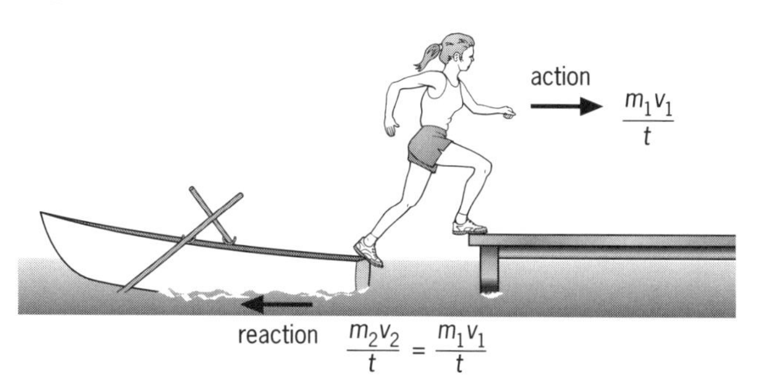

Figure 9.7 A pair of forces

Newton summarised the existence of these pairs of forces in his third law: **when two objects interact, they exert equal and opposite forces on one another**.

When the rower steps off her boat, her momentum changes from zero to m_1v_1. Similarly, the momentum of the boat changes from zero to m_2v_2. If the two forces causing these changes have the same magnitude and act for the same time it follows that

$$m_1v_1 = m_2v_2$$

Note that v_2 is in the opposite direction to v_1.

Example 5

A rower of mass 60 kg steps off her boat with a velocity of $4.0\,m\,s^{-1}$. Calculate the velocity of the boat if it has a mass of 120 kg.

$$m_1v_1 = m_2v_2$$

$$\text{so} \quad v_2 = \frac{m_1v_1}{m_2}$$

$$= \frac{60\,kg \times 4.0\,m\,s^{-1}}{120\,kg}$$

$$= 2.0\,m\,s^{-1} \text{ in the opposite direction}$$

It is this principle of action and reaction on which jet propulsion and rocket propulsion are based. Fast-moving particles are ejected backwards creating an equal and opposite reaction which pushes the craft forwards (Figure 9.8).

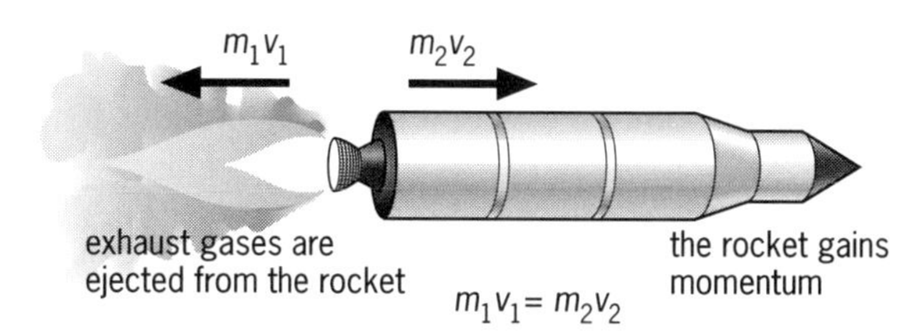

Figure 9.8 Rocket propulsion

Conservation of momentum

When two objects collide the forces they apply to each other cause their momenta to change. By Newton's Third Law the forces acting on the objects during the collision will be equal and opposite and the duration of the collision is the same for both. Therefore the increase in momentum of one of the objects is equal to the decrease in momentum of the second object and there is no change in the total momentum of the system. This is summarised in the Principle of Conservation of Momentum. **When objects in a system interact there is no change to their total momentum, provided no external forces act on the system.**

total momentum before a collision	=	total momentum after a collision

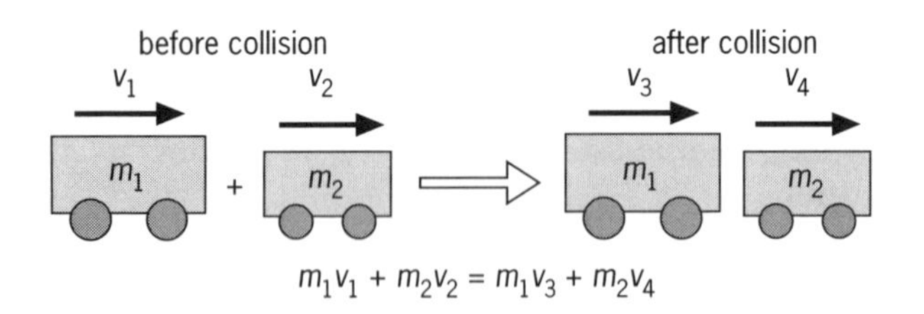

Figure 9.9 Conservation of momentum

Example 6

A car of mass 1000 kg, moving at $25\,\mathrm{m\,s^{-1}}$, collides with a second, stationary car of mass 500 kg. After the collision the first car continues in the same direction with a velocity of $20\,\mathrm{m\,s^{-1}}$. Calculate the new velocity of the second car.

Using the annotation of Figure 9.9 above,

$$m_1v_1 + m_2v_2 = m_1v_3 + m_2v_4$$

$$1000\,\mathrm{kg} \times 25\,\mathrm{m\,s^{-1}} + 500\,\mathrm{kg} \times 0 = 1000\,\mathrm{kg} \times 20\,\mathrm{m\,s^{-1}} + 500\,\mathrm{kg} \times v_4$$

$$\therefore \quad v_4 = \frac{(25\,000 - 20\,000)\,\mathrm{kg\,m\,s^{-1}}}{500\,\mathrm{kg}}$$

$$= 10\,\mathrm{m\,s^{-1}}$$

Momentum is a vector quantity, so its direction needs to be taken account of in calculations.

Example 7

A van of mass 1000 kg, moving at $10\,\mathrm{m\,s^{-1}}$, collides head-on with a van of mass 500 kg which was travelling towards the first van with a velocity of $10\,\mathrm{m\,s^{-1}}$. After the collision the smaller van moves in the opposite direction to its original direction, with a velocity of $8\,\mathrm{m\,s^{-1}}$. Calculate the final velocity of the large van.

Again using

$$m_1v_1 + m_2v_2 = m_1v_3 + m_2v_4$$

and making velocities in the direction of initial travel of the large van positive,

$$1000\,\mathrm{kg} \times 10\,\mathrm{m\,s^{-1}} + 500\,\mathrm{kg} \times (-10\,\mathrm{m\,s^{-1}}) = 1000\,\mathrm{kg} \times v_3 + 500\,\mathrm{kg} \times 8\,\mathrm{m\,s^{-1}}$$

$$v_3 = \frac{(10\,000 - 5000 - 4000)\,\mathrm{kg\,m\,s^{-1}}}{1000\,\mathrm{kg}}$$

$$= 1\,\mathrm{m\,s^{-1}}$$

v_3 is positive, indicating that the large van is still moving in its original direction.

Elastic and inelastic collisions

The Principle of Conservation of Momentum can be applied to both **elastic** and **inelastic** collisions. An elastic collision is one in which there is no loss of kinetic energy (see topic 10) during the collision, so that the sum of the kinetic energies of all the objects before the collision is equal to the sum of the kinetic energies after the collision.

So for an elastic collision:

kinetic energy before = kinetic energy after

$$\tfrac{1}{2}m_1v_1^2 + \tfrac{1}{2}m_2v_2^2 = \tfrac{1}{2}m_1v_3^2 + \tfrac{1}{2}m_2v_4^2$$

as well as

momentum before = momentum after

$$m_1v_1 + m_2v_2 = m_1v_3 + m_2v_4$$

An inelastic collision is one in which there *is* loss of kinetic energy during the collision, so that the sum of the kinetic energies of all the objects before the collision is greater than the sum of the kinetic energies after the collision. The kinetic energy lost during an inelastic collision is changed into other forms of energy such as heat, sound and strain energy.

So for an inelastic collision:

$$\tfrac{1}{2}m_1v_1^2 + \tfrac{1}{2}m_2v_2^2 \neq \tfrac{1}{2}m_1v_3^2 + \tfrac{1}{2}m_2v_4^2$$

but remember that momentum is conserved in all collisions:

$$m_1v_1 + m_2v_2 = m_1v_3 + m_2v_4$$

Example 8

Is the collision between the two cars in Example 6 an elastic or an inelastic collision?

total kinetic energy before collision $= \tfrac{1}{2}m_1v_1^2 + \tfrac{1}{2}m_2v_2^2$

$$= \tfrac{1}{2} \times 1000 \times 25^2 + \tfrac{1}{2} \times 500 \times 0^2$$
$$= \tfrac{1}{2} \times 1000 \times 25^2$$
$$= 312\,500\,\mathrm{J}$$

total kinetic energy after collision $= \tfrac{1}{2}m_1v_3^2 + \tfrac{1}{2}m_4v_2^2$

$$= \tfrac{1}{2} \times 1000 \times 20^2 + \tfrac{1}{2} \times 500 \times 10^2$$
$$= 200\,000 + 25\,000$$
$$= 225\,000\,\mathrm{J}$$

The sum of the kinetic energies of the cars before the collision is greater than the sum of the kinetic energies after the collision. It is therefore an inelastic collision.

Level 1

1 a An unbalanced force of 5.0 N acts on a mass of 4.0 kg. What is the acceleration?

b If the force acts for 25.0 s and the mass was initially stationary, what is the final velocity?

c What distance has been covered while the force acts?

d What would the new velocity be if the mass was initially moving at $2.0\,\mathrm{m\,s^{-1}}$ in the direction of the force?

2 The dummy in a crash test car has a head of mass 3.0 kg. The dummy is strapped in firmly so that the body does not move. The car's forward speed changes from $24\,\mathrm{m\,s^{-1}}$ to zero in 0.25 s.

a What is the force on the dummy's neck?

b Compare this force with the normal weight of the head.

c Why do you think the passenger seats in armed forces' aeroplanes face the rear of the plane?

3 What thrust is exerted by the rotor of a helicopter with mass 8.0 tonne

a to enable it to hover,

b to accelerate it upwards at $1.4\,\mathrm{m\,s^{-2}}$?

(1 tonne = 1000 kg)

4 The motion of a trolley with a glass of water on it is videoed by the science students. One frame is shown in Figure 9.10.

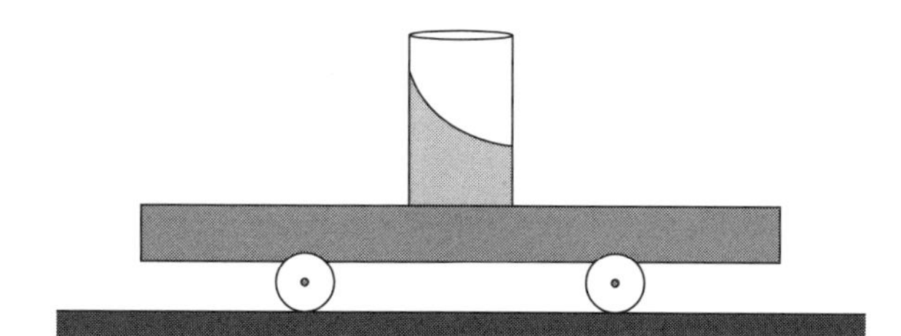

Figure 9.10

a If the trolley had previously been moving to the left, what is now happening?

b If the trolley had previously been stationary, what is now happening?

5 a What is the momentum of a 10 tonne lorry travelling at $22\,\mathrm{m\,s^{-1}}$? (1 tonne = 1000 kg)

b When it is brought to rest, what is the change in its momentum?

c What is the rate of change of momentum if it is brought to rest in

i 3 s

ii 0.3 s

iii 0.03 s?

d What is the significance of the *rate of change of momentum*? (What physical quantity does it tell us about?)

e The quantity *force × time* is called the *impulse*. What can you say about the impulse in this case, if you use the three sets of figures from part c?

6 An alpha particle is emitted from a stationary nucleus at $1.8 \times 10^{7}\,\mathrm{m\,s^{-1}}$. The mass of the particle is 4 u and that of the nucleus is 212 u. (u is a unit of mass equal to 1.66×10^{-27} kg.)

a Find the recoil velocity of the nucleus.

b Explain how you know the direction of movement.

7 A compressed-air gun shoots ten ping-pong balls a second. They travel horizontally at $14\,\mathrm{m\,s^{-1}}$ and have mass of 8.0 g each. They rebound perfectly elastically from a vertical wall. Calculate the force on the wall.

8 A ball of mass 0.20 kg is dropped from a height of 2.5 m and bounces back up to 1.6 m. Taking the acceleration due to gravity as $9.8\,\mathrm{m\,s^{-2}}$, calculate:

a the velocity of the ball as it hits the floor

b the velocity of the ball as it leaves the floor

c the change in momentum caused by the impact

d the average force of the floor on the ball if the impact time is 40 ms.

How do you account for conservation of momentum in this case?

Level 2

9 An aeroplane of mass 20 tonne lands on an aircraft carrier with a horizontal velocity of $80\,\mathrm{m\,s^{-1}}$ (Figure 9.11). A 100 m runway is available.

a What must be the acceleration to bring the aeroplane to rest safely?

b What retarding force is required?

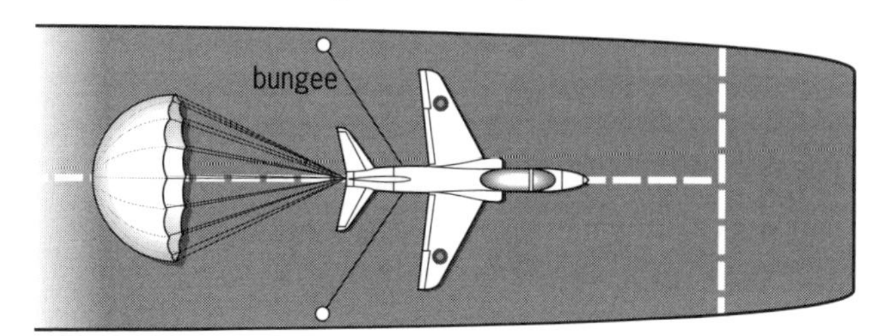

Figure 9.11

10 Looking at a line of wrecked vehicles waiting for collection, your friend remarks that modern cars are not built strongly enough; they just crumple on impact. Comment on this.

11 Two 'gliders' are held at rest at the centre of an air track with a pair of repelling magnets, one fixed to each glider, Figure 9.12a. When they are released the gliders move apart and a stroboscopic photograph is taken. This is drawn to scale in Figure 9.12b.

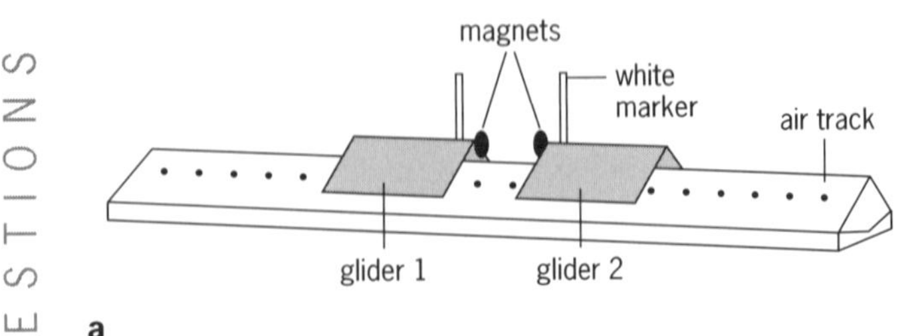

Figure 9.12

The mass of glider 1 is 800 g. What is the mass of glider 2?

12 The graph in Figure 9.13 shows how the force that a cricket *bat* exerts *on a ball* varies with time.

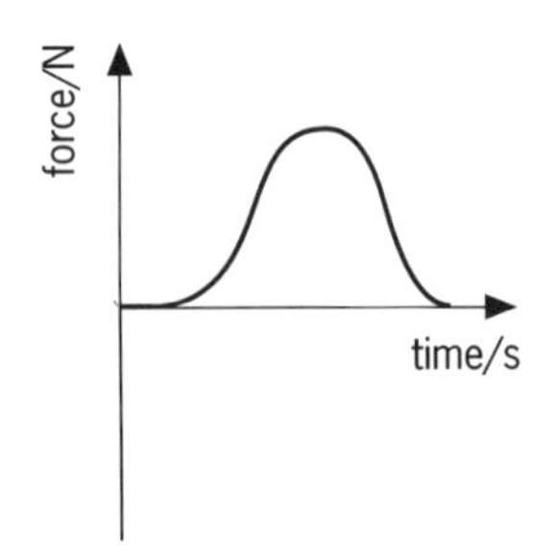

Figure 9.13

a Copy the graph and on the same axes sketch the force that the *ball* exerts *on the bat* over the same time interval.
b How does this illustrate the Principle of Conservation of Momentum?

13 Sand falls vertically, at a rate of $6.0\,\text{kg}\,\text{s}^{-1}$, onto a horizontal conveyor belt which moves steadily at $2.0\,\text{m}\,\text{s}^{-1}$, Figure 9.14. Calculate the average force needed to keep the belt moving.

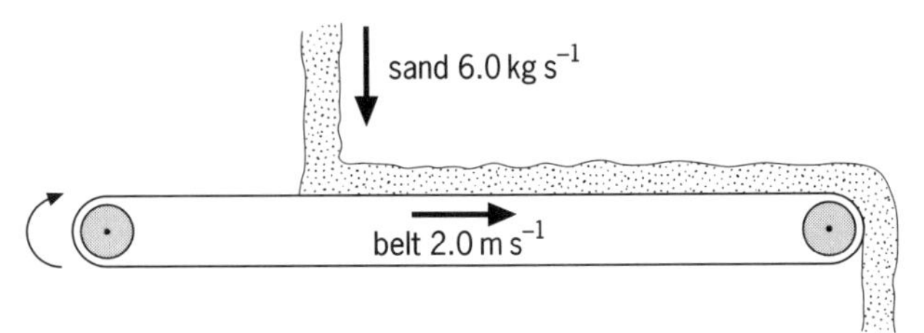

Figure 9.14

Level 3

14 A ball of mass 0.30 kg is dropped off a cliff 75 m high. When it hits the sand at the foot of the cliff it sinks in to a depth of 4.0 cm. Find
a the velocity when it hits the sand
b the retarding force exerted by the sand.
Ignore air resistance. Use $g = 9.8\,\text{m}\,\text{s}^{-2}$.

This can be solved using the equations of motion or by energy considerations (see topic 10); both solutions are given.

15 A rocket ejects gas at high speed. Assuming that the gas is stationary at first, calculate the thrust of a *Saturn V* rocket which ejects 14 tonne of gas per second at a speed of $2400\,\text{m}\,\text{s}^{-1}$.

16 A ball of mass 0.60 kg rolls North at $4.0\,\text{m}\,\text{s}^{-1}$. It meets a slope which causes a force of 0.18 N East. This force lasts for 10 s. Calculate the final velocity of the ball. Neglect any friction effects.

17 Your friends are confused! One says 'I thought momentum was conserved only in elastic collisions.' Another says 'No, it's energy that's conserved in elastic collisions.' What have you got to say about this?

18 Body A, of mass 1.2 kg, moving at velocity $+3.0\,\text{m}\,\text{s}^{-1}$, has a head-on collision with body B, of mass 0.8 kg and velocity $-4.0\,\text{m}\,\text{s}^{-1}$ (Figure 9.15). After the collision, B is going at $+2.6\,\text{m}\,\text{s}^{-1}$ along the original line.

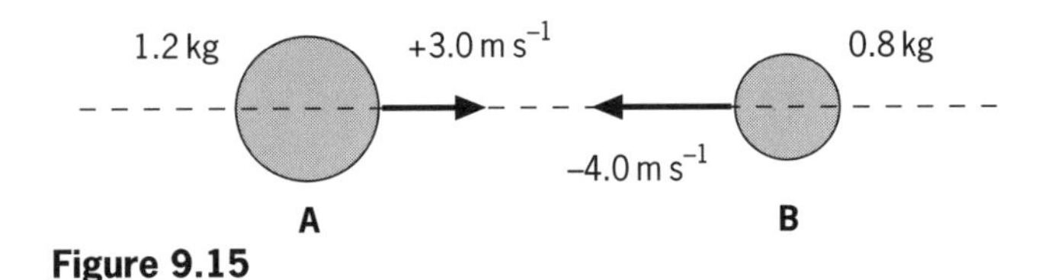

Figure 9.15

a Find the velocity of A after the collision.
b Sketch graphs of momentum against time for each body, covering the time before and after the collision.

19 The graph in Figure 9.16 shows the force on a squash ball when served during a game. Find the mass of the ball if it leaves the racket with a velocity of $40\,\text{m}\,\text{s}^{-1}$. (Assume the ball to be stationary before it is struck.)

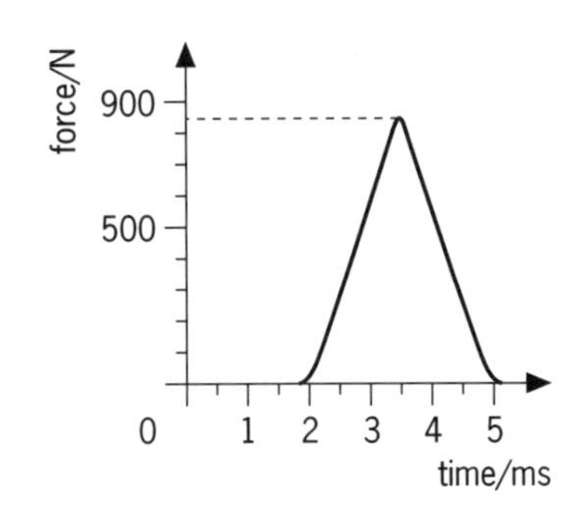

Figure 9.16

Level 1

1 a Force = mass × acceleration or $F = m \times a$

$$\therefore a = \frac{F}{m} = \frac{5.0\,\text{N}}{4.0\,\text{kg}} = \frac{5.0\,\text{kg m s}^{-2}}{4.0\,\text{kg}} = \mathbf{1.25\,m\,s^{-2}}$$

b $v = u + at = 0 + 1.25\,\text{m s}^{-2} \times 25.0\,\text{s} = \mathbf{31.2\,m\,s^{-1}}$

c The distance travelled, s, can be found by multiplying the average speed, v_{av}, by the time.

$$v_{av} = \frac{31.2\,\text{m s}^{-1}}{2} = 15.6\,\text{m s}^{-1}$$

$$\therefore \quad s = 15.6\,\text{m s}^{-1} \times 25.0\,\text{s} = \mathbf{391\,m}$$

Or you can use the equation $s = ut + \frac{1}{2}at^2$, giving

$$s = (0 \times 25.0\,\text{s}) + \tfrac{1}{2} \times 1.25\,\text{m s}^{-2} \times (25.0\,\text{s})^2 = 391\,\text{m}$$

d The change in velocity would be the same whatever the original velocity. So

final velocity = initial velocity + change in velocity
$= 2.0\,\text{m s}^{-1} + 31.2\,\text{m s}^{-1} = \mathbf{33.2\,m\,s^{-1}}$

2 a Acceleration $= \dfrac{\text{change in velocity}}{\text{time taken}}$

$$= \frac{24 - 0\,\text{m s}^{-1}}{0.25\,\text{s}} = 96\,\text{m s}^{-2}$$

The force on the head exerted by the neck to produce this acceleration is

$$F = ma = 3.0\,\text{kg} \times 96\,\text{m s}^{-2} = \mathbf{288\,N}$$

This is also the size of the force on the neck exerted by the head.

b The weight of the head is about 30 N so the force of deceleration is **about 10 times** the force that the head and neck would feel normally.

c A plane is most likely to stop suddenly (crash) when it is travelling forwards. If the seats are facing the rear of the plane, the headrest can support the neck and create the force necessary to stop the head carrying on forwards due to inertia.

3 a To let the helicopter hover the thrust upwards must equal the weight downwards. So

thrust $= W = m \times g = 8.0 \times 10^3\,\text{kg} \times 9.8\,\text{m s}^{-2}$
$= \mathbf{7.8 \times 10^4\,N}$

b To accelerate upwards there must be an extra unbalanced force given by

$$F = ma = 8.0 \times 10^3\,\text{kg} \times 1.4\,\text{m s}^{-2} = 1.1 \times 10^4\,\text{N}$$

so the total force required is
$(7.8 + 1.1) \times 10^4 = \mathbf{8.9 \times 10^4\,N}$.

Alternatively, you could use the change in acceleration needed, compared to free fall, when the engines cut out. This would be

$9.8\,\text{m s}^{-2}$ to overcome gravity + $1.4\,\text{m s}^{-2}$
extra $= 11.2\,\text{m s}^{-2}$

and then use

$$F = ma = 8 \times 10^3\,\text{kg} \times 11.2\,\text{m s}^{-2} = \mathbf{8.96 \times 10^4\,N}$$

4 a If the trolley had been moving to the left then it is now stopping. You can see that inertia of the water makes it carry on towards the left until it hits the sides of the glass.

b If the trolley had been stationary then it is now starting to move to the right. The inertia of the water causes it to remain stationary until it receives the necessary force from the sides of the glass.

5 a Momentum, p = mass × velocity
$= 10 \times 10^3\,\text{kg} \times 22\,\text{m s}^{-1}$
$= \mathbf{2.2 \times 10^5\,kg\,m\,s^{-1}}$

b The change in momentum is the difference between the final momentum and the initial momentum.

$$\Delta p = p_{\text{final}} - p_{\text{initial}}$$

The final momentum, when the lorry is at rest, is zero. So

$\Delta p = 0\,\text{kg m s}^{-1} - 2.2 \times 10^5\,\text{kg m s}^{-1}$
$= \mathbf{-2.2 \times 10^5\,kg\,m\,s^{-1}}$

The change is the same size as the initial momentum but in the opposite direction.

c Rate of change of momentum $= \dfrac{\Delta p}{\text{time taken, } t}$

i For $t = 3\,\text{s}$,

$$\frac{\Delta p}{t} = \frac{-2.2 \times 10^5\,\text{N s}}{3\,\text{s}} = \mathbf{-73\,kN}$$

Note that the unit of momentum can be written as kg m s^{-1} or N s. Check these are equivalent.

ii For $t = 0.3\,\text{s}$,

$$\frac{\Delta p}{t} = \frac{-2.2 \times 10^5\,\text{N s}}{0.3\,\text{s}} = \mathbf{-730\,kN}$$

iii For $t = 0.03\,\text{s}$,

$$\frac{\Delta p}{t} = \frac{-2.2 \times 10^5\,\text{N s}}{0.03\,\text{s}} = \mathbf{-7300\,kN}$$

d The rate of change of momentum is equal to the force causing the change of momentum.

(You can see from the answers to part c that you have to apply a very large force to make momentum change quickly.)

e Impulse = force × time = $F \times t$ but force $= \dfrac{\Delta p}{t}$

$$\therefore \text{impulse} = \frac{\Delta p}{t} \times t = \Delta p$$

In the three cases of part c:

impulse $= 73\,\text{kN} \times 3\,\text{s} = 730\,\text{kN} \times 0.3\,\text{s}$
$= 7300\,\text{kN} \times 0.03\,\text{s}$
$= 2.2 \times 10^5\,\text{N s}$

The impulse is always equal to the change in momentum. For a given change in momentum, the shorter the time, the greater the force required.

6 The situation just after emission is shown in Figure 9.17.

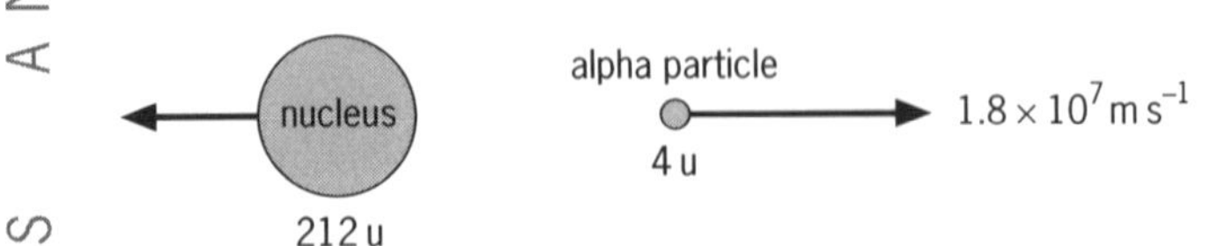

Figure 9.17

a The momentum of the nucleus before the emission is zero, so the total momentum afterwards of the nucleus and the alpha particle will be zero. The particles move in opposite directions, so one velocity must be considered negative.

$$m_{nucleus}\,v_{nucleus} + m_{alpha}\,v_{alpha} = 0$$

$$\therefore v_{nucleus} = \frac{-m_{alpha}\times v_{alpha}}{m_{nucleus}} = \frac{-4\,u\times1.8\times10^7\,m\,s-^1}{212\,u} = \mathbf{-3.4\times10^5\,m\,s^{-1}}$$

Note that you don't need to use the value for u, because it cancels out.

b Because only two particles are involved, they must go in opposite directions for momentum to be conserved. (If three or more particles are involved, they can go almost anywhere.)

7 You need to find the force *on a ball*, by considering what happens to it. The change in velocity of each ball as it bounces back from the wall is from $+14\,m\,s^{-1}$ to $-14\,m\,s^{-1}$ (taking velocity towards the wall as being positive), because there is no loss of energy in the elastic collision. See Figure 9.18.

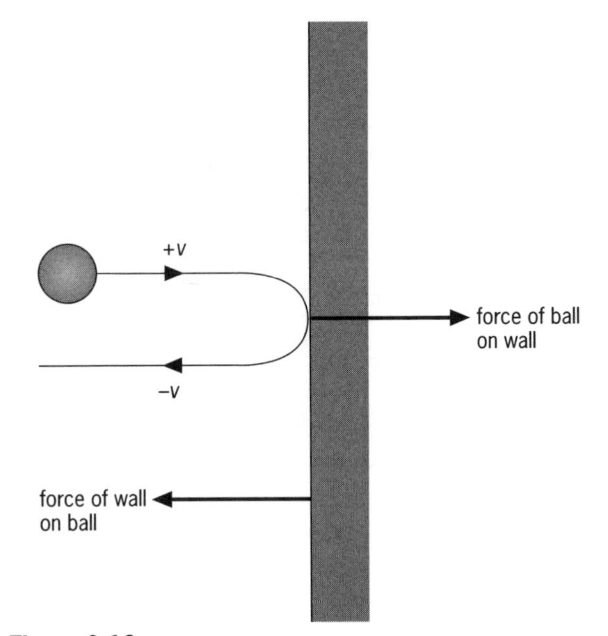

Figure 9.18

$$\Delta v = \text{final velocity} - \text{initial velocity} = -14\,m\,s^{-1} - 14\,m\,s^{-1} = -28\,m\,s^{-1}$$

The change in momentum for each ball is therefore

$$\Delta p = m\Delta v = 0.008\,kg\times(-28\,m\,s^{-1}) = -0.224\,N\,s$$

Because ten balls strike the wall every second, the total rate of change of momentum (total change of momentum per second) is

$$\frac{\Delta p}{t} = \frac{10\times(-0.224\,N\,s)}{1\,s} = -2.2\,N$$

This is the total force of the wall on the balls and it acts away from the wall. The force of the balls on the wall is equal in size but opposite in direction

$$= \mathbf{+2.2\,N}$$

8 a You can use the equation $v^2 = u^2 + 2as$ to calculate the velocity when the ball reaches the floor. The initial velocity, $u = 0$. Choosing movement downwards as being positive, the acceleration, $a = +9.8\,m\,s^{-2}$, and the distance travelled, $s = +2.5\,m$. So

$$v^2 = 0^2 + 2\times9.8\,m\,s^{-2}\times2.5\,m = 49\,m^2\,s^{-2}$$
$$\therefore v = \mathbf{7.0\,m\,s^{-1}}$$

b When the ball has bounced and reached its maximum height, $v = 0$ and $s = -1.6\,m$. (Remember we chose to take movement downwards as positive and movement upwards as negative.) So, using the same equation, $v^2 = u^2 + as$,

$$0 = u^2 + 2\times9.8\,m\,s^{-2}\times(-1.6\,m)$$
$$u^2 = 31.4\,m^2\,s^{-2}$$
$$\therefore u = \mathbf{-5.6\,m\,s^{-1}}$$

You need the negative square root here because the ball is going upwards.

c The velocity changes from $+7.0\,m\,s^{-1}$ (downwards) to $-5.6\,m\,s^{-1}$ (upwards) when the ball bounces. This is a change in velocity of $-12.6\,m\,s^{-1}$; so the change in momentum, given by $\Delta p = m\times\Delta v$, is

$$0.20\,kg\times(-12.6\,m\,s^{-1}) = \mathbf{-2.5\,N\,s}$$

d The ball is in contact with the floor for 40 ms, which is the time it takes to cause this change of momentum. The rate of change of momentum is equal to the force applied to the ball by the floor. So the force is

$$\frac{-2.5\,N\,s}{40\times10^{-3}\,s} = \mathbf{-62.5\,N}$$

The minus sign indicates that the force is upwards.

In order to conserve momentum, the Earth must undergo a change in momentum equal in magnitude but opposite in direction to the change calculated here.

Level 2

9 a The equation that relates acceleration and the distance over which it occurs is $v^2 = u^2 + 2as$, where v is the final velocity and u is the initial velocity. You can rearrange this to give

$$a = \frac{v^2 - u^2}{2s}$$

In this case the initial velocity is $80\,\mathrm{m\,s^{-1}}$ and the final velocity is, we hope, zero; the distance s is 100 m.

$$\therefore a = \frac{0^2 - (80\,\mathrm{m\,s^{-1}})^2}{2 \times 100\,\mathrm{m}}$$

$$= \frac{-6.4 \times 10^3\,\mathrm{m^2\,s^{-2}}}{200\,\mathrm{m}}$$

$$= \mathbf{-32\,m\,s^{-2}}$$

The minus sign means the direction of the acceleration is opposite to the velocity.

b The retarding force must stop a 20 tonne aeroplane:

$$F = m \times a = 20 \times 10^3\,\mathrm{kg} \times (-32\,\mathrm{m\,s^{-2}})$$
$$= \mathbf{-6.4 \times 10^5\,N}$$

10 The crumpling of the metal bodywork increases the time it takes after impact to stop the parts of the car to which you are strapped. This reduces the force you will feel, because the force is equal to the rate of change of momentum; the more slowly your momentum changes the smaller the force will be. So the crumpling is beneficial.

It is also true that materials that are soft, and crumple, are less likely to do you serious injury if you make contact with them as a result of impact.

11 You need to find the ratio of the velocities of the gliders by measuring the positions of the markers in the diagram of the stroboscopic photograph. You should find that

$$\frac{v_1}{v_2} = -3$$

The minus sign is included because they are moving in opposite directions.

The gliders were stationary at the start, so, from conservation of momentum, the sum of the final momenta must also be zero:

$$m_1v_1 + m_2v_2 = 0$$

$$\therefore \quad m_2 = \frac{-m_1 \times v_1}{v_2} = -m_1 \times \frac{v_1}{v_2}$$

$$= -m_1 \times (-3) = m_1 \times 3$$

$$= 0.80\,\mathrm{kg} \times 3 = \mathbf{2.4\,kg}$$

Glider 2 is three times as massive and has a third the velocity of glider 1.

12 a

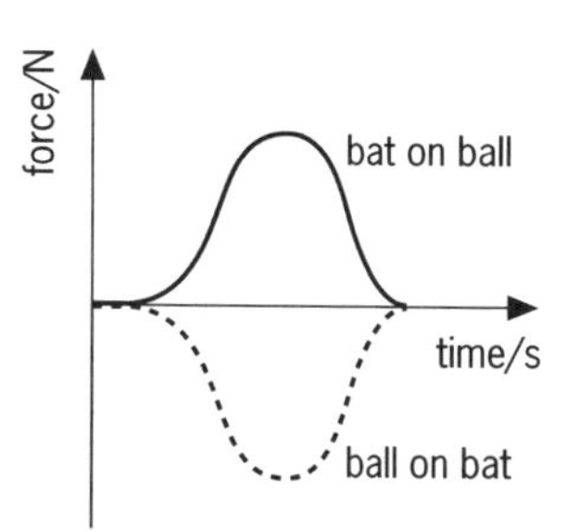

Figure 9.19

From Newton's Third Law, the force of the ball on the bat is the same size as the force of the bat on the ball, but in the opposite direction at all times.

b The area between the time axis and each of these curves is equal to force × time, which is called impulse, and is equal to the change in momentum. The graphs show that the impulses, that is the changes in momentum, are equal and opposite.

13 The conveyor belt has to increase the horizontal component of the velocity of the sand from 0 to $2.0\,\mathrm{m\,s^{-1}}$. It does this for a mass of 6.0 kg every second. The rate of change of momentum (change of momentum per second), which is equal to the force required, is

$$\frac{6.0\,\mathrm{kg} \times 2.0\,\mathrm{m\,s^{-1}}}{1\,\mathrm{s}} = \mathbf{12.0\,N}$$

Level 3

14 a $v^2 = u^2 + 2as$ is the best equation to use, with $u = 0$ and $a = g$. So

$$v^2 = 2 \times 9.8\,\mathrm{m\,s^{-2}} \times 75\,\mathrm{m} = 1470\,\mathrm{m^2\,s^{-2}}$$
$$\therefore v = \mathbf{38\,m\,s^{-1}}$$

b When the ball hits the ground, its velocity is reduced to zero in a distance of 0.040 m. So, using $v^2 = u^2 + 2as$ again and rearranging to make a the subject:

$$a = \frac{v^2 - u^2}{2s}$$

$$= \frac{1470\,\mathrm{m^2\,s^{-2}} - 0}{0.080\,\mathrm{m}} = 1.84 \times 10^4\,\mathrm{m\,s^{-2}}$$

You can now calculate the retarding force using $F = m \times a$:

$$F = 0.30\,\mathrm{kg} \times 1.84 \times 10^4\,\mathrm{m\,s^{-2}} = \mathbf{5510\,N}$$

Alternatively, you could use energy considerations (covered in topic 10) as follows overleaf.

a The gravitational potential energy at the top of the cliff is given by $E = mgh$.

$\therefore \quad E = 0.30\,\text{kg} \times 9.8\,\text{m}\,\text{s}^{-2} \times 75\,\text{m} = 220.5\,\text{J}$

(Note that $kg\,m^2\,s^{-2} \equiv J$.*)*

This is equal to the kinetic energy of the ball ($\frac{1}{2}mv^2$) just before it hits the sand:

$$E = \tfrac{1}{2}mv^2$$

You can rearrange this to make v^2 the subject:

$$v^2 = \frac{2 \times E}{m} = \frac{2 \times 220.5\,\text{J}}{0.30\,\text{kg}} = 1470\,\text{m}^2\,\text{s}^{-2}$$

$\therefore v = \mathbf{38\,m\,s^{-1}}$

b The energy E is also equal to the work done by the sand in bringing the ball to rest, $W = F \times s$, which you can rearrange to give

$$F = \frac{W}{s} = \frac{E}{s} = \frac{220.5\,\text{J}}{0.040\,\text{m}}$$

$$= \frac{5510\,\text{kg}\,\text{m}^2\,\text{s}^{-2}}{\text{m}} = \mathbf{5510\,N}$$

15 A mass of 14×10^3 kg of gas is accelerated from 0 to $2400\,\text{m}\,\text{s}^{-1}$ in each second.

acceleration, $a = 2400\,\text{m}\,\text{s}^{-2}$

The force required to accelerate this mass is given by

$F = m \times a = 14 \times 10^3\,\text{kg} \times 2400\,\text{m}\,\text{s}^{-2} = \mathbf{34\,MN}$

This is the force of the rocket on the gas and it is equal in magnitude and opposite in direction to the force of the gas on the rocket, or the thrust.

You could also solve the problem by considering momentum. In 1 second the change in momentum of the gas is ($m \times v - m \times u$), where v is the final velocity, u is the initial velocity and m is the mass ejected. In this case $v = 2400\,\text{m}\,\text{s}^{-1}$, $u = 0$ and $m = 14$ tonnes (per second). So the change in momentum (per second) is

$$14 \times 10^3\,\text{kg} \times 2400\,\text{m}\,\text{s}^{-1} - 0$$

giving a rate of change of momentum of $34 \times 10^6\,\text{kg}\,\text{m}\,\text{s}^{-2}$, which is equal to the applied force.

$\therefore F = \mathbf{34\,MN}$

16 Rearranging the equation $F = ma$ to make a the subject gives $a = F/m$. So the eastwards acceleration of the ball is

$$\frac{0.18\,\text{N}}{0.60\,\text{kg}} = 0.30\,\text{m}\,\text{s}^{-2}$$

The force lasts for 10 s, so the final component of velocity to the East (initially zero) is given by

$v = at = 0.30\,\text{m}\,\text{s}^{-2} \times 10\,\text{s} = 3.0\,\text{m}\,\text{s}^{-1}$ to the East

The velocity of $4.0\,\text{m}\,\text{s}^{-1}$ North is unaffected by the force. Figure 9.20 shows the velocity triangle.

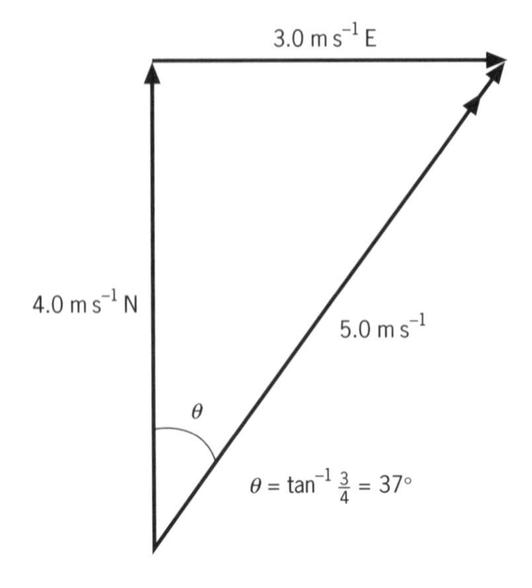

Figure 9.20

You should be able to spot that you have a 3, 4, 5 right-angled triangle and so the final velocity is $\mathbf{5.0\,m\,s^{-1}}$**, at a bearing of 037°** ($\tan^{-1} 3/4$).

17 You should tell them that both total energy and total momentum are *always* conserved, if a complete system is considered. This means that you can account for any changes that occur – a gain in one place is balanced by a loss in another place.

The special thing about an *elastic* collision is that the total *kinetic energy* is the same before and after; none is changed into any other form of energy.

18 a Before the collision, the momentum of body A, p_A, is

$$p_A = m_A v_A = 1.2\,\text{kg} \times (+3.0\,\text{m}\,\text{s}^{-1}) = 3.6\,\text{N}\,\text{s}$$

The momentum of body B, p_B, is

$$p_B = m_B v_B = 0.8\,\text{kg} \times (-4.0\,\text{m}\,\text{s}^{-1}) = -3.2\,\text{N}\,\text{s}$$

So the total initial momentum $= 3.6\,\text{N}\,\text{s} - 3.2\,\text{N}\,\text{s} = +0.4\,\text{N}\,\text{s}$. Because momentum is conserved, this is also the total final momentum.

After the collision the momentum of B is

$$p_B = 0.8\,\text{kg} \times (+2.6\,\text{m}\,\text{s}^{-1}) = 2.1\,\text{N}\,\text{s}$$

The total momentum is therefore

$$p_A + 2.1\,\text{N}\,\text{s} = 0.4\,\text{N}\,\text{s}$$

$$\therefore \quad p_A = 0.4\,\text{N}\,\text{s} - 2.1\,\text{N}\,\text{s} = -1.7\,\text{N}\,\text{s}$$

And so

$$v_A = \frac{p_A}{m_A} = \frac{-1.7\,\text{N}\,\text{s}}{1.2\,\text{kg}} = \mathbf{-1.4\,m\,s^{-1}}$$

b

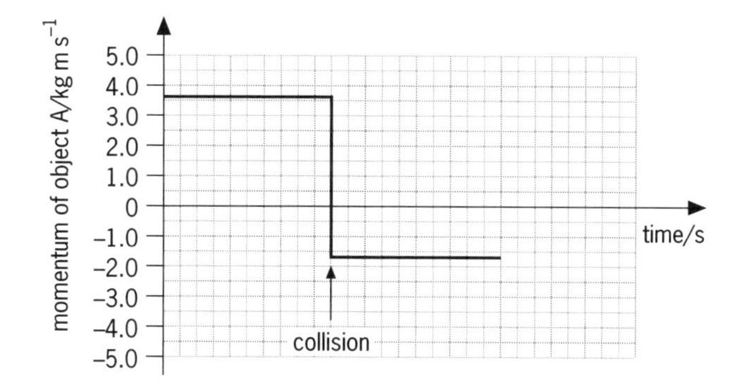

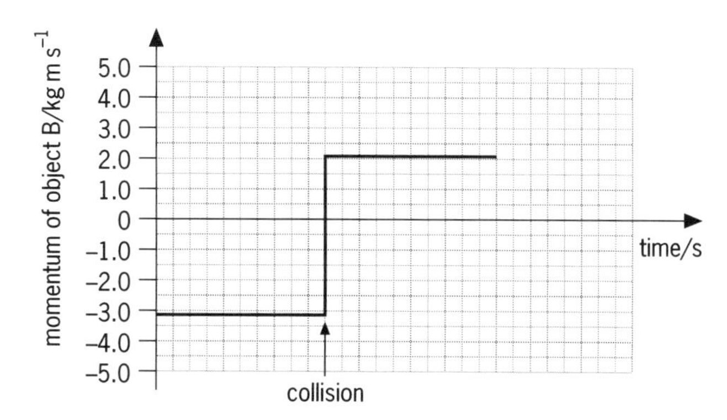

Figure 9.21

19 The area under a force–time graph is equal to the impulse, or change in momentum. The graph shown is very close to being a triangle, and it is acceptable to take the area as $\frac{1}{2}$ base × height. So the change in momentum Δp is, from the graph,

$$\tfrac{1}{2} \times 3 \times 10^{-3}\,\text{s} \times 850\,\text{N} = 1.28\,\text{N s}$$

But the mass of the ball, m, is constant, so

$$\Delta p = m \times \Delta v$$

which you can rearrange to give

$$m = \frac{\Delta p}{\Delta v}$$

$\Delta p = 1.28\,\text{N s}$ and $\Delta v = 40\,\text{m s}^{-1}$

$$\therefore m = \frac{1.28\,\text{N s}}{40\,\text{m s}^{-1}} = \mathbf{0.032\,kg}$$

10 Work, energy and power

Work

Work is done when a force moves its point of application.

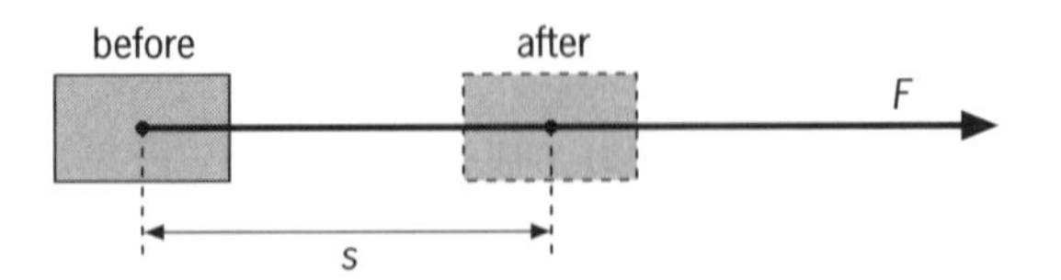

Figure 10.1 Force and displacement

In Figure 10.1, the amount of work done by the force F in moving the box a distance s is $F \times s$, that is, **work done is the product of the force and the distance moved in the direction of the force**. If the force is in newtons and the distance in metres then the unit of work is the joule.

Example 1

Calculate the work done in moving the box in Figure 10.1 a distance of 2.0 m if the applied force is 10 N.

$$W = Fs$$
$$= 10\,\text{N} \times 2.0\,\text{m} = 20\,\text{J}$$

Example 2

Calculate the work done on a box if it is made to move 3.0 m by a 20 N force applied at an angle of 45° to the direction in which it moves (Figure 10.2).

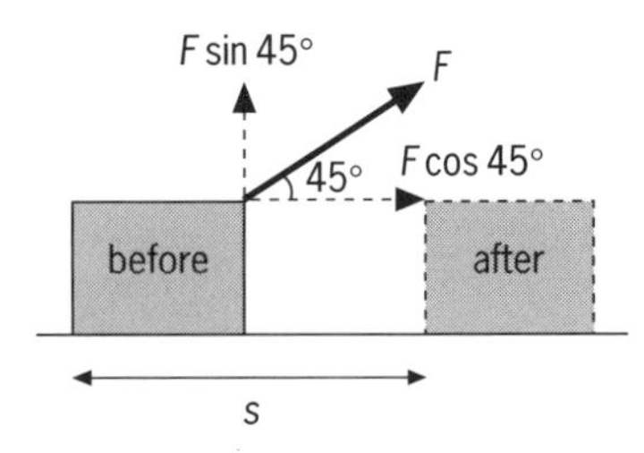

Figure 10.2

We need to use the component of the force which is acting in the direction of motion. This is $20\,\text{N} \times \cos 45°$.

$$W = Fs$$
$$\therefore\ W = 20\,\text{N} \times \cos 45° \times 3.0\,\text{m} = 42\,\text{J}$$

The component of the force perpendicular to the line of motion does no work.

If a graph of force against displacement (in the direction of the force) is plotted, as in Figure 10.3, the area under the graph is equal to the work done by the force.

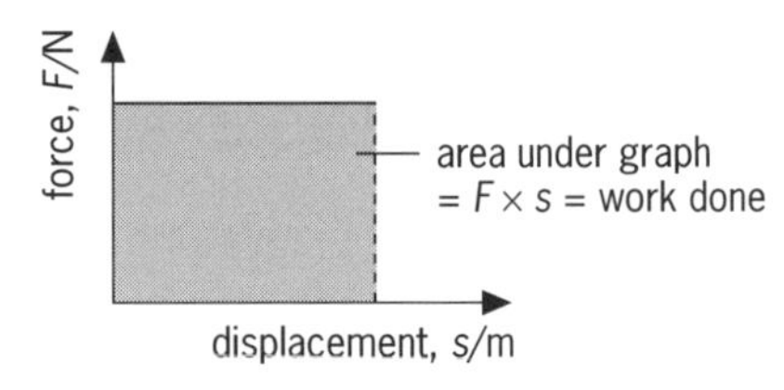

Figure 10.3 Force–displacement graph

Potential energy

Sometimes it is convenient to regard energy as being 'stored work'. For example, the work done in extending a spring is stored in the spring as **elastic potential energy** (see Hooke's Law, page 190).

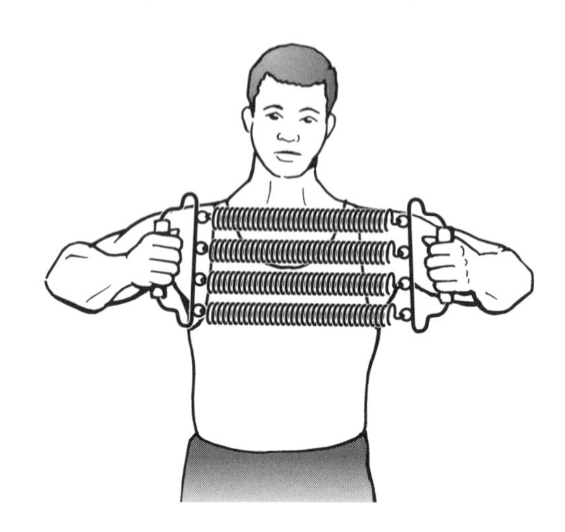

Figure 10.4 As these springs are extended work is done and is stored in the springs as potential energy

Similarly, the work done in lifting a weight from the floor on to a table is stored as **gravitational potential energy**, E_p.

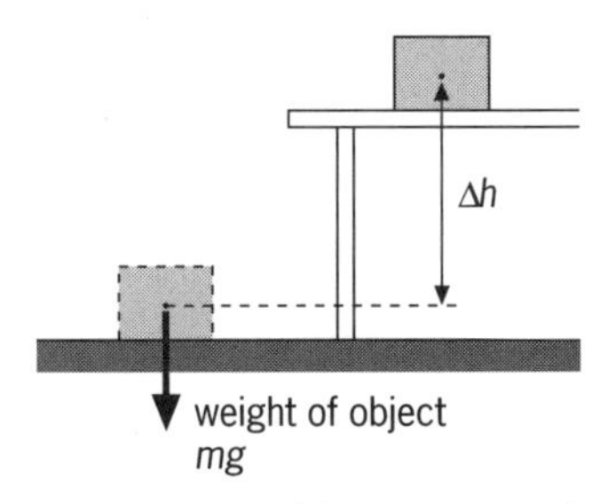

Figure 10.5 As this weight is lifted work is done and is stored in the weight as potential energy

work done on the weight $= \text{force} \times \text{displacement}$

$= mg\,(\text{weight}) \times \Delta h\,(\text{change in height})$

So

change in potential energy, $\Delta E_p = mg\Delta h$

Example 3

Calculate the gain in gravitational potential energy, ΔE_p, of a box of mass 20 kg when it is lifted through a height of 4.0 m. ($g = 10\,\text{m}\,\text{s}^{-2}$)

$$\Delta E_p = mg\Delta h$$
$$= 20\,\text{kg} \times 10\,\text{m}\,\text{s}^{-2} \times 4.0\,\text{m}$$
$$= 800\,\text{J}$$

Kinetic energy

Kinetic energy, E_k, is the energy a body possesses by virtue of its motion. Consider an object of mass m which undergoes acceleration a from rest due to a force F. The work W done by this force is given by

$$W = Fs$$

but, from Newton's Second Law of Motion, $F = ma$, therefore

$$W = mas$$

If the object is undergoing uniform acceleration,

$$v^2 = u^2 + 2as$$

but $\quad u = 0$

$\therefore \quad v^2 = 2as$

so $\quad as = \frac{1}{2}v^2$

and $\quad mas = \frac{1}{2}mv^2$

The work done on the object to get it moving is equal to the kinetic energy of the object. Therefore

$$\text{kinetic energy, } E_k = \tfrac{1}{2}mv^2$$

Example 4

Calculate the kinetic energy of a car of mass 1000 kg travelling with a velocity of 20 m s^{-1} (Figure 10.6).

$$\begin{aligned} E_k &= \tfrac{1}{2}mv^2 \\ &= \tfrac{1}{2} \times 1000\,\text{kg} \times (20\,\text{m s}^{-1})^2 \\ &= 500 \times 400\,\text{kg m s}^{-2} \\ &= 200\,000\,\text{J or } 200\,\text{kJ} \end{aligned}$$

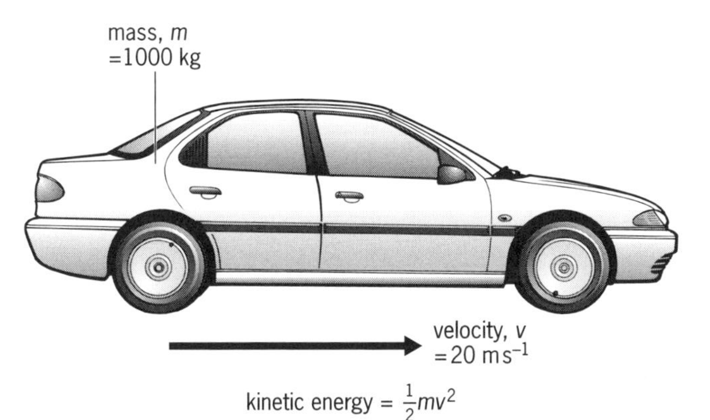

Figure 10.6

Power

There are several ways in which we can describe power. It can be thought of as the rate of doing work, that is

$$\text{power, } P = \frac{\text{work done}}{\text{time taken}}$$

Power is measured in watts (W). A machine has a power of one watt if it is doing one joule of work each second (1 J s^{-1})

Example 5

A fork-lift truck lifts a box weighing 5000 N to a height of 4.0 m in 10 s. Calculate the lifting power of the truck.

$$\begin{aligned} \text{power, } P &= \frac{\text{work done}}{\text{time taken}} \\ &= \frac{5000\,\text{N} \times 4.0\,\text{m}}{10\,\text{s}} \\ &= 2000\,\text{J s}^{-1} \text{ or } 2.0\,\text{kW} \end{aligned}$$

Power can also be thought of as the 'rate of transfer of energy', that is

$$\text{power} = \frac{\text{energy transferred}}{\text{time taken}}$$

Example 6

Water drops 20 m down a waterfall at the rate of 30 kg s^{-1}. Calculate the power of this energy transfer from gravitational potential energy to kinetic energy. ($g = 10\,\text{m s}^{-2}$)

Assuming that all the gravitational potential energy the water loses is changed into kinetic energy,

$$\begin{aligned} \text{power, } P &= \text{rate of loss of gravitational potential energy} \\ &= \frac{\Delta E_p}{t} \\ &= \frac{mg\Delta h}{t} \\ &= \frac{30\,\text{kg} \times 10\,\text{m s}^{-2} \times 20\,\text{m}}{1\,\text{s}} \\ &= 6000\,\text{J s}^{-1} \text{ or } 6\,\text{kW} \end{aligned}$$

Power and velocity

If an object is being moved by a force F through a distance s in a time t, the equation

$$P = \frac{\text{work done}}{\text{time taken}}$$

becomes

$$P = \frac{Fs}{t}$$

But $\frac{s}{t} = v$, the velocity of the object; therefore

$$P = Fv$$

or

$$\text{power} = \text{force} \times \text{velocity}$$

Example 7

Calculate the power of a car's engine which produces a propulsive force of 10 000 N when it is travelling at $40\,\text{m}\,\text{s}^{-1}$.

$$\begin{aligned}\text{power} &= \text{force} \times \text{velocity}\\ &= 10\,000\,\text{N} \times 40\,\text{m}\,\text{s}^{-1}\\ &= 400\,000\,\text{W or } 400\,\text{kW}\end{aligned}$$

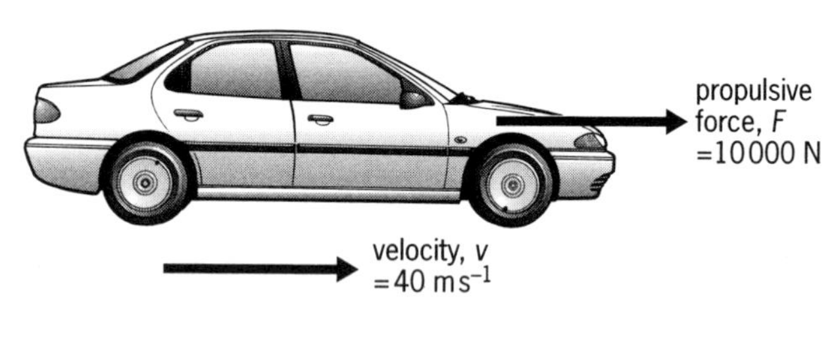

Figure 10.7

Efficiency

When work is done and energy transferred, there is always some energy 'lost'; the energy input is greater than the *useful* energy output. The **efficiency** of an energy transfer can be calculated using the equation

$$(\%)\ \text{efficiency} = \frac{\text{useful energy (or power) output}}{\text{total energy (or power) input}} \times 100\%$$

Example 8

A crane uses 10 000 J of energy to lift a 60 kg load to a height of 10 m. Calculate the efficiency of the crane. ($g = 10\,\text{m}\,\text{s}^{-2}$)

$$\begin{aligned}\text{efficiency} &= \frac{\text{energy gained by load}}{\text{total energy input}} \times 100\%\\ &= \frac{mgh}{\text{total energy input}} \times 100\%\\ &= \frac{60\,\text{kg} \times 10\,\text{m}\,\text{s}^{-2} \times 10\,\text{m}}{10\,000\,\text{J}} \times 100\%\\ &= 60\%\end{aligned}$$

Level 1

1 A car of mass 1000 kg is travelling at $31\,m\,s^{-1}$. It then stops in 75 m, with an acceleration of $-6.4\,m\,s^{-2}$.

a Calculate the work done in stopping the car.

b Show that this work is equal to the kinetic energy the car had when moving.

2

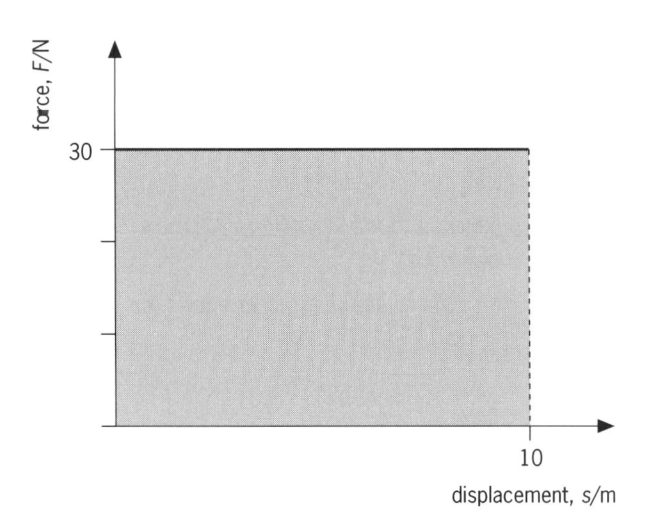

Figure 10.8

A 3.1 kg mass falls 10 m. A force acts on the mass over this distance as shown on the graph, Figure 10.8. The force and displacement are in the same direction.

a Use the graph to calculate the kinetic energy acquired.

b Compare the kinetic energy acquired by the mass with the gravitational potential energy lost. ($g = 9.8\,m\,s^{-2}$)

3 Estimate your power if you do 100 step-ups in 60 seconds.

4 A girder of mass 600 kg is raised to third floor level (10 m) using a motorised crane.

a If the crane motor has a constant output power of 3.0 kW, how long does it take?

b The motor is only 20% efficient. How much energy is actually used?

5 a Guess (without calculating), to the nearest 100 J, the kinetic energy of

i a football in a free kick (mass 0.40 kg, speed $30\,m\,s^{-1}$)

ii a bullet fired from a gun (mass 10 g, speed $300\,m\,s^{-1}$).

b Now calculate their kinetic energies.

c Now try to guess

i the kinetic energy of a male sprinter (100 m in about 10 s), to the nearest kJ

ii the kinetic energy of a lorry at 70 mph (40 tonne, $31\,m\,s^{-1}$), to the nearest MJ.

Level 2

6 The mass of an oil tanker is 6.0×10^{8} kg and it travels at $10\,m\,s^{-1}$.

a Calculate its kinetic energy.

If the engines are put into reverse the total maximum retarding force is 3.0 MN.

b What distance does the tanker take to stop?

7 A skier is towed up a slope with a bar as shown in Figure 10.9. How much work is done moving her 200 m up the slope?

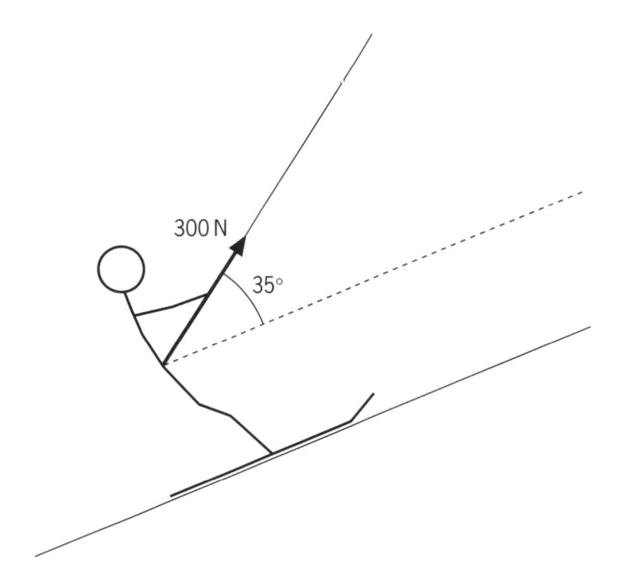

Figure 10.9

Level 3

8 Each time the heart pumps it accelerates about 20 g of blood from $0.20\,m\,s^{-1}$ to $0.34\,m\,s^{-1}$.

a What is the increase in kinetic energy of the blood with each beat?

b Calculate the power of the heart when it beats at about 80 beats per minute.

9 You walk up a 1000 m high mountain, which takes you 5 hours. Walking on the flat consumes about 400 W. You are hoping to lose some of your 70 kg mass with all this exercise but on the way you eat four small chocolate bars, which according to the wrapper will each supply 1130 kJ.

a Calculate the energy you expend during the 5 hour walk, assuming that while you are climbing, your body works with an efficiency of only 15%.

b Calculate the energy that must be obtained from your body after the energy of the chocolate has been used.

c If fat converts to energy at a rate of $38\,MJ\,kg^{-1}$, how much mass are you likely to lose?

10 a Show that power may be expressed as force × velocity.

b A car engine has a maximum power of 110 kW, find the possible tractive (pulling) force at $10\,m\,s^{-1}$ and at $30\,m\,s^{-1}$.

Level 1

1 a Work = force × distance moved (in the same line of action as the force). We are given the distance but we need to find the retarding force.

$$F = ma = 1000\,\text{kg} \times (-6.4\,\text{m s}^{-2}) = -6400\,\text{N}$$

The minus sign shows that the force is in the opposite direction to the movement.

$\therefore$ work = 6400 N × 75 m = **480 kJ**

Work is not a vector so no minus sign is needed here.

b Kinetic energy $= \frac{1}{2}mv^2$

$$= \tfrac{1}{2} \times 1000\,\text{kg} \times (31\,\text{m s}^{-1})^2 = \mathbf{480\,kJ}$$

So work done in stopping is equal to kinetic energy lost.

2 a The area under a force–distance graph is equal to the work done or energy gained. The graph has area 30 N × 10 m = **300 J**. This is the kinetic energy acquired.

b Gravitational potential energy, lost $\Delta E_p = mg\Delta h$

$$= 3.1\,\text{kg} \times 9.8\,\text{m s}^{-2} \times 10\,\text{m} = \mathbf{300\,J}$$

This is equal to the kinetic energy acquired.

3 If your mass is about 60 kg, your weight is about 600 N. The step-up should be about 0.30 m (about 1 ft) each time. You will lift your weight up to this height 100 times.

$$\therefore \text{power} = \frac{\text{work}}{\text{time}} = \frac{\text{force} \times \text{distance}}{\text{time}}$$

$$= \frac{600\,\text{N} \times 0.30\,\text{m} \times 100}{60\,\text{s}} = \mathbf{300\,W}$$

The result you get will depend on your estimate of your mass, but it should not be too different from 300 W.

4 a The work done by the crane is equal to the extra gravitational potential energy of the girder when it is in place.

$$\therefore \text{work done} = mg\Delta h$$
$$= 600\,\text{kg} \times 9.8\,\text{m s}^{-2} \times 10\,\text{m}$$
$$= 5.9 \times 10^4\,\text{J or } 59\,\text{kJ}$$

$$\text{power} = \frac{\text{work}}{\text{time}} \text{ or } P = \frac{W}{t}$$

$$\therefore \quad t = \frac{W}{P} = \frac{5.9 \times 10^4\,\text{J}}{3 \times 10^3\,\text{W}} = \mathbf{20\,s}$$

b If 20% of the energy used is 59 kJ then the total amount converted from the fuel is 100/20 = 5 times this, that is 5 × 59 J = **295 kJ**.

5 a The accuracy of your guesses will depend on whether you have any feeling for the size of 100 J and on whether you have ever kicked a football or fired a gun.

b Using $E_k = \frac{1}{2}mv^2$ (with m in kg and v in m s^{-1})

i $E_k = 0.5 \times 0.40\,\text{kg} \times (30\,\text{m s}^{-1})^2 = \mathbf{180\,J}$

ii $E_k = 0.5 \times 0.010\,\text{kg} \times (300\,\text{m s}^{-1})^2 = \mathbf{450\,J}$

c Check your guesses with these calculations. Again using $E_k = \frac{1}{2}mv^2$

i $E_k = 0.5 \times 70\,\text{kg} \times (10\,\text{m s}^{-1})^2 = 3.5\,\text{kJ}$ (taking the mass as 70 kg) or **4 kJ** to the nearest kJ

ii $E_k = 0.5 \times 40 \times 10^3\,\text{kg} \times (30\,\text{m s}^{-1})^2 = \mathbf{18\,MJ}$

Level 2

6 a The kinetic energy is given by

$$E_k = \tfrac{1}{2}mv^2 = 0.5 \times 6.0 \times 10^8\,\text{kg} \times (10\,\text{m s}^{-1})^2$$
$$= \mathbf{3.0 \times 10^{10}\,J}$$

b The work done to stop the tanker equals this energy, so

$$W = Fs = 3.0 \times 10^{10}\,\text{J}$$

$$\therefore \quad s = \frac{3.0 \times 10^{10}\,\text{J}}{F} = \frac{3.0 \times 10^{10}\,\text{J}}{3.0 \times 10^6\,\text{N}}$$

$$= \mathbf{1 \times 10^4\,m} \text{ or } \mathbf{10\,km}$$

*Think about this distance. It is the **shortest** stopping distance!*

7 You need to calculate the component of the force in the direction of movement of the skier:

$$300\,\text{N} \times \cos 35° = 246\,\text{N}$$

When the skier has moved 200 m, the work done is force × distance,

$$246\,\text{N} \times 200\,\text{m} = \mathbf{4.9 \times 10^4\,J} \text{ or } \mathbf{49\,kJ}$$

Level 3

8 a The change in kinetic energy of the blood for each beat of the heart is given by the equation

$$\text{change in ke} = \Delta E_k = \tfrac{1}{2}mv^2 - \tfrac{1}{2}mu^2 = \tfrac{1}{2}m(v^2 - u^2)$$

where v = final velocity and u = initial velocity of the blood.

$$\therefore \Delta E_k = 0.5 \times 0.020\,\text{kg} \times (0.116 - 0.040)\,\text{m}^2\,\text{s}^{-2}$$
$$= \mathbf{7.6 \times 10^{-4}\,J} \text{ or } \mathbf{0.76\,mJ}$$

b This increase in kinetic energy occurs 80 times in 60 seconds.

$$\text{power} = \frac{\text{work done}}{\text{time taken}}$$

$$= \frac{7.6 \times 10^{-4}\,(\text{J beat}^{-1}) \times 80\,(\text{beat minute}^{-1})}{60\,(\text{s minute}^{-1})}$$

$$= \frac{7.6 \times 10^{-4} \times 80}{60}\,\text{J s}^{-1} = \mathbf{1.0\,mW}$$

9 a 1 watt is 1 joule per second so walking on the flat for 5 h with a power of 400 W would use

$$400\,\mathrm{J\,s^{-1}} \times 60 \times 60 \times 5\,\mathrm{s} = 7.2\,\mathrm{MJ}$$

Climbing the mountain increases your gravitational potential energy by

$$mg\Delta h = 70\,\mathrm{kg} \times 9.8\,\mathrm{m\,s^{-2}} \times 10^3\,\mathrm{m} = 0.69\,\mathrm{MJ}$$

But because your body is only 15% efficient you will expend more energy than this. The actual energy you use to raise yourself is

$$\frac{0.69\,\mathrm{MJ} \times 100\%}{15\%} = 0.69\,\mathrm{MJ} \times 6.67 = 4.6\,\mathrm{MJ}$$

So the total energy required to climb the hill is $7.2\,\mathrm{MJ} + 4.6\,\mathrm{MJ} = \mathbf{11.8\,MJ}$.

b The energy you will get from the chocolate bars is $4 \times 1130\,\mathrm{kJ} = 4.5\,\mathrm{MJ}$.
So the difference between energy intake and output $= 11.8 - 4.5 = \mathbf{7.3\,MJ}$.

c You can calculate the mass of fat that you will have to convert to energy to make up the difference by simply dividing the energy you need by the energy available per kilogram of fat:

$$\text{mass of fat} = \frac{\text{energy required}}{\text{energy available per kg of fat}}$$

$$= \frac{7.3 \times 10^6\,\mathrm{J}}{38 \times 10^6\,\mathrm{J\,kg^{-1}}} = \mathbf{0.192\,kg} \text{ or } \mathbf{192\,g}$$

The energy from the chocolate almost cancelled out the extra work done climbing the mountain.

10 a $\text{Power} = \dfrac{\text{work}}{\text{time}} = \dfrac{\text{force} \times \text{displacement}}{\text{time}}$

where displacement is in the direction of the force.

But $\text{velocity} = \dfrac{\text{displacement}}{\text{time}}$

$\therefore$ power = force × velocity
(in the direction of the force)

b We can rearrange the above to give

$$\text{force} = \frac{\text{power}}{\text{velocity}}$$

So for a power of 110 kW at a velocity of $10\,\mathrm{m\,s^{-1}}$

$$\text{force} = \frac{110 \times 10^3\,\mathrm{W}}{10\,\mathrm{m\,s^{-1}}} = \mathbf{11\,kN}$$

and at $30\,\mathrm{m\,s^{-1}}$

$$\text{force} = \frac{110 \times 10^3\,\mathrm{W}}{30\,\mathrm{m\,s^{-1}}} = \mathbf{3.7\,kN}$$

More tractive force is available at lower velocities.

11 Basic wave properties

A wave motion is a means of transferring energy from place to place but without transferring matter; for example, light waves carry light energy from the Sun to the Earth. There are two large families of waves, namely **transverse waves** and **longitudinal waves**.

A transverse wave, Figure 11.1, vibrates at right angles to the direction in which the energy is being transferred. Light waves and water waves belong to this family.

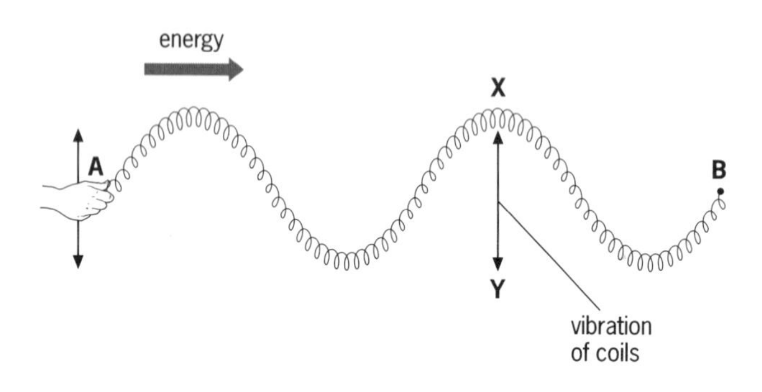

Figure 11.1 The coils of the slinky vibrate in the direction X–Y but the energy is travelling from A to B

A longitudinal wave, Figure 11.2, vibrates along the direction in which the energy is being transferred. Sound waves are longitudinal waves.

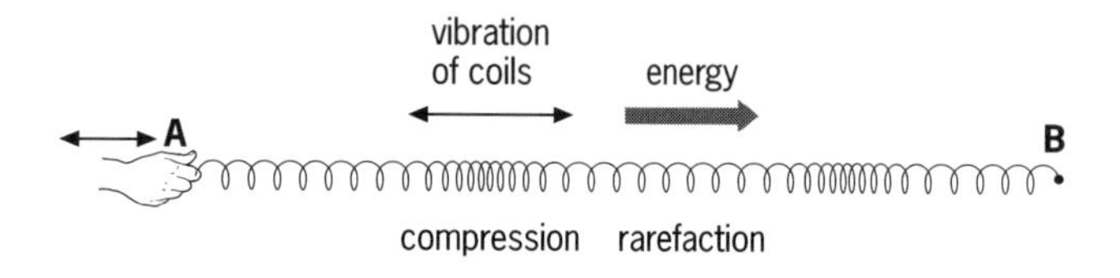

Figure 11.2 The coils of the slinky vibrate in the direction A–B, and the energy is also travelling in this direction

Main features of a wave

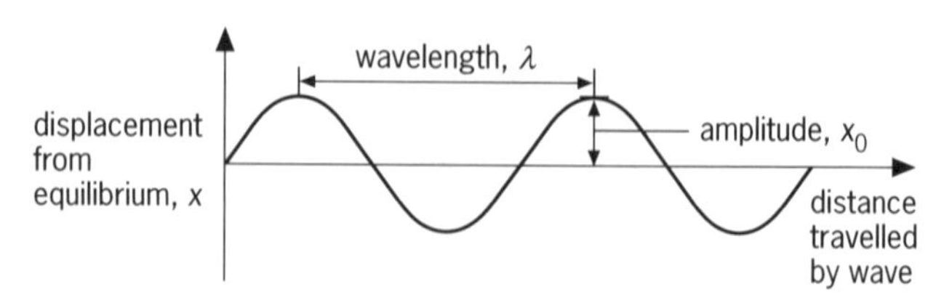

Figure 11.3 The main features of a wave

Wavelength (λ)	The distance from a point on a wave to the identical point on the next wave, for example from the crest of a transverse wave to the next crest of the wave
Amplitude (x_0)	The maximum displacement of a particle from its undisturbed position
Frequency (f)	The number of complete waves created by the source per second *or* the number of complete waves that pass a fixed point each second. The frequency of a wave is measured in hertz (Hz): 1 Hz is one wave or one cycle per second
Period (T)	The time it takes for the source to produce one complete wave *or* the time it takes a vibrating particle to complete one oscillation

The relationship between the period and frequency of a wave is described by the equation

$$T = \frac{1}{f} \quad \text{or} \quad f = \frac{1}{T}$$

For example, a source which produces 10 waves in one second has frequency of 10 Hz and a period of 1/10 or 0.10 s.

Wave velocity

The velocity, v, frequency, f, and wavelength, λ, of a wave are related by the equation

$$v = f \times \lambda$$

Example 1

A tuning fork vibrates with a frequency of 170 Hz. If the sound wave it produces has a wavelength of 2.0 m calculate the speed of this sound wave.

$$\begin{aligned} v &= f \times \lambda \\ &= 170\,\text{s}^{-1} \times 2.0\,\text{m} \\ &= 340\,\text{m}\,\text{s}^{-1} \end{aligned}$$

The electromagnetic spectrum

Visible light is just one small part of a much larger family of waves known as the electromagnetic spectrum. Electromagnetic waves are transverse waves. They consist of magnetic and electric fields vibrating at right angles to each other and to the direction in which the wave is moving. They travel at $3 \times 10^8 \, \mathrm{m\,s^{-1}}$ in vacuo and exhibit all basic wave properties – they can be reflected, refracted, diffracted and brought together to create interference patterns. The electromagnetic spectrum is usually subdivided into a series of regions or groups determined by the wavelengths and frequencies of the waves.

The ripple tank

Many of the basic properties of waves can be demonstrated using a ripple tank.

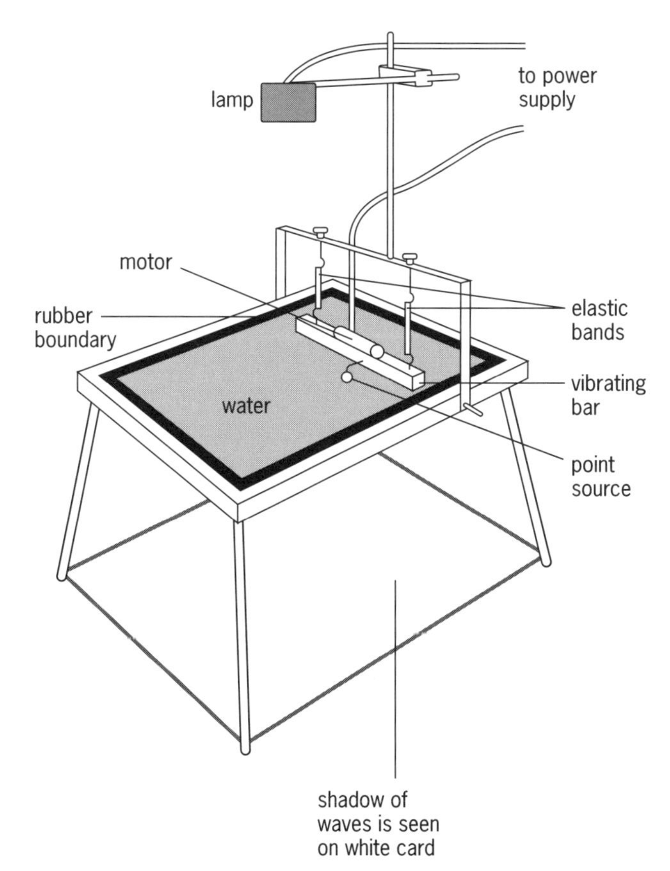

Figure 11.4 A ripple tank

Figures 11.5a–d show the reflection, refraction, diffraction and interference of water ripples.

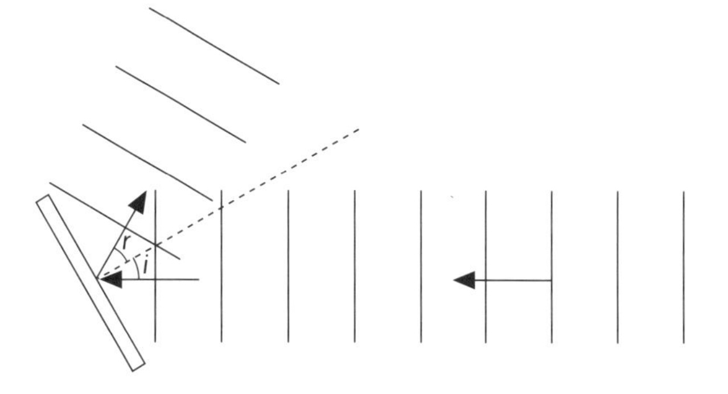

a When the waves strike a plane surface they are reflected such that the angle of incidence and the angle of reflection are equal

b As the waves travel from deeper water to shallow water they slow down and are refracted, that is, they change direction (see page 84)

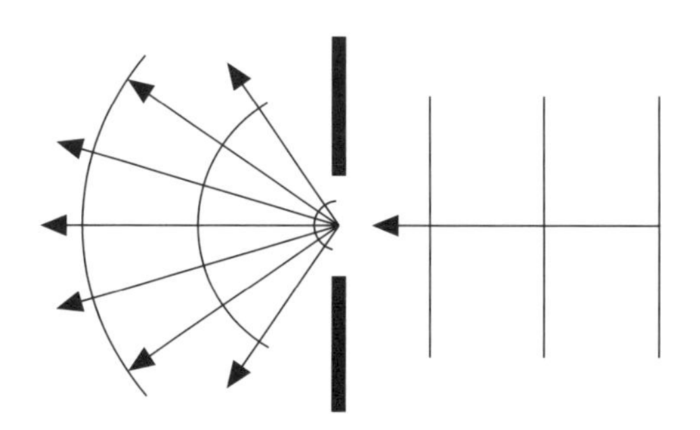

c As the waves travel through a gap they spread out. This 'bending' of the waves is called diffraction (see page 100)

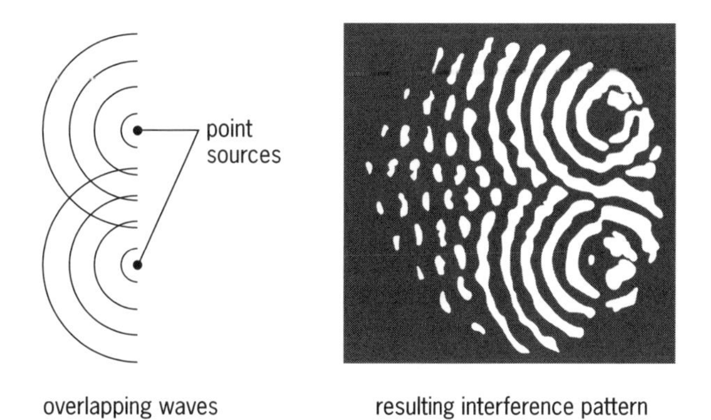

d If two sets of waves overlap they may produce an interference pattern (see page 94)

Figure 11.5

Wave intensity from a point source

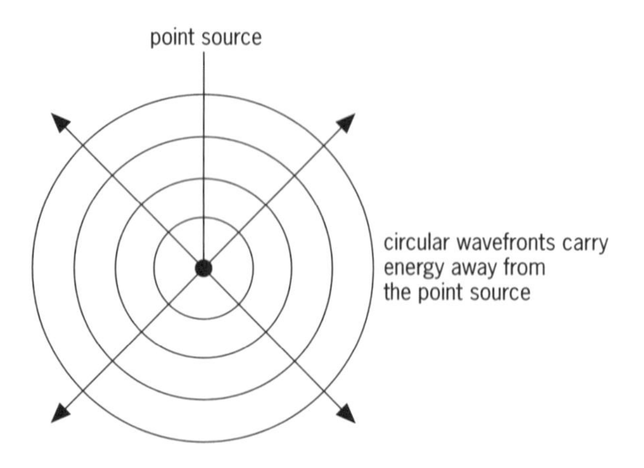

Figure 11.6 Wavefronts from a single point source in a ripple tank

The ripple tank diagram in Figure 11.6 shows us that the energy from a point source of water ripples is carried outwards on circular wavefronts. From a point source of light or sound the energy moves in three dimensions so the wavefronts are spherical, Figure 11.7. The surface area of these spherical wavefronts is $4\pi r^2$, where r is the distance the wave has moved from the source. As the wavefronts move further from the source the energy they carry is spread over a larger surface area.

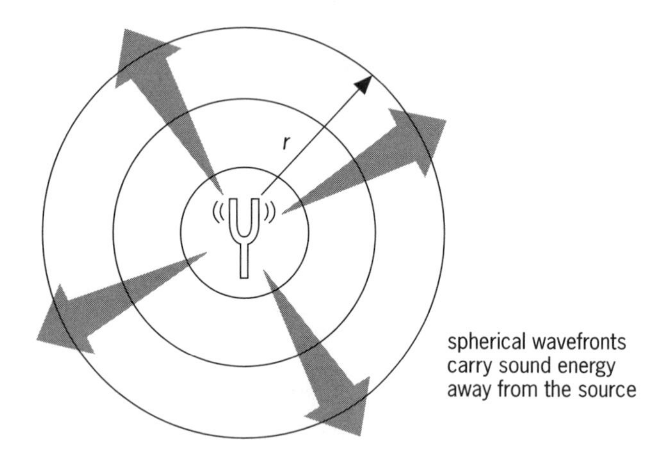

Figure 11.7 Wavefronts from a point source of sound waves

If the waves have travelled a distance r from the source the energy carried per unit area of wavefront will be

$$\frac{E}{4\pi r^2}$$

where E is the total energy carried by a single wavefront. This expression becomes more useful if we consider the energy emitted each second, or the power, P, of the source. We can now write an equation,

$$I = \frac{P}{4\pi r^2}$$

I is known as the **intensity** of the wave at a point distant r from the source, and is measured in $\mathrm{W\,m^{-2}}$.

Example 2

Calculate the intensity of a sound wave 5.0 m from a source of power 30 W.

$$I = \frac{P}{4\pi r^2}$$

$$= \frac{30\,\mathrm{W}}{4\pi \times (5.0\,\mathrm{m})^2}$$

$$= 0.095\,\mathrm{W\,m^{-2}} \text{ or } 95\,\mathrm{mW\,m^{-2}}$$

Level 1

1 A vibrating bar makes waves in a ripple tank. Figure 11.8a shows the displacement of the wave as it travels out from the bar.

The position of a floating cork in the tank varies with time as shown in Figure 11.8b.

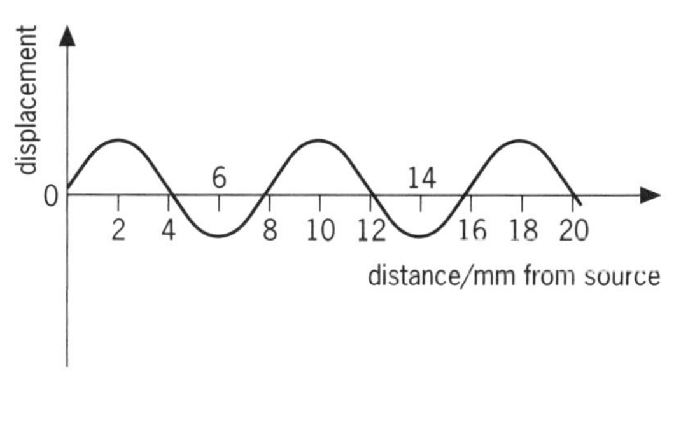

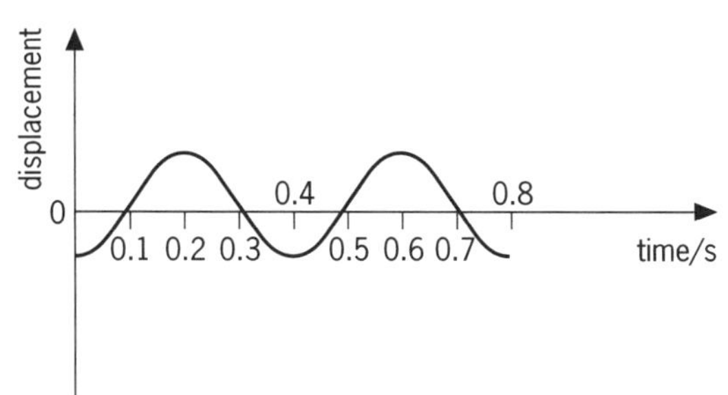

Figure 11.8

From the two graphs determine:

a the wavelength, λ
b the time period, T
c the frequency, f
d the speed of the waves, v.

2 All electromagnetic waves travel through space at a velocity of $3.0 \times 10^8\,\mathrm{m\,s^{-1}}$.

a A microwave demonstration apparatus gives a wavelength of 3.2 cm. What is the frequency of the waves that it produces?
b The long-wave dial on my radio is only marked in wavelengths. Where will I find 198 kHz?

3 A low energy, 13-watt bulb illuminates your desk from a distance of 1.2 m. You can assume that it acts as a point source and that it is 100% efficient. What is the intensity (power per unit area) at the desk (in $\mathrm{W\,m^{-2}}$)?

4 A diagram of the electromagnetic spectrum usually shows the infrared region on the left of the 'visible' region and ultraviolet on the right of it. On which side of the diagram would you find

a radio waves
b X-rays
c microwaves
d gamma rays?

Level 2

5 The intensity of a sound wave is $3.7\,\mathrm{mW\,m^{-2}}$ when the detector is 80 m from a loudspeaker. Estimate the total power output of the speaker and say what assumptions were made.

6 The graph in Figure 11.9 shows the displacement of air molecules from their equilibrium position when subjected to a sound wave.

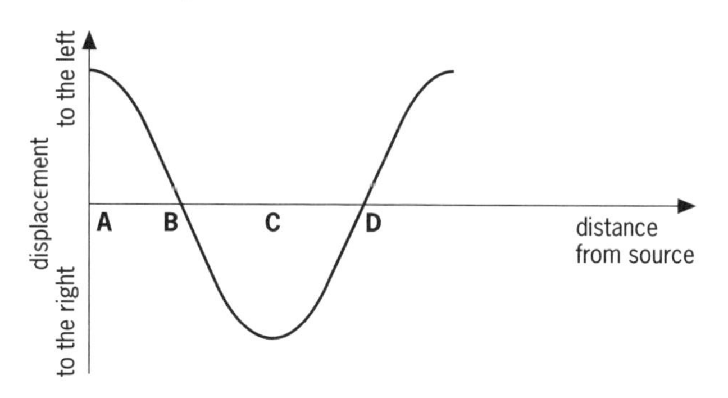

Figure 11.9

Which point will correspond to a compression and which to a rarefaction?

Hint: Draw a line of molecules and then displace them as shown.

7

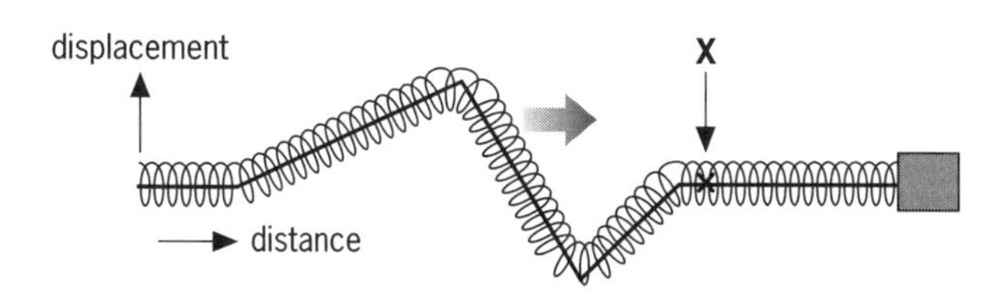

Figure 11.10

The graph in Figure 11.10 shows a saw-tooth pulse travelling along a slinky spring.

a Sketch a graph of the motion of the indicated coil (X) against time.
b Draw the pulse after it has been reflected at the right hand end of the spring, which is fixed.

Level 3

8 The 'mouth' of the parabolic dish of a steerable radio telescope has a radius of 34 m. The central detector needs at least $10^{-9}\,\mathrm{W}$ input. What is the intensity of the weakest signal that can be detected?

Hint: Think about the area of the dish that is at right angles to the incoming beam.

9 The power per unit length of the wavefront of a circular water ripple is $62\,\mathrm{\mu W\,m^{-1}}$ when it is 0.20 m from the point source.

a What is the power that the source transmits to the wave?
b What is the power per unit length of the wavefront when it is 0.50 m away from the source?

Level 1

1 **a** The wavelength, λ, is the length of one complete cycle on the graph of displacement against *distance*. In this case, $\lambda = \mathbf{8.0\,mm}$.

b The time period, T, is represented by one complete cycle on the graph of displacement against *time*. In this case, $T = \mathbf{0.40\,s}$.

c You can calculate the frequency, f, from the equation

$$f = \frac{1}{T} = \frac{1}{0.40\,\text{s}} = \mathbf{2.5\,Hz}$$

d You can calculate the speed, v, from the equation

$$v = f \times \lambda = 2.5\,\text{Hz} \times 0.0080\,\text{m} = \mathbf{0.020\,m\,s^{-1}}$$

2 **a** For any type of wave,

speed = frequency × wavelength or $v = f\lambda$

which you can rearrange to give

$$f = \frac{v}{\lambda}$$

In this case

$$f = \frac{3.0 \times 10^8\,\text{m}\,\text{s}^{-1}}{0.032\,\text{m}} = \mathbf{9.4 \times 10^9\,Hz}$$

b Rearrange the same equation to give

$$\lambda = \frac{v}{f}$$

In this case

$$\lambda = \frac{3.0 \times 10^8\,\text{m}\,\text{s}^{-1}}{198 \times 10^3\,\text{s}^{-1}} = \mathbf{1515\,m}$$

3 The energy from the light bulb spreads out over the surface of a sphere. For a sphere of radius r, the surface area $= 4\pi r^2$. The intensity, I, is defined as power/area so

$$I = \frac{P}{4\pi r^2}$$

$$= \frac{13\,\text{W}}{4 \times \pi \times (1.2\,\text{m})^2} = \mathbf{0.72\,W\,m^{-2}}$$

4 Radio waves (**a**) and microwaves (**c**) will be on the left; like infrared, they have longer wavelengths than visible light. X-rays (**b**) and gamma rays (**d**) will be on the right; like ultraviolet, they are shorter wavelength radiations. See Figure 11.11.

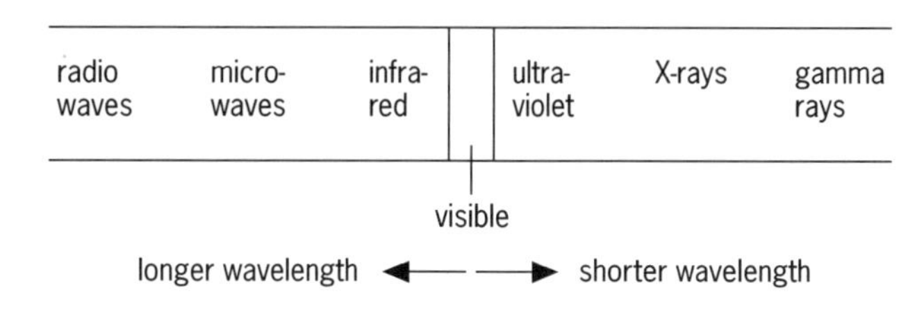

Figure 11.11 The electromagnetic spectrum

Level 2

5 You may assume that the speaker acts as a point source, projecting sound out in all directions, and that there is no absorption by the medium between the speaker and the point in question. The intensity, I, at any distance, r, from the source is given by the equation

$$I = \frac{P}{4\pi r^2}$$

where P is the power of the source. You can rearrange this equation to give

$$P = I \times 4\pi r^2$$

(The total power, P, delivered by the source is equal to the intensity I at distance r from the source, multiplied by the surface area of the sphere of radius r.)

$$\therefore P = 3.7 \times 10^{-3}\,\text{W}\,\text{m}^{-2} \times 4 \times \pi \times (80\,\text{m})^2 = \mathbf{298\,W}$$

6 At **D**, the particles are displaced towards each other on either side of an undisturbed particle. They are all crowded together, so this is a region of compression.

At **B**, the particles on either side are moving away from the undisturbed particle, so this is a region of **rarefaction**. See Figure 11.12.

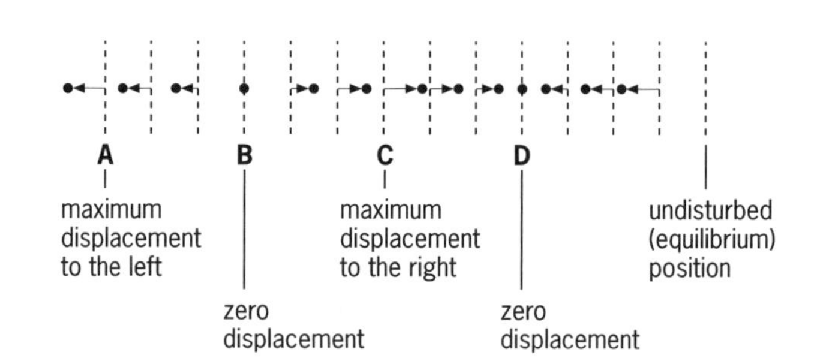

Figure 11.12

7 **a** With the pulse travelling in the direction shown, the coil marked X will first move down then rapidly upwards and then slowly downwards again, as shown in Figure 11.13.

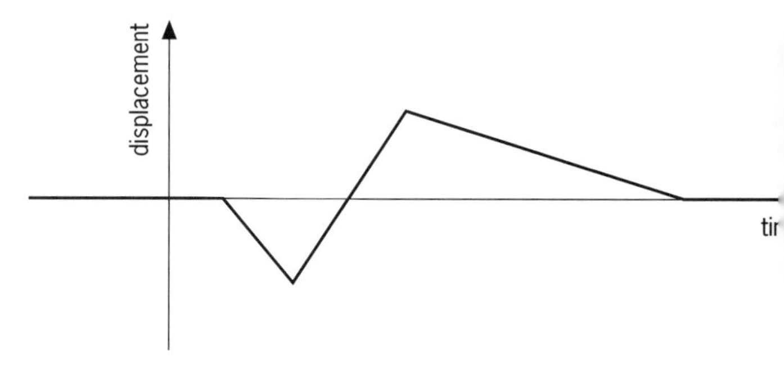

Figure 11.13

b The reflected pulse will be on the opposite side of the spring, the 'sharp' edge of the saw-tooth will still be leading, and the pulse will be travelling in the opposite direction. See Figure 11.14.

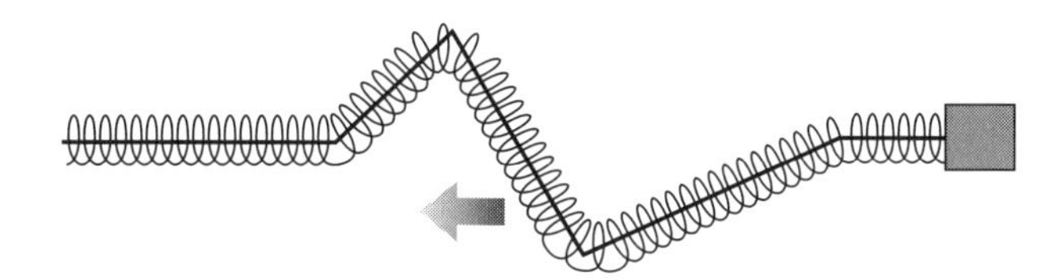

Figuro 11.14

Level 3

8 The dish of the radio telescope is steerable. When it is pointed directly at the source, the area over which it collects signals is at right angles to the beam (Figure 11.15) and is given by the equation $A = \pi r^2$. In this case

$$A = \pi \times (34\,\text{m})^2 = 3.6 \times 10^3\,\text{m}^2$$

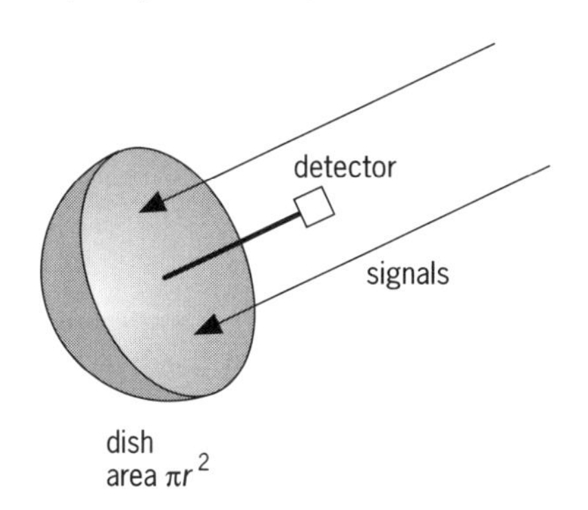

Figure 11.15

Intensity is power per unit area, $I = P/A$, so the power received by the detector is given by the equation

$$\text{power received} = \text{incoming signal intensity} \times \text{effective area of collector}$$

$$P = I \times A$$

so that

$$\text{minimum power received} = \text{minimum signal intensity} \times \text{effective area of collector}$$

which you can rearrange to give

$$\text{minimum signal intensity} = \frac{\text{minimum power received}}{\text{effective area of collector}} = \frac{1.0 \times 10^{-9}\,\text{W}}{3.6 \times 10^3\,\text{m}^2} = \mathbf{2.8 \times 10^{-13}\,W\,m^{-2}}$$

9

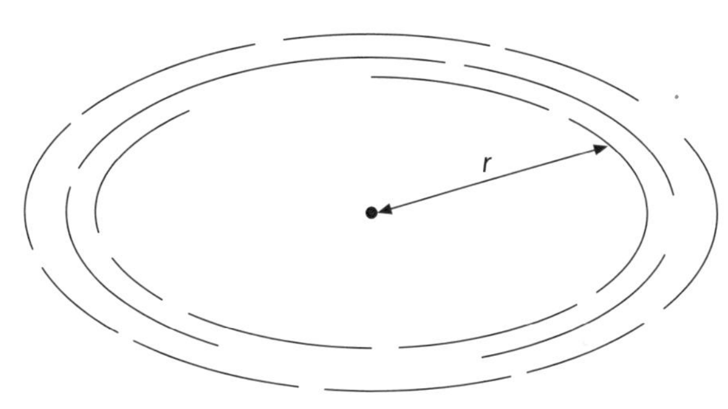

Figure 11.16

This problem is in two dimensions (Figure 11.16) instead of the more usual three-dimensional, spherical case. So the 'power per unit length' is based on the circumference of a circle instead of the surface area of a sphere. We can say

$$\text{intensity} = \text{power per unit length} = \frac{P_{tot}}{2\pi r}$$

a Rearrange the above equation to give

$$\begin{aligned} P_{tot} &= \text{power per unit length} \times 2\pi r \\ &= 62 \times 10^{-6}\,\text{W}\,\text{m}^{-1} \times 2\pi r \\ &= 62 \times 10^{-6}\,\text{W}\,\text{m}^{-1} \times 2\pi \times 0.20\,\text{m} = \mathbf{78\,\mu W} \end{aligned}$$

So (if there has been no absorption) this must be the power transmitted to the wave by the generator.

b Using the first equation,

$$\text{power per unit length} = \frac{P_{tot}}{2\pi r}$$

When $r = 0.50\,\text{m}$,

$$\text{power per unit length} = \frac{78 \times 10^{-6}\,\text{W}}{2 \times \pi \times 0.50\,\text{m}} \approx \mathbf{25\,\mu W\,m^{-1}}$$

12 Refraction and total internal reflection

Refraction: Snell's Law

When a ray of light crosses the boundary between media of different optical densities, such as air and glass, its speed changes. In air light travels at $3 \times 10^8\,\mathrm{m\,s^{-1}}$ whereas in glass it travels at nearer $2 \times 10^8\,\mathrm{m\,s^{-1}}$. This change in speed may cause the ray to alter direction. This phenomenon is called **refraction**. Similarly, when surface water waves move over water of different depths their speeds change. They travel more slowly over shallow water than over deep water. As a consequence the water waves may undergo refraction.

If a ray of light crosses into an optically more dense medium it slows down and bends towards the normal, as in Figure 12.1a. If the ray crosses into an optically less dense medium it speeds up and bends away from the normal.

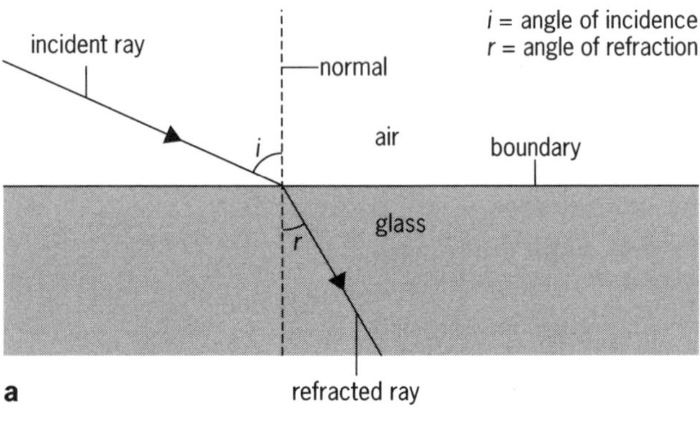

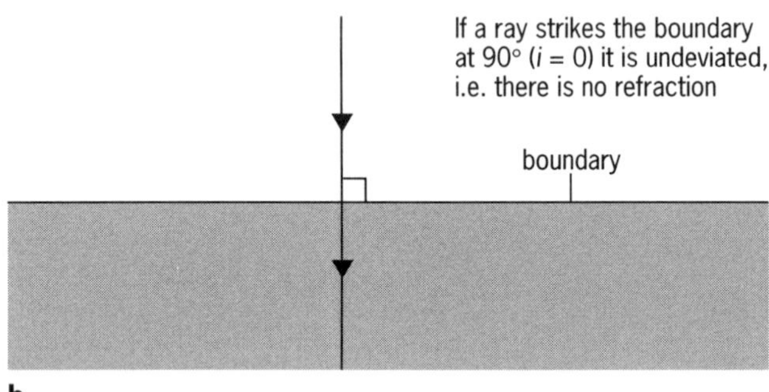

Figure 12.1 Behaviour of a light ray as it enters an optically denser medium

The relationship between the angle of incidence and the angle of refraction is described by the equation

$$\frac{\sin i}{\sin r} = \text{a constant}$$

This relationship is known as **Snell's Law**. The constant is called the **relative refractive index**. In this example, for a ray travelling from air to glass, it is written as ${}_{\mathrm{a}}n_{\mathrm{g}}$. The value of the relative refractive index depends upon the medium the ray is leaving, the medium it enters and the wavelength of the light.

Example 1

Using the information shown in Figure 12.2, calculate the relative refractive index for air and water, ${}_{\mathrm{a}}n_{\mathrm{w}}$.

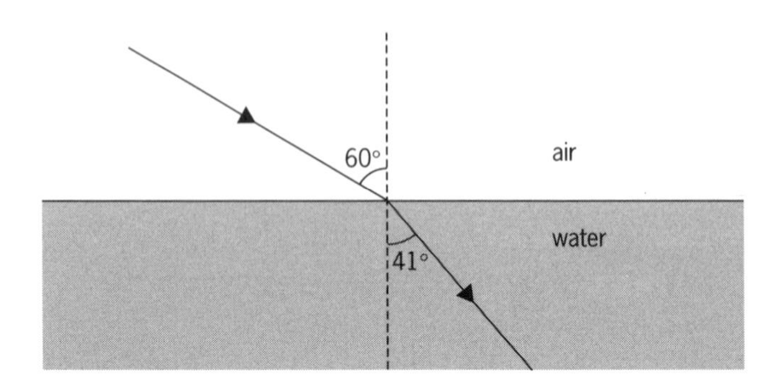

Figure 12.2

$${}_{\mathrm{a}}n_{\mathrm{w}} = \frac{\sin i}{\sin r} = \frac{\sin 60°}{\sin 41°}$$
$$= 1.3$$

Note that refractive index has no unit as it is a ratio.

Relative refractive index and the speed of light

The relative refractive index can also be shown to be the ratio of the speeds of light in the two media:

$${}_{1}n_{2} = \frac{\text{velocity of light in medium 1}}{\text{velocity of light in medium 2}} = \frac{c_1}{c_2}$$

For light travelling from air into glass,

$${}_{\mathrm{a}}n_{\mathrm{g}} = \frac{\text{velocity of light in air}}{\text{velocity of light in glass}} = \frac{c_{\mathrm{a}}}{c_{\mathrm{g}}} = \frac{3 \times 10^8\,\mathrm{m\,s^{-1}}}{2 \times 10^8\,\mathrm{m\,s^{-1}}}$$
$$= 1.5$$

Example 2

Calculate the speed of light in water using the value of ${}_{\mathrm{a}}n_{\mathrm{w}}$ calculated in the previous example.

$${}_{\mathrm{a}}n_{\mathrm{w}} = \frac{\text{velocity of light in air}}{\text{velocity of light in water}} = \frac{c_{\mathrm{a}}}{c_{\mathrm{w}}}$$

$$\therefore\ c_{\mathrm{w}} = \frac{c_{\mathrm{a}}}{{}_{\mathrm{a}}n_{\mathrm{w}}}$$
$$= \frac{3 \times 10^8\,\mathrm{m\,s^{-1}}}{1.3}$$
$$= 2.3 \times 10^8\,\mathrm{m\,s^{-1}}$$

The Principle of Reversibility

Consider a ray of light passing from medium 1 to medium 2, Figure 12.3a.

The relative refractive index, ${}_1n_2$, is given by

$$ {}_1n_2 = \frac{\sin i}{\sin r} $$

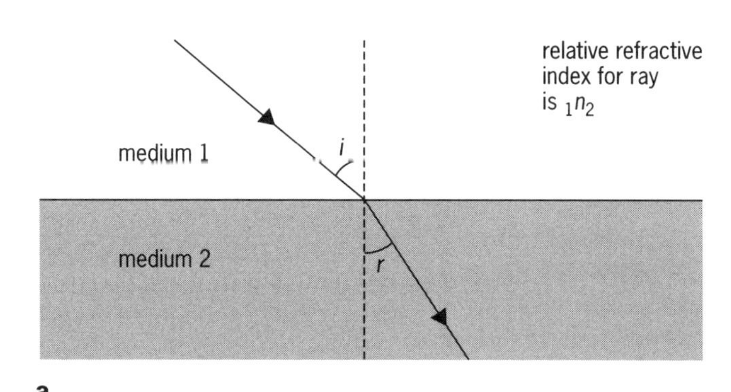

a

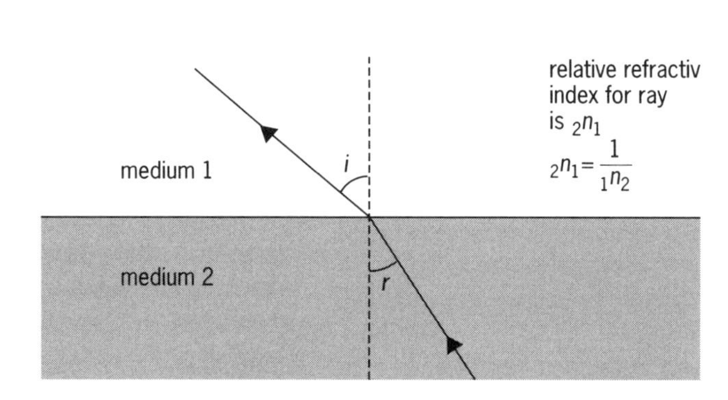

b

Figure 12.3 The Principle of Reversibility

If a ray is travelling in the opposite direction, Figure 12.3b, it traverses the same path and the relative refractive index, ${}_2n_1$, is given by

$$ {}_2n_1 = \frac{\sin r}{\sin i} $$

From these two equations we can see that

$$ {}_2n_1 = \frac{1}{{}_1n_2} $$

Example 3

If the relative refractive index for light travelling from air to water is 1.3, calculate the relative refractive index for a ray travelling from water to air.

$$ {}_\text{w}n_\text{a} = \frac{1}{{}_\text{a}n_\text{w}} $$

$$ \therefore {}_\text{w}n_\text{a} = \frac{1}{1.3} $$

$$ = 0.77 $$

Absolute refractive indices

When values for refractive indices are given they are usually quoted with respect to a vacuum, meaning that light is considered to be passing from vacuum into the new medium. These values are called **absolute refractive indices**. The absolute refractive index for glass is

$$ n_\text{g} = \frac{\text{velocity of light in vacuo}}{\text{velocity of light in glass}} = \frac{c}{c_\text{g}} $$

There is very little difference between the velocity of light in a vacuum and the velocity of light in air. It follows therefore that the absolute refractive index for a medium and its relative refractive index with air are almost identical, so ${}_\text{a}n_1$ can be written as n_1.

Snell's Law in terms of absolute refractive index

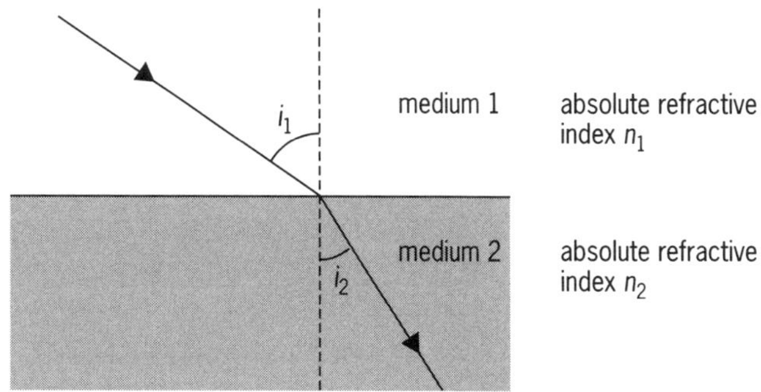

Figure 12.4

For a ray of light travelling from medium 1 into medium 2, Figure 12.4, we can rewrite Snell's Law in terms of the absolute refractive indices of the two substances:

$$ \frac{\sin i_1}{\sin i_2} = {}_1n_2 = \frac{c_1}{c_2} $$

or

$$ \frac{\sin i_1}{\sin i_2} = \frac{c_1}{c} \times \frac{c}{c_2} = \frac{1}{n_1} \times n_2 = \frac{n_2}{n_1} $$

$$ \therefore \quad \boldsymbol{n_1 \sin i_1 = n_2 \sin i_2} $$

or more generally,

$$ n \sin i = \text{constant} $$

where i is the angle in the medium with an absolute refractive index of n.

Example 4

A ray of light travelling from water to glass strikes the boundary at 60° to the normal. If $n_\text{w} = 1.3$ and $n_\text{g} = 1.5$, calculate the angle of refraction of the ray in the glass.

$$ n_\text{w} \sin i_\text{w} = n_\text{g} \sin i_\text{g} $$

$$ \therefore 1.3 \sin 60° = 1.5 \sin i_\text{g} $$

$$ \sin i_\text{g} = \frac{1.3 \sin 60°}{1.5} $$

giving $i_\text{g} = 49°$

Dispersion

White light is composed of a mixture of waves with differing frequencies or colours. When white light passes through a prism the refractive index of the glass varies for each frequency or colour. These variations may cause the colours to become **dispersed**, creating a coloured spectrum.

Total internal reflection

When a ray of light travels from a more dense medium to a less dense medium it is refracted away from the normal. As the incident angle, *i*, is increased the refracted angle, *r*, becomes larger until eventually it reaches 90°. Any further increase in the incident angle results in the ray undergoing **total internal reflection**, and the boundary acts as a mirror. The largest incident angle for which the ray is still partially refracted is called the **critical angle**.

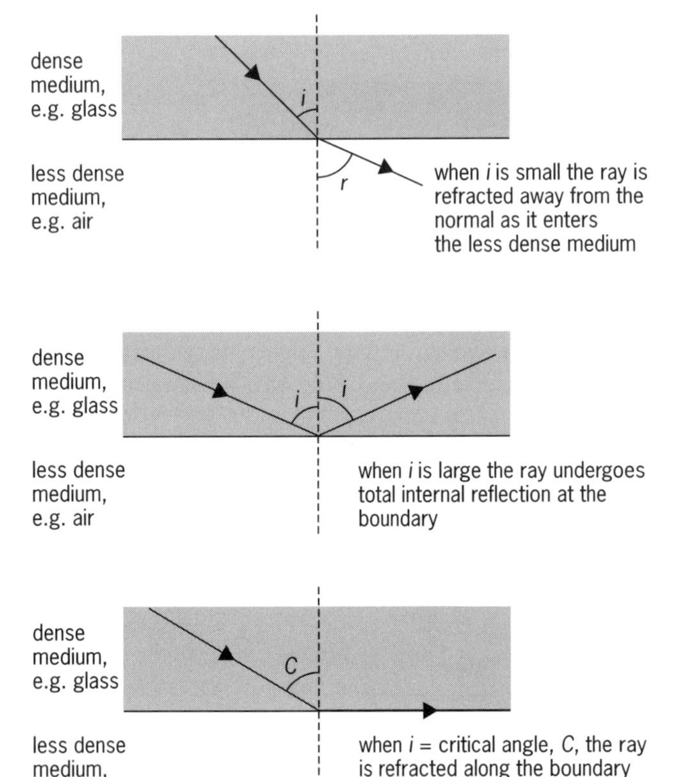

Figure 12.5 Behaviour of a light ray as it enters an optically less dense medium

The critical angle, *C*, is related to the refractive index by the equation

$$n = \frac{1}{\sin C}$$

Note that the critical angle is always in the denser medium and, like absolute refractive index, it is a characteristic of the medium.

Example 5

If the refractive index of glass is 1.5, calculate the critical angle for glass at the glass/air boundary.

$$n = \frac{1}{\sin C}$$

$$\therefore \quad \sin C = \frac{1}{n}$$

$$= \frac{1}{1.5} = 0.66$$

giving $\quad C = 42°$

The critical angle for water is approximately 50°, for Perspex it is 42° and for alcohol it is 47°.

Total internal reflection in prisms

Prismatic periscopes, binoculars and some cameras use total internal reflection in glass prisms to change the direction of a ray of light. This is preferable to using mirrors as prisms give a brighter final image.

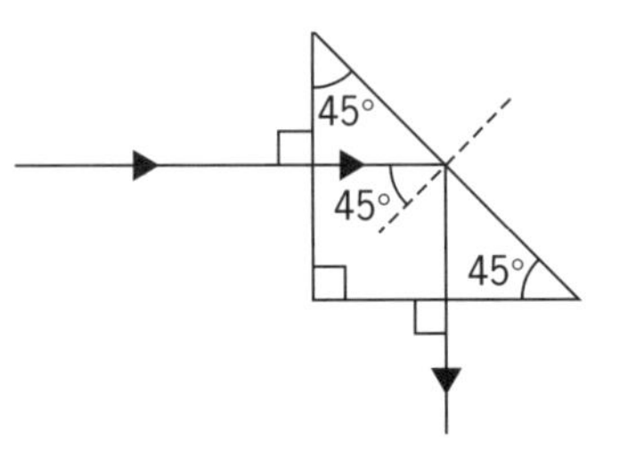

a Using a prism to turn a ray of light through 90°

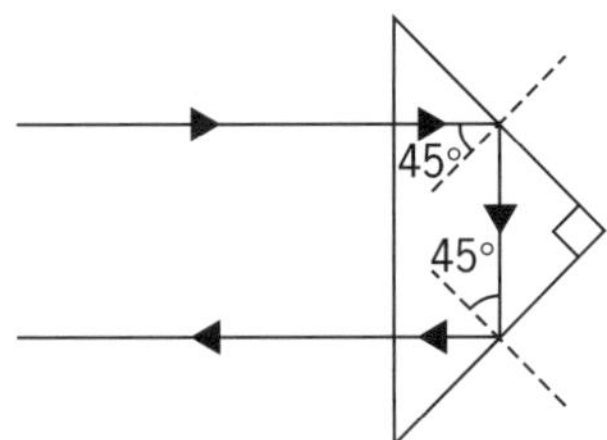

b Using a prism to turn a ray of light through 180°

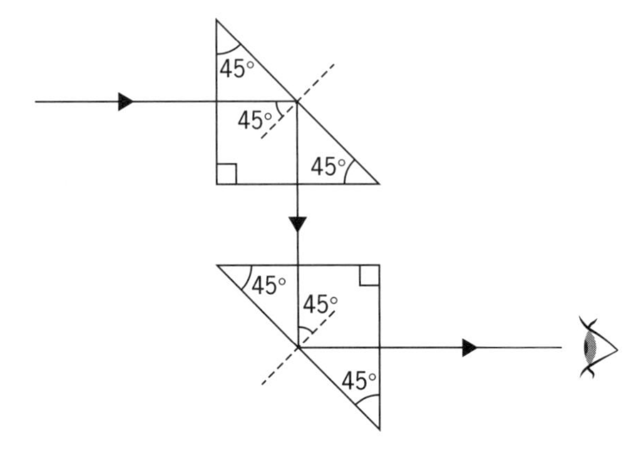

c A simple prismatic periscope which uses two prisms to change the direction of the light

Figure 12.6 Reflection of light by prisms

Optical fibres

Optical fibres also make use of total internal reflection. A single glass fibre consists of a very narrow core of glass surrounded by cladding made from a different type of glass which is optically less dense (Figure 12.7).

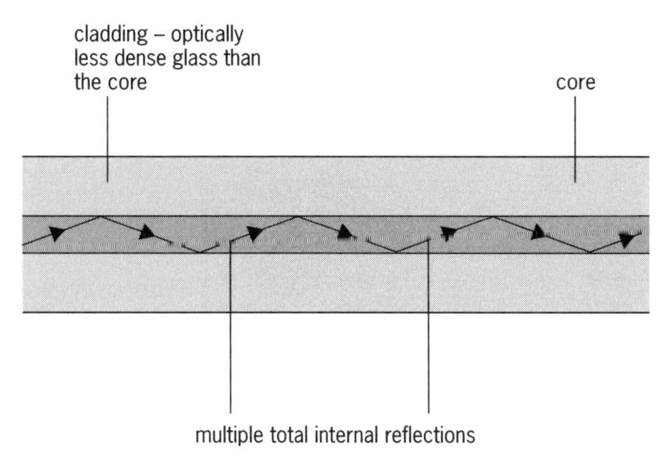

Figure 12.7 Structure of a single optical fibre fibre

Light travelling down the core of a fibre always meets the boundary between the two types of glass at an angle greater than the critical angle and so is totally internally reflected. The fibre thus behaves as a 'light pipe' capable of guiding light along it.

Optical fibres combined with lenses are being used in communication systems to replace copper wire as the medium through which signals are sent. Signals are sent as digital pulses of infrared light. The fibres are cheaper, resistant to chemical attack and are capable of carrying many more signals at any moment in time than conventional copper cables.

Bundles of fibres are also used in optical instruments such as the endoscope, which allow doctors to see inside the body of a patient (Figure 12.8).

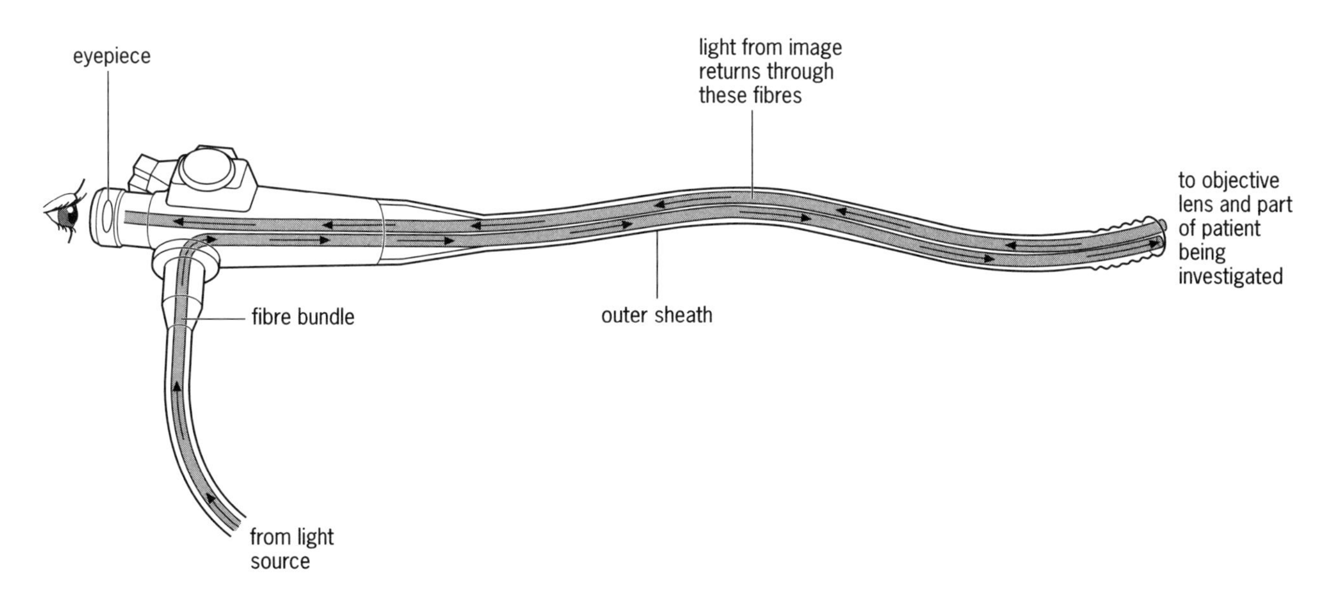

Figure 12.8 The principle of an endoscope

Make sure that you have the correct answer to question 1 before attempting the other questions.

Level 1

1 a What is meant by the *absolute* refractive index of a substance X, n_X?

b What is meant by the *relative* refractive index when a wave goes from substance X to substance Y, ${}_Xn_Y$?

2 a The speed of light in a particular type of glass is $1.9 \times 10^8\,\mathrm{m\,s^{-1}}$. What is the refractive index, n_g, of the glass?

b The refractive index of water, n_w, is 1.3 and that of diamond, n_d, is 2.4. Calculate the speed of light in each substance.

3 A light ray is incident on a block of ice at an angle of 22.0° (measured from the normal as shown in Figure 12.9). This angle is i_a. The ray changes direction slightly at the boundary as it slows in the optically denser substance. The angle with the normal inside the ice (the refracted angle) i_i is 16.6°.

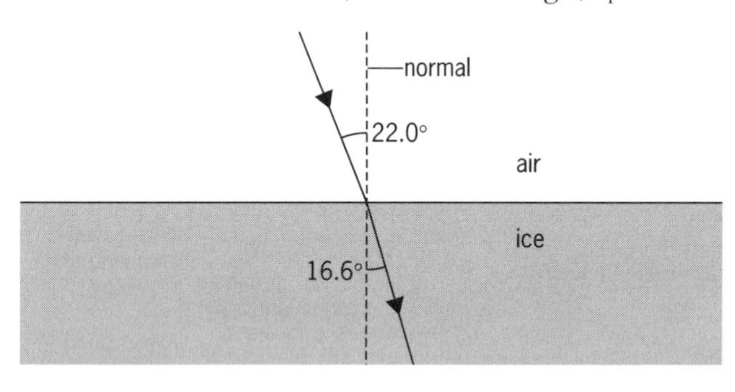

Figure 12.9

What is the refractive index for light going from air to ice, ${}_an_i$?

Note that this will be close to the absolute refractive index of ice, n_i, because the optical density of air is very close to that of a vacuum.

4 The refractive index from air to diamond, ${}_an_d$, is 2.4, very close to the absolute refractive index of diamond, n_d. The angle, i_d, of a ray after refraction at the diamond surface is 12°, measured from the normal (Figure 12.10). At what angle, i_a, in the air, did the ray approach the diamond?

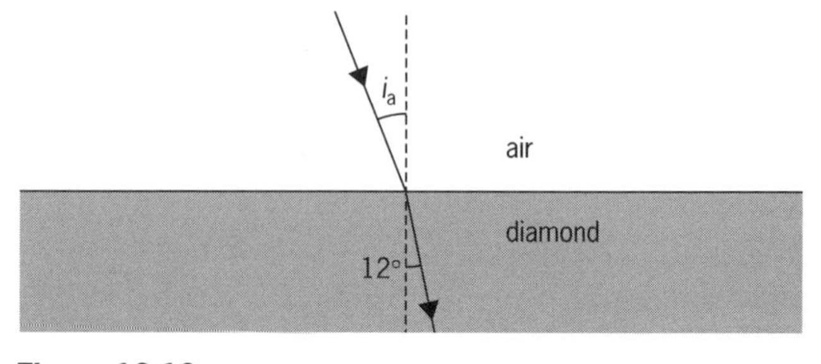

Figure 12.10

5 A semicircular 'D'-block, Figure 12.11, is usually used to demonstrate total internal reflection. Why is this shape chosen?

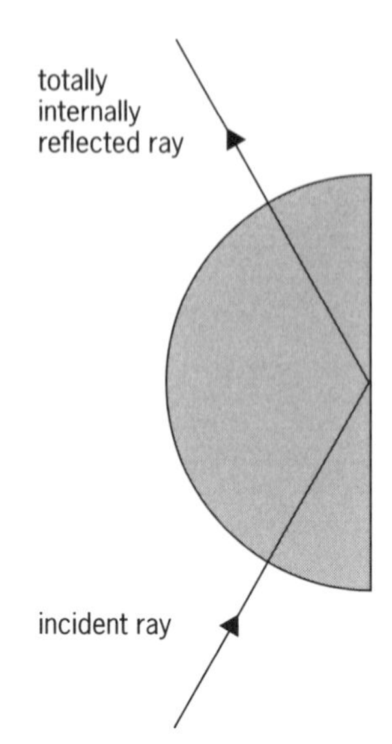

Figure 12.11

6

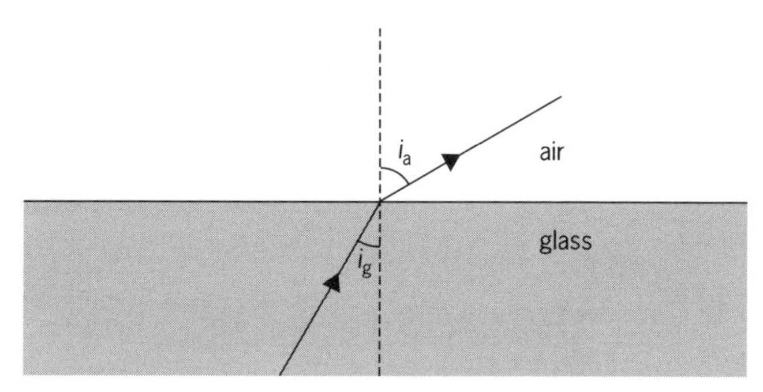

Figure 12.12

The value of ${}_an_g$ for the glass in Figure 12.12 is 1.6. Calculate the refracted angle in air, i_a, if

a $i_g = 35°$
b $i_g = 38°$
c $i_g = 40°$

Comment on your answers and find the critical angle for this glass.

Hint: Use $n_a \sin i_a = n_g \sin i_g$.

7 Find the critical angles for the following substances:

a X with $n_X = 1.31$
b Y with $n_Y = 1.56$
c Z with $n_Z = 2.4$.

Comment on the fact that Z is diamond.

8 Using your answers to question 7, say whether some light is refracted or if it is totally internally reflected at a boundary with air for incident angles of

a 28° in material X
b 38° in Y
c 33° in Z.

Level 2

9 A type of glass has $n_{g,b}$ for blue light = 1.639 and $n_{g,r}$ for red light = 1.621. Calculate the angle between a red ray and a blue ray after refraction in the glass if they were both incident in the air at 60°, Figure 12.13.

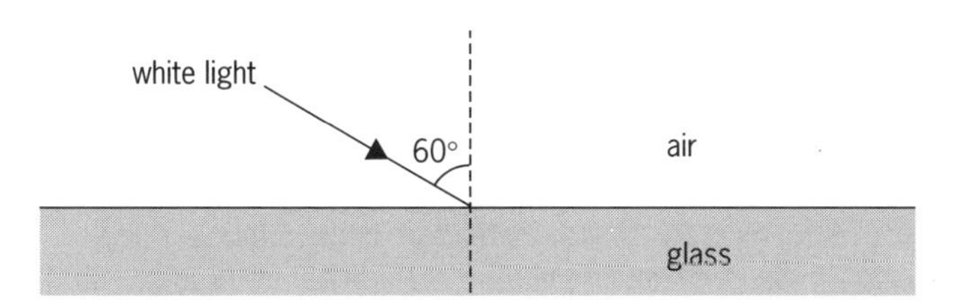

Figure 12.13

10 The *absolute* refractive indices of glass, water and diamond are given as $n_g = 1.5$, $n_w = 1.3$, $n_d = 2.4$. What is the *relative* refractive index

a from glass to water, ${}_gn_w$

b from water to diamond, ${}_wn_d$?

11

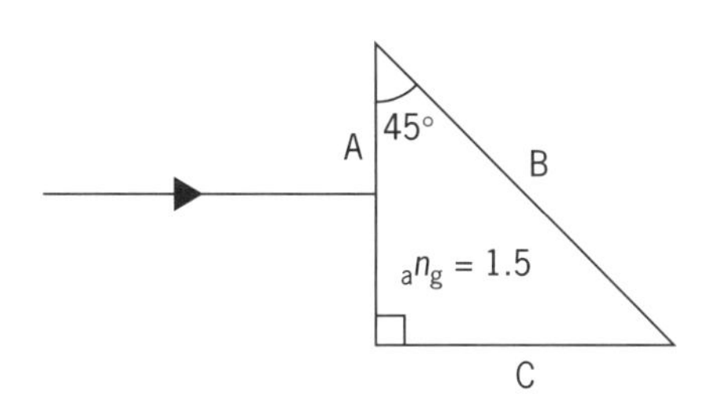

Figure 12.14

Copy Figure 12.14 and complete the path of the ray after it enters face A of the glass prism. Explain what happens when:

a the ray enters the glass

b the ray reaches the far side of the prism, B

c the ray leaves the prism.

12 A batch of fibre optic cable was made with the glass intended for the core used as cladding and vice versa. The core now has $n_g = 1.5$ while the cladding has $n_g = 1.6$. What effect will this have?

(There is no need to calculate anything.)

Level 3

13 Refraction is caused by changes in wave velocity. The speed of sound in air increases with temperature.

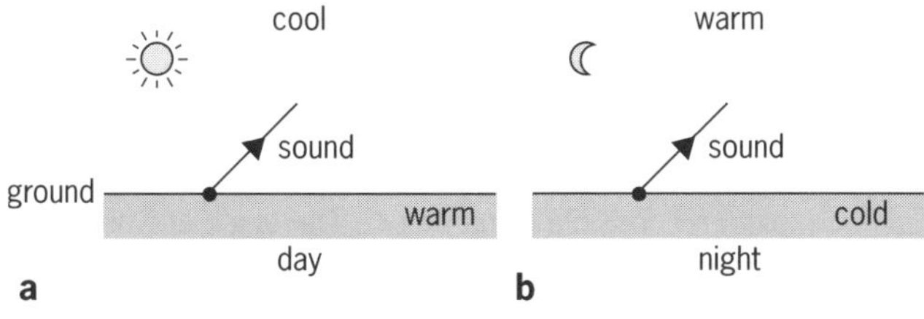

Figure 12.15

Sketch the rough path of the sound under the conditions shown in Figures 12.15a and b.

14 Waves 30 mm apart are generated in a ripple tank by a bar vibrating at 7.0 Hz.

a Show that the speed of these water waves is $210\,\text{mm}\,\text{s}^{-1}$.

The bar is parallel to a step that changes the depth so that the wave speed is reduced to $140\,\text{mm}\,\text{s}^{-1}$.

b Does the depth at the step increase or decrease in the direction in which the waves are travelling?

c What is the new wavelength?

d What is the 'refractive index' of the boundary in the direction in which the waves are travelling?

15 A light-emitting diode (LED) is glued to the end of an optical fibre as shown in Figure 12.16. The refractive index of the glue, $n_{glue} = 1.36$ and of the core glass, $n_{glass} = 1.58$. A ray is incident on the end of the fibre at 10°. At what angle does it hit the core/cladding boundary?

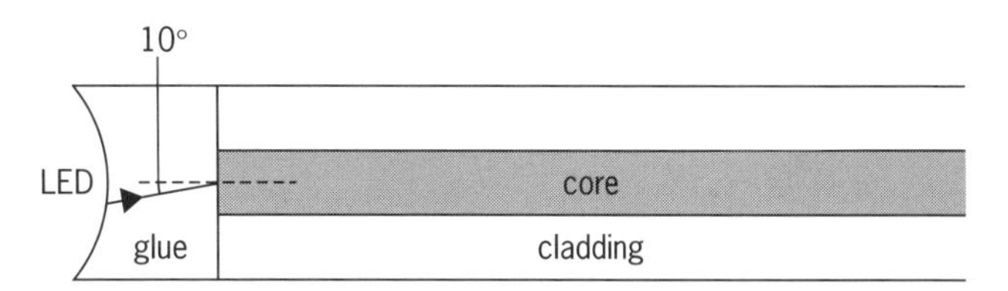

Figure 12.16

16 A pulse of white light is sent straight down the middle of a fibre for which $n_{g,blue} = 1.639$ and $n_{g,red} = 1.621$. What time interval would there be between the blue and red components of the pulse after they have travelled 100 km?

17 a Calculate the critical angle of an optical fibre without any cladding if the glass has ${}_an_g = 1.56$.

b Find the new critical angle when glass cladding with ${}_an_g = 1.48$ is added.

c What advantage is there to this change in critical angle?

18 It is a hot, sunny day.

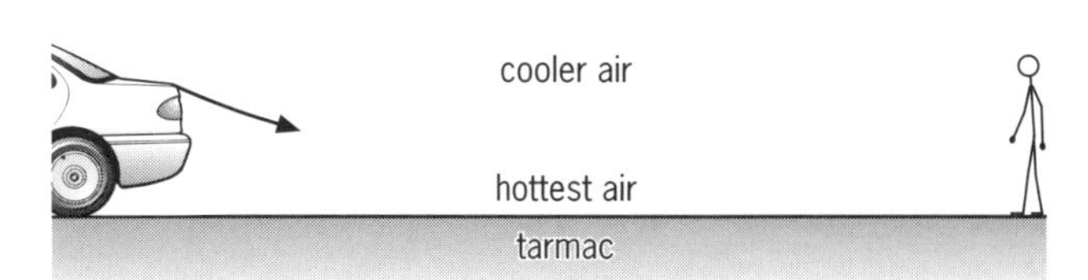

Figure 12.17

a Copy Figure 12.17 and complete the path of the ray of light shown, which reaches the eye of the observer some distance away.

b Explain why it follows this path.

c Describe what the observer sees. (You may add to the diagram if you wish.)

Level 1

1 a The absolute refractive index of a substance, n_X, is equal to the ratio of the speed of electromagnetic radiation in a vacuum, c, to the speed in the substance, c_X:

$$n_X = \frac{c}{c_X}$$

Note that c_X is always less than c, so n_X is always greater than 1.

The absolute refractive index is also equal to the ratio of $\sin i_{vac}$ to $\sin i_X$ where the subscript identifies the substance in which the angle is measured:

$$n_X = \frac{\sin i_{vac}}{\sin i_X}$$

b The relative refractive index, ${}_Xn_Y$, is the ratio of the speeds of a wave (not necessarily electromagnetic) in the two substances X and Y:

$${}_Xn_Y = \frac{v_X}{v_Y}$$

and it can be shown that

$${}_Xn_Y = \frac{n_Y}{n_X} \quad \text{(see question 10)}$$

It is also equal to the ratio of $\sin i_X$ to $\sin i_Y$:

$${}_Xn_Y = \frac{\sin i_X}{\sin i_Y}$$

2 a The refractive index of the glass is the ratio of the speed of light in a vacuum to the speed of light in the glass:

$$n_g = \frac{c}{c_g}$$

where c = speed of light in a vacuum $= 3.0 \times 10^8\,\mathrm{m\,s^{-1}}$. So

$$n_g = \frac{3.0 \times 10^8\,\mathrm{m\,s^{-1}}}{1.9 \times 10^8\,\mathrm{m\,s^{-1}}} = \mathbf{1.6}$$

Refractive index does not have a unit. It is a pure number: the ratio of two values with the same unit.

b From the same equation for the refractive index,

$$n_w = 1.3 = \frac{c}{c_w}$$

$$\therefore c_w = \frac{c}{1.3} = \frac{3.0 \times 10^8\,\mathrm{m\,s^{-1}}}{1.3} = \mathbf{2.3 \times 10^8\,m\,s^{-1}}$$

In the same way,

$$c_d = \frac{c}{2.4} = \frac{3.0 \times 10^8\,\mathrm{m\,s^{-1}}}{2.4} = \mathbf{1.25 \times 10^8\,m\,s^{-1}}$$

The speed of light in water is nearly twice the speed of light in diamond.

3 The rule for refraction is that for any two materials the ratio of the sines of the two angles between the ray and the perpendicular to the surface is a constant, called the (relative) refractive index.

*The ratio of the angles is **not** constant. You cannot cancel the sin functions!*

$${}_an_i = \frac{\sin i_a}{\sin i_i} = \frac{\sin 22.0°}{\sin 16.6°} = \frac{0.375}{0.286} = \mathbf{1.31}$$

If the sines don't come out to these values (to 3 significant figures) on your calculator, check the instructions. Make sure you're not in 'radian' mode instead of 'degree' mode. Check also whether you have to enter the angle before or after you press the sin key.

4 The refractive index is defined as

$${}_an_d = \frac{\sin i_a}{\sin i_d}$$

So ${}_an_d \times \sin i_d = \sin i_a$

(If you have learnt $n_1 \sin i_1 = n_2 \sin i_2$, you just put $n_a = 1.0$.)

$\therefore \quad \sin i_a = 2.4 \times \sin 12° = 2.4 \times 0.208 = 0.499$

giving $\quad i_a = \sin^{-1} 0.499 = \mathbf{29.9°}$

*Remember that $\sin^{-1} x$ means 'the angle which has sine equal to x'. It does **not** mean $1/\sin x$.*

5 The incident ray is aimed through the curved face of the semicircular block, towards the centre of the flat face. It enters the block along a radius and so hits the curved face at 90°. This means that it is not bent at this face and consequently you can concentrate on what happens when the light reaches the flat face of the block.

6 You can use

$${}_an_g = \frac{\sin i_a}{\sin i_g} \quad \text{or} \quad {}_gn_a = \frac{\sin i_g}{\sin i_a}$$

or $\quad n_a \sin i_a = n_g \sin i_g$

to solve this problem. ${}_an_g = 1.6$ and so

$$\sin i_a = 1.6 \times \sin i_g$$

a If $i_g = 35°$, $\sin i_a = 1.6 \times \sin 35° = 0.917$
$i_a = \sin^{-1} 0.917 = \mathbf{66.6°}$

b If $i_g = 38°$, $\sin i_a = 1.6 \times \sin 38° = 0.985$
$i_a = \sin^{-1} 0.985 = \mathbf{80.1°}$

c If $i_g = 40°$, $\sin i = 1.6 \times \sin 40° = 1.028$
$i_a = \sin^{-1} 1.028 =$ ERROR
Your calculator tells you there is an error. Obviously something strange happens between 38° and 40°. The refracted angle reaches 90° and the sine reaches its maximum possible value of 1.0. The angle at which this happens is called the critical angle, often denoted by C.

$$\sin C = \frac{1}{{}_an_g} = \frac{1}{1.6} = 0.625$$

$$C = \sin^{-1} 0.625 = \mathbf{38.7°}$$

7 The critical angle for refraction, C, is given by the equation

$$\sin C = \frac{1}{n}$$

a For X, $n_X = 1.31$, so

$$\sin C = \frac{1}{1.31} = 0.769$$

$$\therefore \quad C = \sin^{-1} 0.769 = \mathbf{50.3°}$$

b For Y, $n_Y = 1.56$, so

$$\sin C = \frac{1}{1.56} = 0.641$$

$$\therefore \quad C = \sin^{-1} 0.641 = \mathbf{39.9°}$$

c For Z, $n_Z = 2.4$, so

$$\sin C = \frac{1}{2.4} = 0.417$$

$$\therefore \quad C = \sin^{-1} 0.417 = \mathbf{24.7°}$$

This shows that the higher the refractive index, the smaller the critical angle. A wave incident at any angle *larger* than the critical angle is internally reflected. There will be much more internal reflection in a diamond than in other substances, hence its sparkling appearance.

8 **a** 28° is smaller than the critical angle for X (50.3°) so some light will be refracted.

b 38° is smaller than the critical angle for Y (39.9°) so some light will be refracted.

c 33° is larger than the critical angle for Z (24.7°) so the light will be totally internally reflected.

Level 2

9 For blue light the refractive index,

$${}_a n_{g,b} = \frac{\sin i_a}{\sin i_g} = 1.639$$

$$\therefore \sin i_g = \frac{\sin 60°}{1.639} = 0.528$$

$$i_g = 31.9°$$

Similarly for red light,

$${}_a n_{g,r} = \frac{\sin i_a}{\sin i_g} = 1.621$$

$$\therefore \sin i_g = \frac{\sin 60°}{1.621} = 0.534$$

$$i_g = 32.3°$$

So the angular separation is 32.3° − 31.9° = **0.4°**. See Figure 12.18.

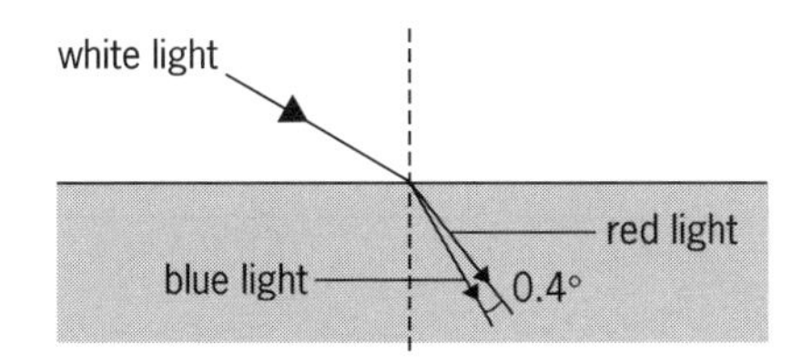

Figure 12.18 Dispersion in glass

10 From the definition of relative refractive index for materials X and Y as

$${}_X n_Y = \frac{c_X}{c_Y}$$

and the definition of the absolute refractive index for material X as

$$n_X = \frac{c}{c_X}$$

you should be able to see that

$${}_X n_Y = \frac{c_X}{c} \times \frac{c}{c_Y} = \frac{1}{n_X} \times n_Y = \frac{n_Y}{n_X}$$

Notice that with speeds it is c_X/c_Y and with refractive indices it is n_Y/n_X.

a $${}_g n_w = \frac{n_w}{n_g} = \frac{1.3}{1.5} = \mathbf{0.87}$$

b $${}_w n_d = \frac{n_d}{n_w} = \frac{2.4}{1.3} = \mathbf{1.8}$$

Notice that ${}_X n_Y$ is <1 if the light is going from a denser substance (X) to a less dense substance (Y), but that n_X is always >1, because there is nothing less dense than a vacuum.

11

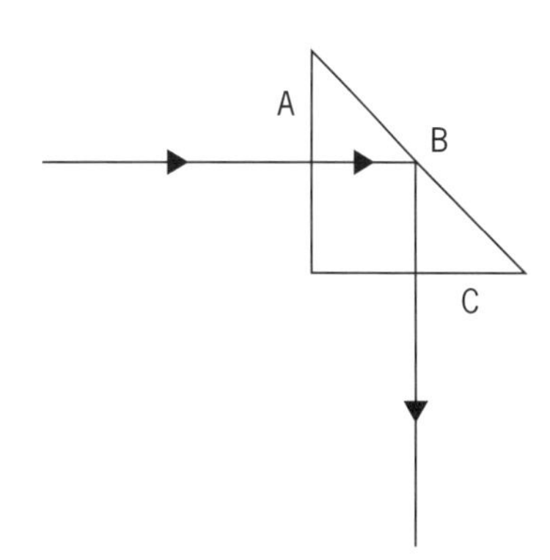

Figure 12.19

Changes in direction occur only at a boundary, unless the density of the glass varies.

a At A, there is no change in direction – no deviation – because the ray strikes the boundary at 90° (the angle of incidence is zero).

b At B, the incident ray strikes the boundary at 45°. You need to find out how this angle compares with the critical angle. The critical angle is given by the equation

$$\sin C = \frac{1}{{}_a n_g} = \frac{1}{1.5} = 0.667$$

$$\therefore \quad C = \sin^{-1} 0.667 = 41.8°$$

So the angle of incidence is greater than C and there will be no refracted ray. The ray is totally internally reflected and turned through an angle of $(2 \times 45°) = 90°$.

c It leaves the prism without refraction because, as when it entered, it strikes the boundary at an angle of 90° (the angle of incidence is zero).

12 There will be no total internal reflection because the light in the core is in the less dense substance. It will be no use as a fibre optic cable because the light will escape by refraction when it strikes the outer glass layer.

Level 3

13

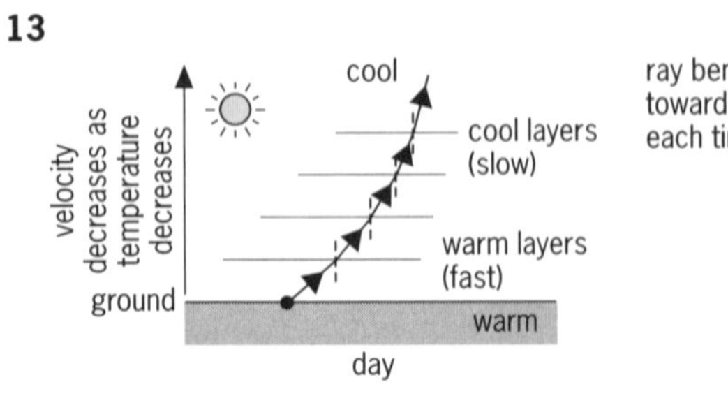

a

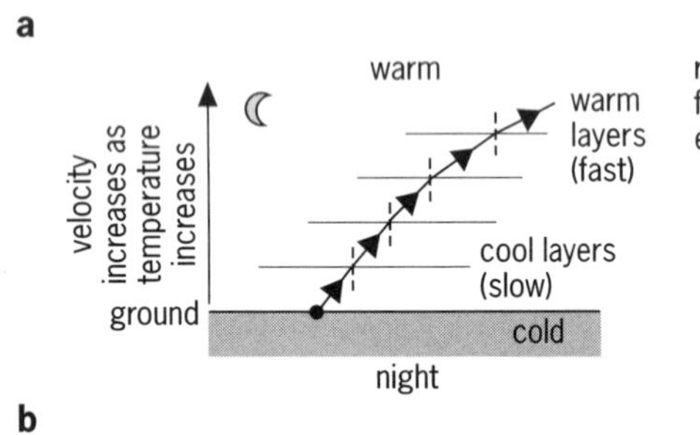

b

Figure 12.20

During the day, when the air near the ground is warmer than the air higher up, sound waves slow down and tend to bend towards the vertical (Figure 12.20a) so that sounds escape upwards. During the night the waves speed up and tend to bend towards the horizontal (Figure 12.20b) so that sounds will carry further along the ground to distant listeners. (This is why more people hear the church bells at midnight on Christmas Eve than at 11 o'clock on a warm Sunday morning.)

Compare this with the behaviour of waves that you are familiar with – light travelling between air and glass. When travelling into a more dense ('slower') medium, the direction of the waves bends towards the normal, Figure 12.21a. When travelling into a less dense ('faster') medium, the direction bends away from the normal, Figure 12.21b.

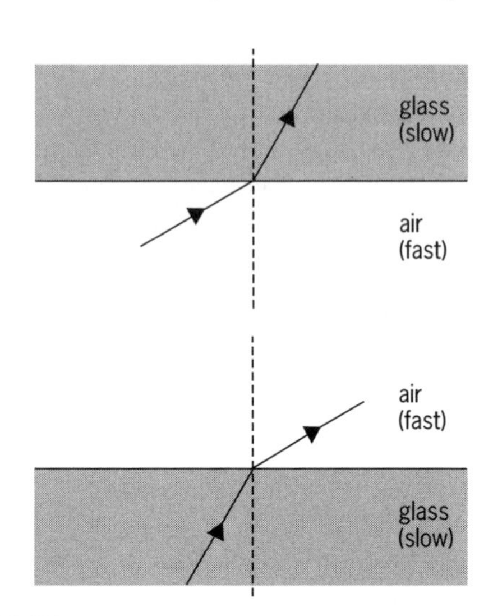

Figure 12.21

14 a For any sort of wave,

velocity = frequency × wavelength or $v = f\lambda$

In this case $v = 7.0\,\text{Hz} \times 30\,\text{mm} = \mathbf{210\,mm\,s^{-1}}$ or $\mathbf{0.21\,m\,s^{-1}}$.

b The velocity of a wave decreases when the water is shallower so the depth decreases at the step.

To remember this, imagine the extra drag of the bottom on the waves, slowing them down.

c The frequency is constant so the wavelength is proportional to the velocity:

$$\frac{v_{\text{shallow}}}{v_{\text{deep}}} = \frac{\lambda_{\text{shallow}}}{\lambda_{\text{deep}}}$$

$$\therefore \lambda_{\text{shallow}} = \lambda_{\text{deep}} \times \frac{v_{\text{shallow}}}{v_{\text{deep}}}$$

$$= 30\,\text{mm} \times \frac{140\,\text{mm}\,\text{s}^{-1}}{210\,\text{mm}\,\text{s}^{-1}} = \mathbf{20\,mm}$$

d The 'refractive index' is given by the ratio of the velocities:

$${}_{\text{deep}}n_{\text{shallow}} = \frac{v_{\text{deep}}}{v_{\text{shallow}}} = \frac{210\,\text{mm}\,\text{s}^{-1}}{140\,\text{mm}\,\text{s}^{-1}} = \mathbf{1.5}$$

15 The simplest way to find the angle in the core is to use

$$n_X \sin i_x = n_Y \sin i_Y$$

so

$$1.58 \sin i_{\text{glass}} = 1.36 \sin i_{\text{glue}}$$

$$\sin i_{\text{glass}} = \frac{1.36 \sin i_{\text{glue}}}{1.58} = \frac{1.36 \sin 10°}{1.58} = 0.150$$

$$i_{\text{glass}} = \sin^{-1} 0.150 = 8.6°$$

The other way is to first find ${}_{\text{glue}}n_{\text{glass}} = 1.16$.

The angle at which the ray will hit the core/cladding boundary will be (Figure 12.22) 90° − 8.6° = **81.4°**.

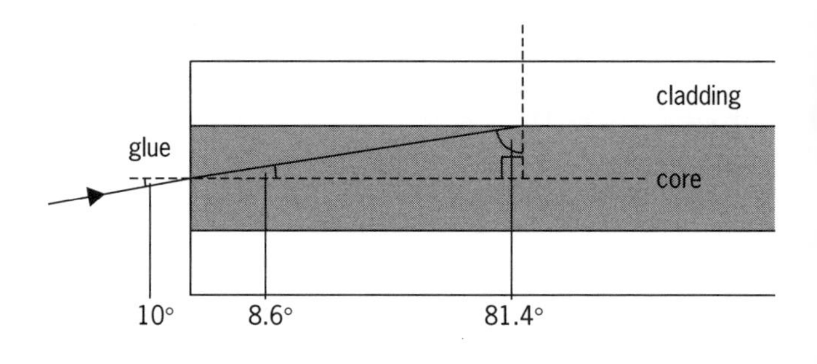

Figure 12.22

16 The absolute refractive index for light of colour 'M' in substance 'G' is defined by

$$n_{G,M} = \frac{c}{c_{G,M}} \quad \text{which you can rearrange to give}$$

$$c_{G,M} = \frac{c}{n_{G,M}}$$

where c = speed of light in a vacuum and $c_{G,M}$ = speed of light of colour 'M' in substance 'G'.

In this question,

$$c_{g,blue} = \frac{3.0 \times 10^8 \,\text{m s}^{-1}}{1.639} = 1.830 \times 10^8 \,\text{m s}^{-1}$$

and

$$c_{g,red} = \frac{3.0 \times 10^8 \,\text{m s}^{-1}}{1.621} = 1.851 \times 10^8 \,\text{m s}^{-1}$$

You can rearrange the equation

$$\text{speed} = \frac{\text{distance}}{\text{time}} \quad \text{to give} \quad \text{time} = \frac{\text{distance}}{\text{speed}}$$

So, for the blue light, the time taken to travel 100 km is

$$t_{blue} = \frac{100 \times 10^3 \,\text{m}}{1.830 \times 10^8 \,\text{m s}^{-1}} = 5.464 \times 10^{-4} \,\text{s}$$

Similarly, for the red light,

$$t_{red} = \frac{100 \times 10^3 \,\text{m}}{1.851 \times 10^8 \,\text{m s}^{-1}} = 5.402 \times 10^{-4} \,\text{s}$$

The time difference between the arrival of the red pulse and blue pulse is therefore

$$5.464 \times 10^{-4} \,\text{s} - 5.402 \times 10^{-4} \,\text{s} = \mathbf{0.062 \times 10^{-4} \,s} \text{ or } \mathbf{6.2 \times 10^{-6} \,s} \text{ or } \mathbf{6.2 \,\mu s}$$

Notice that the values of refractive index in the question are given to 3 decimal places. It is necessary to keep this degree of precision throughout the calculation to achieve a non-zero time interval.

17 a $\sin C = \frac{1}{{}_a n_g} = \frac{1}{1.56} = 0.641$

$C = \sin^{-1} 0.641 = \mathbf{39.9°}$

b $n_{core} \sin i_{core} = n_{cladding} \sin i_{cladding}$

The critical angle, $i_{core} = C$, occurs when $i_{cladding} = 90°$ so $\sin i_{cladding} = 1.0$. Then

$$\sin C = \frac{n_{cladding}}{n_{core}} \times 1.0 = \frac{1.48}{1.56} = 0.949$$

$C = \sin^{-1} 0.949 = \mathbf{71.6°}$

c With a critical angle of only about 40° the distance travelled by a ray that is being reflected close to the critical angle will be very much longer than the straight path (see Figure 12.23). Signals would soon get out of step.

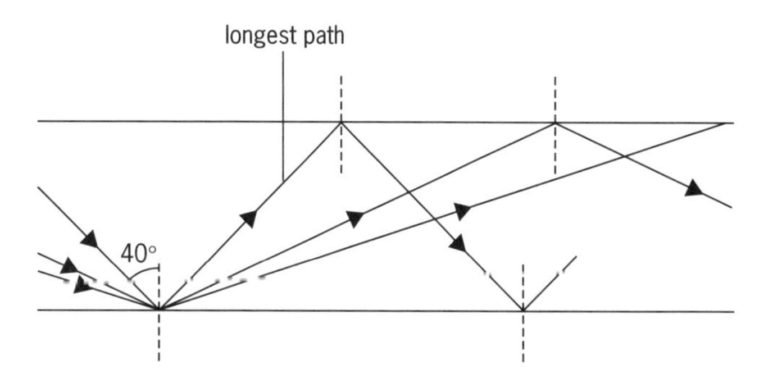

Figure 12.23

It is better to let these rays escape and to allow only rays that hit the walls at a large angle to travel through.

18 a

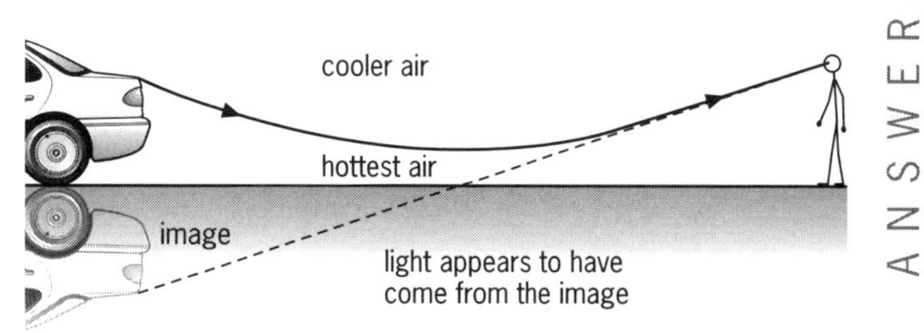

Figure 12.24

b There will be a temperature gradient and the hot air next to the tarmac will be least dense. This will reduce the refractive index so that 'total internal' reflection can take place.

c The observer sees an image of the car upside down and this is interpreted as a reflection in a pool of water – a 'mirage'.

13 Interference between waves

Phase difference

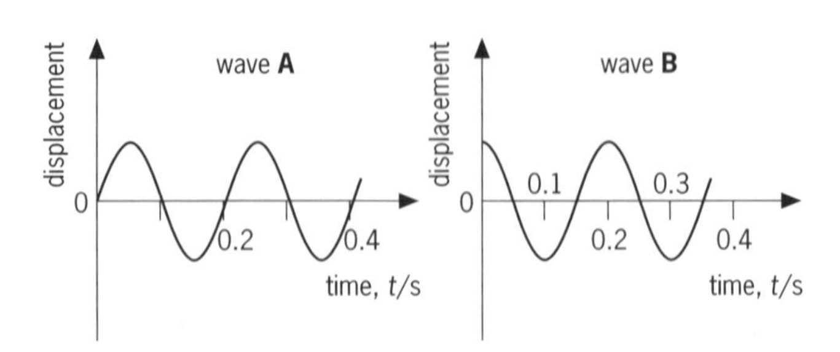

Figure 13.1 These two waves have a phase difference of $T/4$, or a phase angle of 90° or $\pi/2$

The two waves in Figure 13.1 have identical amplitudes, periods and frequencies yet they appear different. This is because the diagrams contain an extra 'ingredient' called **phase difference** or **phase angle**. It answers the questions 'Are the waves synchronised? Are they doing the same thing at the same time?'

In Figure 13.1, wave B is leading wave A. Wave B reaches its peak at a time of $t = 0$ s while A doesn't reach its peak value until $t = 0.05$ s. Wave A is lagging behind or 'out of phase' with wave B. We can express the size of this phase difference by saying that wave A lags behind wave B with a phase difference of $T/4$ (where T is the period) or one quarter of a cycle. This can also be expressed as a phase angle of 90° or $\pi/2$.

Interference

Interference occurs when two or more waves combine to produce a new wave. If the waves combine to produce a resultant with a higher amplitude than the originals they are said to have interfered **constructively**, Figure 13.2a. If however the resultant has a smaller amplitude, the waves are said to have interfered **destructively**, Figure 13.2b

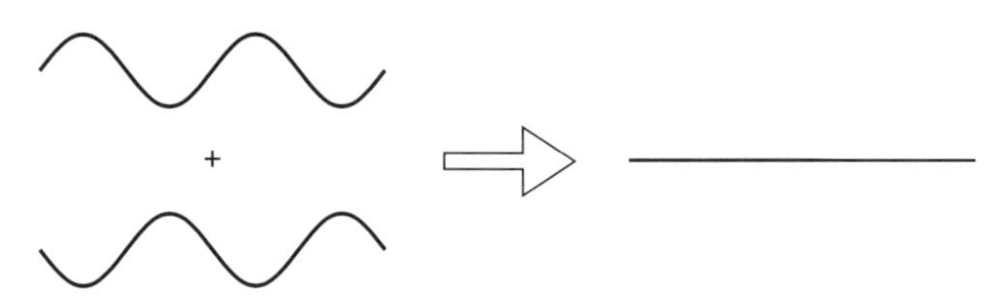

a Constructive interference

b Destructive interference

Figure 13.2

Constructive and destructive interference are examples of the Principle of Superposition of waves. This principle states that when two waves overlap the total displacement at any point is equal to the vector sum of their individual displacements at that point.

Path difference and interference

If the two wave sources A and B in Figure 13.3 are in phase then waves from these sources which meet at point X in the diagram will have travelled the same distance and so will meet in phase. They will therefore interfere constructively, producing a resultant wave with maximum amplitude. Constructive interference will also occur at those points where the path difference between the waves is a whole number of wavelengths, for example point Z.

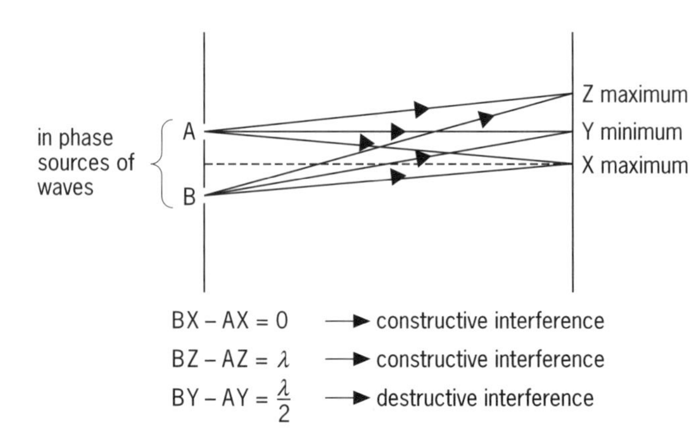

Figure 13.3 Interference between two sources of waves

At a point between X and Z, such as Y, the waves from A will have travelled a shorter distance than those from B and the waves here will not be in phase. If the path difference between AY and BY is half a wavelength, $\lambda/2$, the waves will interfere destructively, producing a resultant wave with a minimum amplitude. Destructive interference will also occur at those points where the path difference between the waves is an odd number of half wavelengths, for example $3\lambda/2$, $5\lambda/2$. If the amplitudes of the waves from A and B are equal, the resultant will have zero amplitude at these points.

Interference patterns

Figure 11.5d on page 79 shows the interference pattern created by two point sources of water waves. In order to produce an observable interference pattern the two sources must:

- produce waves of approximately the same amplitude and frequency (or wavelength); and
- be **coherent** – this means that their phase relationship must not change.

It is relatively easy to produce coherent sources with water waves and sound waves, and so create interference patterns; with light waves it is a little more complicated.

Interference between light waves

It is impossible to produce an observable interference pattern using the light from two independent light sources. This is because light waves are emitted from their source in short pulses and between each pulse there may be an abrupt phase change (Figure 13.4). As a consequence the phase relationship between waves from different sources continuously changes. The sources are incoherent, so no sustained pattern is produced.

Figure 13.4 Interruption of the smooth oscillation of a monochromatic light wave creates an abrupt phase change

Coherent light sources can only be created by separating the light from a single monochromatic source, so that a constant phase relationship is maintained. A monochromatic light source emits light of just one wavelength.

When light waves interfere they produce bands of darkness where they interfere destructively and bright bands of light where they interfere constructively. These bands are called **interference fringes**.

Young's slits experiment

The aim of this experiment is to produce observable interference fringes with light waves.

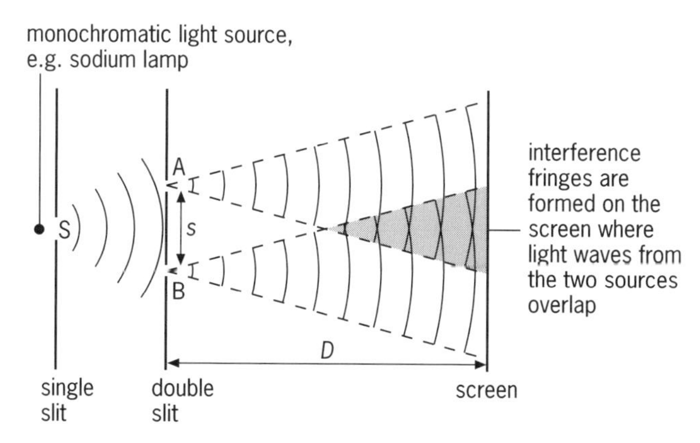

Figure 13.5 Young's slits arrangement

A monochromatic light source is placed behind a screen with a narrow slit in it, S. Light from this single source spreads out (due to diffraction) and passes through two narrow, closely spaced slits in a second screen. This creates two coherent sources, A and B. The waves from these sources spread out and overlap to produce an interference pattern on the viewing screen.

If the two sources are in phase, constructive interference will occur at the screen where the path difference between the waves from A and B is zero or a whole number of wavelengths, that is $n\lambda$. At these places a bright fringe will be seen. Where the path difference is $(n+\frac{1}{2})\lambda$, destructive interference will occur and a dark fringe will be seen. Figure 13.6 shows the resulting interference pattern.

Figure 13.6 Interference fringes seen in a Young's slits experiment

The distance between successive bright fringes, called the fringe width, x, and the wavelength of the light source are related by the equation

$$\text{fringe width, } x = \frac{\lambda D}{s}$$

where D = distance of sources from screen
s = distance between slits

Example 1

Calculate the fringe width of an interference pattern created by two coherent sources 1.0 mm apart and emitting light of wavelength 600 nm. The pattern is observed on a screen 50 cm from the sources.

$$\text{fringe width, } x = \frac{\lambda D}{s}$$

$$= \frac{600 \times 10^{-9}\,\text{m} \times 50 \times 10^{-2}\,\text{m}}{1.0 \times 10^{-3}\,\text{m}}$$

$$= 3.0 \times 10^{-4}\,\text{m}$$

Practical details

1. The light source is usually a monochromatic sodium light or a laser.
2. The interference pattern is rather faint and must be observed in a darkened room.
3. For reasonably wide fringes the screen distance (D) should be approximately 30–50 cm and the slit separation no more than 1 mm.
4. To produce sharp fringes the slits should be as narrow as possible.
5. The width of 10 or 20 fringes should be measured rather than a single fringe width. These and the distance between the slits are usually measured using a travelling microscope. D can be measured with a ruler.

Level 1

1 The diagrams in Figure 13.7 show two pulses created at opposite ends of a rope.

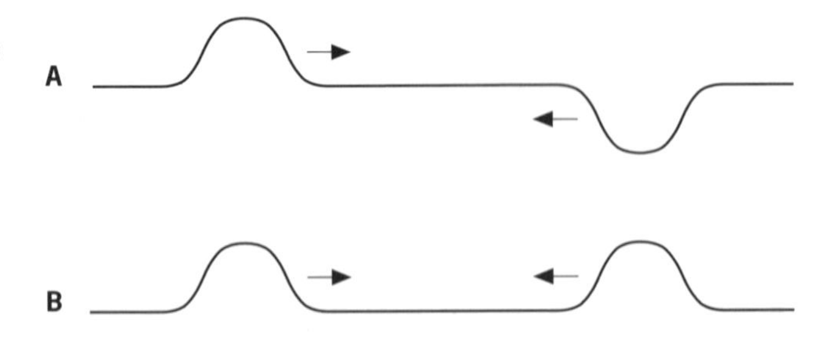

Figure 13.7

a Draw diagrams to show, in each case, what the rope looks like at the instant when the pulses meet.
b Which of your diagrams shows 'constructive interference'?

2 Sketch the wave in Figure 13.8 and, on the same axes, draw a wave that is 90° out of phase with it.

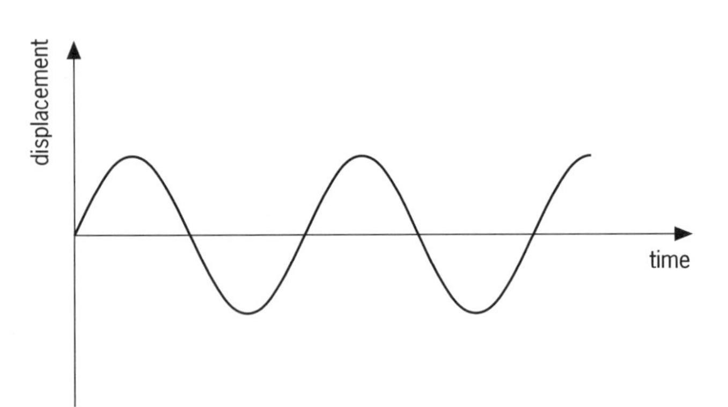

Figure 13.8

3 a Explain why you do not see any interference between the light from two light bulbs connected to the same power supply.
b Explain why you *can hear* interference between sound from two loudspeakers that are connected to the same source.

Level 2

4 Two small loudspeakers are arranged as shown in Figure 13.9. They are both connected to the same signal generator, which is set to give an output at a frequency of 6.6 kHz. You should take the velocity of sound in air as being $330\,\mathrm{m\,s^{-1}}$.

Figure 13.9

a What is the wavelength of the sound?
b The equation

$$\lambda = \frac{xs}{D}$$

applies here, where x is the spacing of the 'fringes'. What do s and D represent and what are their values?
c Rearrange the equation to make x, the spacing of the 'fringes', the subject.
d What will an observer hear at A?
e Calculate the separation of the 'fringes'.
f How far would an observer have to walk, from A, to get to a position of minimum sound?

5 A point source of vibrations, at position S in Figure 13.10, makes waves with a wavelength, $\lambda = 5.0\,\mathrm{cm}$ in a ripple tank. There is a barrier that reflects the waves and a motion sensor floating at D.

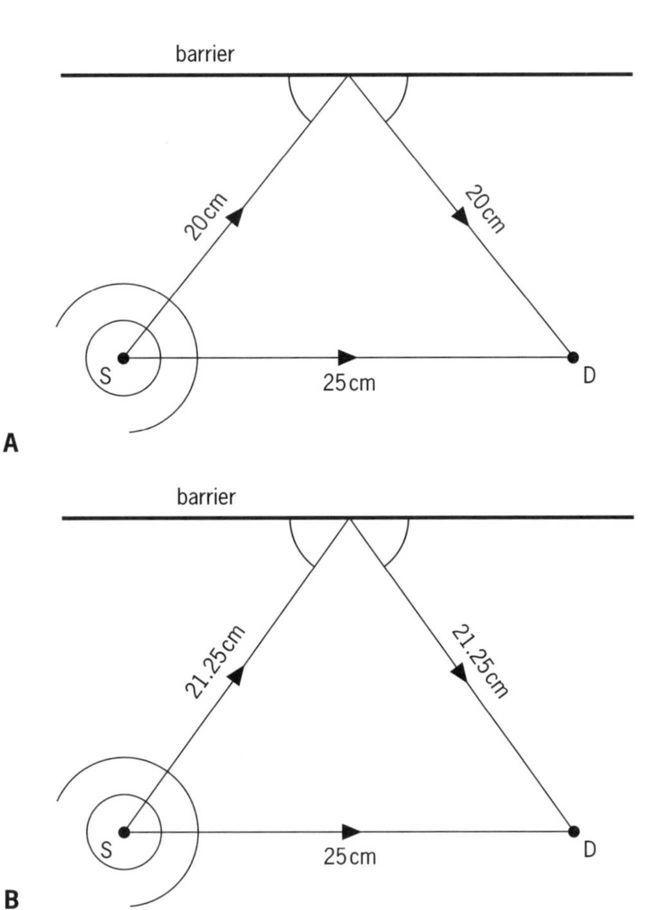

Figure 13.10

a What sort of motion will be detected at D in diagram **A**? Explain your answer.
b What sort of motion will be detected at D in diagram **B**? Explain.
c The barrier forms an image of the source. Draw a diagram showing the position of this image relative to the barrier and the real source.
d How could you calculate the spacing of the maxima that you would expect as the detector is moved at right angles to the barrier?

6 In Figure 13.11, S_1 and S_2 are point sources of ripples. They are on the same bar and so vibrate together with the same frequency and amplitude. Maxima can be seen at A and C, and minima at B and D.

Figure 13.11

How will the pattern change if:

a the two sources are moved closer together, without changing the frequency?

b the frequency of the vibrations is increased, without changing the position of the sources? How would you do this?

c The velocity of the waves is reduced? How would you do this?

Level 3

7 You are giving a presentation on interference. You want to use a laser with $\lambda = 690\,\text{nm}$ and a pair of slits ruled 0.10 mm apart on a slide. Estimate the distance from the slits that the screen should be so that the bright dots (maxima) are clearly separated.

8 Using the same laser as the previous question, you find an unmarked pair of slits and decide to check their spacing and then label for future use. The screen is now 3.0 m away and the maxima are 1.2 cm apart. What is the separation of the slits?

Level 1

1 a See Figure 13.12. When the two pulses in **A** meet they cancel each other out because they are of the same amplitude but opposite phase. When the two pulses in **B** meet they produce a pulse with double the amplitude.

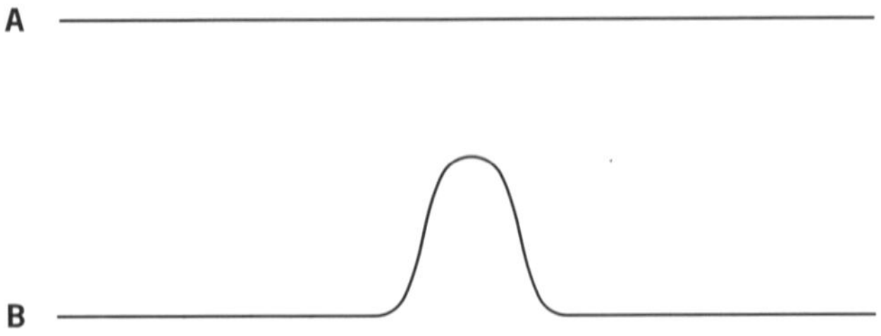

A

B

Figure 13.12

b Diagram **B** shows constructive interference.

2 See Figure 13.13. Both the dotted waves shown are 90° out of phase with the original: one is 90° ahead; the other is 90° behind.

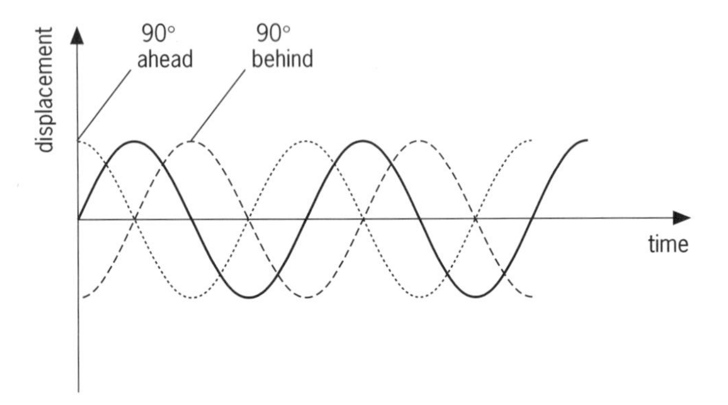

Figure 13.13

3 a The wave trains of light from a filament bulb are emitted in random bursts of different lengths and with different time intervals between them. There can never be a constant phase relationship between the light from two such bulbs, even if the light emitted has the same frequency (colour).

b The pattern of sound vibrations from each speaker will be identical so interference patterns will occur.

Level 2

4 a You can rearrange the equation

$$v = f\lambda \quad \text{to give} \quad \lambda = \frac{v}{f}$$

In this case

$$\lambda = \frac{330\,\mathrm{m\,s^{-1}}}{6.6\times10^{3}\,\mathrm{s^{-1}}} = \mathbf{0.05\,m}$$

b *s* is the separation of the sources: ***s*** **= 0.80 m**.
D is the distance from the sources to the line (parallel to the line joining the speakers) along which the observer moves, so ***D*** **= 25 m**.

c
$$\lambda = \frac{xs}{D}$$
$$\therefore D\times\lambda = x\times s$$
and so
$$x = \frac{D\lambda}{s}$$

d The sound will be **loud** at A, because it is equidistant from the two speakers and so the waves will be in phase and will interfere constructively.

e The separation of the 'fringes' is given by the equation

$$x = \frac{D\lambda}{s}$$
$$= \frac{25\,\mathrm{m}\times0.050\,\mathrm{m}}{0.80\,\mathrm{m}} = \mathbf{1.56\,m}$$

f The distance between two adjacent maxima or two adjacent minima is 1.56 m, so the distance from a maximum to the next minimum will be half this, which is **0.78 m**.

5 a The two routes between the source and the detector in **A** are:
i the 'direct' route, a distance of 25 cm, which is equal to $5\times\lambda$;
ii the 'reflected' route, a distance of $2\times20 = 40$ cm, which is equal to $8\times\lambda$.
Each route has a length equal to a whole number of wavelengths so the signals will arrive at the detector in phase and will interfere constructively to give **maximum** oscillatory motion.

b The length of the 'direct' route is still equal to $5\times\lambda$. The 'reflected' route in **B** is a distance of 42.5 cm, which is equal to $8\frac{1}{2}\times\lambda$. This means that when the signals arrive at the detector they will be completely out of phase, with a phase difference of 180° or π radians. These signals will interfere destructively to produce **minimum** oscillatory motion.

c

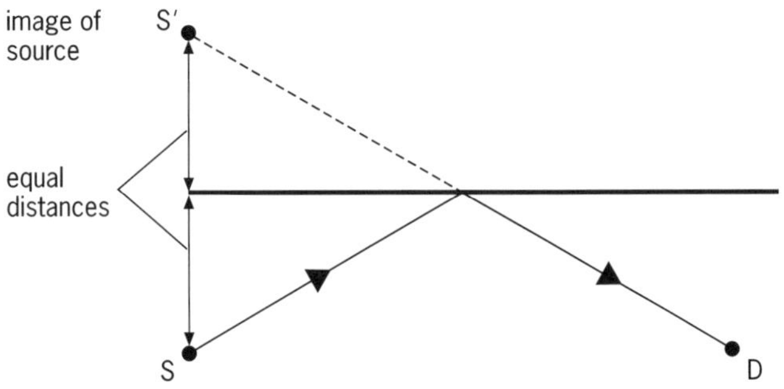

image of source
S′
equal distances
S
D

Figure 13.14

d To use the formula for fringe spacing,

$$x = \frac{\lambda D}{s}$$

you need to know the separation between the real source, S, and its image, S′, to provide *s*. Measure from the source to the barrier and then double it. *D* is the distance between source, S, and detector, D.

6 Look at the equation for the fringe spacing,

$$x = \frac{\lambda D}{s}$$

where s is the separation of (distance between) the sources and D is the distance of the fringes from the line that joins the sources.

a As s gets smaller, x will get larger (x is inversely proportional to s), so the spacing of maxima and minima **increases** as you move the sources together.

b As the frequency, f, increases, the wavelength, λ, decreases ($f \times \lambda =$ the wave velocity, which is constant), which means that x, thc fringe separation, also decreases. So the spacing of maxima and minima **decreases** if you increase the frequency. You can do this by making the motor that drives the vibrating bar rotate faster, usually by increasing the voltage of the supply.

c If v is reduced but f stays the same then λ must get smaller. But as λ decreases the spacing decreases, so the spacing of maxima and minima **decreases** when the velocity is reduced. To reduce the velocity you need to reduce the depth of the water.

Level 3

7 For clarity, the separation of the maxima should be somewhere between 0.50 cm and about 10 cm. The required screen distance is given by the equation

$$D = \frac{xs}{\lambda}$$

So for $x = 0.50\,\text{cm}$ ($5.0 \times 10^{-3}\,\text{m}$),

$$D = \frac{5.0 \times 10^{-3}\,\text{m} \times 0.1 \times 10^{-3}\,\text{m}}{690 \times 10^{-9}\,\text{m}} = 0.72\,\text{m}$$

This is too close for demonstration to a class.

For $x = 10\,\text{cm}$ (0.10 m),

$$D = \frac{0.10\,\text{m} \times 0.1 \times 10^{-3}\,\text{m}}{690 \times 10^{-9}\,\text{m}} = 14\,\text{m}$$

Note that to increase the spacing by a factor of 20 you need to increase the distance to the screen by a factor of 20 too.

This is too far away. A value somewhere in between would be suitable.

8 The equation needed here is

$$s = \frac{\lambda D}{x}$$

In this case,

$$s = \frac{3.0\,\text{m} \times 690 \times 10^{-9}\,\text{m}}{0.012\,\text{m}}$$

$$= \mathbf{1.7 \times 10^{-4}\,m} \text{ or } \mathbf{0.17\,mm}$$

14 Diffraction and polarisation

Diffraction

When waves pass through an aperture, or their path is partially obstructed by an object, the waves spread out and some will enter the 'shadow' zone. This effect is called **diffraction**; the waves are said to have been diffracted.

The extent to which the waves spread out depends upon the wavelength of the waves and the size of the aperture, Figure 14.1.

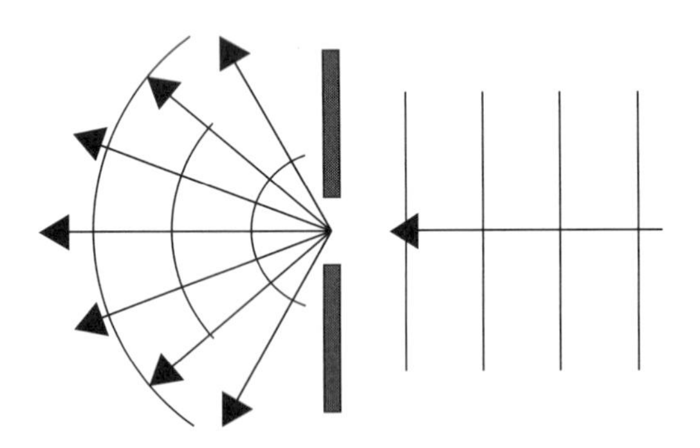

a Diffraction is most noticeable if the size of the aperture is approximately equal to the wavelength of the waves

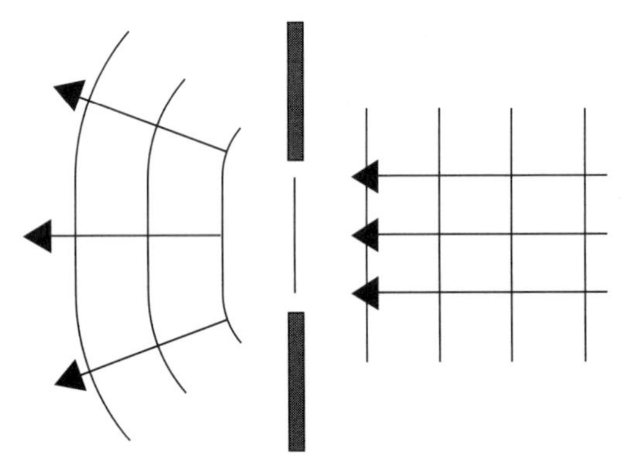

b Diffraction is less obvious if the aperture is much bigger than the wavelength of the waves

Figure 14.1

Single slit diffraction

Some of the diffracted waves emerging from an aperture such as a slit will overlap, creating an interference pattern or 'diffraction pattern'. This consists of maxima where the waves have interfered constructively and minima where they have interfered destructively, Figure 14.2.

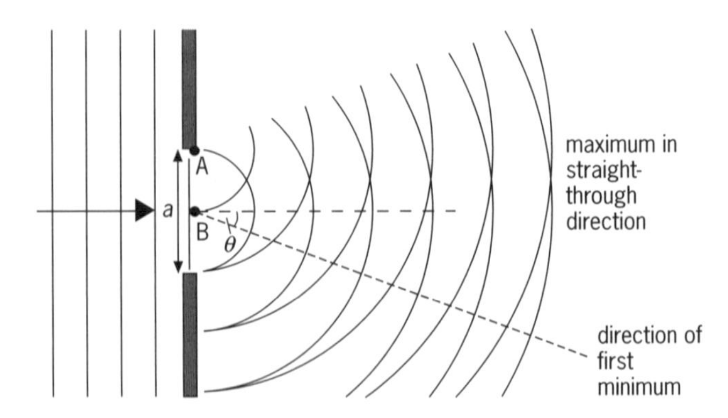

Figure 14.2 Diffracted waves from two points within the slit, such as A and B, overlap and interfere

The positions of the minima in the diffraction pattern depend upon the width of the slit and the wavelength of the waves. The narrower the slit, the wider the pattern. The longer the wavelength, the wider the pattern. The relationship for the angular position of a minimum is described by the equation

$$\sin\theta = \frac{n\lambda}{a}$$

where θ is the angular position of a minimum measured from the slit (Figure 14.2)

n is the minimum being considered, for example first minimum ($n = 1$), second minimum ($n = 2$), and so on

λ is the wavelength of the incident waves

a is the slit width

Example 1

Calculate the angular position of the first minimum for a diffraction pattern created when light of wavelength 600 nm passes through a single rectangular slit of width 0.15 mm.

$$\sin\theta = \frac{n\lambda}{a}$$

$$= \frac{1 \times 600 \times 10^{-9}\,\text{m}}{0.15 \times 10^{-3}\,\text{m}} = 4.0 \times 10^{-3}$$

$$\therefore \quad \theta = \sin^{-1}(4.0 \times 10^{-3}) = 0.23°$$

The diffraction grating

In the Young's slits experiment, page 95, we saw how diffracted light from two coherent sources can produce an interference pattern of light and dark fringes. If these two slits are replaced by multiple slits – a 'diffraction grating' – the interference pattern changes to one containing fewer but sharper intensity maxima. These changes occur because increasing the number of coherent sources decreases the number of directions in which the waves interfere constructively and increases the number of directions in which they interfere destructively.

Consider two rays of light from two adjacent slits, Figure 14.3.

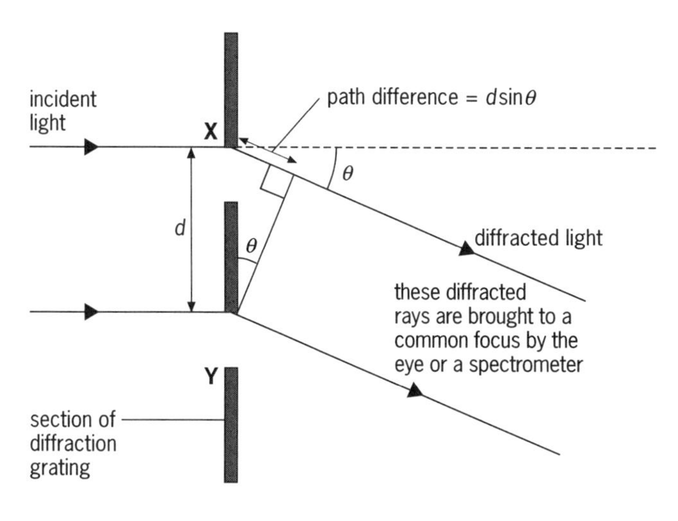

Figure 14.3 Diffraction from two adjacent slits of a diffraction grating

For constructive interference to take place the path difference between waves from similar points in adjacent slits, such as X and Y, must be a whole number of wavelengths. From geometry we can see that the path difference of rays from X and Y is $d \sin\theta$. So, for the production of an intensity maximum,

$$d \sin\theta = n\lambda$$

where d is the slit separation
θ is the angular direction in which constructive interference is taking place
n is the number of the maximum, for example when $n = 1$ this is called the first order maximum

Note that when $n = 0$ this is the zero order maximum, created in the straight-through direction.

Example 2

Calculate the angular position of the second order maximum created when monochromatic light of wavelength 700 nm passes through a diffraction grating of spacing 3.5 µm.

$$d\sin\theta = n\lambda$$

so $$\sin\theta = \frac{n\lambda}{d}$$

$$= \frac{2 \times 700 \times 10^{-9}\,\text{m}}{3.5 \times 10^{-6}\,\text{m}}$$

$$\therefore \quad \theta = \sin^{-1}0.4 = 24°$$

With the wavelengths of visible light it is possible to see three or even four maxima.

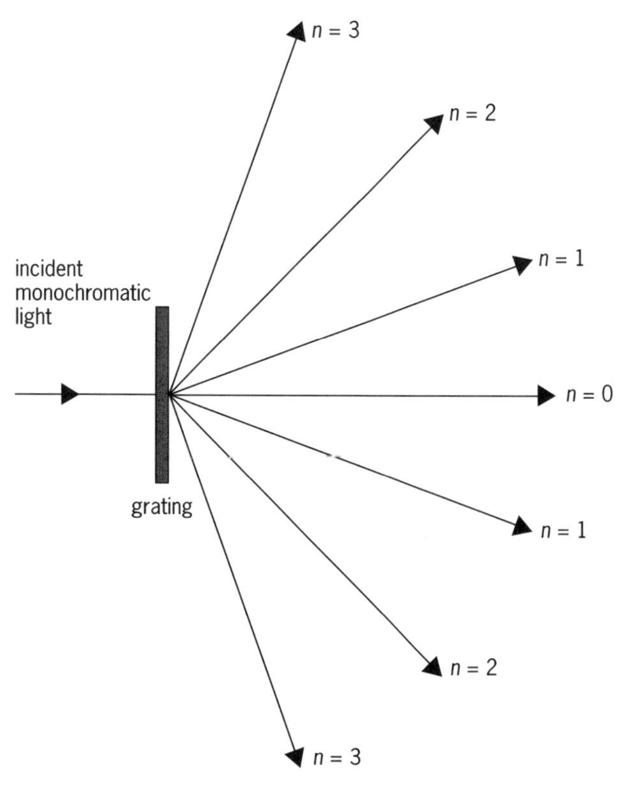

Figure 14.4 Orders of maxima created by an optical grating

Polarisation of electromagnetic waves

Light waves, X-rays, radio waves and microwaves are all examples of electromagnetic waves (see Figure 11.11, page 82). They are transverse waves, that is, their vibrations are at right angles to the direction in which the wave is transferring energy. If the vibrations occur in many different planes the wave is said to be unpolarised, Figure 14.5a. If the vibrations take place in just one plane then the wave is said to be **plane polarised**, Figure 14.5b.

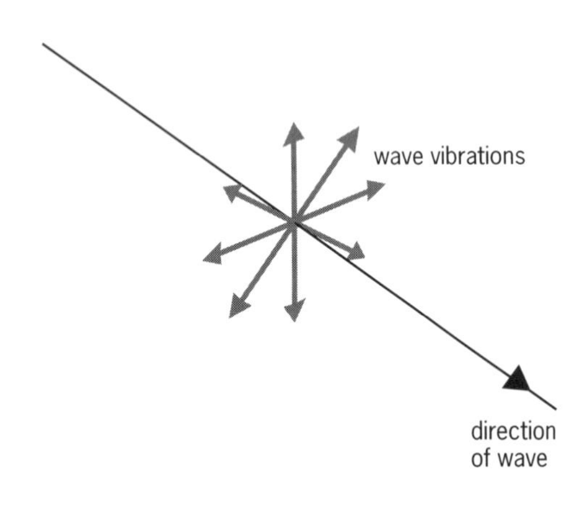

a In an unpolarised wave the vibrations occur in all planes

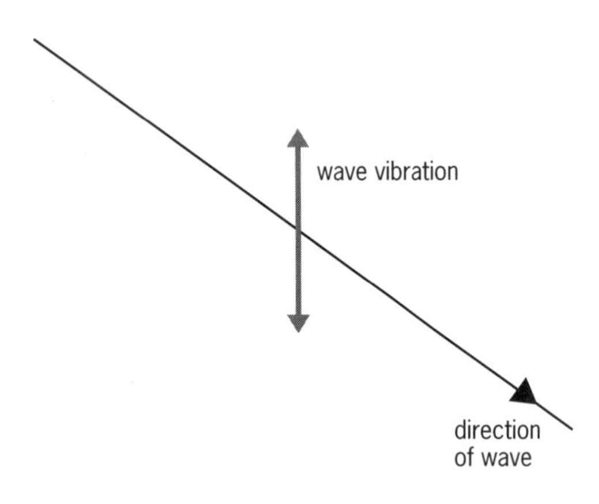

b In a polarised wave the vibrations occur in just one plane

Figure 14.5

Unpolarised light waves can be polarised using a polariser such as a sheet of Polaroid. A Polaroid sheet contains long chains of organic molecules aligned parallel to each other. The molecules allow waves with vibrations in one particular plane to pass through the sheet but absorb all vibrations which are at right angles to this plane, Figure 14.6.

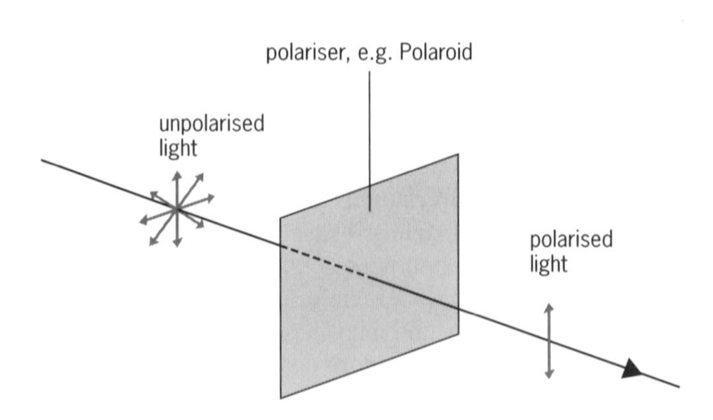

Figure 14.6 Polaroid has the ability to allow particular components of an electromagnetic wave to pass through it while 'blocking' others

If the polarised light is viewed through a second piece of Polaroid, light will not reach the observer if the Polaroids are 'crossed', Figure 14.7.

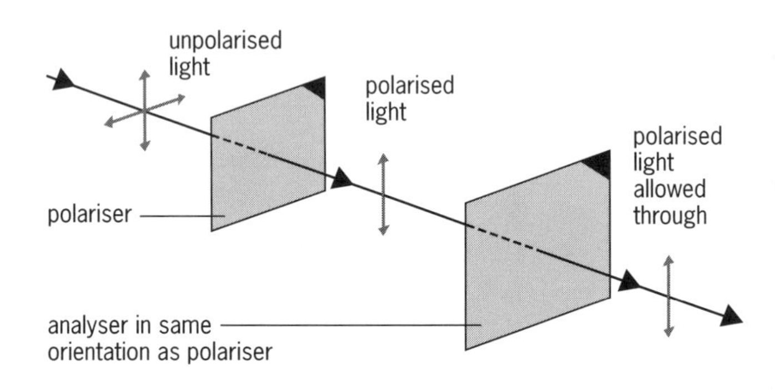

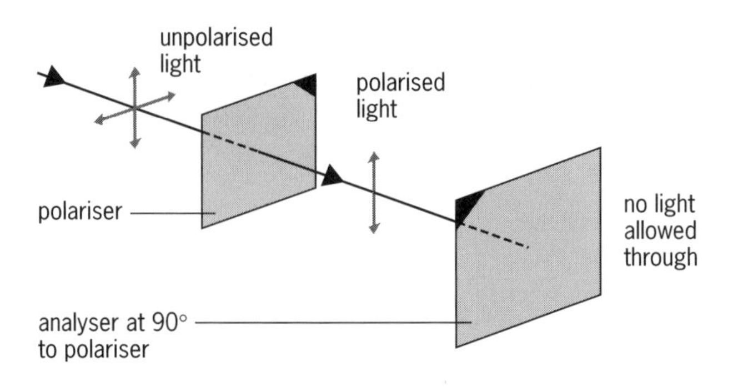

Figure 14.7 Light passes through parallel Polaroids but is unable to pass through 'crossed' Polaroids

The fact that light waves can be polarised is direct proof that they are transverse. Longitudinal waves, such as sound waves, cannot be polarised since their vibrations are along the direction in which the wave is transferring energy.

Level 1

1 Copy and complete the diagrams in Figure 14.8 to show the shape of the wavefronts after they have passed through the gap. The waves have the same speed.

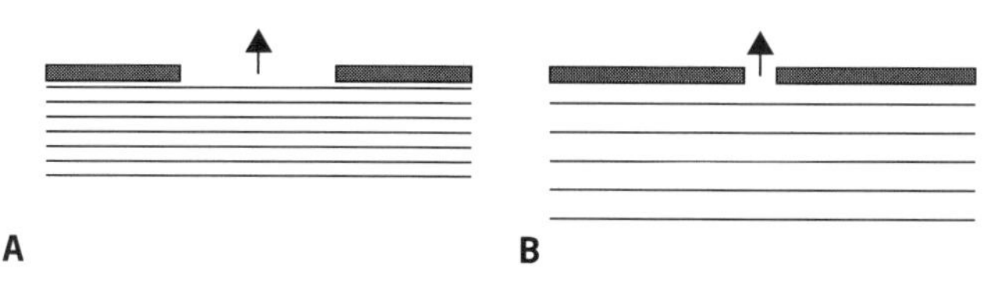

Figure 14.8

a Which wave is diffracted more?
b Which wave has the higher frequency?

2 Explain why you can hear a radio through an open door to the next room, even if you can't see the radio.

Hint: You should mention wavelength.

3 Using sound and light as examples, describe some properties of each that reveal the difference between longitudinal and transverse waves.

4 Explain, using a diagram, how a pair of crossed Polaroids cut out the light.

Level 2

5 Your uncle looks at a photograph in your textbook of diffraction caused by a small circular obstacle and thinks that it is ridiculous to say that it would be bright in the centre of a shadow. How do you explain this to him?

6 A footpath is edged with iron railings that could diffract sound. Roughly what frequency sounds would be diffracted?

7

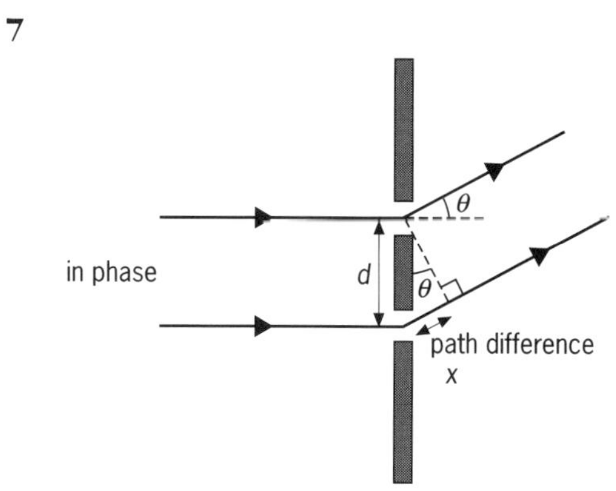

Figure 14.9

After passing through a grating of slits, each separated by distance d, the waves of a single wavelength λ reinforce if their path difference is a whole number (n) of wavelengths. Use the diagram, Figure 14.9, to show that

$$n \times \lambda = d \times \sin\theta$$

for a maximum at angle θ.

Level 3

8 When you look at a CD, you see different colours at different angles because the regular lines act as a reflection grating. Figure 14.10 illustrates that, at normal incidence, the angles for a maximum for each colour are given by the same equation as for a transmission grating.

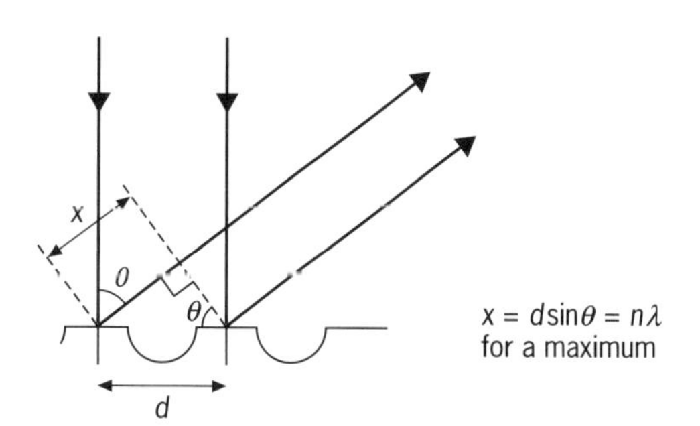

Figure 14.10

If you see red light with, $\lambda = 700\,\text{nm}$, at 25° from the normal, what is the spacing of the lines on the CD?

9

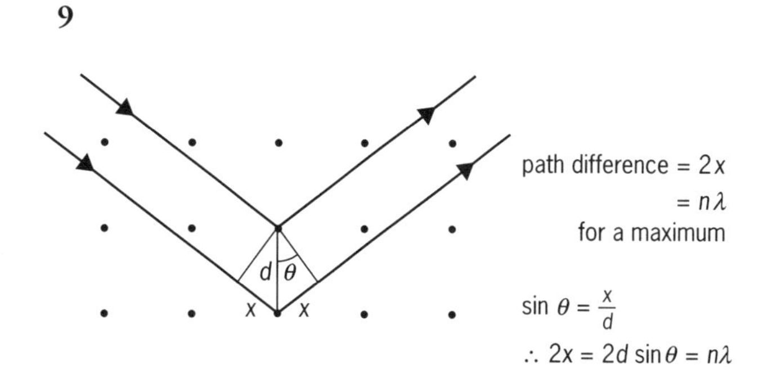

Figure 14.11

Figure 14.11 illustrates that the path difference for X-rays reflected from adjacent planes of atoms is twice that of the diffraction grating already met:

$$2d\sin\theta = n\lambda \quad \text{for a maximum}$$

where d is the spacing of the planes of atoms and θ is the angle between the rays and the layers of atoms.

Find the separation of the planes of atoms if, when X-rays with $\lambda = 0.05\,\text{nm}$ are reflected from a crystal of sodium chloride, the first order diffraction maximum is found at an angle of 5.4° to the crystal planes.

10 A first order spectrum is obtained using an optical grating with $500\,\text{lines}\,\text{m}^{-1}$. What is the angular separation of red, with $\lambda = 700\,\text{nm}$, and violet, with $\lambda = 400\,\text{nm}$?

11 Use your knowledge of diffraction and light to explain why:
a there is no visible interference pattern with a two-slit arrangement if the slits are too wide, even if they are close together;
b no pattern is formed if the slits are too far apart, even if they are narrow.

Level 1

1

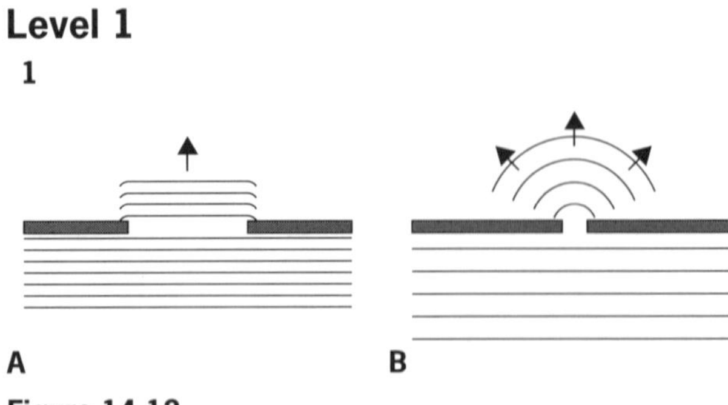

Figure 14.12

a Diagram **B** shows diffraction because the gap is the same order of size as the wavelength, and so alters the direction of the wave.

b The waves in diagram **A** have the higher frequency. You can see that the wavelength is smaller. The velocities of the two waves are the same, so from $v = f\lambda$ their frequencies are inversely proportional to their wavelength:

$$\frac{f_A}{f_B} = \frac{\lambda_B}{\lambda_A}$$

2 Because the wavelength of audible sounds is roughly the same size as the gap, diffraction will occur and the sound will bend around the corners.

The wavelength of the note middle C, which has a frequency of 256 Hz, is about 1 m in air.

Even the longest wavelength of visible light at 700 nm is *tiny* compared with the size of a doorway, so no visible diffraction takes place and the light 'travels in straight lines'. See Figure 14.13.

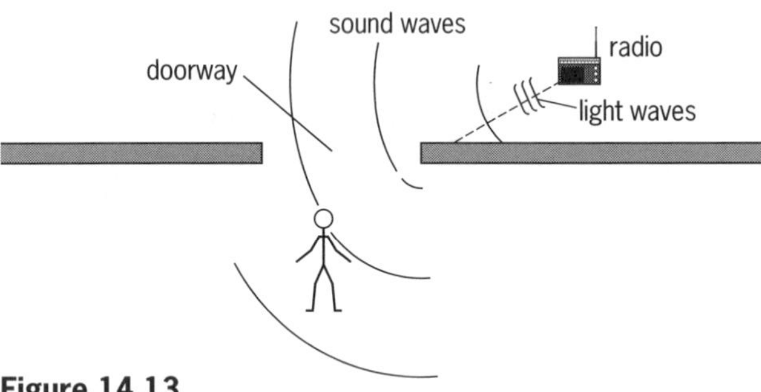

Figure 14.13

3 One property that distinguishes a longitudinal wave from a transverse one is plane polarisation: the confinement of vibrations to one plane at right angles to the direction of travel. If there is no vibration at right angles to the direction of travel, there can be no plane polarisation.

When a sound wave passes through air the vibrations of the molecules are to and fro in the direction of propagation of the wave. This type of wave is longitudinal and cannot show polarisation.

Light is a transverse vibration of electric and magnetic fields. This means the fields oscillate at right angles to the direction of propagation of the wave. Light can be plane polarised using a Polaroid sheet, which absorbs all but one direction of vibration (of the electric field). A second sheet at right angles then cuts out this vibration.

4 Unpolarised light has vibrations in planes in all possible directions at right angles to the direction of motion of the wave. A Polaroid sheet absorbs all but one plane of vibration so the light coming through is 'plane polarised'. If a second Polaroid sheet is placed in the path of this light with its direction of polarisation at right angles to the first, then the plane polarised light is absorbed and no light can get through at all. See Figure 14.14.

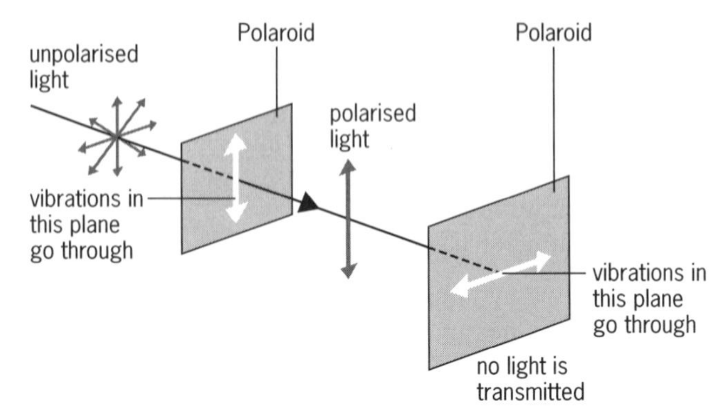

Figure 14.14

Level 2

5 Draw him a picture showing how the waves bend round the back of the obstacle and then interfere constructively. See Figure 14.15.

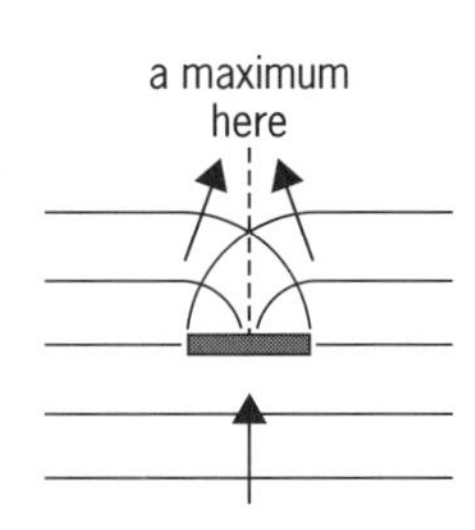

Figure 14.15

6 A row of railings could act as a diffraction grating. The railings are likely to be about 0.1 to 0.3 m apart. Remember that waves are diffracted most by obstacles that are about the same size as their wavelength. So the row of railings will diffract waves with wavelengths of 0.1 to 0.3 m.

$$\text{frequency} = \frac{\text{speed}}{\text{wavelength}}$$

So, in this case, taking the speed of sound to be $330\,\text{m s}^{-1}$, the frequency of the waves that will be diffracted will be between

$$\frac{330\,\text{m s}^{-1}}{0.1\,\text{m}} = \mathbf{3.3\,kHz} \quad \text{and} \quad \frac{330\,\text{m s}^{-1}}{0.3\,\text{m}} = \mathbf{1.1\,kHz}$$

Alternatively, you could use the equation for the diffraction grating,

$$n\lambda = d \sin\theta$$

For a first order angle of between 10° and 80° you get the same lower frequency but the upper end goes up to 16 kHz.

7 After passing through the grating, the difference between the distances travelled by the two rays shown is x. If x is equal to a whole number of wavelengths then the waves will be in phase and will interfere constructively. There will be a maximum at that angle. So, for constructive interference to take place, $x = n\lambda$ where n is a whole number.

From Figure 14.9,

$$\sin\theta = \frac{x}{d}$$

$$\therefore x = d \times \sin\theta$$

and so

$$x = n \times \lambda = d \times \sin\theta$$

Level 3

8 You can rearrange the equation $d\sin\theta = n\lambda$ to give

$$d = \frac{n\lambda}{\sin\theta}$$

Taking $n = 1$, then, for red light with a wavelength of 700 nm,

$$d = \frac{\lambda}{\sin\theta}$$

$$= \frac{700 \times 10^{-9}\,\text{m}}{0.4266}$$

$d = \mathbf{1.66 \times 10^{-6}\,m \approx 1.7\,\mu m}$

The 'grooves' on the CD are 1.7 μm apart.

9 After reflection from adjacent layers of atoms, for constructive interference the path difference must be equal to an integral number of wavelengths:

$$2d\sin\theta = n\lambda$$

The first order diffraction maximum is defined by $n = 1$ and, in this case, $\theta = 5.4°$,

$$\therefore 2d\sin 5.4° = \lambda$$

which you can rearrange to give

$$d = \frac{\lambda}{2\sin 5.4°}$$

$$= \frac{0.05 \times 10^{-9}\,\text{m}}{0.188}$$

$$= \mathbf{2.7 \times 10^{-10}\,m}$$

This is the separation of the planes of atoms.

10 There are 500 lines to the mm so the spacing of the lines on the grating is

$$d = \frac{1 \times 10^{-3}\,\text{m}}{500} = 2.0 \times 10^{-6}\,\text{m}$$

The equation for a grating is

$$d\sin\theta = n\lambda$$

In this case $n = 1$, so you can rearrange the equation to give

$$\sin\theta = \frac{\lambda}{d}$$

For red light

$$\sin\theta_r = \frac{700 \times 10^{-9}\,\text{m}}{2 \times 10^{-6}\,\text{m}} = 0.35$$

and so $\theta_r = \sin^{-1} 0.35 = 20.49°$.

For blue light

$$\sin\theta_b = \frac{400 \times 10^{-9}\,\text{m}}{2 \times 10^{-6}\,\text{m}} = 0.20$$

and so $\theta_b = \sin^{-1} 0.20 = 11.54°$.

The difference in the angles is $20.49° - 11.54° = \mathbf{8.95°}$.

11 a The wavelength of light is very small so there will only be diffraction at very narrow slits. If there is no diffraction there is no overlap of the waves from each slit so there can be no interference; see Figure 14.16.

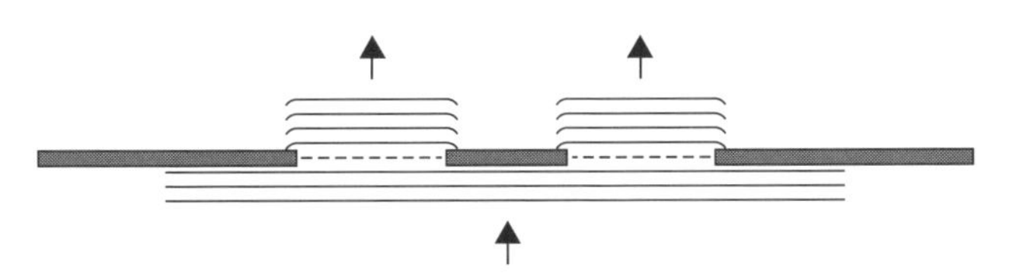

Figure 14.16

b If the slits are too far apart then the path difference between the wave trains from each slit gets too large. There will no longer be a constant phase difference between the two wave trains because of changes of phase within each wave train; see Figure 14.17.

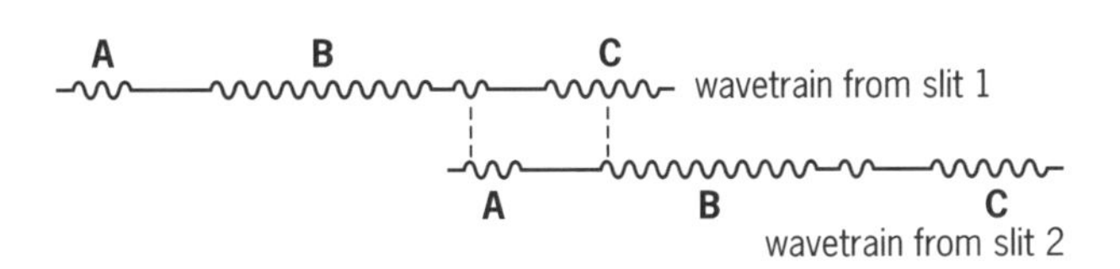

C from slit 1 meets with B from slit 2;
B from slit 1 meets with A from slit 2.
The overlapping waves do not have a constant phase relationship.

Figure 14.17

15 Standing or stationary waves

A water wave moving away from the disturbance which has created it is an example of a **progressive wave**. It transfers energy away from the disturbance. The positions of its peaks and troughs change with time.

Under certain circumstances it is possible to create a motion which oscillates like a wave but the positions of its peaks and troughs do not change. Waves like this are called **standing** or **stationary waves**. They are produced when interference takes place between progressive waves of the same frequency and amplitude travelling in opposite directions along the same line. If the string of a guitar is plucked at its centre, for example, progressive waves travel outwards to the fixed ends and are reflected. These reflected waves overlap to produce standing or stationary waves.

When the waves on a string interfere there are several possible standing waves which can be produced. The simplest of these is called the **fundamental mode of vibration**, Figure 15.1.

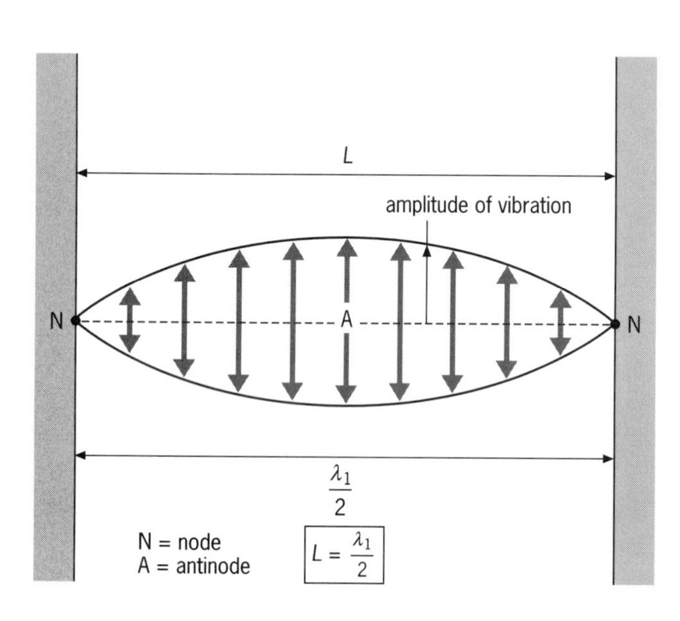

Figure 15.1 The fundamental mode of vibration of a string

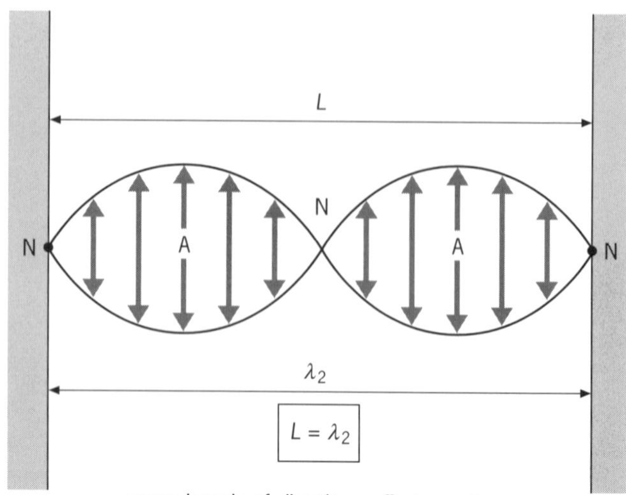

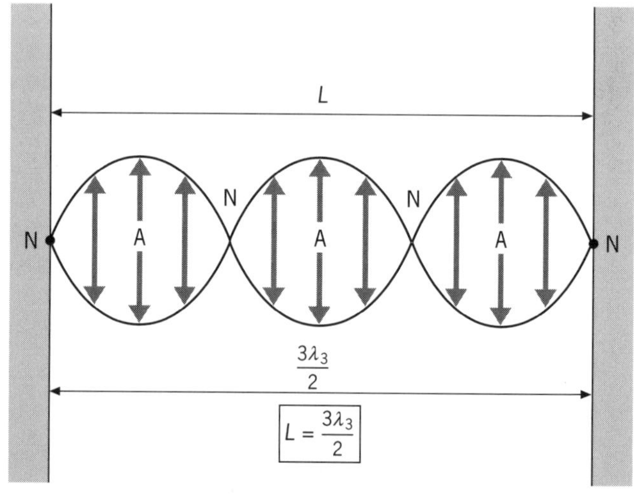

Figure 15.2 Overtones on a string

The next simplest modes are called the **first overtone**, the **second overtone**, and so on, Figure 15.2.

It can be seen that

- there are certain points along the string where there is *no* vibration: these points are known as **nodes**;
- there are places along the string where maximum amplitude of vibration occurs: these are called **antinodes**;
- the positions of the nodes and antinodes are unchanged with time: these are stationary waves;
- the distance between successive nodes is $\lambda/2$ for each mode of vibration.

Differences between progressive and standing waves

1. In a progressive wave all particles have the same maximum displacement, or amplitude. In a standing wave the size of the maximum displacement, or amplitude of vibration, varies with the position of the particle.
2. In a progressive wave there is a phase difference between all the particles – adjacent particles are, at any instant in time, at different points in their vibration. In a standing wave *all* the particles between two adjacent nodes are vibrating in phase.
3. Progressive waves transfer energy; standing waves do not.

Standing waves in air

Closed pipes

If sound waves travelling down a pipe closed at one end interfere with reflected waves travelling up the pipe, they can produce standing waves. As in the above example of the vibrating string there are several ways, or modes, in which the air in the pipe can vibrate.

At the closed end it is not possible for the air molecules to vibrate freely so a node is produced here. At the open end of the pipe where the molecules are least restricted an antinode will be produced. Figure 15.3 shows the first three modes of vibration possible.

Note that standing waves in pipes are longitudinal sound waves in the air, but they are drawn as transverse waves in diagrams in order to show how the amplitude of vibration varies with position in the pipe.

Open pipes

It is also possible to produce standing waves in pipes that are open at both ends. The air at both ends of an open pipe is able to vibrate freely and therefore antinodes – positions of maximum vibration (amplitude) – occur at the ends. Figure 15.4 shows the first three modes of vibration.

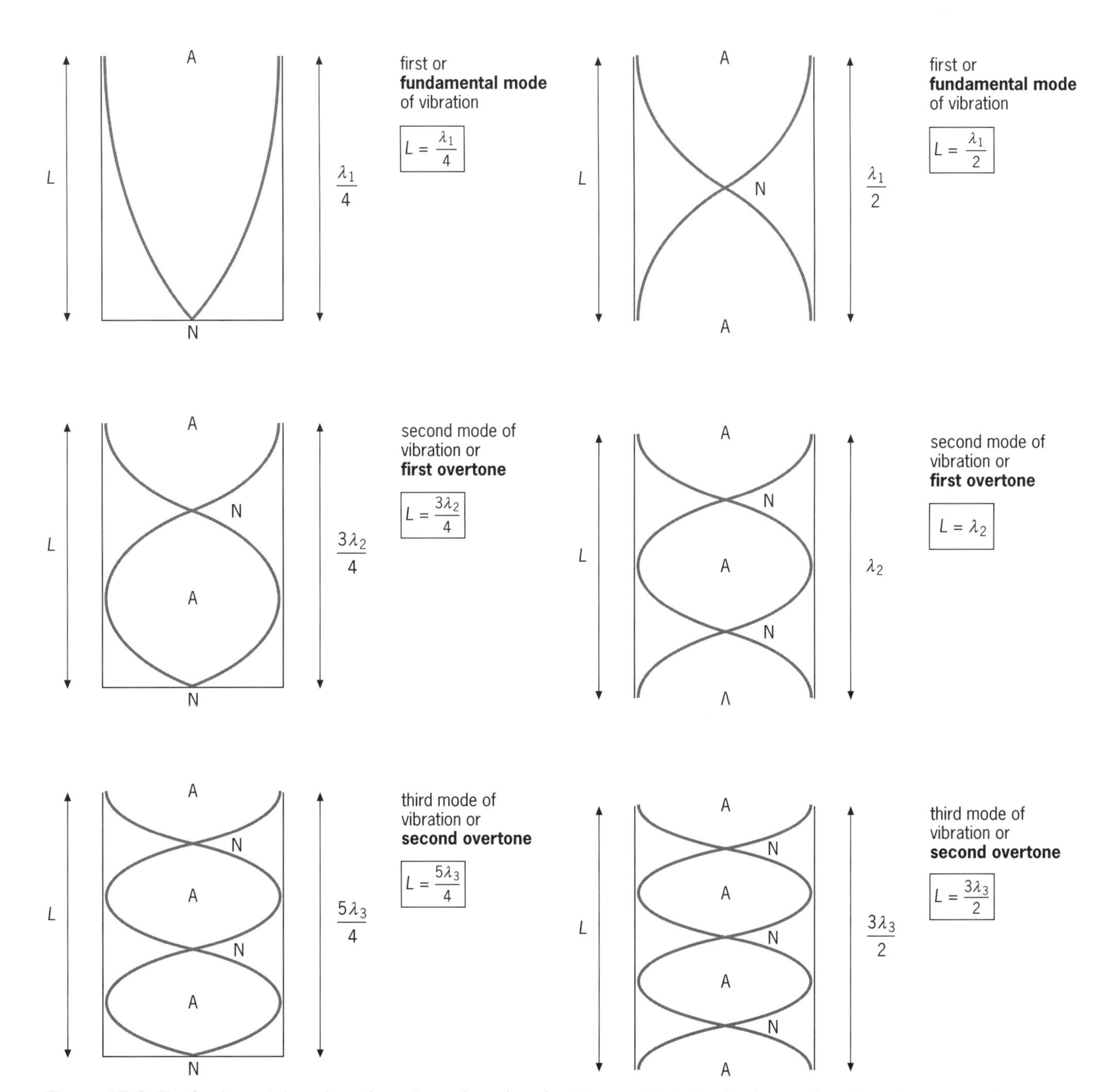

Figure 15.3 The fundamental mode and overtones in a closed pipe

Figure 15.4 The fundamental mode and overtones in an open pipe

Determining the velocity of sound using standing waves in a resonance tube

Procedure

A vibrating tuning fork, or a small loudspeaker connected to a signal generator, is held above the apparatus, Figure 15.5. The tube is raised, varying the length of the air column, until the air column **resonates**, that is, the note is heard much louder. The length of the pipe, L_1, is noted.

The tube is gradually raised further until the next resonant position is found and the new length, L_2, is noted.

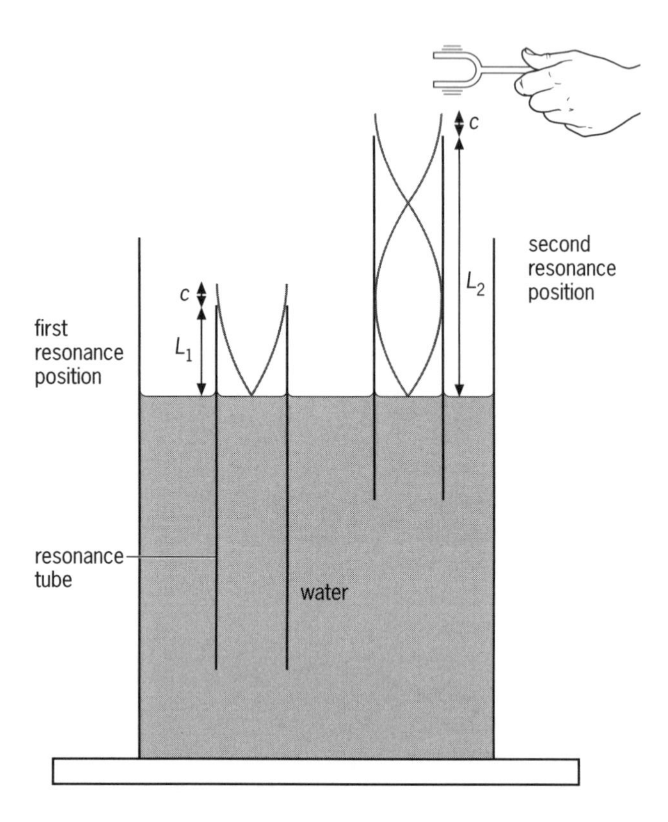

Figure 15.5 Experimental setup

Theory

Resonance is heard when a standing sound wave is created by the note.

At the first resonant position

$$\frac{\lambda}{4} = L_1 + c$$

where c is the small amount the standing wave protrudes from the top of the pipe, called the 'end correction'.

At the second resonant position

$$\frac{3\lambda}{4} = L_2 + c$$

Subtracting these two equations,

$$\frac{\lambda}{2} = L_2 - L_1$$

If the frequency of the tuning fork or signal generator is known then, using the equation $v = f\lambda$, the velocity of sound is given by

$$v = 2f(L_2 - L_1)$$

Determining the velocity of sound using standing waves in free air

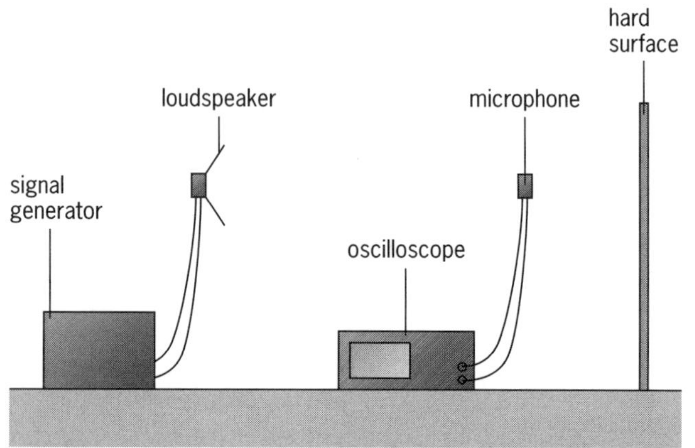

Figure 15.6 Experimental setup

The apparatus is set up as shown in Figure 15.6, with the time base of the cathode ray oscilloscope (CRO) turned off. Waves reflected from the hard surface combine with waves from the loudspeaker to produce a standing wave. The microphone is moved along the line between the loudspeaker and the hard surface and the height of the vertical trace on the CRO is seen to change. The position of a minimum on the oscilloscope is found. The microphone is again moved until the position of the next minimum is found. The distance between two successive minima is $\lambda/2$. If the frequency of the sound waves is known the velocity can be found using the equation $v = f\lambda$.

Level 1

1 Draw diagrams to show three different, simple standing waves on a stretched string plucked in the centre. Write an expression for the wavelength in each diagram.

2 The fundamental standing wave for sound is set up in a tube 0.60 m long, open at both ends.
a Draw a diagram showing the position of the nodes and antinodes.
b Draw a diagram to show how the amplitude of vibration of the air varies along the length of the tube.
c What would be the frequency of this note? (Take the velocity of sound as $330\,\mathrm{m\,s^{-1}}$ and assume, for simplicity, that the nodes or antinodes are exactly at the ends of the tube.)

3 You are one of a line of people in a pool with a wave machine. The line is in the direction that the waves travel. What difference would you notice between a region with progressive waves and a region where stationary waves are set up? (Describe your motion and that of your neighbours in each case.)

4 a How is a stationary wave formed? Illustrate your answer with reference to water waves hitting a barrier at 90°, as in Figure 15.7.

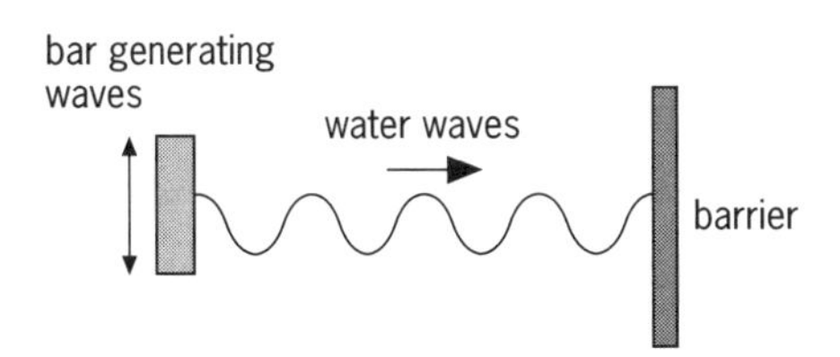

Figure 15.7

b How far apart are the nodes and antinodes in a stationary wave?
c A microwave transmitter faces an aluminium screen and a detector picks up strong signals every 1.6 cm between the transmitter and the screen, Figure 15.8. What is the wavelength of the radiation?

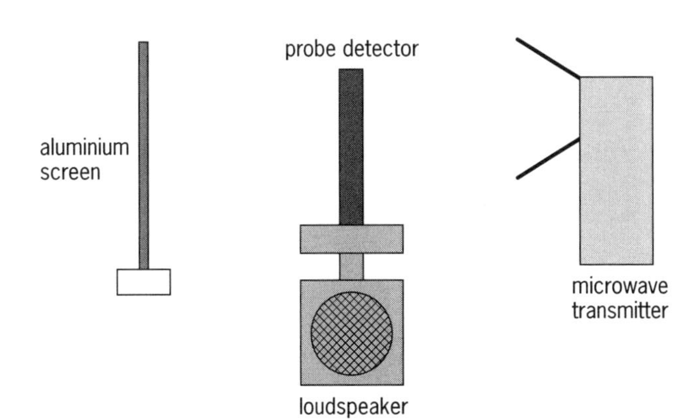

Figure 15.8

Level 2

5 Children in the playground tied a rope to the railings and waggled the end. The rope was 2.0 m long and made a pattern of three and a half loops. When a pulse was timed it took 5.0 s to travel to the railings and back. With what frequency were they waggling the end of the rope?

6 You are asked to demonstrate the standing wave inside a 1.0 m long glass tube. There is a loudspeaker connected to a signal generator at one end of the tube and a cork in the other end. Along the bottom of the tube there is a thin layer of powder, which will be thrown away from the antinodes but will lie in piles at the nodes, so that the wavelength of the standing wave may be seen. Figure 15.9 shows the setup.

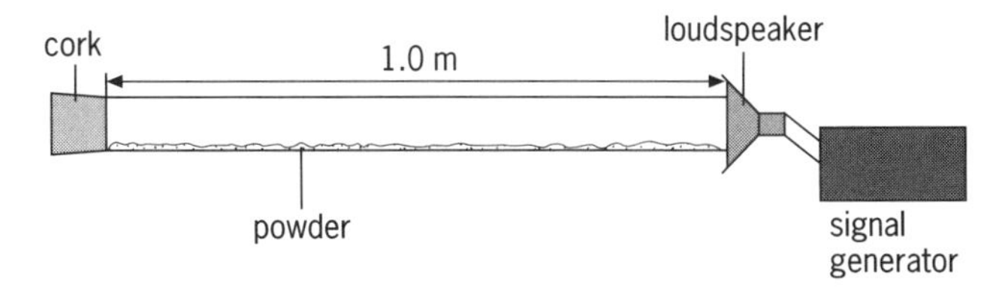

Figure 15.9

What frequency would you choose on the signal generator to make the nodes and antinodes clearly visible? (Take the velocity of sound in air as $330\,\mathrm{m\,s^{-1}}$.)

Level 3

7

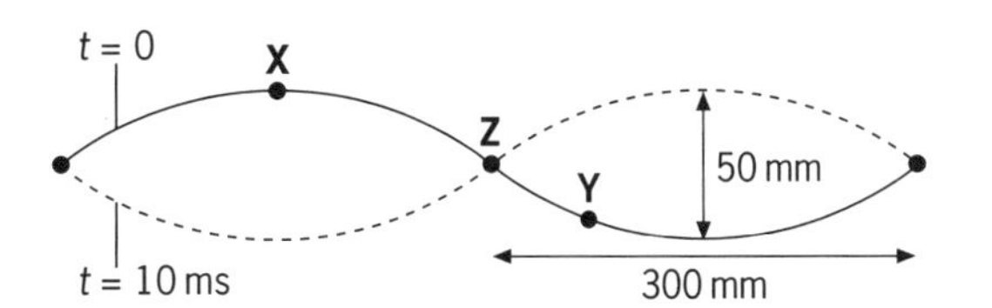

Figure 15.10

Figure 15.10 shows part of a standing wave on a string at times $t = 0$ and $t = 10\,\mathrm{ms}$, during which time the string has moved from the first position to the second.
a What is the time period of the oscillation?
b Sketch the appearance of the string at 5 ms.
c Describe the motion of points X and Y in the diagram, including their frequency and amplitude.
d What is the phase difference between points X and Y?
e Why is point Z stationary? (Give an explanation rather than a simple statement.)
f Calculate the speed of transverse waves on this string.

8 You have a loudspeaker connected to a signal generator. The loudspeaker is facing a large wall and is 4.0 m away from it. You want to demonstrate that a stationary sound wave is formed when the sound is reflected. The class will move around in the space between the loudspeaker and the wall. Estimate a suitable frequency to choose on the signal generator to give a good separation of loud and quiet regions.

9

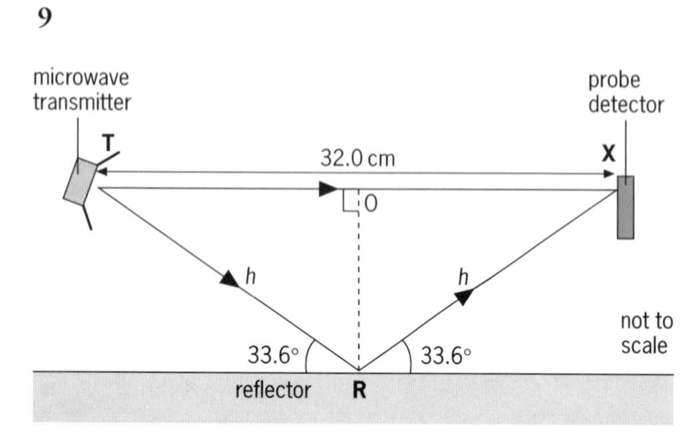

Figure 15.11

A microwave transmitter is set up as shown in Figure 15.11. The wavelength of the microwave radiation is 3.2 cm.

a What will be detected at point X if the phase of the wave is not altered by reflection?

b What will be detected at point X if there is a phase change of 180° on reflection?

10 When you measure the velocity of sound using a resonance tube you should do measurements at two different lengths or at two different frequencies. Why?

11 Two navigation beacons, 60 km apart, transmit radio waves of wavelength 4000 m. The signals start with the same phase and amplitude. A ship is midway between them and is sailing straight towards one of them. Describe what happens to the signals received on board the ship over the next 6 km.

Level 1

1 There must be an antinode at the centre of the string, because that is where the vibrations are created, and there must be nodes at each end, because the string is fixed there. If the length of the string is l, the first three modes and their wavelengths are as shown in Figure 15.12.

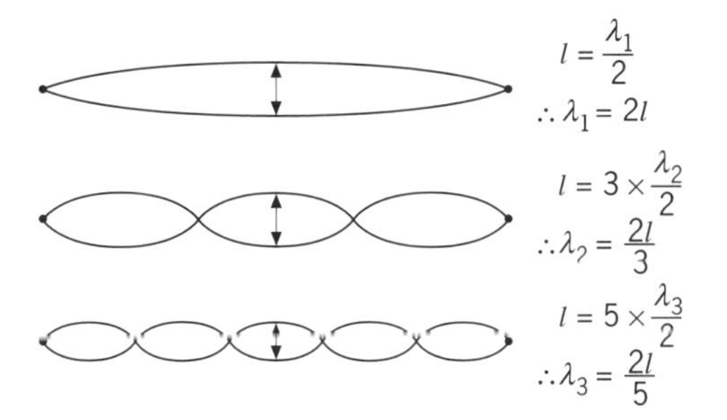

Figure 15.12

2 a The disturbance must be created at one end of the tube, for instance by blowing across the top so that at this end there is an antinode. The sound is reflected at the opposite open end, where there must be another antinode because the air is free to vibrate. All the nodes must be within the tube. The simplest case is shown in Figure 15.13.

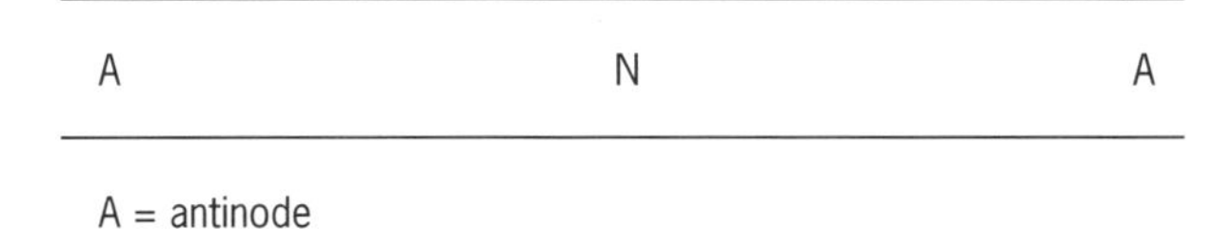

A = antinode
N = node

Figure 15.13

b There is zero displacement at a node and maximum displacement at an antinode.

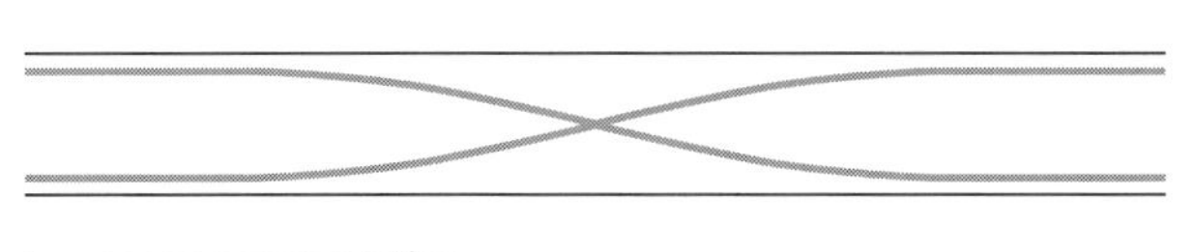

transverse representation of amplitude

Figure 15.14

c The length of the tube is equal to half the length of the standing wave, so

$$\frac{\lambda}{2} = 0.60\,\text{m} \quad \text{and} \quad \lambda = 1.2\,\text{m}$$

You can rearrange the equation $v = f\lambda$ to give

$$f = \frac{v}{\lambda}$$

$$= \frac{330\,\text{m}\,\text{s}^{-1}}{1.2\,\text{m}} = \mathbf{275\,Hz}$$

3 In a region of **progressive** waves, the people nearer the wave generator receive the disturbance first. A neighbour between you and the generator will go up before you go up and then down before you go down, Figure 15.15. People who are further from the generator will go up and down after you. The **extent of movement** will be the **same** for everyone. They will have different phases, but the same amplitude.

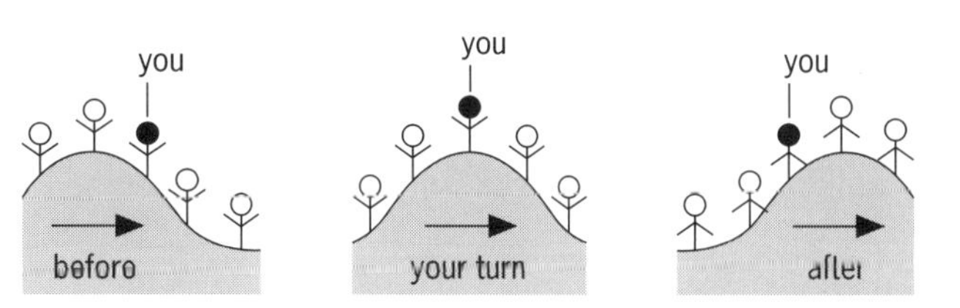

Figure 15.15 A progressive wave

In a region of **stationary** waves some people may hardly move at all (at a node) but others will have a large up and down motion (at an antinode). People on one side of a node will all **go up together**, but to **different heights**, Figure 15.16. At the same time, on the other side of the node people will be going down together but by different amounts. These movements will be in phase, over each half wavelength, but with varying amplitude.

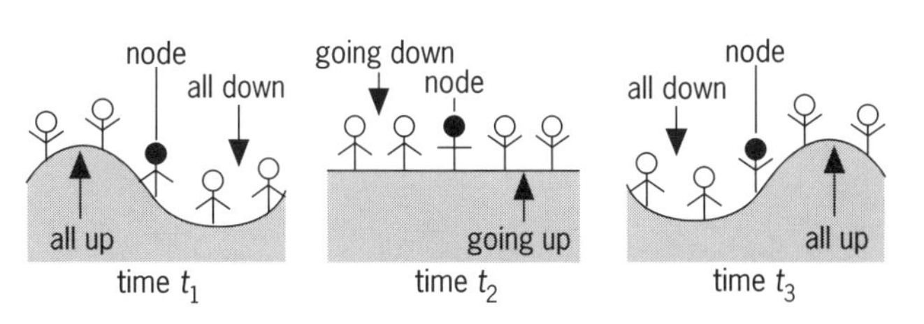

Figure 15.16 A stationary wave

4 a One way to set up a stationary wave is to combine two waves travelling in opposite directions with equal frequency and roughly equal amplitude. An easy way to achieve this is by **reflection**. The incident and reflected waves superpose (add up) and, as they move, at some positions the disturbances always cancel (nodes). At points between the nodes the disturbances combine to give vibrations which may be up to twice the amplitude of the individual waves (antinodes).

b The nodes are always $\lambda/2$ apart and the antinodes are also $\lambda/2$ apart; see Figure 15.17.

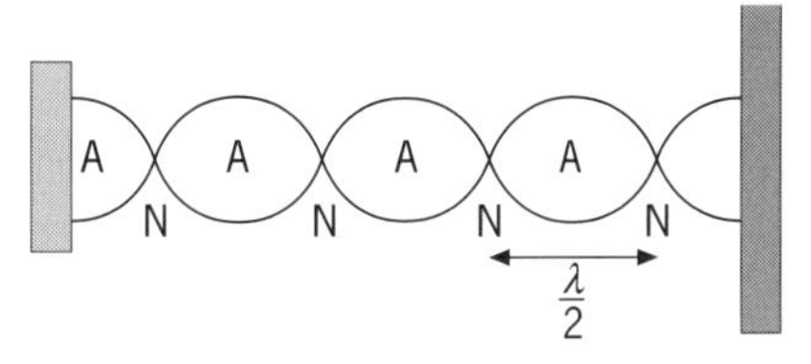

Figure 15.17 The resultant stationary wave

c The wavelength is always double the separation of the nodes or antinodes. In this case the antinodes are 1.6 cm apart, so the wavelength is equal to $2 \times 1.6\,\text{cm} = \mathbf{3.2\,cm}$.

Level 2

5 The situation is shown in Figure 15.18.

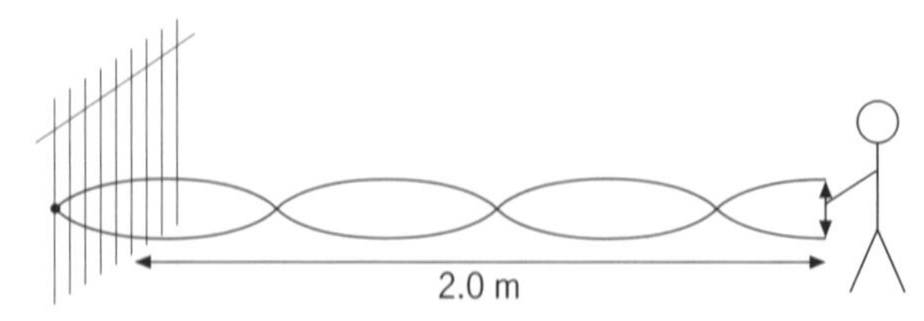

Figure 15.18

Each 'loop' has a length of $\lambda/2$, so $3\frac{1}{2}$ loops have a length of $3.5 \times \lambda/2$. The rope is 2.0 m long, so

$$3.5 \times \frac{\lambda}{2} = 7 \times \frac{\lambda}{4} = 2.0\,\text{m}$$

(Or, just by looking at the rope for whole wavelengths, you have $1\frac{3}{4}\lambda = \frac{7\lambda}{4} = 2.0\,\text{m}$.*)*

$$\therefore \lambda = 2.0\,\text{m} \times \frac{4}{7} = 1.14\,\text{m}$$

To calculate the frequency of the oscillations ('waggling'), you need to use the equation

$$f = \frac{v}{\lambda}$$

You can calculate the speed at which waves travel along the rope from the data about the timing of a pulse:

$$\text{speed, } v = \frac{\text{distance pulse travels along the rope and back}}{\text{time}}$$

$$= \frac{2 \times 2.0\,\text{m}}{5.0\,\text{s}} = 0.80\,\text{m s}^{-1}$$

$$\therefore f = \frac{v}{\lambda} = \frac{0.80\,\text{m s}^{-1}}{1.14\,\text{m}} = \mathbf{0.70\,Hz}$$

6 You should aim to produce several nodes and antinodes, so you need to choose something like the fourth overtone, as shown in Figure 15.19. The speaker causes the disturbance so there will be an antinode at that end. The cork prevents the air moving so there will always be a node there.

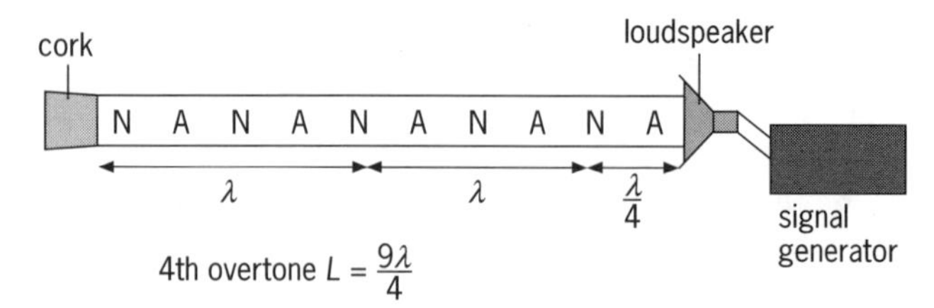

Figure 15.19

The tube is 1.0 m long and has nine node-to-antinode wave segments. Each node-to-antinode has a length of $\lambda/4$, so $9\lambda/4 = 1.0\,\text{m}$.

$$\therefore \lambda = \frac{4}{9} \times 1.0\,\text{m} = 0.44\,\text{m}$$

and

$$f = \frac{v}{\lambda} = \frac{330\,\text{m s}^{-1}}{0.44\,\text{m}} = \mathbf{742\,Hz}$$

You could choose other values, such as **412 Hz** (second overtone), **579 Hz** (third overtone), or **907 Hz** (fifth overtone).

Level 3

7 a It will take another 10 ms for the string to get back to its initial position, so the time period,

$T = 2 \times 10\,\text{ms}$

$= \mathbf{20\,ms}$.

b 5 ms is $\frac{1}{4}$ of the time period; the string will be half way between its positions at $t = 0$ and $t = 10\,\text{ms}$.

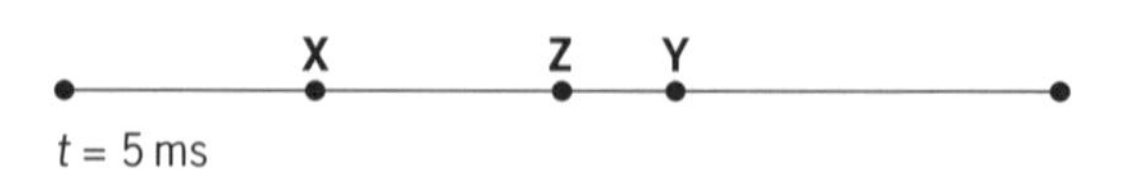

Figure 15.20

c **At point X** the frequency is

$$f = \frac{1}{T} = \frac{1}{20 \times 10^{-3}\,\text{s}}$$

$$= \mathbf{50\,Hz}$$

The amplitude is equal to *half* the displacement from peak to trough, that is

$$\frac{50\,\text{mm}}{2} = \mathbf{25\,mm}$$

At point Y the frequency is also **50 Hz**, but the amplitude is **less than 25 mm**.

d Y is 180° out of phase with X; when X is going up Y is going down and vice versa.

With a standing wave points can only be in phase or 180° out of phase.

e Z is stationary because it is at a point where the two progressive waves that create the standing wave always cancel. Point Z is called a **node**.

Note it is not an ***explanation*** *simply to say it is a node.*

f The wavelength of this standing wave,

$\lambda = 600\,\text{mm} = 0.60\,\text{m}$.

$v = f\lambda = 50\,\text{s}^{-1} \times 0.60\,\text{m}$

$= \mathbf{30\,m\,s^{-1}}$

8 When walking towards the wall you want the sound to go from loud (an antinode) to quiet (a node) fairly rapidly, say every 0.2 m. The distance between a node and the next antinode is $\lambda/4$. This makes λ about 0.8 m, so the required frequency is

$$f = \frac{v}{\lambda} = \frac{330\,\text{m s}^{-1}}{0.8\,\text{m}}$$

$$\approx \mathbf{400\,Hz}$$

You could perhaps use a node-to-antinode distance, $\lambda/4$, as low as 0.05 m, giving a wavelength of 0.2 m and frequency of 1.6 kHz; or as high as 1 m, giving a wavelength of 4 m and frequency of 80 Hz.

9 a The distance travelled by the reflected beam will be twice the length of the hypotenuse, h, of the triangle $\triangle$TRO in Figure 15.11.

$$\frac{h}{16.0\,\text{cm}} = \cos 33.6°$$

$$\therefore \quad h = \frac{16.0\,\text{cm}}{\cos 33.6°} = \frac{16.0\,\text{cm}}{0.833} = 19.2\,\text{cm}$$

So, the reflected path length $= 2 \times 19.2\,\text{cm} = 38.4\,\text{cm}$.

The microwave wavelength $= 3.2\,\text{cm}$. So the reflected path length is equivalent to

$$\frac{38.4\,\text{cm}}{3.2\,\text{cm}} = 12 \times \text{wavelength}$$

and the direct path length is equivalent to

$$\frac{32.0\,\text{cm}}{3.2\,\text{cm}} = 10 \times \text{wavelength}$$

With no phase changes upon reflection the two waves will arrive in phase, and there will be a maximum at X.

b If there is a phase change of 180° on reflection, the waves will arrive at point X out of phase and destructive interference will occur, so the resultant signal will have an amplitude of zero. (At other positions between T and X, the direct and the phase-shifted reflected waves will arrive in phase, and at these positions a maximum would be detected.)

10 This method usually involves measuring the length of a tube that resonates with a signal of a known frequency, such as a tuning fork. The shortest resonant length occurs at $\lambda/4$. You can then use the equation $v = f\lambda$ to calculate the speed of sound. However, your measurement of the resonant length will involve an error because the antinode at the end of an open tube is never exactly at the end of the tube, but at a small distance outside the tube. You must find, or eliminate from your calculations, the value of the 'end correction'.

The procedure described on page 108 uses two different lengths of tube and eliminates the end correction; it therefore gives a relationship for the speed of sound that does not include the end correction.

Alternatively, if you take measurements at two different resonant frequencies, and assume that the end correction is constant, you can calculate its value. For instance, suppose you make measurements at frequencies f_1 and f_2 and measure the shortest resonant lengths to be l_1 and l_2. If the end correction is c then the corresponding wavelengths are equal to $4 \times (l_1 + c)$ and $4 \times (l_2 + c)$, respectively. Using the equation $v = f\lambda$ for each measurement, you will see that

$$v = 4 \times (l_1 + c) \times f_1 = 4 \times (l_2 + c) \times f_2$$

You should be able to rearrange the two right hand elements of this equation to give

$$c = \frac{(l_1 \times f_1) - (l_2 \times f_2)}{f_2 - f_1}$$

You can calculate this value from your recorded results and add it to the measured lengths so that you can calculate an accurate value for the speed of sound.

11 There is a maximum at the midway point because the signals from each beacon have travelled the same distance and are in phase. When the boat has travelled 2 km further, the waves are in phase again, because one signal has travelled $28\,\text{km} = 7 \times \lambda$ and the other has travelled $32\,\text{km} = 8 \times \lambda$. There is a maximum every 2 km and a minimum between each maximum. For instance, where one has travelled 29 km ($7\frac{1}{4}\lambda$) and the other has travelled 31 km ($7\frac{3}{4}\lambda$), they will be out of phase.

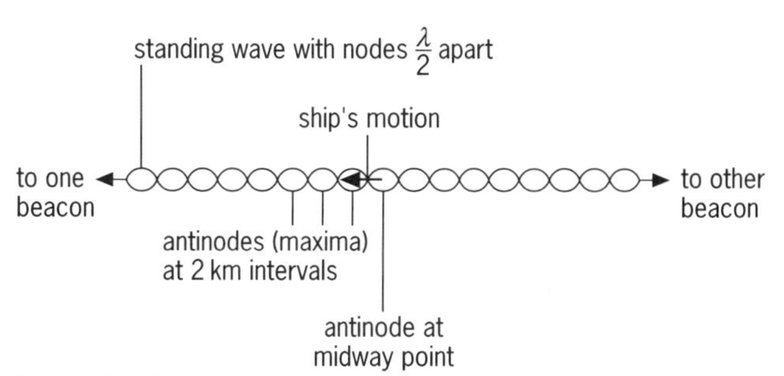

Figure 15.21

The resulting signal is a standing wave with nodes and antinodes at intervals of half a wavelength.

16 Photoelectric effect and wave–particle duality

Electromagnetic radiation incident on the surface of a metal can cause electrons to be emitted. This phenomenon is called **photoelectric emission**.

Although in physics we usually study the effect of electromagnetic radiation on atoms, biologists will know that the release of electrons from organic molecules by incoming radiation is the basis of all life.

Results of photoelectric experiments

Experiments show that a particular metal will only emit electrons from its atoms if the frequency of the light is above a certain value, called the **threshold frequency**, f_0, for that element. If the frequency is below this value electrons will not be emitted, however high the intensity (brightness) of the light (Figure 16.1). If the frequency is equal to or above this value electrons will be emitted, no matter how low the intensity of the incident light.

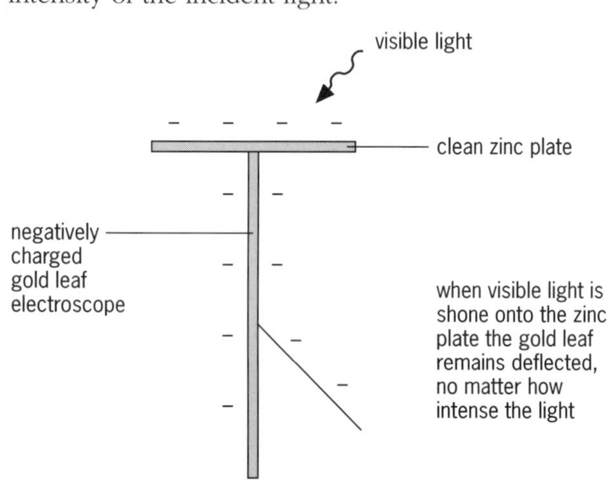

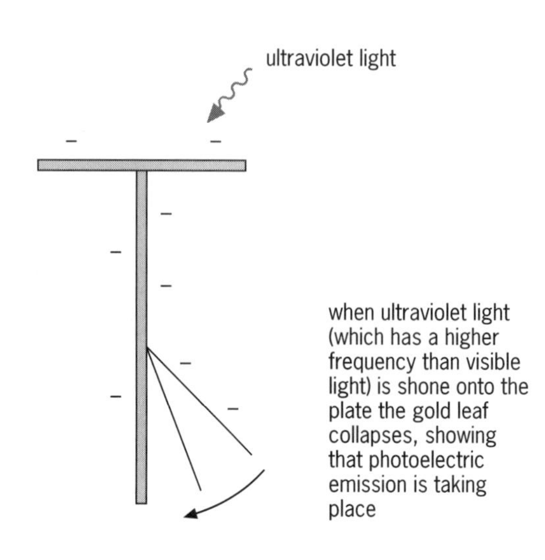

Figure 16.1 Photoelectric experiments with a charged gold leaf electroscope

To explain these discoveries a new theory of the nature of light was proposed by Einstein. He suggested that light should be considered not as a wave but as a stream of particles called **photons**. These photons

- move at the speed of light;
- carry a **quantum** or packet of energy equal to hf, where f is the frequency of the light and h is a constant called the Planck constant and has a value of 6.6×10^{-34} J s;
- interact individually with one electron.

When a photon collides with an electron it gives it energy equal to hf.

Before an electron can be liberated it must receive sufficient energy to overcome the attractive forces of the metal. The minimum amount of energy needed to do this is called the **work function** (ϕ). Different metals have different work functions, as shown in Table 16.1.

Table 16.1 Values of the work function for some metals

Metal atom	**Work function (ϕ) /** electronvolts (eV)
barium	2.49
calcium	2.70
tungsten	4.49
thorium	3.50

1 eV = 1.6×10^{-19} J

If the energy of a single photon of incident radiation (hf) is less than ϕ the electron cannot escape and no emission takes place. At the threshold frequency (f_0) the energy of the photon (hf_0) is equal to the work function of the metal. An electron will be emitted but it will have no kinetic energy once it has escaped the surface of the metal.

$$\phi = hf_0$$

At frequencies greater than the threshold frequency the energy of the photon is greater than the work function of the metal and the electron will escape with excess energy, seen as the kinetic energy of the electron after it has escaped the surface of the metal. If an electron is emitted without losing any energy in interactions with surrounding atoms, its kinetic energy will be a maximum for that frequency of radiation and that metal (Figure 16.2):

$$hf = \phi + \tfrac{1}{2}mv^2_{\text{max}}$$

This relationship is shown graphically in Figure 16.3.

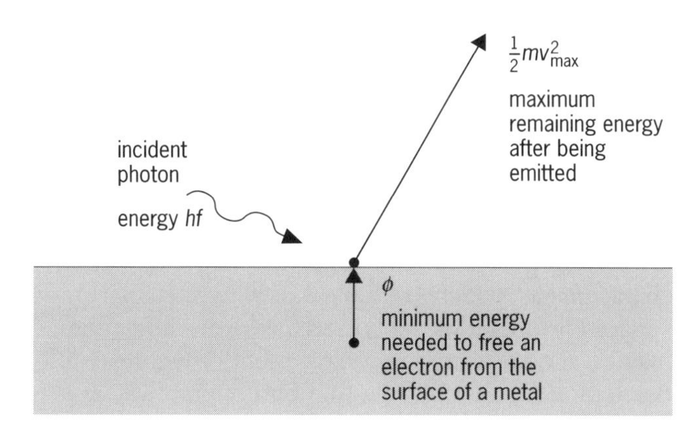

Figure 16.2 Energy transfer involved in photoelectric emission

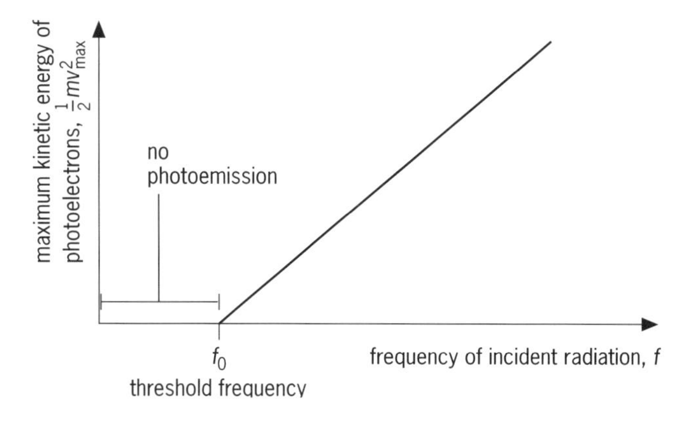

Figure 16.3 Relationship between maximum emission energy and frequency of light

---

Example 1

When light of frequency 7.5×10^{14} Hz is incident on the surface of a metal, photoelectrons with a maximum kinetic energy of 1.6×10^{-19} J are emitted. Calculate the threshold frequency of this metal. ($h = 6.6 \times 10^{-34}$ J s)

$$hf = \phi + \tfrac{1}{2}mv^2_{max}$$

but $$\phi = hf_0$$

$\therefore$ $$hf = hf_0 + \tfrac{1}{2}mv^2_{max}$$

Rearranging,

$$hf_0 = hf - \tfrac{1}{2}mv^2_{max}$$

$$f_0 = \frac{(hf - \frac{1}{2}mv^2_{max})}{h}$$

$$= \frac{(6.6 \times 10^{-34}\,\text{J s} \times 7.5 \times 10^{14}\,\text{s}^{-1} - 1.6 \times 10^{-19}\,\text{J})}{6.6 \times 10^{-34}\,\text{J s}}$$

$$= \frac{(4.95 - 1.6) \times 10^{-19}\,\text{J}}{6.6 \times 10^{-34}\,\text{J s}}$$

$$= 5.1 \times 10^{14}\,\text{Hz}$$

The *intensity* of a beam of light is determined by the number of photons it contains and not by the energy carried by each photon. The intensity of a beam of light therefore controls the *number* of electrons that are emitted, but not whether they will be emitted.

Measuring the energies of photoelectrons

Using the circuit shown in Figure 16.4 it is possible to investigate the energies of photoelectrons. A photocell consists of two electrodes sealed in an evacuated tube. One of the electrodes, the cathode, is usually curved and is made from a photoemissive metal. Radiation incident on this electrode with a sufficiently high frequency will cause photoelectrons to be emitted. Some of the emitted electrons will travel across the vacuum to the anode; the incident radiation has caused current to flow across the evacuated tube. If the incident radiation is interrupted the flow of current ceases.

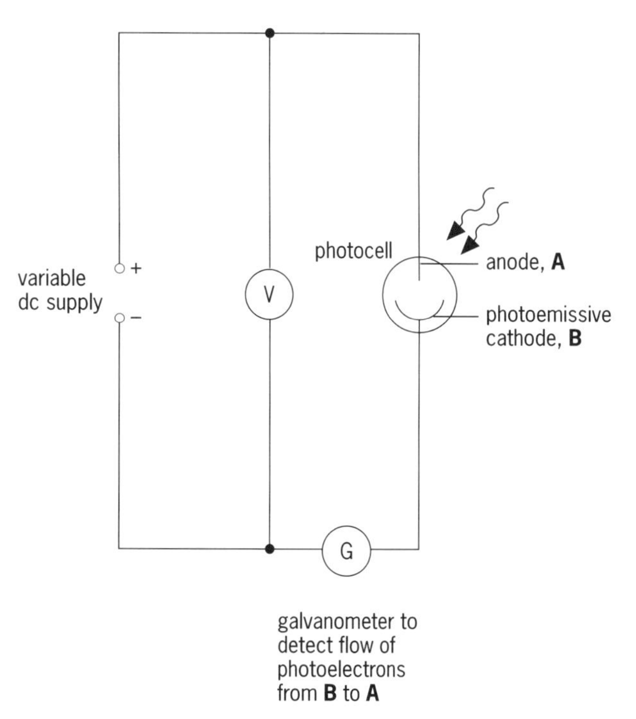

Figure 16.4 A photocell circuit for determining stopping potential

In this investigation, the 'anode' A is made *negative* with respect to the 'cathode' B, so photoelectrons travelling from B to A will have to do work against the opposing potential V. The potential difference is gradually increased, that is A is made more negative, until no photoelectrons have sufficient energy to cross the gap between B and A and the galvanometer reads zero. This potential is called the **stopping potential**, V_s. The maximum energy of the emitted electrons is then equal to the work done against the opposing pd:

$$\tfrac{1}{2}mv^2_{max} = eV_s$$

The photoelectric equation can then be written as

$$hf = \phi + eV_s$$

or

$$hf = hf_0 + eV_s$$

This is shown graphically in Figure 16.5.

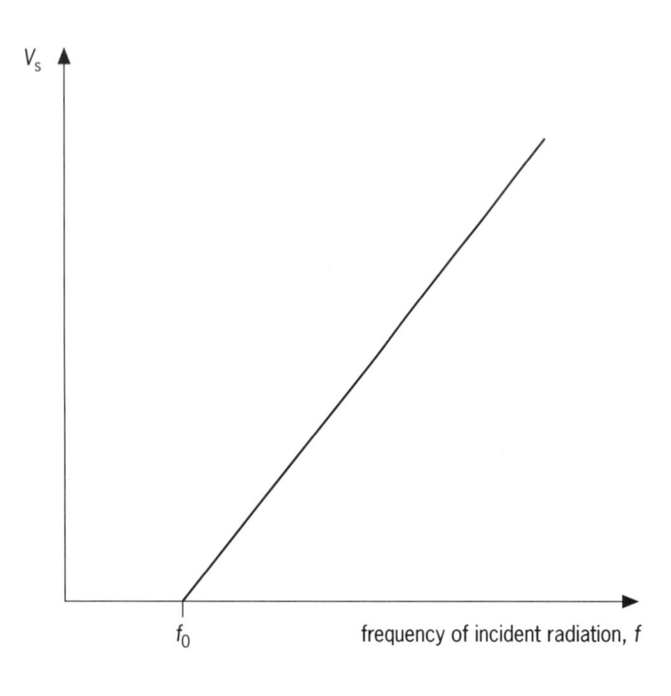

Figure 16.5 Relationship between stopping potential and frequency of light

Example 2

Calculate the stopping potential for photoelectrons which are emitted when incident light of frequency 1.0×10^{15} Hz strikes the surface of a metal with a work function of 2.5 eV. ($h = 6.6 \times 10^{-34}$ J s, $e = -1.6 \times 10^{-19}$ C)

$$hf = \phi + eV_s$$

$$\phi = 2.5\,\text{eV or } 2.5 \times 1.6 \times 10^{-19}\,\text{J}$$

$$V_s = \frac{(hf - \phi)}{e}$$

$$= \frac{(6.6 \times 10^{-34}\,\text{J s} \times 1.0 \times 10^{15}\,\text{s}^{-1} - 2.5 \times 1.6 \times 10^{-19}\,\text{J})}{1.6 \times 10^{-19}\,\text{C}}$$

$$= \frac{(6.6 \times 10^{-19} - 4.0 \times 10^{-19}\,\text{J})}{1.6 \times 10^{-19}\,\text{C}}$$

$$= 1.6\,\text{V}$$

Wave–particle duality

In 1924 a French physicist called Louis de Broglie suggested that because light has a dual nature, that is it sometimes behaves as a wave and sometimes as a particle (or photon, as in the photoelectric effect), perhaps objects which have previously been regarded as consisting of particles might show some characteristics which are typical of waves. His suggestion proved to be correct. Electron diffraction may be shown using the arrangement of Figure 16.6. A beam of electrons strikes a thin film of graphite on a metal grid just beyond a hole in the anode. The electrons are clearly exhibiting a property normally associated with waves.

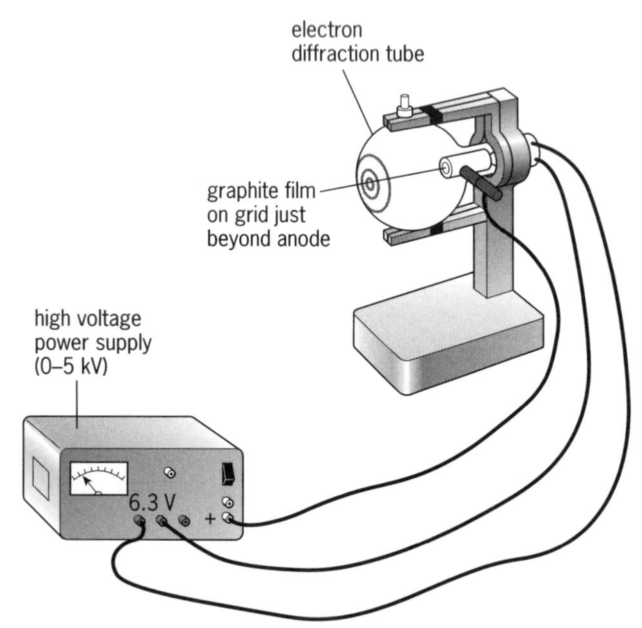

Figure 16.6 Electron diffraction

De Broglie suggested that p, the momentum, and λ, the wavelength, of a particle are related by the equation

$$p = \frac{h}{\lambda}$$

where h is the Planck constant (6.6×10^{-34} J s).

Example 3

Calculate the wavelength of an electron moving at $1.0 \times 10^{8}\,\text{m s}^{-1}$. (mass of an electron $= 9.1 \times 10^{-31}$ kg)

$$\text{momentum, } p = mv = \frac{h}{\lambda}$$

$$\therefore \quad \lambda = \frac{h}{mv}$$

$$= \frac{6.6 \times 10^{-34}\,\text{J s}}{9.1 \times 10^{-31}\,\text{kg} \times 1.0 \times 10^{8}\,\text{m s}^{-1}}$$

$$= 7.3 \times 10^{-12}\,\text{m}$$

Level 1

1 Light of frequency 7.0×10^{14} Hz shines on a metal plate. The work function of the metal, ϕ, which is the energy needed to free an electron from the surface of the metal, is 4.0×10^{-19} J.

a How much energy, in joules, does one photon of light of this frequency have?

b Will this light eject any electrons?

c What maximum kinetic energy would the electrons have?

2 Explain what is meant by the terms

a stopping potential,

b threshold frequency,

in connection with the photoelectric effect.

3 An experiment with a photocell is set up as in Figure 16.7.

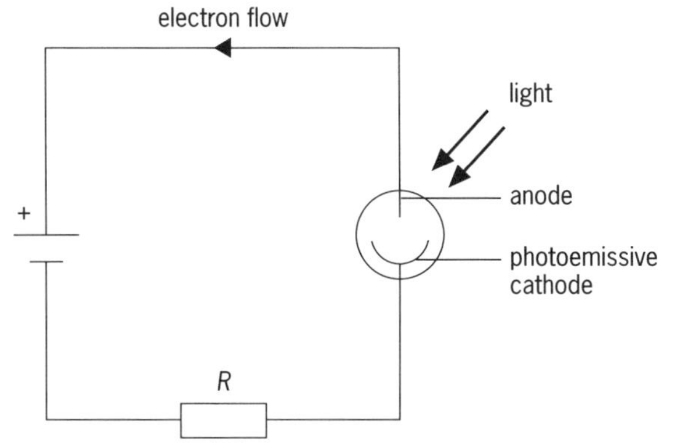

Figure 16.7

a What instrument would you use to 'count' the electrons ejected per second?

b What instrument would you use to measure the kinetic energy of the electrons?

4 a Explain to a non-scientist friend what you mean by a photon.

b In a beam of photons, which factor determines the intensity and which the energy?

Level 2

5 Ultraviolet light of wavelength 300 nm is shining onto a caesium target that has a work function of 3.2×10^{-19} J. Calculate:

a the frequency of the light

b the energy of one photon (in joules)

c the maximum kinetic energy of an ejected electron in joules

d the maximum kinetic energy in eV

e the stopping potential.

6 Electrons are ejected by light shining on potassium. Explain what happens to

a the number of electrons

b the energy of the electrons,

when

i the frequency of the light is raised (for example from red to blue)

ii the intensity ($W m^{-2}$) of the light is doubled.

7 A freshly cleaned zinc plate is connected to a coulombmeter and charged negatively using a polythene rod. The meter shows that there is a charge of -46 nC on the plate. An ultraviolet lamp is then arranged to shine on the zinc (taking care that no one can look into the lamp). The reading on the coulombmeter gradually decreases to zero. Explain this.

8 The wave behaviour of the electron is the basis of the electron microscope, which can give us information on a smaller scale than can a visible light microscope.

In an electron microscope an electron is accelerated through 5.0 kV.

a Calculate the kinetic energy gained by the electron in joules.

b The electron was originally stationary. Calculate its final speed.

c Calculate the electron's final momentum.

d What wavelength is associated with this momentum?

e Is this wavelength longer or shorter than that of 'visible light'?

Level 3

9 An electron in a hydrogen atom has a ground-state energy of -13.6 eV. An ultraviolet photon with energy 20.0 eV is absorbed and ejects the electron. What is the maximum possible velocity of the ejected electron? (mass of an electron $m_e = 9.1 \times 10^{-31}$ kg)

10 Explain what is meant by 'threshold wavelength' in the photoelectric effect.

11 A photographic exposure meter may have a photocell with the cathode coated with a material which emits electrons when visible light shines on it.

a When light of wavelength 550 nm shines on it, electrons are emitted with negligible kinetic energy.

i What is the threshold frequency of this material?

ii What is the work function, ϕ, in joules?

b The wavelength of the light is changed to 400 nm, at the same intensity.

i What is the maximum possible kinetic energy of an ejected electron?

ii How does this change of frequency affect the number of electrons emitted each second?

c The intensity is then doubled. What is the effect on

i the kinetic energy of the ejected electrons

ii the number of electrons emitted per second?

12 In a photoelectric experiment, the stopping potential was measured for incident light of different frequencies by using coloured filters, Figure 16.8. The following results were obtained.

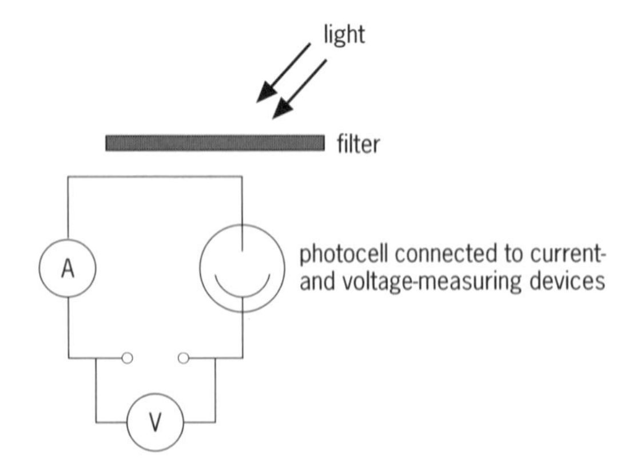

Figure 16.8

Potential/V	2.0	1.6	1.2	0.78	0.36
Frequency/$\times 10^{14}$ Hz	10.0	9.0	8.0	7.0	6.0

Use these figures to find a value for the Planck constant.

13 The momentum of elementary particles is often given in units of GeV/c, where the prefix G means $\times 10^9$, eV is the electronvolt, and c is the velocity of light.

a Show that the base units corresponding to GeV/c are $\mathrm{kg\,m\,s^{-1}}$.

b What similar unit could be used for mass?

Level 1

1 a The energy of a photon is given by $E = hf$, where h is the Planck constant $= 6.6 \times 10^{-34}$ J s. So, for light of this frequency,

$$E = 6.6 \times 10^{-34}\,\text{J s} \times 7.0 \times 10^{14}\,\text{s}^{-1} = \mathbf{4.6 \times 10^{-19}\,J}$$

b **Yes**. Electrons will be ejected because the photon energy is greater than the value of the work function (4.0×10^{-19} J).

c Subtract the work function, ϕ, from the total photon energy to get the maximum possible energy of the electron:

$$ke_{max} = hf - \phi = (4.6 - 4.0) \times 10^{-19}\,\text{J} = \mathbf{0.6 \times 10^{-19}\,J}$$

2 a 'Stopping potential' is the potential difference needed to bring an electron of a certain kinetic energy to rest.

energy = charge × potential difference $= e \times V_s = \frac{1}{2}mv^2$

b 'Threshold frequency' is the lowest frequency of light that will release electrons from a particular substance. Light at any lower frequency does not have enough energy. At the threshold frequency the energy of a photon is equal to the work function so that when an electron is released there is no energy left over and the kinetic energy of the electron is zero. For light of any higher frequency there is some energy available and the electrons will have some kinetic energy.

3 a You use a sensitive **ammeter** to measure the 'current' of electrons flowing in the circuit. Electrons must be crossing the gap between the electrodes.

b You use a **voltmeter** to measure the energy. Put a potential difference across the gap to make the anode negative instead of positive, so that the potential difference opposes the flow of electrons. Steadily increase the potential difference. Monitor the potential difference as the current gradually decreases and record the potential difference at which no electrons are able to get across the gap – when the current falls to zero. Because the potential difference is equal to 'the energy per unit charge', this tells you how much energy the electrons had. Numerically, the kinetic energy in electronvolts is equal to the potential difference when the current is zero. This value of the potential difference is called the 'stopping potential'.

4 a A photon is like a package of energy. Electromagnetic radiation, such as light, microwaves or X-rays, may be thought of as a beam of photons (like a stream of bullets) travelling at the speed of light.

b The **brighter** (or greater the intensity of) the light the **more photons** there are in the beam. The **energy** of a photon increases directly with the **frequency** of the radiation. (The frequency of the light also determines its colour.) If you double the frequency you double the energy. X-ray photons are much more energetic and therefore much more harmful than photons of light or microwaves.

Level 2

5 a For electromagnetic radiation the frequency, f, and wavelength, λ, are linked by the equation $c = f \times \lambda$ where c is the velocity of light. Rearranging the equation,

$$f = \frac{c}{\lambda}$$

So, for ultraviolet light of wavelength 300 nm,

$$f = \frac{3.0 \times 10^{8}\,\text{m s}^{-1}}{300 \times 10^{-9}\,\text{m}} = \mathbf{1.0 \times 10^{15}\,Hz}$$

b The quantum of energy for light of this frequency is given by

$$\begin{aligned} E &= h \times f \\ &= 6.6 \times 10^{-34}\,\text{J s} \times 1.0 \times 10^{15}\,\text{Hz} \\ &= \mathbf{6.6 \times 10^{-19}\,J} \end{aligned}$$

c To calculate the maximum kinetic energy of an ejected electron, subtract the energy needed to release the electron (the work function) from the energy of the photon. The remainder is the maximum possible kinetic energy. So

$$ke_{max} = hf - \phi = (6.6 - 3.2) \times 10^{-19} = \mathbf{3.4 \times 10^{-19}\,J}$$

d 1 eV $= 1.6 \times 10^{-19}$ J, so the maximum kinetic energy in eV is

$$\frac{3.4 \times 10^{-19}\,\text{J}}{1.6 \times 10^{-19}\,\text{J eV}^{-1}} = \mathbf{2.1\,eV}$$

e The potential needed to stop a 2.1 eV electron is **2.1 V**.

6 a **i** Increasing the frequency does not affect the number of incoming photons so, provided the threshold frequency has been reached, the same number of electrons are ejected and the current stays the same.
ii When the intensity of the light is doubled, twice as many photons are incident on the potassium so twice as many electrons are freed; the current will be twice as big.

b **i** When the frequency of the light increases, the incident photon energy increases so the kinetic energy of the ejected electrons increases. The stopping potential is higher if the electrons are more energetic.
ii The energy of each incident photon is the same so the maximum kinetic energy of each electron is the same as before and the stopping potential does not change.

7 Light from the ultraviolet lamp releases electrons from the zinc; they are repelled by the overall negative charge and they leave. The overall negative charge decreases. (When the plate is slightly positively charged, the electrons may not have enough energy to overcome the electrostatic attraction.)

8 a The kinetic energy gained by one electron going through 5.0 kV is

$$E_k = qV = 5000 \text{ electronvolts} = 5.0\,\text{keV}$$

To change this to joules you must multiply by 1.6×10^{-19} joules per eV, giving

$$E_k = 5.0 \times 10^3\,\text{eV} \times 1.6 \times 10^{-19}\,\text{C} = \mathbf{8.0 \times 10^{-16}\,J}$$

b The kinetic energy of the electron is given by the equation $E_k = \frac{1}{2}m_e v^2$, where m_e is its mass and v is its velocity. You can rearrange this equation to give

$$v = \sqrt{\frac{2E_k}{m_e}}$$

$$= \sqrt{\frac{2 \times 8.0 \times 10^{-16}\,\text{J}}{9.1 \times 10^{-31}\,\text{kg}}} = \mathbf{4.2 \times 10^7\,m\,s^{-1}}$$

c Momentum = mass × velocity, or

$$p = m_e v = 9.1 \times 10^{-31}\,\text{kg} \times 4.2 \times 10^7\,\text{m}\,\text{s}^{-1}$$
$$= \mathbf{3.8 \times 10^{-23}\,N\,s}$$

(Note that $kg\,m\,s^{-1} = kg\,m\,s^{-2}\,s = N\,s$.)

d The wavelength associated with a particle by virtue of its motion is called the 'de Broglie' wavelength and is given by the equation $\lambda = h/p$ where h is the Planck constant and p is the momentum of the particle. In this case,

$$\lambda = \frac{6.6 \times 10^{-34}\,\text{J}\,\text{s}}{3.8 \times 10^{-23}\,\text{N}\,\text{s}} = \mathbf{1.7 \times 10^{-11}\,m}$$

e This is considerably **shorter** than the wavelength of 'visible light', which is around 600 nm (6×10^{-7} m). This means that an electron wave is more energetic and has a greater resolving power than a normal light wave.

Level 3

9 The surplus energy available to the electron after it has been released from the atom, E, is 20.0 eV – 13.6 eV = 6.4 eV.

$$1\,\text{eV} = 1.6 \times 10^{-19}\,\text{J}$$

$$\therefore E = 6.4\,\text{eV} \times 1.6 \times 10^{-19}\,\text{J}\,\text{eV}^{-1} = 1.02 \times 10^{-18}\,\text{J}$$

If *all* this energy is carried off as kinetic energy by the electron then

$$E = \tfrac{1}{2}m_e v_{max}^2$$

and so

$$v = \sqrt{\frac{2E}{m_e}} = \sqrt{\frac{2 \times 1.02 \times 10^{-18}}{9.1 \times 10^{-31}}} = \mathbf{1.5 \times 10^6\,m\,s^{-1}}$$

(This is the maximum possible velocity.)

10 The threshold *frequency* is the lowest frequency (minimum energy) of radiation that will cause an electron to be ejected/released from a material. Because wavelength varies inversely with frequency, the threshold wavelength is the **longest wavelength** at which emission will occur. Any wavelength longer than this does not have enough energy.

11 a If the ejected electrons have negligible kinetic energy then this wavelength is the threshold wavelength, λ_0

i The threshold frequency may be found from

$$f_0 = \frac{c}{\lambda_0} = \frac{3.0 \times 10^8\,\text{m}\,\text{s}^{-1}}{550 \times 10^{-9}\,\text{m}} = \mathbf{5.45 \times 10^{14}\,Hz}$$

ii The work function is the energy of a photon of the threshold frequency.

$$\phi = h \times f_0 = 6.6 \times 10^{-34}\,\text{J}\,\text{s} \times 5.4 \times 10^{14}\,\text{s}^{-1}$$
$$= \mathbf{3.60 \times 10^{-19}\,J}$$

b i The new frequency is given by

$$f_2 = \frac{c}{\lambda_2} = \frac{3.0 \times 10^8\,\text{m}\,\text{s}^{-1}}{400 \times 10^{-9}\,\text{m}} = 7.50 \times 10^{14}\,\text{Hz}$$

The surplus energy is given by

$$hf_2 - hf_0 = h \times (f_2 - f_0) = h \times \Delta f$$

where f_0 = the threshold frequency.

$$h \times \Delta f = 6.6 \times 10^{-34}\,\text{J}\,\text{s} \times 2.05 \times 10^{14}\,\text{s}^{-1}$$
$$= \mathbf{1.35 \times 10^{-19}\,J}$$

This will be the maximum possible kinetic energy of the electrons.

Alternatively, you could work out the new photon energy, $h \times f_2 = 4.95 \times 10^{-19}$ J, and find the difference between that and the work function, $(4.95 - 3.60) \times 10^{-19}$ J.

ii The change in frequency will not change the number of electrons emitted.

c i Doubling the intensity does not change the energy of the electrons.

ii Doubling the intensity will double the number of photons per second and will therefore double the number of electrons released per second.

12 The equation describing the photoelectric effect involves the Principle of Conservation of Energy. A photon with energy $E = hf$ releases an electron by supplying energy equal to the work function, ϕ, any energy left over appearing as the kinetic energy of the electron.

$$hf = \phi + eV_s$$

where e is the charge on the electron and V_s is the 'stopping potential'.

If this equation is rearranged to make V_s the subject, you have

$$V_s = \frac{hf}{e} - \frac{\phi}{e}$$

As this is of the form $y = mx + c$ (where V is equivalent to y, and x is equivalent to f), if you plot V_s against f you will get a straight line with

$$\text{gradient} = \frac{h}{e} \quad \text{and} \quad \text{intercept} = -\frac{\phi}{e}$$

Because you know the value of *e* you can find *h* from the gradient. See Figure 16.9.

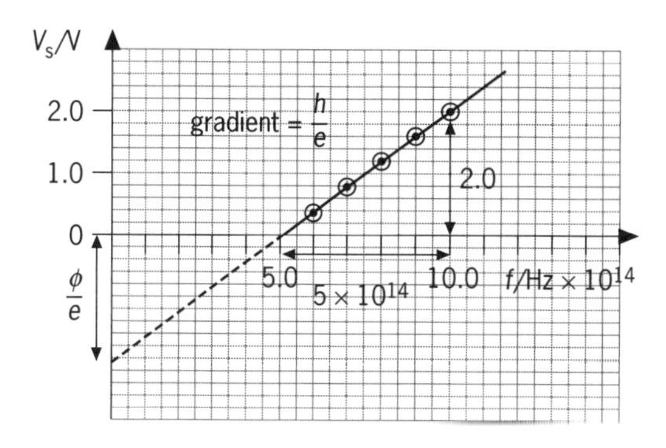

Figure 16.9

The gradient (= *h/e*) is approximately

$$\frac{2.0\,\text{V}}{5.0 \times 10^{14}\,\text{Hz}} = 4.0 \times 10^{-15}\,\text{V Hz}^{-1} (\equiv \text{V s})$$

The Planck constant, *h*, is then given by

$$h = \text{gradient} \times e = 4.0 \times 10^{-15}\,\text{V s} \times 1.6 \times 10^{-19}\,\text{C}$$
$$= \mathbf{6.4 \times 10^{-34}\,J\,s}$$

13 a Looking at the numerator (top line), GeV, meaning giga-electronvolts, is a unit of energy.

$$\text{energy (or work)} = \text{force} \times \text{distance}$$
$$= \text{mass} \times \text{acceleration} \times \text{distance}$$

which has units of

$$\text{kg} \times \text{m s}^{-2} \times \text{m} = \text{kg m}^2\,\text{s}^{-2}$$

The denominator (bottom line), *c*, is a velocity, in m s^{-1}.

∴ GeV/*c* has base units of

$$\frac{\text{kg m}^2\,\text{s}^{-2}}{\text{m s}^{-1}} = \text{kg m s}^{-1}$$

which is a unit of mass × velocity = momentum.

b From part a,

$$\frac{\text{energy}}{\text{velocity}} = \text{momentum} = \text{mass} \times \text{velocity}$$

so energy/velocity^2 = mass and therefore units of GeV/c^2 could be used for mass.

17 Emission and absorption spectra

Electron energy levels

The electrons orbiting the nucleus of an atom have energies which are **quantised**, that is there are only certain energy levels which are permitted, Figure 17.1. A single electron may have any of these energies but it cannot have values between them.

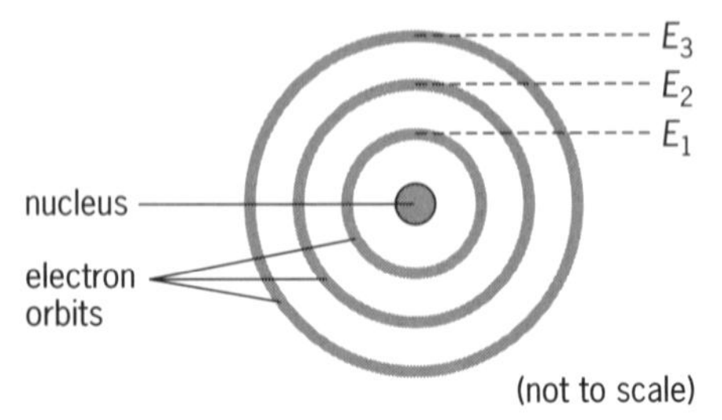

Figure 17.1 Allowed electron orbits with discrete energies

We can represent these permitted energy values as an energy level diagram, Figure 17.2. Only a certain number of electrons can occupy a certain level, described by its **principal quantum number**.

zero — $n = \infty$
−0.38 — $n = 6$
−0.54 — $n = 5$
−0.85 — $n = 4$
−1.51 — $n = 3$
−3.39 — $n = 2$
−13.6 — $n = 1$
energy in eV
n = principal quantum number

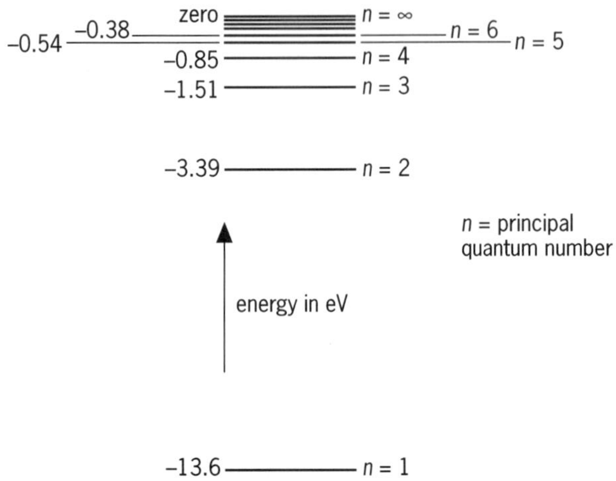

Figure 17.2 Energy level diagram for a hydrogen atom

Within an atom electrons will tend to occupy the lowest energy levels available to them. If there are no unoccupied energy levels below an electron, the atom is said to be in its **ground state**. At temperatures above 0 K (see topic 18) it is possible for individual electrons to absorb energy and jump to a higher energy level. The atom is then said to be **excited**.

After a short time the electron raised to a higher energy level will return to the lower energy level, but before it can do so it must lose energy equal to the difference between the two energy levels, say $E_2 - E_1$. It does this by emitting a photon. The frequency of the quantum of energy emitted is related to the energy levels by the equation

$$hf = E_2 - E_1$$

These movements between energy levels are called **transitions**.

Example 1

Calculate the frequency of a photon which is emitted when an electron in a hydrogen atom jumps from an energy level of −3.4 eV to an energy level of −13.6 eV. ($h = 6.6 \times 10^{-34}$ J s, $e = -1.6 \times 10^{-19}$ C)

$$hf = E_2 - E_1$$

$$\text{so} \quad f = \frac{(E_2 - E_1)}{h}$$

The energy values in this equation must be in joules, not electronvolts.

$$1\,\text{eV} = 1.6 \times 10^{-19}\,\text{J}$$

$$\therefore f = \frac{[-3.4 - (-13.6)] \times 1.6 \times 10^{-19}\,\text{J}}{6.6 \times 10^{-34}\,\text{J s}}$$

$$= \frac{10.2\,\text{eV} \times 1.6 \times 10^{-19}\,\text{J eV}^{-1}}{6.6 \times 10^{-34}\,\text{J s}}$$

$$= 2.5 \times 10^{15}\,\text{Hz}$$

The frequency of the photon emitted by an electron as it moves to a lower energy level may lie inside or outside the visible part of the electromagnetic spectrum. In the above example the photon emitted would not be visible but ultraviolet (see Figure 17.3).

If an electron absorbs sufficient energy to reach the zero energy level (principal quantum number $n = \infty$), it escapes the atom and the atom is said to be **ionised**. For hydrogen, this happens when the energy absorbed is greater than or equal to 13.6 eV.

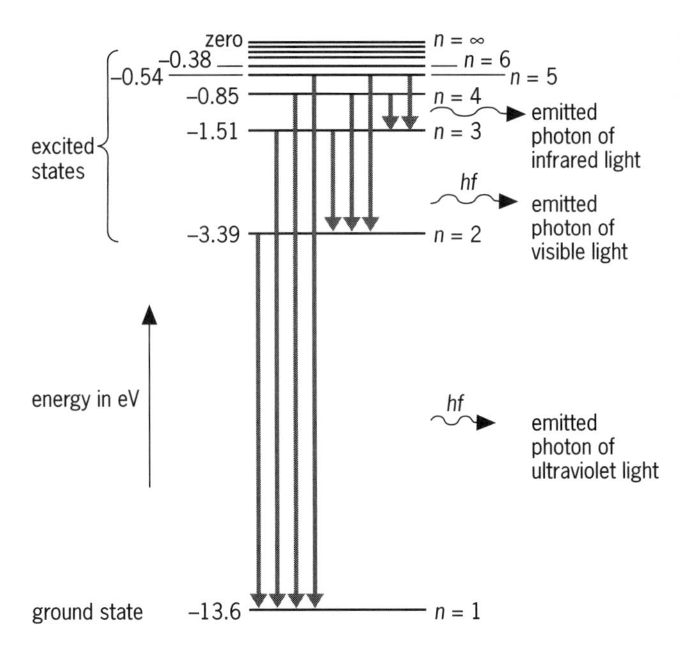

Figure 17.3 Principal transitions of the electron in a hydrogen atom

Emission spectra

As can be seen from Figure 17.3, there may be several transitions that an electron can make within an atom. Each of these transitions will result in photons of a different frequency being emitted.

Each element has its own specific electron energy levels. Therefore the frequencies of the quanta emitted by electrons as they fall to lower energy levels are characteristic of that atom and, like a fingerprint, they can be used to detect its presence. The light radiated in this way by a source is called an **emission spectrum**. There are three main types of emission spectrum.

Line, band and continuous spectra

If the light emitted from a hydrogen atom, by electrons making transitions between the different energy levels, is passed through a slit into a prism or a spectrometer an image consisting of a series of bright lines is seen. This is called a **line spectrum**, the lines being images of the slit illuminated by the different colours of light dispersed by the prism. Figure 17.4 shows some of the lines in the spectrum of hydrogen.

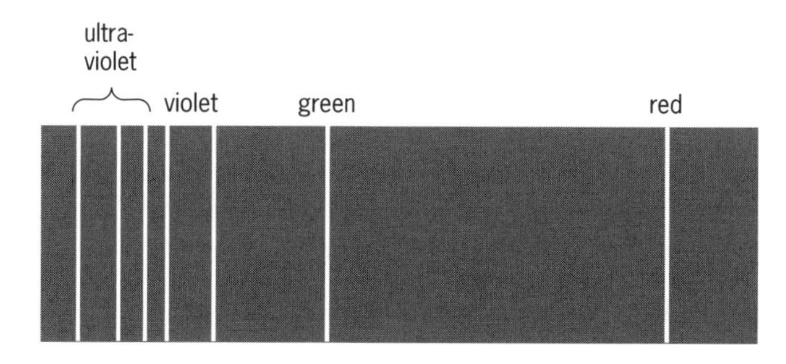

Figure 17.4 Line spectrum characteristic of hydrogen

Line spectra are produced by isolated atoms, for example the atoms of a gas where there is no interaction between the electrons of neighbouring atoms.

Where there is some interaction between electrons, such as when atoms combine to form molecules, a greater number of energy levels is available and spectra are produced which consist of lots of lines packed very close together, Figure 17.5. These are called **band spectra**.

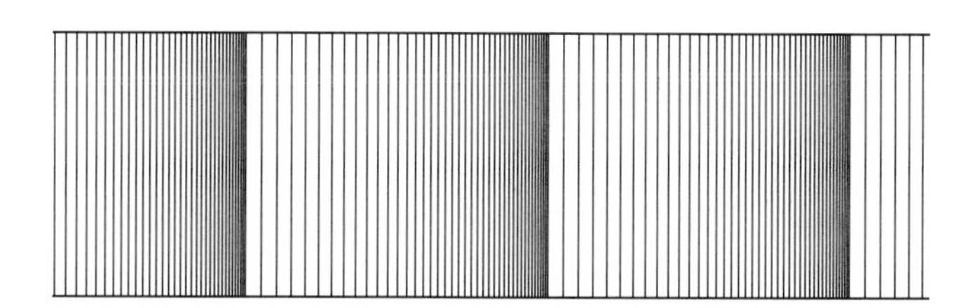

Figure 17.5 Diagrammatic representation of a band spectrum

Where neighbouring atoms are very close together, for example in a solid, a liquid or a gas at high pressure, there is a great deal of interaction between orbiting electrons and *all* energies within a range are available to the electrons. As a result all frequencies of light within this range are emitted and the spectrum seen is called a **continuous spectrum**, Figure 17.6.

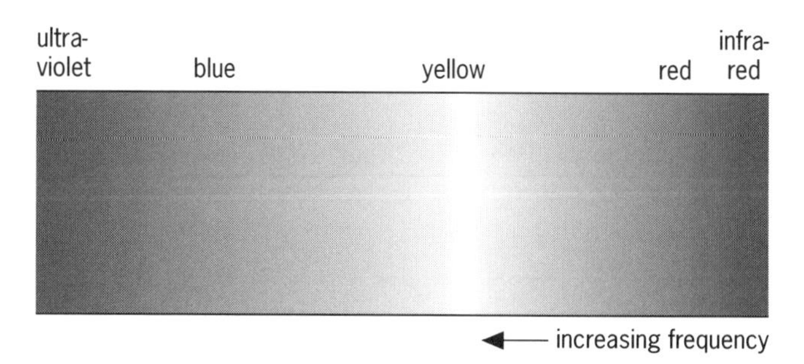

Figure 17.6 Representation of a continuous spectrum

Absorption spectra

If white light passes through a cloud of gas some frequencies of light may be selectively absorbed. These frequencies correspond to transitions that take place within the gas atoms when electrons absorb energy and jump to higher energy levels. The spectrum of the emerging light shows these missing frequencies as dark lines. This is called an **absorption spectrum** and again can be used to identify the presence of elements within the gas.

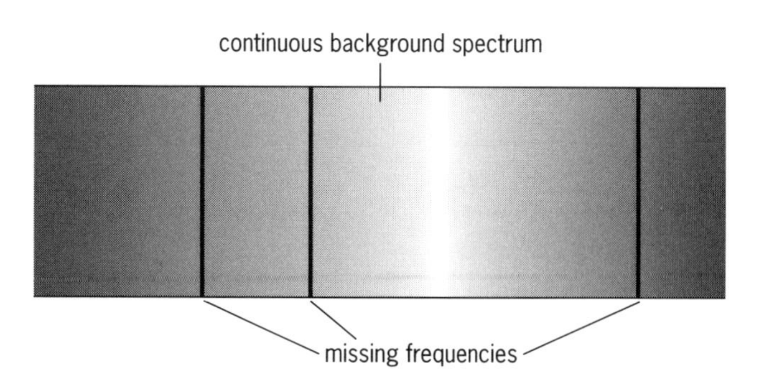

Figure 17.7 Absorption spectrum characteristic of hydrogen

1 eV = 1.6×10^{-19} J Mass of an electron, $m_e = 9.1 \times 10^{-31}$ kg Planck constant, $h = 6.6 \times 10^{-34}$ J s

Level 1

1 Electron A is at its lowest atomic energy level as shown in Figure 17.8.

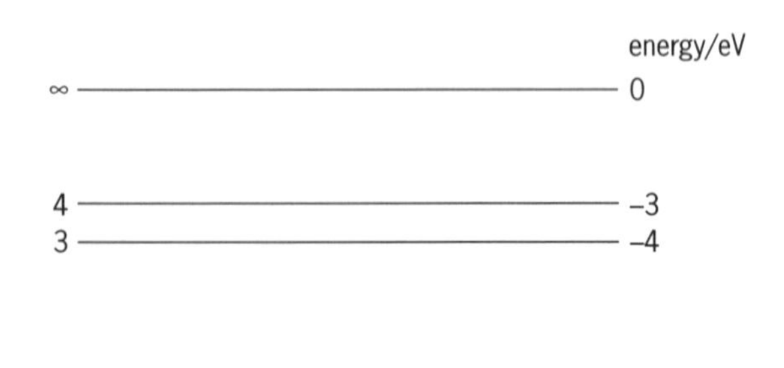

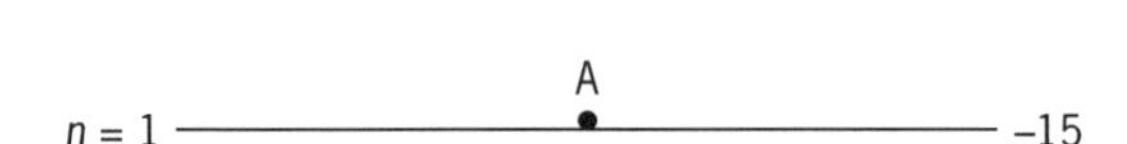

Figure 17.8

If each of the following interacted with the atom, say whether you think *ionisation*, *excitation* or *nothing* will probably happen:

a a 7.0 eV photon
b a 4.0 eV electron
c a 22 eV electron
d an 11 eV photon
e an 11.5 eV photon.

2 Some of the energy levels of an electron in a hydrogen atom are shown in Figure 17.9.

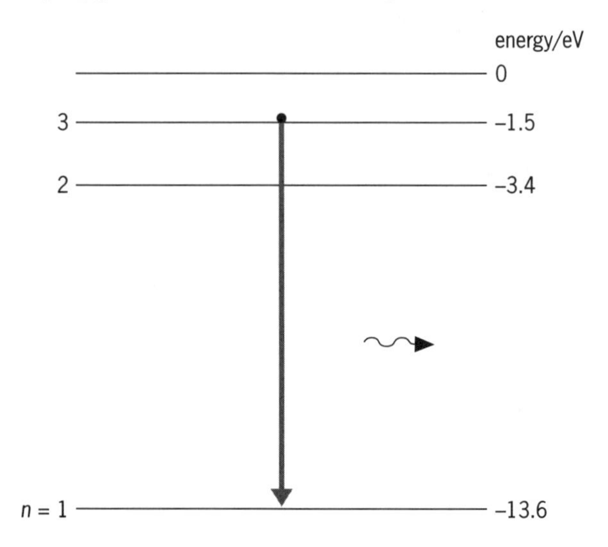

Figure 17.9

An electron has been excited to the −1.5 eV level and then drops back to the ground state, emitting a photon.

a How much energy, in eV, is released from this transition?
b Where does this energy go?
c What is this energy in joules?
d Find the frequency, f, of the photon emitted (its energy = hf).
e Find the wavelength of the light emitted.
f Will it be visible to a human?

3 What is meant by the word 'quantised' when talking about the energy levels of an electron in an atom?

Level 2

4 In a helium–neon laser, the electrons in the neon atoms drop down from their excited state at −4.026 eV to −5.990 eV. What is the wavelength of the light emitted?

5 In an X-ray tube, electrons are accelerated through 100 kV. They then crash into the target where they may excite the electrons of the target atoms. An excited electron then drops down to a lower energy level, emitting a photon as it does so.

a What energy, in joules, do the accelerated electrons acquire?
b What is the maximum possible energy of a photon leaving the target?
c What frequency would this photon have?
d What is the minimum photon wavelength possible from this tube?

6 Explain briefly how forensic scientists can use emission line spectra and absorption line spectra.

Level 3

7 Light of wavelength 656 nm is emitted when an electron drops from −1.51 eV to a lower energy level.

a What is the frequency of the light?
b What is the energy change in joules?
c Find the final energy level in eV.

8 The spectral lines emitted by transitions to the ground state of the hydrogen atom are found in the ultraviolet region, while the transitions to the first excited state are found in the visible region and those to the second excited state are found in the infrared. Give a qualitative explanation for this.

9 The wavelength of light emitted by a sodium lamp is 590 nm. The time taken for the electron transition, which produces the photon, is of the order of nanoseconds. How many oscillations of the electromagnetic wave will occur during this time?

Level 1

1 **a** A photon with energy 7.0 eV will **excite** electron A to level 2.

b An electron energy of 4.0 eV will have **no effect** on electron A.

c An electron with energy 22 eV will **ionise** the atom and the atomic electron will leave with surplus energy of 7.0 eV.

d A photon with energy 11 eV will **excite** electron A to level 3.

e A photon with energy 11.5 eV will have **no effect** on the electron because 11.5 eV is not *exactly* equal to the difference between the energy of any two levels.

2 **a** The difference in energy between the levels −1.5 eV and −13.6 eV is

$$-1.5\,\text{eV} - (-13.6\,\text{eV}) = +\mathbf{12.1\,eV}$$

b This will be the energy of the photon that is emitted.

c The electronvolt (eV) is the energy gained or lost when the charge on an electron goes through a potential difference of 1 volt.

energy = charge × potential difference

so $1\,\text{eV} = 1.6 \times 10^{-19}\,\text{C} \times 1\,\text{V} = 1.6 \times 10^{-19}\,\text{J}$

and $12.1\,\text{eV} = 1.6 \times 10^{-19} \times 12.1 = \mathbf{1.94 \times 10^{-18}\,J}$

d Energy of the photon, $E = hf$

$$\therefore f = \frac{E}{h} = \frac{1.94 \times 10^{-18}\,\text{J}}{6.6 \times 10^{-34}\,\text{J s}^{-1}} = \mathbf{2.9 \times 10^{15}\,Hz}$$

e For electromagnetic waves $c = f\lambda$; rearranging this equation gives

$$\lambda = \frac{c}{f}$$

$$= \frac{3.0 \times 10^{8}\,\text{m s}^{-1}}{2.9 \times 10^{15}\,\text{s}^{-1}} = \mathbf{1.02 \times 10^{-7}\,m} \text{ or } \mathbf{102\,nm}$$

f **No**. It will be in the ultraviolet region, which is not the visible to the human eye. The 'visible' spectrum goes from about 700 nm to 400 nm.

3 The statement that the energy levels are 'quantised' means that only certain energies are allowed. The electrons cannot have an energy of any 'in-between' value.

Level 2

4 The energy change of each electron is (5.990 − 4.026) eV = 1.964 eV. This is equal to the energy of the photons that are emitted, E.
To change eV to joules, multiply by $1.6 \times 10^{-19}\,\text{J eV}^{-1}$.

$$E = 1.964\,\text{eV} \times 1.6 \times 10^{-19}\,\text{J eV}^{-1} = 3.14 \times 10^{-19}\,\text{J}$$

For an electromagnetic wave,

$$E = hf = \frac{hc}{\lambda}$$

$$\therefore \lambda = \frac{hc}{E}$$

$$= \frac{6.6 \times 10^{-34}\,\text{J s} \times 3.0 \times 10^{8}\,\text{m s}^{-1}}{3.14 \times 10^{-19}\,\text{J}}$$

$$= \mathbf{6.3 \times 10^{-7}\,m} \text{ or } \mathbf{630\,nm}$$

(This is in the red region of the visible spectrum.)

5 **a** The energy gained by an electron that is accelerated through a voltage of 100 kV is *defined* as 100 keV. To change eV to joules, multiply by $1.6 \times 10^{-19}\,\text{J eV}^{-1}$. So,

energy gained by the electrons

$$= 100 \times 10^{3}\,\text{eV} \times 1.6 \times 10^{-19}\,\text{J eV}^{-1}$$

$$= \mathbf{1.6 \times 10^{-14}\,J}$$

b The maximum photon energy will occur when the energy of the incoming electron excites a single electron in the target and this electron subsequently makes a transition between two adjacent energy levels, giving out a single photon. Maximum possible photon energy, $E_{max} = \mathbf{1.6 \times 10^{-14}\,J}$

c You can rearrange the equation $E_{max} = hf_{max}$ to give

$$f_{max} = \frac{E_{max}}{h} = \frac{1.6 \times 10^{-14}\,\text{J}}{6.6 \times 10^{-34}\,\text{J s}} = \mathbf{2.4 \times 10^{19}\,Hz}$$

d The minimum wavelength occurs when the frequency and energy are maximum and you can find it using the equation $c = f\lambda$, which you can rearrange to give

$$\lambda_{min} = \frac{c}{f_{max}} = \frac{3.0 \times 10^{8}\,\text{m s}^{-1}}{2.4 \times 10^{19}\,\text{s}^{-1}} = \mathbf{1.2 \times 10^{-11}\,m}$$

6 Line spectra match the pattern of energy levels in atoms. Only photons with energies equal to the difference between two levels can be emitted or absorbed. The spectra are unique to each element and molecule so that a line spectrum 'fingerprint' can be used to identify a sample of a substance.

Level 3

7 **a** The wavelength, $\lambda = 656\,\text{nm} = 656 \times 10^{-9}\,\text{m}$. You can rearrange the equation $c = f\lambda$ to give

$$f = \frac{c}{\lambda} = \frac{3.0 \times 10^{8}\,\text{m s}^{-1}}{656 \times 10^{-9}\,\text{m}} = \mathbf{4.6 \times 10^{14}\,Hz}$$

b The energy change that gave rise to the photon is given by the equation

$$\Delta E = hf$$
$$= 6.6 \times 10^{-34}\,\text{J s} \times 4.6 \times 10^{14}\,\text{s}^{-1}$$
$$= \mathbf{3.0 \times 10^{-19}\,J}$$

c $1\,\text{eV} = 1.6 \times 10^{-19}\,\text{J}$

So, to convert energy in joules to energy in eV, divide by $1.6 \times 10^{-19}\,\text{J eV}^{-1}$.

$$\Delta E = \frac{3.0 \times 10^{-19}\,\text{J}}{1.6 \times 10^{-19}\,\text{J eV}^{-1}} = 1.88\,\text{eV}$$

This is equal to the difference in energy levels,

$$\Delta E = E_{upper} - E_{lower} = 1.88\,\text{eV}$$

so the lower energy level must be at

$$E_{lower} = E_{upper} - 1.88 = (-1.51) - 1.88 = \mathbf{-3.39\,eV}$$

See Figure 17.10.

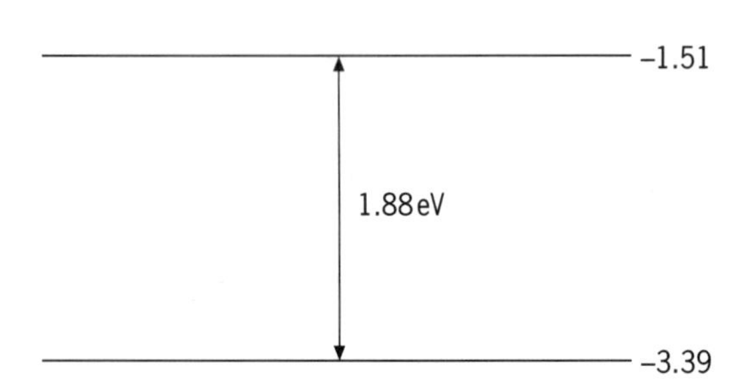

Figure 17.10

8 The difference in energy levels is greatest for the transition to the ground state so this transition produces the photons with the highest energy, and highest frequency ($E = hf$). The 'light' from these transitions will be in the ultraviolet region of the electromagnetic spectrum. When the final state is an excited state the energy change will be smaller so that the spectral lines will be in regions of lower energy, lower frequency and longer wavelength; the 'light' emitted will be in the visible and the infrared regions of the spectrum. See Figure 17.11.

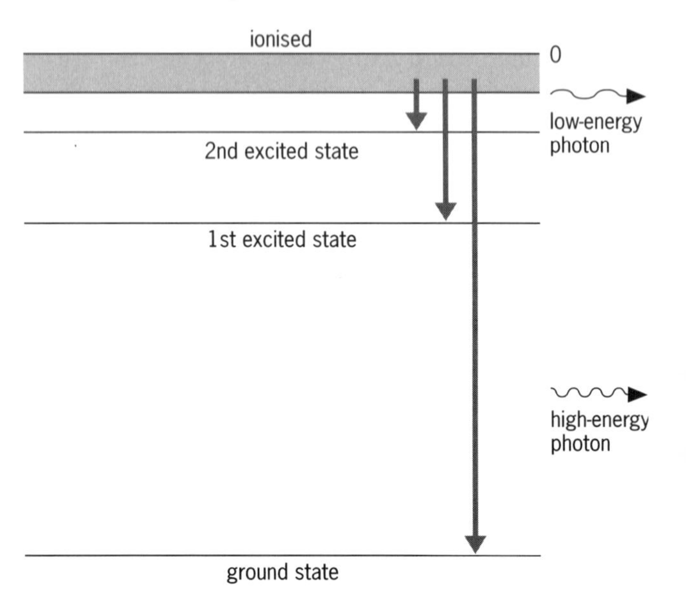

Figure 17.11

9 You can find the frequency of the light using the equation $c = f\lambda$, which you can rearrange to give

$$f = \frac{c}{\lambda} = \frac{3.0 \times 10^{8}\,\text{m s}^{-1}}{590 \times 10^{-9}\,\text{m}} = 5.1 \times 10^{14}\,\text{Hz}$$

This is the number of oscillations in 1 second. So in 1 ns (1×10^{-9} s) there will be about

$$5 \times 10^{14}\,\text{s}^{-1} \times 1 \times 10^{-9}\,\text{s} = \mathbf{10^{5}\ oscillations}$$

18 Temperature change

Temperature and thermometers

Some important points about temperature, heat and energy:

- Temperature is a measure of the hotness of an object.
- The hotter an object is the more internal energy it possesses.
- The internal energy of an object is the sum of the kinetic energy and potential energy of all its particles.
- The average kinetic energy of the molecules is proportional to the absolute temperature.
- Heat is the flow of energy caused by a difference in temperature.

We measure the temperature of an object using a thermometer. Thermometers are based on physical properties that vary with temperature, for example the length of a column of liquid, the resistance of a piece of wire, or the pressure or volume of a gas. To establish a scale of temperature two fixed points are needed. The two fixed points which define the centigrade scale are the **ice point** and the **steam point**. The ice point defines 0 °C, which is the temperature at which pure ice melts. The steam point defines 100 °C, which is the temperature of steam at standard atmospheric pressure.

To calibrate a thermometer on the centigrade scale the temperature-sensitive property (thermometric property) is measured when the thermometer is at the ice point (X_0) and then again when it is at the steam point (X_{100}). The change in the property corresponds to a temperature change of 100 °C. If the thermometric property is now measured at some unknown temperature θ this temperature can be found using the equation

$$\theta = \frac{(X_\theta - X_0)}{(X_{100} - X_0)} \times 100\,°\text{C}$$

Example 1

Calculate the value of the temperature θ measured by a mercury-in-glass thermometer if the length of the mercury column at this temperature is 6.2 cm. The length of the mercury column at the ice point is 3.3 cm and at the steam point 12.0 cm.

$$\theta = \frac{(X_\theta - X_0)}{(X_{100} - X_0)} \times 100\,°\text{C}$$

$$= \frac{(6.2 - 3.3)}{(12.0 - 3.3)} \times 100\,°\text{C} = \frac{290\,°\text{C}}{8.7}$$

$$= 33\,°\text{C}$$

Calibrating a mercury-in-glass thermometer

The uncalibrated thermometer is placed in pure melting ice to discover the length of the mercury column at the ice point (Figure 18.1a). It is then placed in a steam jacket at atmospheric pressure to determine the length at the steam point (Figure 18.1b).

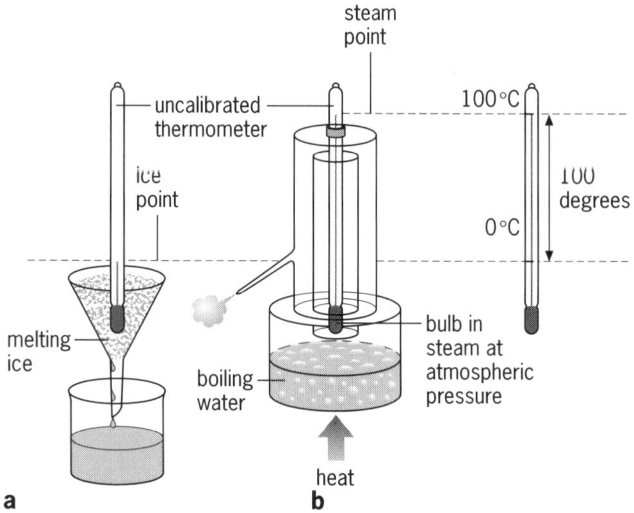

Figure 18.1 Calibration apparatus

The absolute thermodynamic scale of temperature

Thermometers based on different thermometric properties often give slightly different values for the same unknown temperature. The **absolute thermodynamic scale** of temperature does not depend upon physical properties, like a thermometric scale, but is based upon the energy flow in heat engines (see topic 22). It uses as its two fixed points:

1. **absolute zero** – this is the lowest temperature possible and is the temperature at which all substances have minimum internal energy; and
2. the **triple point of water** – this is the temperature at which ice, water and water vapour can exist in equilibrium.

The unit of the absolute scale of temperature is the **kelvin**, K. Absolute zero is defined as 0 K while the triple point of water is defined as 273.16 K.

The relationship between temperature on the absolute thermodynamic scale of temperature and temperature on the **Celsius scale** is described by the equation

absolute temperature/K = Celsius temperature/°C + 273
(to 3 significant figures)

A temperature interval of 1 kelvin equals 1 degree Celsius.

The Celsius scale is for all practical purposes identical to the centigrade scale but it is based upon the thermodynamic scale of temperature, whereas the centigrade scale is based upon a thermometric property measured at the ice point and steam point of water as described above.

Specific heat capacity

When energy is transferred to an object the increase in its internal energy may be indicated by a rise in temperature. The size of the temperature rise depends upon the amount of energy transferred, the mass of the object and the material from which the object is made. The **specific heat capacity** of a material, *c*, is defined as being the energy required to increase the temperature of 1 kg of it by 1 K or 1 °C. So

$$c = \frac{\text{energy}}{\text{mass} \times \text{temperature change}}$$

The unit of specific heat capacity is joule per kilogram per kelvin, or $\text{J kg}^{-1}\text{K}^{-1}$.

The relationship between the energy supplied, ΔQ, to an object and its subsequent rise in temperature, $\Delta\theta$, is described by the equation

$$\Delta Q = mc\Delta\theta$$

where m is the mass of the object in kg
c is the specific heat capacity of the material from which the object is made
$\Delta\theta$ is the increase in temperature caused by the energy supplied.

Example 2

How much energy is needed to raise the temperature of 5.0 kg of water from 10 °C to 30 °C? The specific heat capacity of water is $4200\,\text{J kg}^{-1}\text{K}^{-1}$.

Energy needed, $\Delta Q = m \times c \times \Delta\theta$

$= 5.0\,\text{kg} \times 4200\,\text{J kg}^{-1}\text{K}^{-1} \times 20\,\text{K}$

$= 420\,000\,\text{J}$ or $420\,\text{kJ}$

Molar heat capacities

If the object absorbing the energy is a gas it is more usual to measure the quantity of the gas in moles rather than kilograms. Gases have two molar heat capacities, a molar heat capacity for the gas when it is heated at *constant pressure* (when some of the energy supplied does external work) and a molar heat capacity for a gas when it is heated at *constant volume*.

Heat capacity

While it is important to know the specific heat capacity of individual materials, objects are often composed of more than one material. Under these circumstances it can be more useful to consider the amount of energy necessary to warm the whole object. The **heat capacity** of an object is defined as being the energy required to raise the temperature of an object by 1 K or 1 °C. The unit of heat capacity is J K^{-1}.

If the object is made from a single material, the relationship between the heat capacity of the object and the specific heat capacity of the material is described by the equation

heat capacity = mass × specific heat capacity

$$C = mc$$

Example 3

A kettle made completely of aluminium has a mass of 1.5 kg. Calculate **a** the heat capacity of the kettle and **b** the energy required to increase the temperature of the kettle by 20 °C. The specific heat capacity of aluminium is $900\,\text{J kg}^{-1}\text{K}^{-1}$.

a Heat capacity = mass × specific heat capacity

$= 1.5\,\text{kg} \times 900\,\text{J kg}^{-1}\text{K}^{-1}$

$= 1350\,\text{J K}^{-1}$ or $1.35\,\text{kJ K}^{-1}$

b Energy required = mass × specific heat capacity × rise in temperature

$= 1.5\,\text{kg} \times 900\,\text{J kg}^{-1}\text{K}^{-1} \times 20\,\text{K}$

$= 27\,000\,\text{J}$ or $27\,\text{kJ}$

or

energy required = heat capacity × rise in temperature

$= 1350\,\text{J K}^{-1} \times 20\,\text{K}$

$= 27\,000\,\text{J}$ or $27\,\text{kJ}$

Measurement of specific heat capacity

Electrical methods for solids

The apparatus is shown in Figure 18.2.

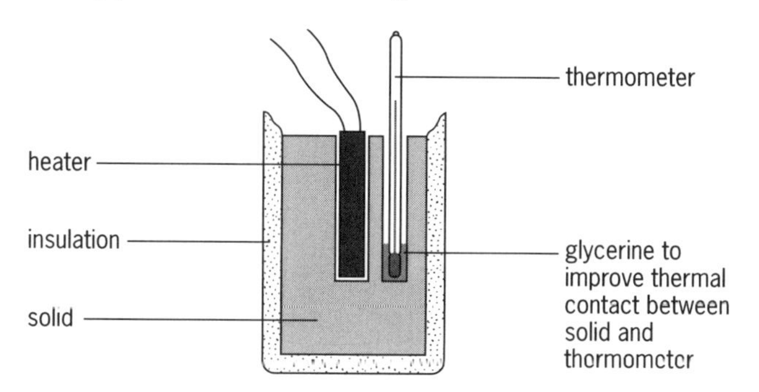

Figure 18.2 Measuring the specific heat capacity of a solid

The mass of the solid, m, and its initial temperature, θ_1, are measured. The heating coil is turned on and the pd across the ends of the coil, V, and the current flowing through it, I, are noted. (A joulemeter may also be used.) At the same instant that the heater is turned on a timer is started. When the temperature of the solid has risen by approximately 10 °C the heater is turned off and the time, t, noted. The highest temperature, θ_2, reached by the solid after the coil has been turned off is also noted.

Assuming there are no energy losses to the surroundings,

$$\text{electrical energy supplied by heater} = \text{energy gained by solid}$$

$$IVt = mc(\theta_2 - \theta_1)$$

which can be rearranged to give

$$c = \frac{IVt}{m(\theta_2 - \theta_1)}$$

Note:

1 Some energy will be lost to the surroundings but this can be kept to a minimum by **a** surrounding the solid with a good insulator such as expanded polystyrene and **b** not allowing the temperature of the solid to rise to more than 10 °C above room temperature.

2 The heat capacity of the heater and the thermometer should be small so that the amount of heat they absorb is negligible.

Example 4

A 2.0 kg piece of copper was heated electrically so that its temperature increased by 10 °C. The pd across the heater was 5.0 V, the current passing through it was 2.0 A and the heater was turned on for 800 s. Calculate the specific heat capacity of copper.

$$c = \frac{IVt}{m(\theta_2 - \theta_1)}$$

$$= \frac{2.0\,\text{A} \times 5.0\,\text{V} \times 800\,\text{s}}{2.0\,\text{kg} \times 10\,\text{K}}$$

$$= 400\,\text{J}\,\text{kg}^{-1}\,\text{K}^{-1}$$

Electrical methods for liquids

A similar method to that described above can be used to determine the specific heat capacity of a liquid. The apparatus is shown in Figure 18.3.

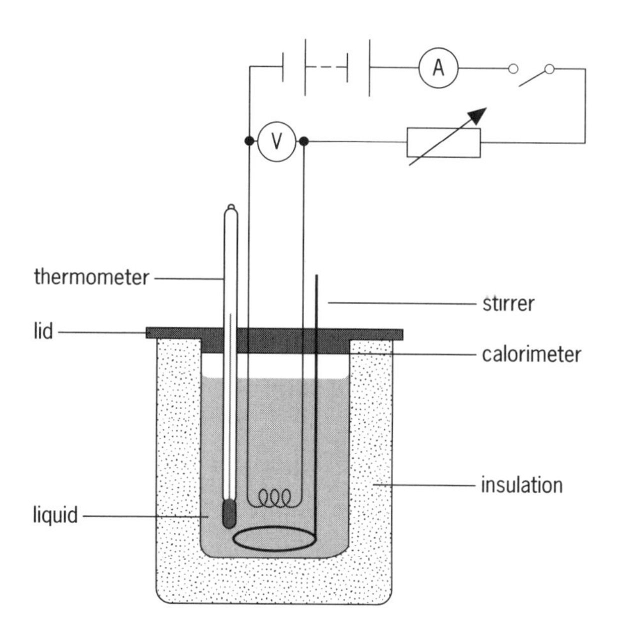

Figure 18.3 Measuring the specific heat capacity of a liquid by an electrical method

While the liquid is being heated it must be stirred to ensure an even temperature throughout. Energy will be absorbed by the container (more usually called a calorimeter); this must be taken into account in the calculation.

Assuming there are no energy losses to the surroundings,

$$\text{electrical energy supplied by heater} = \text{energy gained by liquid} + \text{energy gained by calorimeter}$$

$$IVt = mc(\theta_2 - \theta_1) + m_c c_c(\theta_2 - \theta_1)$$

where m_c is the mass of the calorimeter and c_c is the specific heat capacity of the material from which the calorimeter is made.

Example 5

Calculate the energy supplied to 0.12 kg of water in a copper calorimeter of mass 0.050 kg if the water and the container increase in temperature from 15 °C to 22 °C. The specific heat capacity of water is $4200\,\text{J}\,\text{kg}^{-1}\,\text{K}^{-1}$. The specific heat capacity of copper is $400\,\text{J}\,\text{kg}^{-1}\,\text{K}^{-1}$.

$$\begin{aligned}\text{Energy supplied} &= mc(\theta_2 - \theta_1) + m_c c_c(\theta_2 - \theta_1)\\ &= 0.12\,\text{kg} \times 4200\,\text{J}\,\text{kg}^{-1}\,\text{K}^{-1} \times 7\,\text{K}\\ &\quad + 0.050\,\text{kg} \times 400\,\text{J}\,\text{kg}^{-1}\,\text{K}^{-1} \times 7\,\text{K}\\ &= 3528\,\text{J} + 140\,\text{J}\\ &= 3668\,\text{J or } 3.7\,\text{kJ}\end{aligned}$$

Continuous flow method for liquids

A steady stream of liquid is passed through a tube with a heating coil, Figure 18.4.

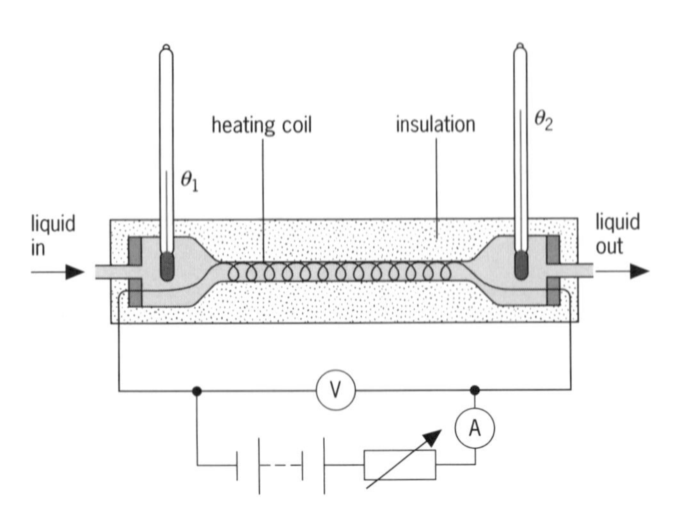

Figure 18.4 Measuring the specific heat capacity of a liquid by a continuous flow method

The current through the coil, I, the pd across its ends, V, and the liquid flow rate are adjusted so that when the thermometer readings become steady the temperature difference $(\theta_2 - \theta_1)$ is approximately 5 °C. The liquid passing through the tube in a given time t s is collected and its mass measured, m.

electrical energy supplied by heater = energy gained by liquid + energy lost to surroundings (H)

$$IVt = mc(\theta_2 - \theta_1) + H$$

If the experiment is repeated over the same length of time, t, using different values of current, I', and voltage, V', but with the flow rate adjusted so that the temperatures θ_1 and θ_2 are at the same values, the energy lost to the surroundings will be the same as that lost in the first experiment.

$$\therefore \quad I'V't = m'c(\theta_2 - \theta_1) + H$$

where m' is the mass of liquid collected in time t.

If these two equations are subtracted the energy lost to the surroundings is eliminated from the calculation:

$$(IV - I'V')t = (m - m')c(\theta_2 - \theta_1)$$

Although this method of determining the specific heat capacity requires large quantities of liquid there are several advantages:

1. It is not necessary to know the heat capacity of the apparatus.
2. The temperature readings are constant and very sensitive thermometers can be used.
3. Energy loss to the surroundings is eliminated.
4. The method can also be used to determine the specific heat capacity (molar heat capacity) of a gas.

Example 6

A liquid flows through a continuous flow calorimeter at the rate of 0.010 kg s^{-1}. With a pd of 200 V across the heating coil and a current of 1.0 A flowing through it a temperature difference of 4.5 °C is maintained. The same temperature difference is achieved with a liquid flow rate of 0.005 kg s^{-1}, a pd of 60 V and a current of 2.0 A. Calculate the specific heat capacity of the liquid.

$$(IV - I'V')t = (m - m')c(\theta_2 - \theta_1)$$

Rearranging,

$$c = \frac{(IV - I'V')t}{(m - m')(\theta_2 - \theta_1)}$$

For $t = 1$ s,

$$c = \frac{(1.0\,\text{A} \times 200\,\text{V} - 2.0\,\text{A} \times 60\,\text{V}) \times 1\,\text{s}}{(0.010 - 0.005)\,\text{kg} \times 4.5\,\text{K}}$$

$$= \frac{80\,\text{J}}{0.0225\,\text{kg}\,\text{K}}$$

$$= 3600\,\text{J}\,\text{kg}^{-1}\,\text{K}^{-1}$$

Specific heat capacities, c:	water	$4200\,\text{J kg}^{-1}\text{K}^{-1}$
	copper	$385\,\text{J kg}^{-1}\text{K}^{-1}$

Level 1

1 a How much energy is needed to raise the temperature of 0.80 kg of copper from 20 °C to 90 °C?

b How much energy is released when 2.2 kg of water cools from 80 °C to 30 °C?

c 1.4 kg of copper is given 5.0 kJ of energy. What is the temperature increase?

d Some water is given 6.2 kJ of energy and the temperature rises by 4.0 K. What is the mass of the water?

2 820 g of liquid is kept in an insulated calorimeter of negligible heat capacity. A 12 V, 3.5 A electric heater, immersed in the liquid, is switched on for 10 minutes and the temperature rise is observed to be 28.7 K. Calculate the specific heat capacity of the liquid.

3 A thermocouple consists of two junctions between two metals; when one junction is at a higher temperature than the other, an emf is generated.

The hot junction and the cold junction of a thermocouple are at 373 K and 273 K, respectively. The emf generated is 1.0 mV. Show that the emf changes to 0.85 mV when the hot junction is moved to a water bath at 358 K, if the emf generated varies linearly with the temperature difference between the junctions.

4 1.0 kg of water per second flows through the cooling system of a diesel engine. The difference between the temperatures of the incoming and outgoing water is 5.0 K. How much energy is being removed each second?

Level 2

5 An air-cooled engine needs 8.0 kW of surplus energy removed to keep the rise in air temperature within its limit of 40 K. Calculate the air flow which will achieve this. (Take the specific heat capacity of air under these conditions to be $1000\,\text{J kg}^{-1}\text{K}^{-1}$.)

6 A tube, 1.0 m long, contains 120 g of lead shot at room temperature. The tube is inverted rapidly 20 times and then the lead shot is emptied into a polystyrene cup. The specific heat capacity of lead is $130\,\text{J kg}^{-1}\text{K}^{-1}$.

a Each time the tube is inverted the lead shot loses gravitational potential energy, which produces a rise in temperature as the shot crashes at the bottom. What is the maximum possible temperature rise?

b Would the rise in temperature be greater if the mass of shot was larger?

c Why is it a good idea to use lead in this experiment?

7 A gas water-heater can raise the temperature of 2.2 kg of water by 45 K in 1 minute.

a What is the power of the heater?

b What practical difficulties might there be in using an *electric* water heater with the same performance?

Hint: How large is the current that would be required?

Level 3

8 Power cables normally operate at temperatures up to about 70 °C. If a short circuit occurs, considerable heat may be evolved which cannot escape quickly. The temperature of the conductor must never rise above 250 °C or the electrical insulation will be damaged.

Under short-circuit conditions, a current of 7000 A lasts for 1.0 s in a cable that has an effective area of cross-section $50\,\text{mm}^2$, resistance per unit length $0.40\,\Omega\,\text{km}^{-1}$ and density $8.9 \times 10^3\,\text{kg m}^{-3}$.

a Calculate the heat evolved in 1.0 m of cable during this short circuit.

b Calculate the rise in temperature of the cable.

9 The cold air falls out of an upright freezer every time you open the door. This does not happen with a chest freezer. If the cost of electricity is 7.4p for 3.6 MJ, comment on the extra cost incurred running an upright freezer of capacity $0.50\,\text{m}^3$. Take the inside temperature to be −18 °C, the specific heat capacity of air under these conditions to be $600\,\text{J kg}^{-1}\text{K}^{-1}$ and the density of air to be $1.2\,\text{kg m}^{-3}$.

10 A power station gets rid of 600 MW of surplus energy into a nearby river. The river is 20 m wide, 3.0 m deep and flows at $1.2\,\text{m s}^{-1}$. Calculate the temperature rise.

Level 1

1 a Temperature change, $\Delta\theta = 90\,°\mathrm{C} - 20\,°\mathrm{C} = 70\,\mathrm{K}$

energy needed, $\Delta Q = mc\Delta\theta$
$= 0.80\,\mathrm{kg} \times 385\,\mathrm{J\,kg^{-1}\,K^{-1}} \times 70\,\mathrm{K}$
$= \mathbf{22\,kJ}$

b Temperature change, $\Delta\theta = 80\,°\mathrm{C} - 30\,°\mathrm{C} = 50\,\mathrm{K}$

energy released, $\Delta Q = mc\Delta\theta$
$= 2.2\,\mathrm{kg} \times 4200\,\mathrm{J\,kg^{-1}\,K^{-1}} \times 50\,\mathrm{K}$
$= \mathbf{460\,kJ}$

(You use the same equation whether the object is heating up or cooling down.)

c You can rearrange the equation $\Delta Q = mc\Delta\theta$ to make the change in temperature the subject of the equation:

$$\Delta\theta = \frac{\Delta Q}{mc}$$

$\therefore$ change in temperature of copper, $\Delta\theta$

$$= \frac{5.0 \times 10^3\,\mathrm{J}}{1.4\,\mathrm{kg} \times 385\,\mathrm{J\,kg^{-1}\,K^{-1}}} = \mathbf{9.3\,K}$$

d You can also rearrange the equation to make the mass the subject:

$$m = \frac{\Delta Q}{c\Delta\theta}$$

$\therefore$ mass of water,

$$m = \frac{6.2 \times 10^3\,\mathrm{J}}{4200\,\mathrm{J\,kg^{-1}\,K^{-1}} \times 4.0\,\mathrm{K}} = \mathbf{0.37\,kg}$$

2 The energy supplied by the electric heater is given by

ΔQ = current $\times$ potential difference $\times$ time (in seconds)

$\Delta Q = IVt = 3.5\,\mathrm{A} \times 12\,\mathrm{V} \times 10 \times 60\,\mathrm{s} = 25\,200\,\mathrm{J}$

Rearrange the equation, $\Delta Q = mc\Delta\theta$, to make the specific heat capacity the subject:

$$c = \frac{\Delta Q}{m\Delta\theta} = \frac{25\,200\,\mathrm{J}}{0.82\,\mathrm{kg} \times 28.7\,\mathrm{K}} = \mathbf{1070\,J\,kg^{-1}\,K^{-1}}$$

3 The initial temperature difference of 100 K produced an emf of 1.0 mV. So, by proportion, because the new temperature difference is $(358 - 273) = 85\,\mathrm{K}$, the new emf E is given by

$$\frac{E}{1.0\,\mathrm{mV}} = \frac{85\,\mathrm{K}}{100\,\mathrm{K}}$$

$$\therefore \quad E = \frac{85}{100} \times 1.0\,\mathrm{mV} = \mathbf{0.85\,mV}$$

4 The basic equation relating energy change, mass, specific heat capacity and temperature change is

$$\Delta Q = mc\Delta\theta$$

If the *rates of change* of these quantities are *constant*, it is useful to divide both sides of the equation by time to get

$$\frac{\Delta Q}{t} = \frac{mc\Delta\theta}{t}$$

This equation says that the *rate of change of energy* is proportional to the *rate of change of mass* (mass flow rate), $\Delta Q/t = (m/t) \times c \times \Delta\theta$, if the temperature change is constant; it is proportional to the *rate of change of temperature*, $\Delta Q/t = m \times c \times \Delta\theta/t$, if the mass is constant.

So for a flow rate of $1\,\mathrm{kg\,s^{-1}}$ through the cooling system and a temperature change of 5.0 K, the energy removed per second is

$$\frac{\Delta Q}{t} = \left(\frac{m}{t}\right) \times c \times \Delta\theta = \left(\frac{1.0\,\mathrm{kg}}{1\,\mathrm{s}}\right) \times 4200\,\mathrm{J\,kg^{-1}\,K^{-1}} \times 5.0\,\mathrm{K}$$
$$= \mathbf{21\,kJ\,s^{-1}}$$

Level 2

5 Power = rate of change of energy
= rate of change of (mass $\times$ specific heat capacity $\times$ temperature difference)

or, if these rates of change are constant,

$$\frac{\Delta Q}{t} = \frac{mc\Delta\theta}{t}$$

In this question, the temperature difference is specified. What you need to find is the mass of air that has to have a temperature rise of 40 K, each second, to remove energy from the engine at a rate of 8.0 kW

Rewrite the equation as

$$\frac{\Delta Q}{t} = \left(\frac{m}{t}\right) \times c \times \Delta\theta$$

and then rearrange it to give

$$\frac{m}{t} = \frac{\Delta Q/t}{c \times \Delta\theta} = \frac{8.0 \times 10^3\,\mathrm{J\,s^{-1}}}{1000\,\mathrm{J\,kg^{-1}\,K^{-1}} \times 40\,\mathrm{K}}$$

$$= \mathbf{0.20\,kg\,s^{-1}}$$

(This is the minimum air flow needed to keep the engine cool.)

6 a The total gravitational potential energy lost is

$$\Delta E_p = 20 \times mg\Delta h$$

where m is the mass of the lead shot, g is the acceleration due to gravity and Δh is the vertical distance through which the lead shot falls.

This is equal to the change in energy associated with the maximum possible rise in temperature,

$$\Delta Q = mc\Delta\theta$$

where m is still the mass of the lead shot, c is the specific heat capacity of lead and $\Delta\theta$ is the maximum increase in temperature.

So

$$20 \times mg\Delta h = mc\Delta\theta$$

You can rearrange this equation to give the increase in temperature:

$$\Delta\theta = \frac{20 \times g\Delta h}{c} = \frac{20 \times 9.8\,\mathrm{m\,s^{-2}} \times 1.0\,\mathrm{m}}{130\,\mathrm{J\,kg^{-1}\,K^{-1}}} = \mathbf{1.5\,K}$$

b The mass of lead shot does **not affect** the temperature rise because the more mass there is, the more energy to lift the shot *and* to raise the temperature. Both the energy released and the energy needed are proportional to the mass.

c Lead has a low specific heat capacity so that a measurable temperature rise is achieved more easily. Lead also collides inelastically so that the gravitational potential energy is more easily converted to internal energy.

7 a The power of the water-heater is given by the equation

$$\frac{\Delta Q}{t} = \frac{mc\Delta\theta}{t} = \frac{22\,\mathrm{kg} \times 4200\,\mathrm{J\,kg^{-1}\,K^{-1}} \times 45\,\mathrm{K}}{60\,\mathrm{s}}$$

$$= \mathbf{6.9\,kW}$$

b The power of an electric heater is given by the equation

power = voltage × current, $P = VI$

which you can rearrange to give

$$\text{current, } I = \frac{P}{V}$$

Using a 230 V mains to provide 6.9 kW requires a current of

$$I = \frac{6.9 \times 10^3\,\mathrm{J\,s^{-1}}}{230\,\mathrm{V}} = \mathbf{30\,A}$$

This is too large for a standard 13 A socket. The heater would have to be connected to a circuit designed for larger currents, such as the electric cooker circuit.

Level 3

8 a The heat evolved in a cable is equal to the square of the current, I, multiplied by the resistance, R, and the time in seconds, t.

$\Delta Q = I^2Rt$ (see topic 4)

For the cable in the question, the current = 7000 A. The resistance is given as 0.40 Ω per kilometre, so the resistance of 1.0 m = $0.40 \times 10^{-3}\,\Omega$. The time is 1.0 second. So

$$\Delta Q = (7000\,\mathrm{A})^2 \times 0.40 \times 10^{-3}\,\Omega \times 1.0\,\mathrm{s} = \mathbf{19.6\,kJ}$$

b The temperature rise, $\Delta\theta$, for an energy increase, ΔQ, is given by the equation

$$\Delta\theta = \frac{\Delta Q}{mc}$$

For the 1.0 m length of cable, $\Delta Q = 19.6\,\mathrm{kJ}$, $c = 385\,\mathrm{J\,kg^{-1}\,K^{-1}}$. You need to find the mass of the cable, m.

mass = density × volume

$m = \rho V$

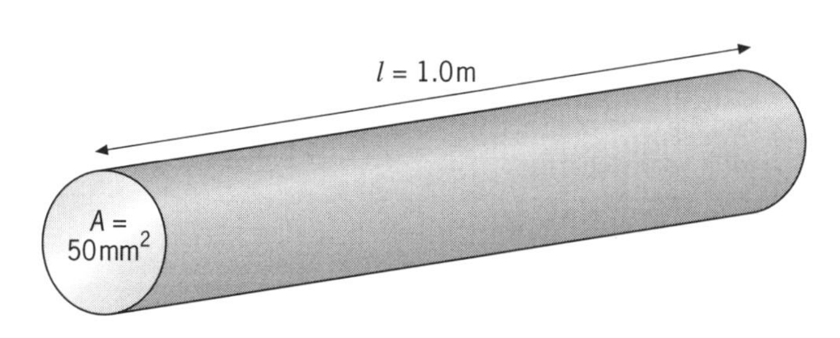

Figure 18.5

volume of cable, V = cross-sectional area × length

$= A \times l$

Note that the cross-sectional area is given in mm²: 50 mm² = 50 × 10⁻⁶ m².

$$\therefore \text{mass, } m = \rho V = \rho Al$$

$$= 8.9 \times 10^3\,\mathrm{kg\,m^{-3}} \times 50 \times 10^{-6}\,\mathrm{m^2} \times 1.0\,\mathrm{m}$$

$$= 0.445\,\mathrm{kg}$$

$$\text{and } \Delta\theta = \frac{\Delta Q}{mc} = \frac{19.6 \times 10^3\,\mathrm{J}}{0.445\,\mathrm{kg} \times 385\,\mathrm{J\,kg^{-1}\,K^{-1}}}$$

$$= \mathbf{114\,K}$$

This means that the temperature will reach 70 °C + 114 K = 184 °C, which is within the specified limit.

9 The empty upright freezer has a volume of $0.50\,m^3$ and contains air at a temperature of $-18\,°C$ with a density of $1.2\,kg\,m^{-3}$.

mass of air in freezer $=$ density $\times$ volume
$= 1.2\,kg\,m^{-3} \times 0.50\,m^3 = 0.60\,kg$

The air at $-18\,°C$ is replaced by air at room temperature, which you can take as $22\,°C$.

(This is on the high side but you might as well look at the worst-case scenario!)

So each time the freezer is opened it is necessary to cool $0.60\,kg$ of air from $22\,°C$ to $-18\,°C$, a temperature difference of $40\,K$. The energy required for this is

$$\Delta Q = mc\Delta\theta = 0.60\,kg \times 600\,J\,kg^{-1}\,K^{-1} \times 40\,K = 14\,kJ$$

If you opened the door of the freezer four times every day, in one year you would use

$$14 \times 10^3\,J \times 4 \times 365 = 2.04 \times 10^7\,J$$

$= 20.4\,MJ$ of electricity

The cost of electricity is quoted as 7.4p for 3.6 MJ.

In fact electricity is sold in 'units' of 1 kWh, which you should be able to show is the same as 3.6 MJ.

So, per MJ of energy the electricity costs

$$\frac{7.4\,p}{3.6\,MJ} = 2.06\,p\,MJ^{-1}$$

and the cost of opening the freezer door four times a day for a year would be

$$20.4\,MJ \times 2.06\,p\,MJ^{-1} = \mathbf{42p}$$

This is not a lot to worry about!

10 The power station needs to dump energy at a rate, $\Delta Q/t$, of 600 MW ($MJ\,s^{-1}$).

The volume flow rate of water flowing past the power station, V/t,

$= $ cross-sectional area (m^2) $\times$ velocity ($m\,s^{-1}$)
$= 20\,m \times 3.0\,m \times 1.2\,m\,s^{-1} = 72\,m^3\,s^{-1}$
(see Figure 18.6)

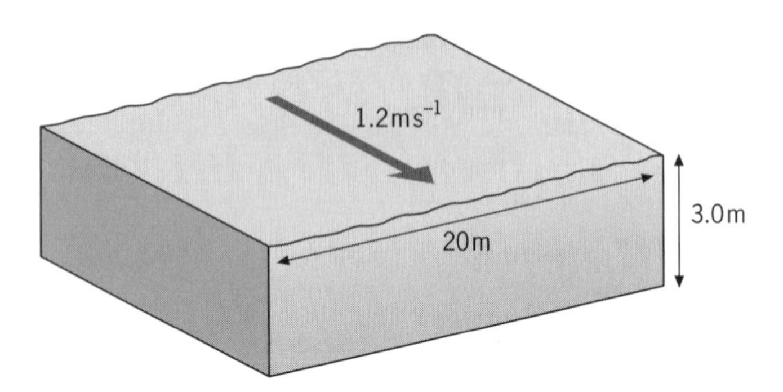

Figure 18.6

The mass flow rate of water flowing past the power station, m/t,

$=$ density $\times$ volume flow rate
$= 1000\,kg\,m^{-3} \times 72\,m^3\,s^{-1} = 72 \times 10^3\,kg\,s^{-1}$

You can rearrange the equation

$$\frac{\Delta Q}{t} = \frac{mc\Delta\theta}{t}$$

to give

$$\Delta\theta = \frac{\Delta Q/t}{(m/t) \times c}$$

$$= \frac{600 \times 10^6\,J\,s^{-1}}{72 \times 10^3\,kg\,s^{-1} \times 4200\,J\,kg^{-1}\,K^{-1}} = \mathbf{1.98\,K}$$

This would have a significant impact on the environment.

19 Latent heat

An object may absorb or lose internal energy and yet its temperature may remain constant. This may happen if the object changes state (phase), that is if a solid melts to become a liquid or a liquid boils to become a gas. The energy which is absorbed or lost during the change in state is called the **latent heat**. The internal potential energy of the molecules changes and work is done involving intermolecular forces.

The **specific latent heat** of a substance is defined as being the energy necessary to change the state of 1 kg of that substance without any change to its temperature. The unit of latent heat is $\mathrm{J\,kg^{-1}}$.

Heating a solid

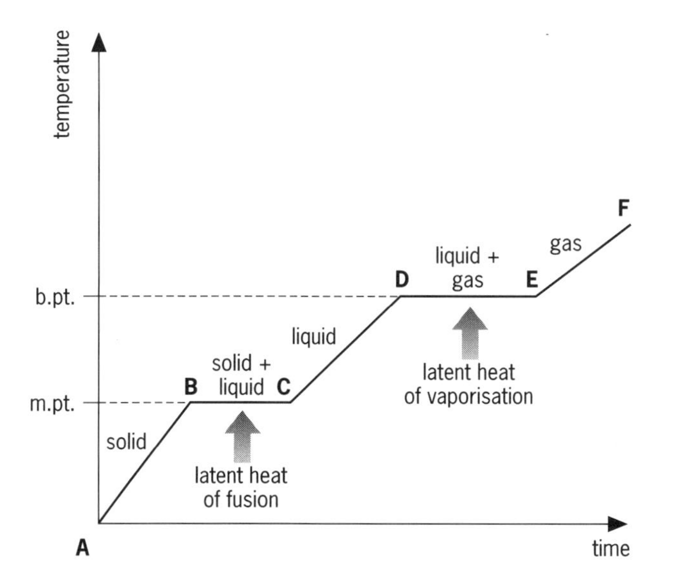

Figure 19.1 Change of temperature with time as a solid is heated

If you heat a solid at a constant rate, its temperature rises according to a graph like that in Figure 19.1. The graph can be divided into the following distinct sections:

AB As the solid absorbs energy, the kinetic energy of its molecules increases and its temperature rises.

BC The energy being absorbed now causes the solid to melt with no increase in temperature. The temperature at which this occurs is the melting point of the solid.

CD When all of the solid has melted the absorption of more energy causes a further increase in the kinetic energy of the molecules and the temperature of the liquid rises.

DE The energy now being absorbed causes the liquid to boil with no increase in temperature. The temperature at which this occurs is the boiling point of the liquid.

EF When all of the liquid has vaporised the absorption of more energy causes the kinetic energy of the gas molecules to increase and the temperature of the gas rises.

As the graph shows, a substance takes in latent heat twice in the course of heating, and may have two values of specific latent heat:

- the **specific latent heat of fusion** of a substance, L_f, is defined as being the amount of energy necessary to change the state of 1 kg of that substance from its solid state into a liquid without any change in temperature;
- the **specific latent heat of vaporisation** of a substance, L_v, is defined as being the amount of energy necessary to change the state of 1 kg of that substance from its liquid state into a gas without any change in temperature.

The relationship between the energy supplied and the mass of the substance, m, which changes state is described by the equation

$$\text{energy, } \Delta Q = mL$$

and so

$$L = \text{energy per unit mass} = \frac{\Delta Q}{m}$$

The unit of L is $\mathrm{J\,kg^{-1}}$.

Example 1

Calculate the energy necessary to **a** change 3.0 kg of ice at 0 °C into water at the same temperature and **b** change 4.0 kg of water at 100 °C into steam at the same temperature. The specific latent heat of fusion of water, L_f, is $330\,000\,\mathrm{J\,kg^{-1}}$ and the specific latent heat of vaporisation of water, L_v, is $2\,260\,000\,\mathrm{J\,kg^{-1}}$.

a $\Delta Q = mL_f$
$= 3.0\,\mathrm{kg} \times 330\,000\,\mathrm{J\,kg^{-1}}$
$= 990\,000\,\mathrm{J}$ or $990\,\mathrm{kJ}$

b $\Delta Q = mL_v$
$= 4.0\,\mathrm{kg} \times 2\,260\,000\,\mathrm{J\,kg^{-1}}$
$= 9\,040\,000\,\mathrm{J}$ or $9.0\,\mathrm{MJ}$

Measurement of the latent heat of vaporisation of a liquid

The apparatus is shown in Figure 19.2.

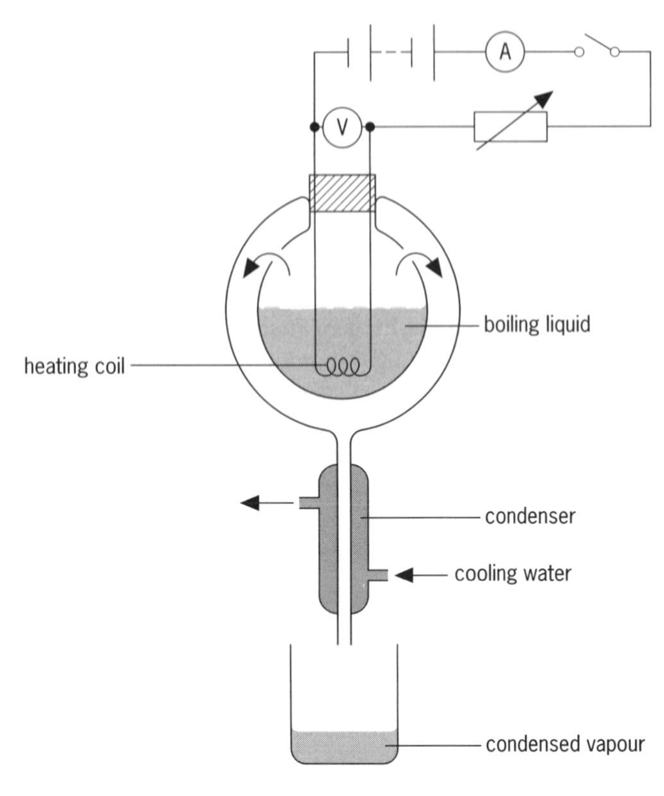

Figure 19.2 Measuring the latent heat of vaporisation of a liquid

The liquid is heated until it is boiling freely. A container is placed under the condenser and the condensed liquid collected over a known period of time, t. The mass of the liquid collected in this time, m, is measured. The current through the heating coil, I, and the pd across its ends, V, are noted.

electrical energy supplied by heater = energy required to vaporise liquid + energy lost to surroundings (H)

$$IVt = mL_v + H$$

If the experiment is repeated with different values of current and pd, I' and V', but the vaporised liquid collected over the same period of time, the heat lost to the surroundings in both parts of the experiment will be the same because the temperature of the vessel is the same.

$$\therefore \quad I'V't = m'L_v + H$$

where m' is the mass of liquid collected in time t.

If these two equations are subtracted the energy lost to the surroundings is eliminated from the calculation:

$$(IV - I'V')t = (m - m')L_v$$

Example 2

A student carried out the experiment described above to determine the specific latent heat of vaporisation of a liquid.

In his first experiment the heater voltage was 20 V and the current 3.0 A. The mass of liquid collected in 10 minutes was 0.085 kg.

In his second experiment the heater voltage was 10 V and the current 2.0 A. The mass of liquid collected in 10 minutes was 0.025 kg.

Calculate the specific latent heat of vaporisation of the liquid.

$$(IV - I'V')t = (m - m')L_v$$

Rearranging,

$$L_v = \frac{(IV - I'V')t}{(m - m')}$$

$$= \frac{(3.0\,\text{A} \times 20\,\text{V} - 2.0\,\text{A} \times 10\,\text{V})10 \times 60\,\text{s}}{(0.085 - 0.025)\,\text{kg}}$$

$$= \frac{(60 - 20) \times 600\,\text{J}}{0.060\,\text{kg}}$$

$$= \frac{24\,000}{0.060}\,\text{J}\,\text{kg}^{-1}$$

$$= 400\,000\,\text{J}\,\text{kg}^{-1}$$

Specific latent heat of fusion of water, $L_f = 330\,kJ\,kg^{-1}$
Specific latent heat of vaporisation of water, $L_v = 2.26\,MJ\,kg^{-1}$
Specific heat capacity of water, $c = 4200\,J\,kg^{-1}\,K^{-1}$
Density of water, $\rho = 1000\,kg\,m^{-3}$ or $1.0\,g\,ml^{-1}$

Level 1

1 a Sketch a graph of the temperature rise of a solid heated at a constant rate through its melting and boiling points. Label all the important aspects clearly.

b Which type of internal energy (kinetic or potential) could be used as a label on the vertical axis?

c Indicate the times when the potential energy is increasing but not the kinetic energy.

2 a What mass of ice at 0 °C needs 6.0 MJ of energy to melt it?

b What mass of water at 100 °C needs 6.0 MJ of energy to change it to vapour at the same temperature?

c Explain why the energy needed to *melt* a given mass of a solid is generally much less than that needed to *vaporise* the same mass of a liquid.

3 A well-insulated 1.2 kW jug kettle comes to the boil and fails to switch off. Over the next 4 minutes the mass of water decreases by 127 g. Calculate a value for the specific latent heat of vaporisation of water.

4 How long does it take a refrigerator to freeze 1.0 kg of water at 0 °C if the refrigerator is running at 85 W?

5 Finely crushed ice at 0 °C surrounds an electric heater placed in a filter funnel, as shown in Figure 19.3.

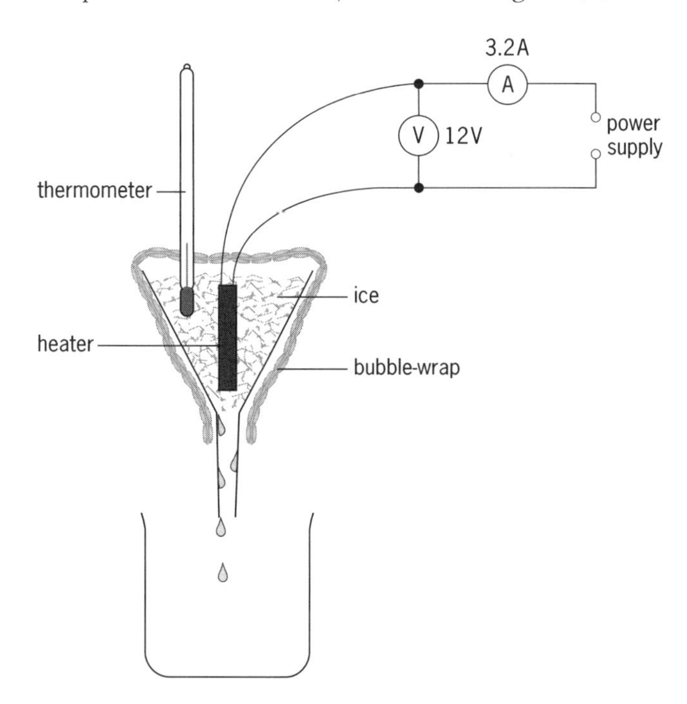

Figure 19.3

The whole apparatus is wrapped in bubble-wrap so you may assume no heat is gained from or lost to the surroundings. The water that drips out of the funnel in 5 minutes weighs 36 g. The heater is working at 12 V and 3.2 A.

a Calculate a value for the specific latent heat of fusion of ice from these measurements.

b If the insulation is not perfect will the calculated value be too high or too low?

6 Water is brought to the boil by an electric heater in a vacuum flask fitted with a condenser, Figure 19.4. A joulemeter is used to measure the energy supplied. What should it read when 10 ml of water has collected in the measuring cylinder?

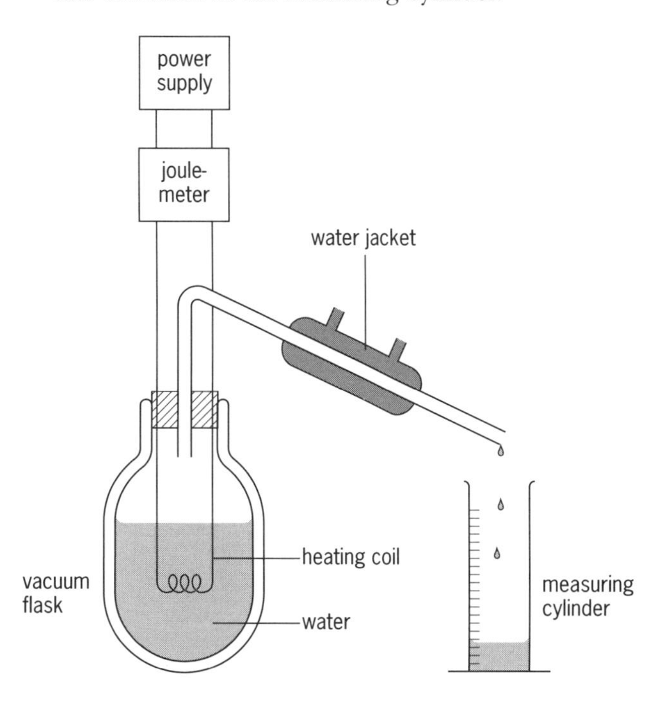

Figure 19.4

Level 2

7 When steam is added to heat a cup of coffee, it first condenses, giving out energy, and then it cools down, from 100 °C to the final drink temperature of 85 °C.

a How many joules are released per gram of steam condensing at 100 °C?

b How many joules are released when 1 gram of water cools from 100 °C to 85 °C?

c How many joules per gram are released in total?

d What temperature rise would this energy produce in a cup holding 150 ml of water?

e Roughly what mass of steam is required to bring each cup of coffee to the required temperature?

8 To cool cooked vegetables quickly from 100 °C, ready for freezing, they may be put into a mixture of ice and water at 0 °C. What mass of ice is melted when 5.0 kg of vegetables with specific heat capacity $2200\,J\,kg^{-1}\,K^{-1}$ are cooled in this way? (Assume that the cooling mixture stays at 0 °C.)

Level 3

9 A person running generates 800 W. The heat capacity of a typical human body is equivalent to 60 kg of water.

a Assuming that no energy is lost, how much will the runner's body temperature rise per minute?

In fact most of the heat generated is lost. If the runner is not going to overheat, 500 W must be lost, by sweating.

b What mass of perspiration per minute needs to evaporate to produce the desired cooling? (Perspiration can be considered to be just water.)

10 In a candle factory, a cylindrical tank of wax is melted and maintained above melting point, 56 °C. It has a diameter of 1.2 m and is filled to a depth of 1.5 m, Figure 19.5.

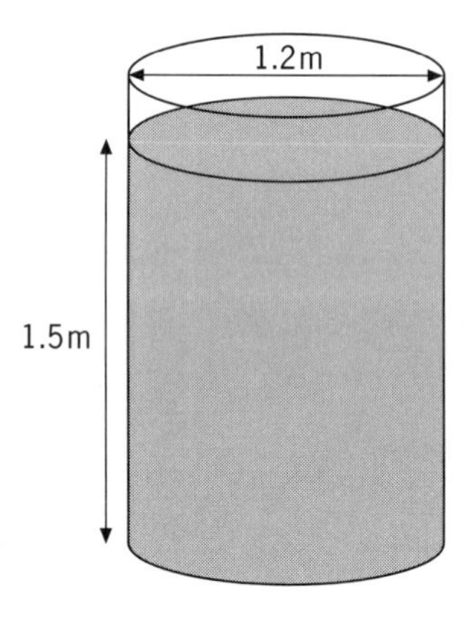

Figure 19.5

The specific heat capacity of the wax, both solid and liquid, is $2900\,\mathrm{J\,kg^{-1}\,K^{-1}}$. The latent heat of fusion of the wax, $L_f = 146\,\mathrm{kJ\,kg^{-1}}$ and its density, $\rho = 900\,\mathrm{kg\,m^{-3}}$.

a What mass of wax is there in the tank?

b How much energy is needed to melt it once it has reached its melting point?

c What power heater is required if the wax has to be melted and its temperature raised to 60 °C, starting at 12 °C, in just 12 hours? (You should assume that the tank is well insulated and that its heat capacity is negligible.)

Level 1

1 a See Figure 19.1, p. 135.

b The temperature is proportional to the average internal **kinetic energy** of the particles, so the vertical axis could be labelled with this.

c The temperature does not rise when the potential energy alone increases, so the horizontal parts BC and DE could be labelled 'potential energy increasing'.

2 a The energy gained or lost in any change of state is given by the equation

$\Delta Q = \text{mass} \times \text{the appropriate specific latent heat}$

When a material is changing from a solid, such as ice, to a liquid, such as water, you need to use the specific latent heat of fusion:

$\Delta Q = mL_f$

In this case you must rearrange the equation to make the mass, m, the subject:

$$m = \frac{\Delta Q}{L_v}$$

$$= \frac{6.0 \times 10^6\,\text{J}}{3.3 \times 10^5\,\text{J kg}^{-1}} = \mathbf{18.2\,kg} \text{ of ice}$$

b In this case you can use the same equation, but because the material is changing from a liquid to a vapour you must use the specific latent heat of vaporisation:

$\Delta Q = mL_v$

So

$$m = \frac{\Delta Q}{L_v}$$

$$= \frac{6.0 \times 10^6\,\text{J}}{2.26 \times 10^5\,\text{J kg}^{-1}} = \mathbf{2.65\,kg} \text{ water}$$

c Think of the increase in separation of the molecules. There is little difference between the spacing between molecules in solids and in liquids. It is much larger for vapours, when the separation increases tenfold. More work must be done against the intermolecular forces when a liquid is vaporised, so the specific latent heat of vaporisation is greater than that of fusion.

3 The heater in the kettle supplies energy at a rate of 1.2 kW, which is equal to $1.2 \times 10^3\,\text{J s}^{-1}$, for 4 minutes.

$$\begin{aligned}\text{total energy supplied} &= \text{power} \times \text{time}\\ &= 1.2 \times 10^3\,\text{J s}^{-1} \times 4 \times 60\,\text{s}\\ &= 288\,\text{kJ}\end{aligned}$$

This energy was sufficient to vaporise 127 grams of water.

The equation governing vaporisation is $\Delta Q = mL_v$, which you can rearrange to give the latent heat of vaporisation,

$$L_v = \frac{\Delta Q}{m}$$

$$= \frac{288 \times 10^3\,\text{J}}{0.127\,\text{kg}} = \mathbf{2.27 \times 10^6\,J\,kg^{-1}}$$

4 The energy gained or lost in any change of state is given by the equation

$\Delta Q = \text{mass} \times \text{the appropriate specific latent heat}$

In the case of water freezing the appropriate value is the specific latent heat of fusion; so

$\Delta Q = mL_f$

The energy to be removed in order to freeze 1 kg (already at 0 °C)

$= 1.0\,\text{kg} \times 330 \times 10^3\,\text{J kg}^{-1} = 3.3 \times 10^5\,\text{J}$

The refrigerator runs at $85\,\text{W} = 85\,\text{J s}^{-1}$.

You can rearrange the equation

$$\text{power} = \frac{\text{work}}{\text{time}}$$

to give the time required to do a specified amount of work at a given rate of working,

$$\text{time} = \frac{\text{work}}{\text{power}}$$

So, to provide 3.3×10^5 J at a rate of $85\,\text{J s}^{-1}$ (85 W) will take a time

$$t = \frac{3.3 \times 10^5\,\text{J}}{85\,\text{J s}^{-1}} = \mathbf{3882\,s} \quad \text{(which is a little over 1 hour)}$$

5 a Total energy supplied by the electric heater is given by the equation

$$\begin{aligned}\Delta Q &= \text{current} \times \text{potential difference} \times \text{time (s)}\\ &= IVt\\ &= 3.2\,\text{A} \times 12\,\text{V} \times 5.0 \times 60\,\text{s} = 11\,520\,\text{J}\end{aligned}$$

This is the energy used to melt the 36 grams of ice, for which

$\Delta Q = \text{mass} \times \text{specific latent heat of fusion of ice} = mL_f$

You can rearrange this equation to give

$$L_f = \frac{\Delta Q}{m}$$

$$= \frac{11\,520\,\text{J}}{0.036\,\text{kg}} = \mathbf{320\,000\,J\,kg^{-1}}$$

b If the insulation is not perfect then heat will be gained from the surroundings, which are at a higher temperature. So more ice will be melted than should be melted by the energy from the electric heater. So the calculated value for the specific latent heat of fusion will be too low.

6 The experiment produces 10 ml of water ($10 \times 10^{-6}\,\text{m}^3$), which will have a mass given by

$$\begin{aligned}\text{mass} = \text{density} \times \text{volume} &= 1000\,\text{kg m}^{-3} \times 10 \times 10^{-6}\,\text{m}^3\\ &= 10 \times 10^{-3}\,\text{kg}\end{aligned}$$

The energy needed to vaporise this amount of water, already at 100 °C, is given by

$\Delta Q = mL_v = 0.010\,\text{kg} \times 2.26 \times 10^6\,\text{J kg}^{-1} = \mathbf{22\,600\,J}$

The joulemeter should show that this quantity of energy has been supplied.

Level 2

7 a Latent heat of vaporisation will be released when the steam condenses. The quantity released per kilogram will be equal to the specific latent heat of vaporisation of water, $L_v = 2.26\,\text{MJ kg}^{-1}$. So, dividing this by 1000, each gram of steam condensing will release **2.26 kJ**.

b When water is cooled the quantity of energy released per kilogram per K is defined as its specific heat capacity, $c = 4200\,\text{J kg}^{-1}\text{K}^{-1}$. The general equation describing the process is

$$\Delta Q = mc\Delta\theta$$

For a temperature change from 100 °C to 85 °C, which equals 15 K, and a mass of 1 gram,

$$\Delta Q = 0.0010\,\text{kg} \times 4200\,\text{J kg}^{-1}\text{K}^{-1} \times 15\,\text{K} = \mathbf{63\,J}$$

This is small compared with the energy involved in vaporising or condensing 1 gram of water.

c The total energy released during the condensation and cooling to the lower temperature per gram of steam is

$$2260\,\text{J} + 63\,\text{J} = \mathbf{2320\,J} \quad \text{or} \quad \mathbf{2.32\,kJ}$$

d Using the density of water given at the start of this section of questions, which you might be expected to know, you should be able to calculate the mass of 150 ml of water to be 0.15 kg.

$$\Delta Q = mc\Delta\theta$$

$$\therefore \Delta\theta = \frac{\Delta Q}{mc} = \frac{2.32 \times 10^3\,\text{J}}{0.15\,\text{kg} \times 4200\,\text{J kg}^{-1}\text{K}^{-1}} = \mathbf{3.7\,K}$$

(If we ignore the extra mass of the added steam, then the increase in temperature of the water does not depend on its initial temperature.)

e To get a cup of coffee up to 'drinking temperature' from a room temperature of 15 °C requires a temperature rise of

$$85\,°\text{C} - 15\,°\text{C} = 70\,\text{K}$$

So, if the steam gives a temperature rise of $3.7\,\text{K g}^{-1}$, you are going to need

$$\frac{70\,\text{K}}{3.7\,\text{K g}^{-1}} = \mathbf{19\,g} \text{ of steam}$$

8 The vegetables start at 100 °C and cool to 0 °C. The mixture of ice and water stays at 0 °C, but some of the ice must melt. So the energy released by the vegetables will be equal to the energy used to melt the ice.

energy released by vegetables, ΔQ_{veg}

$$= m_{veg}c_{veg}\Delta\theta$$
$$= 5.0\,\text{kg} \times 2200\,\text{J kg}^{-1}\text{K}^{-1} \times 100\,\text{K} = 1.1\,\text{MJ}$$

This is equal to the energy required to melt the ice,

$$\Delta Q_{veg} = \Delta Q_{ice} = m_{ice}L_f$$

You can rearrange this to give

$$m_{ice} = \frac{\Delta Q_{ice}}{L_f} = \frac{\Delta Q_{veg}}{L_f}$$

$$= \frac{1.1 \times 10^6\,\text{J}}{3.3 \times 10^5\,\text{J kg}^{-1}} = \mathbf{3.3\,kg}$$

(If all the ice melted, the temperature of the cooling mixture would rise above 0 °C.)

Level 3

9 a $800\,\text{W} = 800\,\text{J s}^{-1}$, so in 1 minute the runner will generate

$$800\,\text{J s}^{-1} \times 1 \times 60\,\text{s} = \mathbf{48\,000\,J}$$

Assume that all this energy is used to raise the runner's temperature. The process is governed by the equation $\Delta Q = mc\Delta\theta$, with mc equal to the *equivalent heat capacity* of the runner, which is that of 60 kg of water with a specific heat capacity of $4200\,\text{J kg}^{-1}\text{K}^{-1}$.

You can rearrange this equation to give the increase in temperature, in 1 minute:

$$\Delta\theta = \frac{\Delta Q}{mc} = \frac{48 \times 10^3\,\text{J}}{60\,\text{kg} \times 4200\,\text{J kg}^{-1}\text{K}^{-1}} = \mathbf{0.19\,K}$$

b $500\,\text{W} = 500\,\text{J s}^{-1}$, so in 1 minute the runner must lose

$$500\,\text{J s}^{-1} \times 60\,\text{s} = 30\,000\,\text{J}$$

If this energy is to be lost, in 1 minute, by vaporising water (sweat) then the process will be governed by the equation $\Delta Q = mL_v$, which you can rearrange to give the mass that must be vaporised each minute,

$$m = \frac{\Delta Q}{L_v} = \frac{30 \times 10^3\,\text{J}}{2.26 \times 10^6\,\text{J kg}^{-1}}$$
$$= \mathbf{0.013\,kg} \quad \text{or} \quad \mathbf{13\,g}$$

10 a The volume of wax in the cylindrical tank is given by the equation $V = \pi r^2 h$ where r, the radius of the tank $= \frac{1}{2} \times$ diameter $= 0.6\,\text{m}$, and h, the height of the wax $= 1.5\,\text{m}$.

$$\therefore V = \pi \times (0.6\,\text{m})^2 \times 1.5\,\text{m} = 1.7\,\text{m}^3$$

You can rearrange the equation

$$\text{density, } \rho = \frac{m}{V}$$

to give

$$m = V \times \rho = 1.7\,\text{m}^3 \times 900\,\text{kg m}^{-3} = \mathbf{1530\,kg}$$

b The energy needed to melt the wax, once it has reached its melting point, is given by

$$\Delta Q = mL_f = 1530\,\text{kg} \times 146 \times 10^3\,\text{J kg}^{-1} = \mathbf{223\,MJ}$$

c The total energy needed is equal to the energy needed to raise the temperature of the wax to its melting point, 56 °C, plus the energy needed to actually melt the wax, plus the energy needed to raise the temperature of the melted wax from 56 °C to the final temperature, 60 °C.

To get the wax from 12 °C to 56 °C, a difference of 44 K, requires

$$\Delta Q_1 = mc\Delta Q = 1530\,\text{kg} \times 2900\,\text{J kg}^{-1}\,\text{K}^{-1} \times 44\,\text{K} = 195\,\text{MJ}$$

To melt the wax at 56 °C, from part **b**, requires

$$\Delta Q_2 = 223\,\text{MJ}.$$

To get the melted wax from 56 °C to 60 °C, a difference of 4 K, requires

$$\Delta Q_3 = mc\Delta Q = 1530\,\text{kg} \times 2900\,\text{J kg}^{-1}\,\text{K}^{-1} \times 4\,\text{K} = 18\,\text{MJ}$$

So, the total energy needed = 195 MJ + 223 MJ + 18 MJ = 436 MJ.

In fact, because the specific heat capacity is the same before and after melting, in this case you could just calculate the energy needed to raise the temperature of the wax from 12 °C to 60 °C, a difference of 48 K.

To supply this energy in 12 hours you would need to supply it at a rate of

$$\frac{436\,\text{MJ}}{12\,\text{h}} = \frac{436\,\text{MJ}}{12 \times 60 \times 60\,\text{s}} = 10\,\text{kJ s}^{-1} = \mathbf{10\,kW}$$

20 Kinetic theory of matter

The **kinetic theory** seeks to explain the basic properties of solids, liquids and gases in terms of the motions of their atoms and molecules.

Solids have a definite shape and volume. Their atoms are held in place by mainly electrostatic forces but are able to vibrate about an equilibrium position. At low temperatures these vibrations are slight but as the temperature of the solid increases the amplitude of the vibrations increases.

Liquids do not have a definite shape. They are able to flow and take the shape of the container in which they are placed. But, at a given temperature they do, like solids, have a definite volume. The atoms of solids and liquids are closely packed together and, as a result, they are dense materials and very difficult to compress. In general, solids and liquids are regarded as being incompressible.

The atoms or molecules of a gas are not closely packed. Gases, therefore, have much lower densities than solids or liquids and are compressible.

Nearly all the mass of an atom is in the very small volume of its nucleus, which is therefore very dense. Density and pressure are important concepts in the kinetic theory.

Density

The density of an object is defined as being its mass per unit volume and can be calculated using the equation

$$\text{density} = \frac{\text{mass}}{\text{volume}} \quad \text{or} \quad \rho = \frac{m}{V}$$

Example 1

Calculate the density of **a** a $2.3 \times 10^{-4}\,\text{m}^3$ block of copper with a mass of 2.0 kg, **b** a sample of air which has a volume of $3.0\,\text{m}^3$ and a mass of 3.9 kg.

Using density, $\rho = \dfrac{m}{V}$:

a $\rho = 2.0\,\text{kg}/2.3 \times 10^{-4}\,\text{m}^3 = 8.7 \times 10^3\,\text{kg}\,\text{m}^{-3}$

b $\rho = 3.9\,\text{kg}/3.0\,\text{m}^3 = 1.3\,\text{kg}\,\text{m}^{-3}$

Pressure

The fact that solids and liquids are incompressible means that forces or pressure can be transmitted through them. Pressure is defined as force per unit area and is measured in the unit **pascal** (Pa). 1 Pa is the equivalent of $1\,\text{N}\,\text{m}^{-2}$. Pressure can be calculated using the equation

$$\text{pressure} = \frac{\text{force}}{\text{area}} \quad \text{or} \quad p = \frac{F}{A}$$

Example 2

A 40 N weight is placed on top of a vertical wooden cylinder of uniform cross-sectional area $5.0 \times 10^{-4}\,\text{m}^2$. Assuming that the weight of the wooden cylinder is negligible, calculate the pressure at its base.

$$\text{pressure, } p = \frac{F}{A}$$

$$= \frac{40\,\text{N}}{5.0 \times 10^{-4}\,\text{m}^2} = 8.0 \times 10^4\,\text{Pa}$$

In the above example the pressure at the top of the block is the same as at the bottom, but if the cross-sectional area of the cylinder had varied then the *pressure* transmitted would vary. The *force* transmitted would be constant at 40 N.

Kinetic theory of gases

It is assumed that the molecules of a gas are completely free and, unlike the molecules of a solid or liquid, experience no intermolecular forces. Their velocities, even at room temperature, are very high, for example $500\,\text{m}\,\text{s}^{-1}$. Support for these ideas is found in two simple experiments.

Brownian motion experiment

If smoke particles are introduced into a glass cell and observed using a microscope, Figure 20.1a, they are seen to move around in a random, haphazard fashion, Figure 20.1b. It is believed that they do this because they are being randomly jostled by the much smaller but fast-moving air molecules. This confirms that the molecules of a gas are in continuous, random motion, called **Brownian motion**.

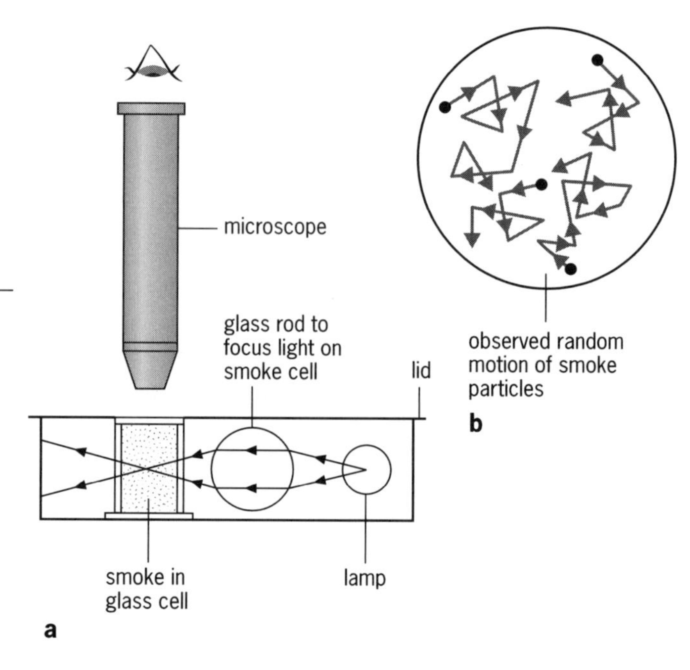

Figure 20.1 Viewing Brownian motion

Diffusion experiment

The apparatus is shown in Figure 20.2.

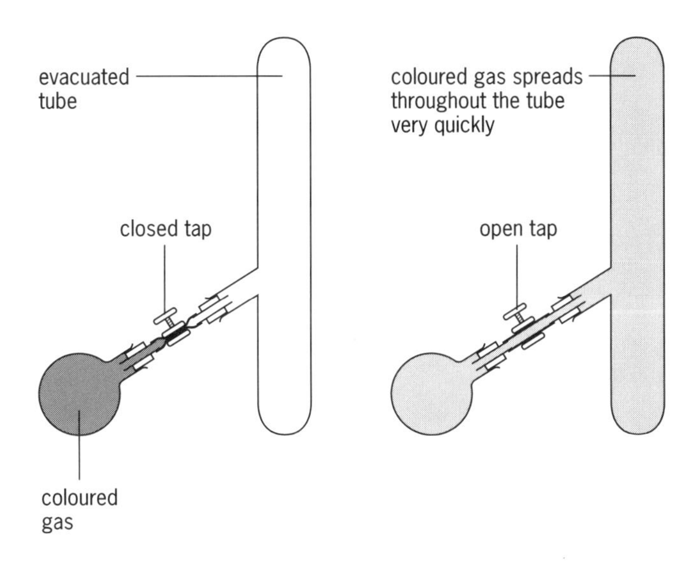

Figure 20.2 Demonstrating diffusion

When the tap is opened the coloured gas is seen to fill the tube almost instantaneously, confirming that the molecules of a gas are not restricted by intermolecular forces and that they have high velocities. If the experiment is repeated but with a tube filled with air rather than a vacuum, the coloured gas again spreads throughout the tube but it takes much longer. This is because of collisions between the coloured gas and air molecules. The spreading of one substance through another is called **diffusion**.

Gas Laws

Further evidence on which we can base our model of gas behaviour comes from the **Gas Laws**. These are three laws which have been discovered through experiments.

Boyle's Law

If the pressure and volume of a fixed mass of gas at constant temperature are measured over a range of values and the data plotted on a graph, results such as those in Figure 20.3 are obtained.

Pressure, *p*/Pa	**Volume, *V*/cm^3**
1.0×10^5	120
2.0×10^5	60
3.0×10^5	40
4.0×10^5	30
5.0×10^5	24
6.0×10^5	20

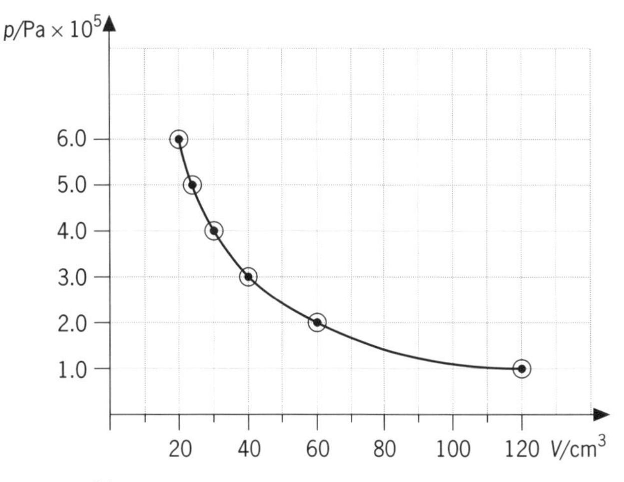

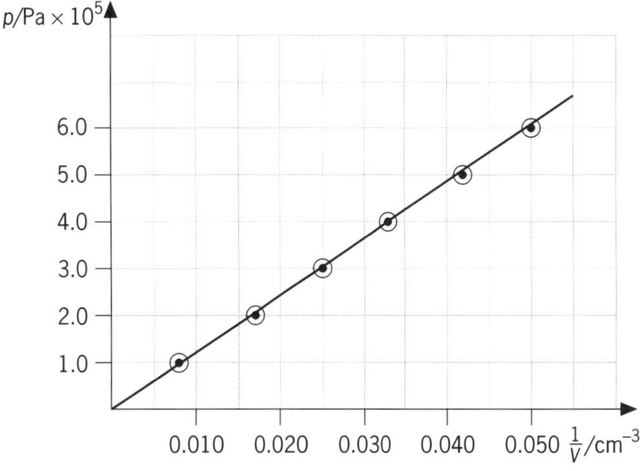

Figure 20.3 Investigating the relationship between pressure and volume of a gas at constant temperature

The shapes of the graphs indicate there is an *inverse relationship* between p and V, that is $p \propto 1/V$. If we double p we halve V. This relationship can be written in the form of an equation:

$$p \times V = \text{a constant} \quad \text{or} \quad p_1V_1 = p_2V_2$$

These results are summarised by **Boyle's Law**, which states that **the volume of a fixed mass of gas is inversely proportional to the pressure provided the temperature remains constant**.

Example 3

A sample of gas whose volume is $1.5\,\text{m}^3$ and pressure $3.0 \times 10^5\,\text{Pa}$ is compressed slowly so that its temperature remains constant until it occupies a volume of $0.50\,\text{m}^3$. Calculate the new pressure of the gas.

$$p_1V_1 = p_2V_2$$

$$\therefore \quad p_2 = \frac{p_1V_1}{V_2}$$

$$= \frac{3.0 \times 10^5\,\text{Pa} \times 1.5\,\text{m}^3}{0.50\,\text{m}^3}$$

$$= 9.0 \times 10^5\,\text{Pa}$$

Pressure Law

If the pressure and temperature of a fixed mass of gas at constant volume are measured over a range of values and the data plotted on a graph, results such as those in Figure 20.4 are obtained.

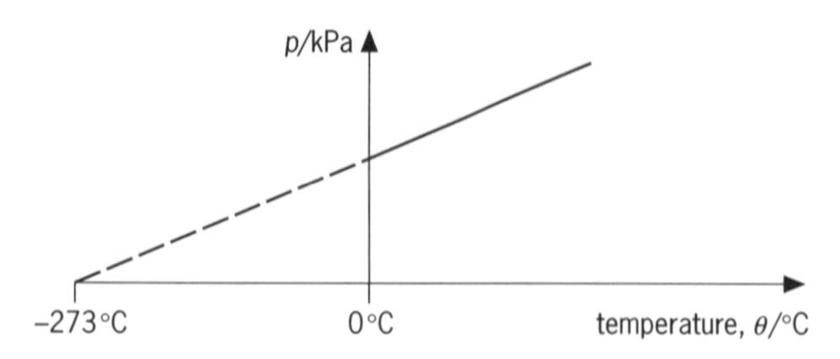

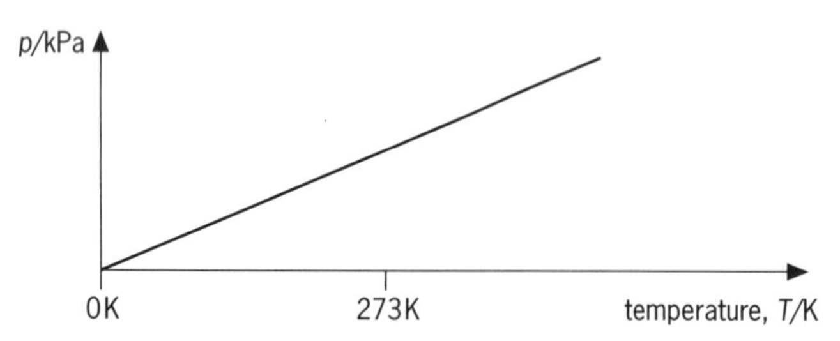

Figure 20.4 The relationship between pressure and temperature of a gas at constant volume

The shapes of the graphs indicate there is a *direct proportionality* between p and T, providing the temperature is the absolute temperature (see topic 18), measured in kelvin. This relationship can be written in the form of an equation:

$$\frac{p}{T} = \text{a constant} \quad \text{or} \quad \frac{p_1}{T_1} = \frac{p_2}{T_2}$$

These results are summarised by the **Pressure Law**, which states that **the pressure of a fixed mass of gas is proportional to the absolute temperature provided the volume remains constant**.

Example 4

A sample of gas in a container of fixed volume is heated from 27 °C to 227 °C. Calculate the final pressure of the gas if its original pressure was 3.0×10^5 Pa.

$$\frac{p_1}{T_1} = \frac{p_2}{T_2}$$

$$\therefore \quad p_2 = \frac{p_1 T_2}{T_1}$$

Remember the temperatures must be in kelvin. Add 273 to the temperature in °C.

$$p_2 = \frac{3.0 \times 10^5\,\text{Pa} \times (227 + 273)\,\text{K}}{(27 + 273)\,\text{K}}$$

$$= 3.0 \times 10^5\,\text{Pa} \times \frac{500}{300}$$

$$= 5.0 \times 10^5\,\text{Pa}$$

Charles' Law

If the volume and temperature of a fixed mass of gas at constant pressure are measured over a range of values and the data plotted on a graph, results such as those in Figure 20.5 are obtained.

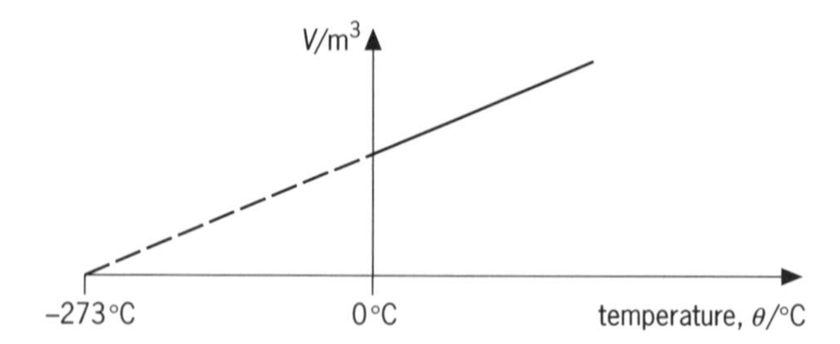

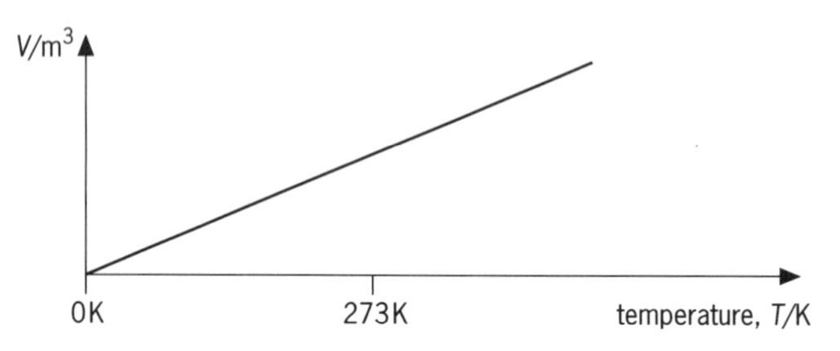

Figure 20.5 The relationship between volume and temperature of a gas at constant pressure

The shapes of the graphs indicate there is a direct proportionality between V and T, providing the temperature is the absolute temperature, measured in kelvin. This relationship can be written in the form of an equation:

$$\frac{V}{T} = \text{a constant} \quad \text{or} \quad \frac{V_1}{T_1} = \frac{V_2}{T_2}$$

These results are summarised by **Charles' Law**, which states that **the volume of a fixed mass of gas is proportional to the absolute temperature provided the pressure remains constant**.

Example 5

A sample of gas expands as it is heated at constant pressure from 77 °C to 227 °C. Calculate the final volume of the gas if its original volume was $0.70\,\text{m}^3$.

$$\frac{V_1}{T_1} = \frac{V_2}{T_2}$$

$$\therefore \quad V_2 = \frac{V_1 T_2}{T_1}$$

$$= \frac{0.70\,\text{m}^3 \times (227 + 273)\,\text{K}}{(77 + 273)\,\text{K}}$$

$$= 0.70\,\text{m}^3 \times \frac{500}{350}$$

$$= 1.0\,\text{m}^3$$

Ideal gas equation

The three Gas Laws: Boyle's Law, the Pressure Law and Charles' Law, can be combined into one equation:

$$\frac{p_1 V_1}{T_1} = \frac{p_2 V_2}{T_2}$$

where the subscripts 1 and 2 refer to the states of a fixed mass of gas before and after a change (Figure 20.6).

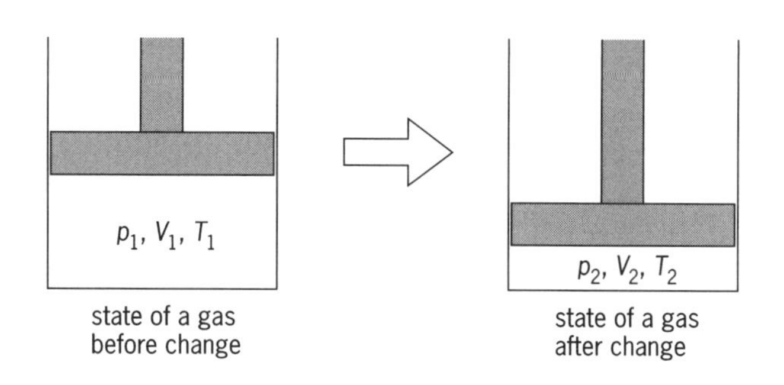

Figure 20.6

Example 6

$10.0\,m^3$ of gas at $27\,°C$ and a pressure of $3.0 \times 10^5\,Pa$ is heated until its temperature has increased to $627\,°C$ and its volume increased to $15.0\,m^3$. Calculate the new pressure of the gas.

$$\frac{p_1 V_1}{T_1} = \frac{p_2 V_2}{T_2}$$

$$\therefore \quad p_2 = \frac{p_1 V_1 T_2}{V_2 T_1}$$

$$= \frac{3.0 \times 10^5\,Pa \times 10.0\,m^3 \times (627 + 273)\,K}{15.0\,m^3 \times (27 + 273)\,K}$$

$$= 6.0 \times 10^5\,Pa$$

Alternatively, the equation can be written as

$$\frac{pV}{T} = \text{constant}$$

For one mole of gas this constant is R, the **universal molar gas constant**, which has the value $8.31\,J\,mol^{-1}\,K^{-1}$

So, for one mole of gas,

$$pV = RT$$

For n moles of gas the equation becomes

$$pV = nRT$$

This is called the **ideal gas equation**, or the **equation of state** for an ideal gas.

Example 7

Calculate the pressure of 8.0 moles of a gas which at a temperature of $177\,°C$ occupy a volume of $0.10\,m^3$.

$$pV = nRT$$

$$\therefore \quad p = \frac{nRT}{V}$$

$$177\,°C = (177 + 273)\,K = 450\,K$$

$$p = \frac{8.0\,mol \times 8.31\,J\,mol^{-1} \times 450\,K}{0.10\,m^3}$$

$$= 3.0 \times 10^5\,Pa$$

The mole

The **mole** is the unit for measuring the amount of a substance. One mole (1 mol) of a substance contains 6.02×10^{23} particles. The value $6.02 \times 10^{23}\,mol^{-1}$ is known as the **Avogadro constant**. Knowing the Avogadro constant and the number of moles of a substance allows us to calculate the number of atoms or molecules it contains. In the above example the 8.0 moles of gas consists of

$$8.0\,mol \times 6.02 \times 10^{23}\,mol^{-1} = 4.8 \times 10^{24}\ \text{particles}$$

The **molar mass** of a substance is that mass which contains 6.02×10^{23} particles; when expressed in $g\,mol^{-1}$ it is numerically equal to the relative atomic or relative molecular mass of the substance. For example:

- The relative molecular mass of oxygen is 32; its molar mass is $32\,g\,mol^{-1}$ and therefore a sample of 32 g of oxygen will contain 6.02×10^{23} molecules of oxygen.
- The relative atomic mass of sodium is 23; its molar mass is $23\,g\,mol^{-1}$ and therefore a sample of 23 g of sodium will contain 6.02×10^{23} atoms of sodium.

Ideal and real gases

A gas which obeys the equation pV/T = a constant is described as being an **ideal gas**. In practice most real gases obey this equation only if

1 the density of the gas is low; and
2 the temperature of the gas is well above its boiling point.

Under these conditions there are almost no intermolecular forces between the particles of the gas, and the volume of the particles of gas is negligible compared with the volume of the gas as a whole. At lower temperatures and high pressures the intermolecular forces between the gas particles, and their finite volume, cause them not to behave as ideal gases.

Molar gas constant, $R = 8.3\,\mathrm{J\,K^{-1}\,mol^{-1}}$
Molar mass of hydrogen $= 2.0\,\mathrm{g\,mol^{-1}}$
Molar mass of helium $= 4.0\,\mathrm{g\,mol^{-1}}$
Molar mass of oxygen $= 32.0\,\mathrm{g\,mol^{-1}}$
Avogadro constant, $N_A = 6.0 \times 10^{23}\,\mathrm{mol^{-1}}$
Density of water $= 1000\,\mathrm{kg\,m^{-3}}$
Density of air $= 1.3\,\mathrm{kg\,m^{-3}}$
Density of lead $= 11.3 \times 10^3\,\mathrm{kg\,m^{-3}}$

Level 1

1 The pressure of a fixed mass of ideal gas is proportional to the absolute temperature if the volume remains constant.

A bubble is trapped in resin at atmospheric pressure and 300 K and is unable to expand.

a What temperature is 300 K in degrees C?
b Roughly how many pascal is atmospheric pressure?

Hint: Use a pressure of '1 atmosphere' in your calculations if you don't know this.

c The temperature doubles to 600 K. What is the new pressure?
d The temperature trebles to 900 K. What is the new pressure?
e What is this final temperature in degrees C?

2 The bubble in the previous question has a volume of 1 mm^3 at atmospheric pressure and 300 K. It is now able to expand and the pressure stays constant.

a What will be the volume at 600 K?

Hint: Don't bother changing to m^3.

b What will be the volume at 900 K?

3 If a fixed mass of ideal gas is kept at constant temperature, pressure × volume is constant.

An airship has a volume of $1.0 \times 10^4\,\mathrm{m^3}$ at ground level. It rises to a height where the pressure is 80% of atmospheric but the temperature is the same as at ground level. What is the new volume of the airship?

4 At the start of a journey a tyre has a pressure of 270 kPa at a temperature of 15 °C. At the end of the journey the pressure is 300 kPa. Find the new temperature. (Assume that the tyre stays at constant volume.)

5 a Calculate the pressure exerted on the floor by a person weighing 800 N who has 0.050 m^2 area of feet in contact with the floor.
b The pressure of the atmosphere is about 100 kPa.
i Estimate the force on the top surface of a table measuring 1.0 m by 1.5 m.
ii Why don't the legs collapse under this force?
c i What area would have a force of 12 N on it due to the pressure of the atmosphere?
ii Change your answer to mm^2 and give the dimensions of a rectangle of this area.

6 a Write down an estimate of the mass of 1.0 m^3 (numerically, the density) of
i carbon dioxide gas
ii ice at 0 °C
iii solid gold.

Hint: Refer to the densities of substances listed at the beginning of the questions.

b What is the average density of a box 0.10 m × 0.10 m × 0.10 m in size which has mass 2.5 kg?
c What is the mass of the water in a swimming pool 25 m × 10 m × 3 m?
d What volume of air has a mass of 10 tonne?

Level 2

7 A helium balloon has a volume of $2.0 \times 10^{-3}\,\mathrm{m^3}$ at a pressure of 110 kPa and a temperature of 295 K.

a Calculate the number of moles of gas molecules in the balloon.

Hint: Use $pV = nRT$.

b What is the mass of the gas in the balloon?

8 Hydrogen and oxygen are stored in gas cylinders with the same volume at the same temperature and pressure.

a Will each cylinder contain the same number of molecules?
b Will each cylinder contain the same mass of gas?

9 8.0 g of hydrogen and 8.0 g of helium are mixed together in a container that has a volume of 0.30 m^3. The temperature is 300 K.

a How many moles of hydrogen are there?
b How many moles of helium are there?
c What is the total number of moles?
d Calculate the pressure inside the container.
e What is the pressure due to the hydrogen alone?

10 2.0 litre of an inert gas is heated in a closed vessel and the pressure recorded as the temperature increases, with the results shown in Figure 20.7. Use the data in the graph to calculate:

a how many moles of gas there are in the vessel

b how many molecules of gas there are in the vessel.

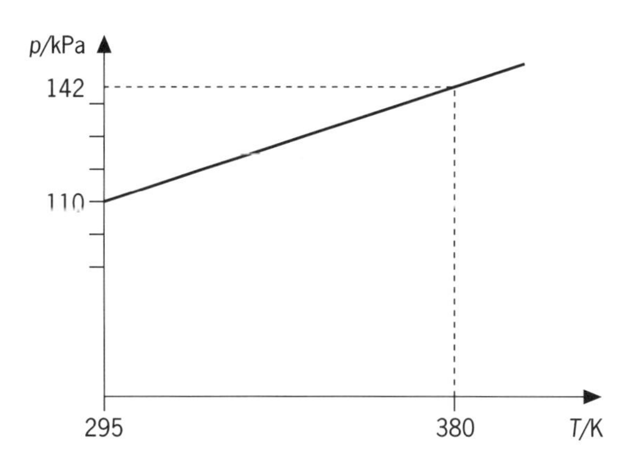

Figure 20.7

11 The lift produced by an aeroplane wing is caused by a pressure difference between the upper and lower wing surfaces as it moves through the air. What must be the difference in pressure for a jet plane in straight and level flight if it has a loaded weight of 3.0 MN and the total wing area is about $400\,m^2$?

12 All nuclei have approximately the same density. The mass of a copper nucleus is 1.1×10^{-25} kg and its radius is 4.8×10^{-15} m.

a Calculate the density of this nucleus, taking it to be spherical.

b Find the radius of a lead nucleus of mass 3.5×10^{-25} kg.

Level 3

13 a Justify the suggestion that a single mole of any gas will occupy the same volume under the same conditions of temperature and pressure.

b Find the volume of 1.0 mole of a gas at 300 K and 100 kPa.

c The volume of a particular gas molecule is $2.0 \times 10^{-29}\,m^3$. What percentage of the molar volume is the volume occupied by the molecules?

d Comment on the treatment of the gas as ideal, with reference to part c.

Level 1

1 a The absolute zero of temperature, 0 K, is equal to −273 °C.

To convert temperatures in K to their equivalent in °C, subtract 273 from the value in K.

∴ 300 K in °C is (300 − 273) = **27 °C**.

b Atmospheric pressure is roughly **100 kPa**.

You should remember this value.

c For a fixed volume, the pressure will double when the absolute temperature doubles, since

$$\frac{p_1}{T_1} = \frac{p_2}{T_2}$$

In this case 1 atmosphere (=100 kPa) will become **2 atmospheres** = **200 kPa**.

d The pressure will treble when the absolute temperature trebles, so 100 kPa will become 3 × 100 kPa = **300 kPa** = **3 atmospheres**.

e 900 K in °C is (900 − 273) = **627 °C**.

2 a When the absolute temperature doubles at constant pressure, the volume of the bubble will also double since

$$\frac{V_1}{T_1} = \frac{V_2}{T_2}$$

The new volume will be **2 mm³**.

b When the absolute temperature trebles, the volume of the bubble will also treble, to **3 mm³**.

3 Because pressure × volume is constant,

$$p_1V_1 = p_2V_2$$

With the airship on the ground, p_1 = 1 atm = 100 kPa, $V_1 = 1.0 \times 10^4\,\text{m}^3$.
With the airship aloft, p_2 = 80% of 1 atm = 0.8 atm = 80 kPa.

You can rearrange the above equation to give the new volume of the airship:

$$V_2 = \frac{p_1V_1}{p_2} = \frac{100\,\text{kPa} \times 1.0 \times 10^4\,\text{m}^3}{80\,\text{kPa}} = \mathbf{1.25 \times 10^4\,m^3}$$

4 For the fixed mass of gas (air) inside the tyre,

$$\frac{p_1V_1}{T_1} = \frac{p_2V_2}{T_2}$$

Remember that all the temperatures in these calculations must be in K.

The temperature at the start of the journey, T_1 = 15 °C = (273 + 15) = 288 K, and the pressure, p_1 = 270 kPa. At the end of the journey, p_2 = 300 kPa.

You have been told that the volume of the tyre stays constant and you have to assume that it doesn't leak!

Because the volume, $V_1 = V_2$, is constant, the equation is simplified to

$$\frac{p_1}{T_1} = \frac{p_2}{T_2}$$

which you can rearrange to give

$$T_2 = \left(\frac{p_2}{p_1}\right)T_1$$

$$= \frac{300 \times 10^3\,\text{Pa}}{270 \times 10^3\,\text{Pa}} \times 288\,\text{K}$$

= **320 K** or (320 − 273) °C = **47 °C**

5 a $\text{Pressure} = \frac{\text{force}}{\text{area}} = \frac{800\,\text{N}}{0.050\,\text{m}^2}$

$= 16\,000\,\text{N}\,\text{m}^{-2} = \mathbf{16\,kPa}$

The pascal, Pa, is one newton per square metre.

b i Area of table = 1.5 m × 1.0 m = 1.5 m². The equation pressure = force/area can be rearranged to give

force = pressure × area = $100\,\text{kN}\,\text{m}^{-2} \times 1.5\,\text{m}^2$
= **150 kN**

ii There is almost exactly the same force pushing *up* on the under surface of the table.

At the same depth in a fluid pressure acts equally in all directions and the difference in pressure because of the thickness of the table top will be negligible.

c i Pressure = force/area can be rearranged to give

$$\text{area} = \frac{\text{force}}{\text{pressure}} = \frac{12\,\text{N}}{100\,\text{kN}\,\text{m}^{-2}} = \mathbf{1.2 \times 10^{-4}\,m^2}$$

ii There are $10^6\,\text{mm}^2$ in 1 m²,

so $1.2 \times 10^{-4}\,\text{m}^2 = 120 \times 10^{-6}\,\text{m}^2 = \mathbf{120\,mm^2}$

A rectangle 12 mm by 10 mm gives this area, so it is about one square centimetre.

6 a i Carbon dioxide is used in fire extinguishers because it sinks, covering the fire. It is denser than air, so your answer should be greater than 1.3 kg. In fact the density is $1.98\,\text{kg}\,\text{m}^{-3}$ so 1 m³ has mass of approximately **2 kg**.
ii Ice is less dense than water since a block of ice floats in water. The mass of 1 m³ is about **920 kg**.
iii Gold is very dense, even denser than lead (at $11\,300\,\text{kg}\,\text{m}^{-3}$). Its density is $19\,300\,\text{kg}\,\text{m}^{-3}$ so 1 m³ has mass 19 300 kg or 19.3 tonne.

b Volume of box = 0.10 m × 0.10 m × 0.10 m = $1.0 \times 10^{-3}\,\text{m}^3$

$$\text{density} = \frac{\text{mass}}{\text{volume}} = \frac{2.5\,\text{kg}}{1.0 \times 10^{-3}\,\text{m}^3} = \mathbf{2500\,kg\,m^{-3}}$$

c Volume of water in pool = 25 m × 10 m × 3 m = 750 m³

mass = density × volume = $1000\,\text{kg}\,\text{m}^{-3} \times 750\,\text{m}^3$
= $\mathbf{7.5 \times 10^5\,kg}$ or 750 tonne

d $\text{Volume} = \frac{\text{mass}}{\text{density}} = \frac{10 \times 10^3\,\text{kg}}{1.3\,\text{kg}\,\text{m}^{-3}} = \mathbf{7700\,m^3}$

At a volume of nearly 8000 m³ this would be a cube of side 20 m, obtained by using the cube root function on the calculator.

Level 2

7 a You can rearrange the equation $pV = nRT$, to make the number of moles, n, the subject:

$$n = \frac{pV}{RT}$$

In this case, the pressure in the balloon is 110 kPa, its volume is $2.0 \times 10^{-3}\,m^3$ and the temperature is 295 K. The molar gas constant $R = 8.3\,J\,K^{-1}\,mol^{-1}$. So

$$n = \frac{110 \times 10^3\,Pa \times 2.0 \times 10^{-3}\,m^3}{8.3\,J\,K^{-1}\,mol^{-1} \times 295\,K} = 0.090\,Pa\,m^3\,J^{-1}\,mol$$

The units of this expression look a bit complicated! But remember that $Pa \equiv N\,m^{-2}$ and that $J \equiv N\,m$.

$\therefore\ Pa\,m^3\,J^{-1}\,mol \equiv N\,m^{-2}\,m^3\,N^{-1}\,m^{-1}\,mol \equiv mol$

So the answer is simply **0.090 mol**.

b 1 mole of helium has mass 4.0 g, so the mass of 0.090 mol is

$0.090\,mol \times 4.0\,g\,mol^{-1} = \mathbf{0.36\,g}$ or $\mathbf{3.6 \times 10^{-4}\,kg}$

8 a For any (ideal) gas $pV = nRT$. You can rearrange this equation to give $n = pV/RT$.

The cylinders of oxygen and hydrogen are at the same temperature and pressure and have the same volume, and the value of R is a constant that does not depend on the type of gas. So for each cylinder the number of moles, $n = pV/RT$, must be the same.

The Avogadro constant, N_A, the number of molecules in a mole, is a constant for all substances. So the number of molecules in each cylinder is equal to $n \times N_A$, and is the same for the two cylinders. This is **Avogadro's Law**.

b Although the cylinders contain the same number of molecules, a molecule of oxygen is heavier than a molecule of hydrogen, so the total mass of gas in the oxygen cylinder will be the greater.

9 a The molar mass of hydrogen is 2.0 g per mol. So 8.0 g of hydrogen molecules is 4.0 mol.

b The molar mass of helium is 4.0 g per mol. So 8.0 g of helium molecules is 2.0 mol.

c The total number of moles, $n = 4.0\,mol + 2.0\,mol = 6.0\,mol$.

d The general equation for ideal gases applies: $pV = nRT$. You can rearrange this equation to give

$$p = \frac{nR}{V}T = \frac{6.0\,mol \times 8.3\,J\,K^{-1}\,mol^{-1}}{0.30\,m^3} \times 300\,K$$

$$= 50\,000\,J\,m^{-3} = 50\,000\,N\,m\,m^{-3}$$
$$= 50\,kN\,m^{-2} = \mathbf{50\,kPa}$$

e When the equation for ideal gases applies, the pressure is proportional to the number of moles (or the number of molecules). Four-sixths (two-thirds) of the molecules in the container are hydrogen molecules. So hydrogen contributes two-thirds of the total pressure:

$$p_{hydrogen} = \frac{2}{3} \times 50\,kPa = \mathbf{33\,kPa}$$

10 a You can rearrange the equation for an ideal gas, $pV = nRT$, to give

$$p = \frac{nR}{V}T$$

This equation shows that if you plot P against T, with V held constant, you should get a straight line with a gradient, $m = nR/V$. Once you have found the gradient m, you can find n because you can rearrange this equation to give $n = m \times V/R$.

From the graph (Figure 20.7), for a change in temperature $\Delta T = 380\,K - 295\,K = 85\,K$ there is a change in pressure $\Delta p = 142\,kPa - 110\,kPa = 32\,kPa$. So the gradient

$$m = \frac{\Delta p}{\Delta T} = \frac{32 \times 10^3\,kPa}{85\,K} = 376\,Pa\,K^{-1}$$

Now you have to multiply the gradient by V/R, but you have been given V in litres and you must convert it to m^3.

How do YOU remember how to change litres to m^3? One way is to remember that a cubic box of side 10 cm holds 1 litre and that there would be ten of these boxes along each side of a cubic metre so there would be $10 \times 10 \times 10 = 1000$ litre in $1\,m^3$. ***Or*** *you can remember that 1 litre is $1000\,cm^3$, and that there are 100 cm along each side of a 1 m cube so that $1\,m^3 = 100\,cm \times 100\,cm \times 100\,cm = 10^6\,cm^3$ or 10^3 litre.*

Whichever way you get to the answer you should find that 2.0 litre $= 2.0 \times 10^{-3}\,m^3$. So

$$n = m \times \frac{V}{R} = 376\,Pa\,K^{-1} \times \frac{2.0 \times 10^{-3}\,m^3}{8.3\,J\,K^{-1}\,mol^{-1}}$$

$$= 0.091\,Pa\,m^3\,J^{-1}\,mol$$

These units look complicated, but remember that $Pa = N\,m^{-2}$ and $J = N\,m$, so the units become $N\,m^{-2}\,m^3\,N^{-1}\,m^{-1}\,mol$, which is just mol.

$\therefore\ n = \mathbf{0.091\,mol}$

b The number of 'particles' in 1 mole of absolutely anything is equal to the Avogadro constant, $N_A = 6.0 \times 10^{23}\,mol^{-1}$.

The Avogadro constant defines a mole.

So, because there are 0.091 moles of the gas in the vessel the number of molecules is

$0.091\,mol \times 6.0 \times 10^{23}\,mol^{-1} = \mathbf{5.5 \times 10^{22}}$

11 If the plane is flying straight and level, the lift must be equal to the weight of the plane, $W = 3.0 \times 10^6\,N$.

lift = pressure difference × wing area

so

$$\text{pressure difference} = \frac{\text{lift}}{\text{wing area}}$$
$$= \frac{30 \times 10^6\,N}{400\,m^2} = \mathbf{7500\,Pa}$$

12 a Taking the nucleus as a sphere with volume, $V = \frac{4}{3}\pi r^3$,

volume of copper nucleus $= \frac{4}{3}\pi(4.8 \times 10^{-15})^3$
$= 4.6 \times 10^{-43}\,\text{m}^3$

$$\text{density} = \frac{\text{mass}}{\text{volume}} = \frac{1.1 \times 10^{-25}\,\text{kg}}{4.6 \times 10^{-43}\,\text{m}^3}$$

$= \mathbf{2.4 \times 10^{17}\,kg\,m^{-3}}$

b We are assuming that the densities of all nuclei are the same, so we can calculate the radius of the lead nucleus by finding the volume of the nucleus by proportion:

$$\frac{m_{Cu}}{V_{Cu}} = \frac{m_{Pb}}{V_{Pb}}$$

Rearranging this equation,

$$V_{Pb} = \frac{m_{Pb}}{m_{Cu}} V_{Cu}$$

$$= \frac{3.5 \times 10^{-25}\,\text{kg} \times 4.6 \times 10^{-43}\,\text{m}^3}{1.1 \times 10^{-25}\,\text{kg}}$$

$$= 1.5 \times 10^{-42}\,\text{m}^3$$

$$V_{Pb} = \tfrac{4}{3}\pi \times (r_{Pb})^3$$

$$\therefore (r_{Pb})^3 = V_{Pb} \times 3/4\pi = \frac{1.5 \times 10^{-42}\,\text{m}^3 \times 3}{4 \times \pi}$$

$$= 3.6 \times 10^{-43}\,\text{m}^3$$

giving

$$r_{Pb} = \mathbf{7.1 \times 10^{-15}\,m}$$

Level 3

13 a You can rearrange the basic ideal gas equation to give

$$V = \frac{nRT}{p}$$

R is a constant, so for a given set of values for n, T and P (including $n = 1$) the volume, V will always be the same.

This is a result of Avogadro's Law – at the same volume, same temperature and same pressure, ideal gases contain the same number of molecules.

b For 1.0 mole at 300 K and 100 kPa,

$$V = \frac{1.0\,\text{mol} \times 8.3\,\text{J}\,\text{K}^{-1}\,\text{mol}^{-1} \times 300\,\text{K}}{1.0 \times 10^5\,\text{Pa}}$$

$= \mathbf{0.025\,m^3}$ (25 litres)

Check that you can show that $J\,Pa^{-1} = m^3$.

c There are the Avogadro constant, N_A, molecules in every mole. So the volume of the molecules in 1.0 mole is

$$N_A \times V_{molecule} = 6.0 \times 10^{23} \times 2.0 \times 10^{-29}\,\text{m}^3$$
$$= 1.2 \times 10^{-5}\,\text{m}^3$$

The rest of the molar volume (0.025 m³ in total, from part **b**) is just empty space. The % of the molar volume that is actually occupied by the molecules is

$$\frac{1.2 \times 10^{-5}\,\text{m}^3}{0.025\,\text{m}^3} \times 100\% \approx \mathbf{0.05\%}$$

d One of the assumptions of the ideal gas model is that the volume of the particles should be negligible. This is fulfilled in this case.

21 Mathematical description of the behaviour of gases

To describe mathematically the behaviour of gases we have to make certain assumptions. These are:

1. gases consist of small particles that are in continuous and random motion;
2. the volume of the gas particles is negligible compared with the volume of the gas itself;
3. all collisions are elastic, that is there is no loss of kinetic energy;
4. there are no interactions between the particles except during collisions.

Consider a particle of gas of mass m, moving with velocity c, in a cubic container of side l. Let the component of the particle's velocity at right angles to the side ABCD be u. See Figure 21.1.

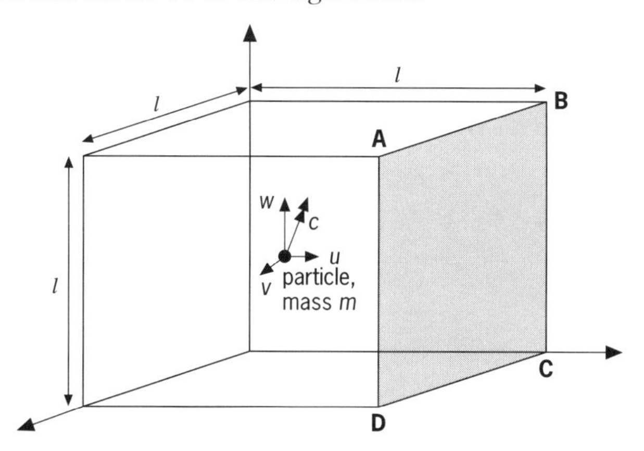

Figure 21.1 Motion of a single gas particle

There is no loss of energy when the particle strikes the sides of the container so the magnitude of the particle's velocity (its speed) will be unchanged. The change in momentum, when the particle strikes the side ABCD is

$$mu-(-mu)=2mu$$

The time between each collision with this wall is

$$\frac{\text{distance travelled}}{\text{speed}}=\frac{2l}{u}$$

Therefore the number of times this particle will strike ABCD per second is

$$\frac{u}{2l}$$

The resulting change in momentum per second is

$$\frac{u}{2l}\times 2mu=\frac{mu^2}{l}$$

By Newton's Second Law (page 62), force = rate of change of momentum, so the force exerted on the side ABCD by the particle is

$$\frac{mu^2}{l}$$

The pressure exerted on the wall by the particle is

$$\frac{\text{force}}{\text{area}}=\frac{mu^2/l}{l^2}=\frac{mu^2}{l^3}=\frac{mu^2}{V}$$

where V is the volume of the container (that is, of the gas).

If there are n identical particles in the cube with different velocity components in this direction u_1, u_2, u_3, and so on, the total pressure, p, exerted by these particles on the side ABCD will be given by

$$p=\frac{mu_1^{\,2}}{V}+\frac{mu_2^{\,2}}{V}+\frac{mu_2^{\,2}}{V}+\ldots+\frac{mu_n^{\,2}}{V}$$

$$=\frac{m}{V}(u_1^{\,2}+u_2^{\,2}+u_3^{\,2}+\ldots+u_n^{\,2})$$

The mean of the squares of the speeds along this direction is

$$\langle u^2\rangle=\frac{(u_1^{\,2}+u_2^{\,2}+u_3^{\,2}+\ldots+u_n^{\,2})}{n}$$

so

$$p=\frac{nm\langle u^2\rangle}{V}$$

Note that the mean square speed $\langle u^2\rangle$ is sometimes written as $\overline{u^2}$.

Similarly, along the other axes (see Figure 21.1) we can write

$$p=\frac{nm\langle v^2\rangle}{V} \quad\text{and}\quad p=\frac{nm\langle w^2\rangle}{V}$$

As we have seen (Figure 21.1) the velocity, c, of one particle is related to its three perpendicular components u, v, w by

$$c^2=u^2+v^2+w^2$$

Similarly, for the mean square speeds,

$$\langle c^2\rangle=\langle u^2\rangle+\langle v^2\rangle+\langle w^2\rangle$$

If there are a large number of particles moving randomly, the mean square speeds in the three perpendicular directions will, on average, be equal:

$$\langle u^2\rangle=\langle v^2\rangle=\langle w^2\rangle$$

It follows, therefore, that

$$\tfrac{1}{3}\langle c^2\rangle=\langle u^2\rangle=\langle v^2\rangle=\langle w^2\rangle$$

The pressure, therefore, exerted on any of the faces of the cube will be the same and given by

$$p = \frac{1}{3}\frac{nm\langle c^2\rangle}{V}$$

or

$$pV = \tfrac{1}{3}nm\langle c^2\rangle$$

where nm is the mass of the gas and so nm/V is the density, ρ.

Therefore another way of writing the equation is

$$p = \tfrac{1}{3}\rho\langle c^2\rangle$$

The **root mean square (rms)** speed of a gas is the square root of the mean of the squares of the speeds, that is

$$c_{rms} = \sqrt{\frac{c_1^2 + c_2^2 + c_3^2 + \ldots + c_n^2}{n}}$$

$$= \sqrt{\langle c^2\rangle}$$

Example 1

Calculate the rms speed of the atoms in a sample of helium which has a density of $1.2\,\text{kg}\,\text{m}^{-3}$ and a pressure of $1.0 \times 10^5\,\text{N}\,\text{m}^{-2}$.

Rearrange the equation

$$p = \tfrac{1}{3}\rho\langle c^2\rangle$$

$$\langle c^2\rangle = \frac{3p}{\rho}$$

$$= \frac{3 \times 1.0 \times 10^5\,\text{N}\,\text{m}^{-2}}{1.2\,\text{kg}\,\text{m}^{-3}}$$

$$= 250\,000\,\text{m}^2\,\text{s}^{-2}$$

$$c_{rms} = \sqrt{\langle c^2\rangle} = 500\,\text{m}\,\text{s}^{-1}$$

Comparison of the derived equation with empirical laws

If the kinetic theory equation derived above is to have any value it must confirm those laws which have been derived from experimental data.

Boyle's Law

Boyle's Law states that the volume of a fixed mass of gas is inversely proportional to its pressure providing the temperature remains constant, that is pV= constant.

The kinetic theory equation states that

$$pV = \tfrac{1}{3}nm\langle c^2\rangle$$

For a given mass of gas nm is a constant, and for a fixed temperature the mean square speed $\langle c^2\rangle$ is a constant (see the next section); it follows therefore that

$$pV = \text{constant}$$

and there is agreement between the kinetic theory equation and Boyle's Law.

Avogadro's Law

Avogadro's Law states that equal volumes of gases under the same conditions of temperature and pressure contain the same number of molecules.

Consider two gases of equal volume and pressure:

$$p_1V_1 = p_2V_2$$

Therefore, from the kinetic theory equation,

$$\tfrac{1}{3}n_1m_1\langle c_1^2\rangle = \tfrac{1}{3}n_2m_2\langle c_2^2\rangle$$

If the temperatures of the gases are the same their kinetic energies will also be the same (see the next section):

$$\tfrac{1}{2}m_1\langle c_1^2\rangle = \tfrac{1}{2}m_2\langle c_2^2\rangle$$

From these two equations it follows that $n_1 = n_2$: the gases contain equal numbers of molecules and there is agreement between Avogadro's Law and the kinetic theory equation.

Relationship between the kinetic energy of a gas and its temperature

One mole of any gas contains 6.02×10^{23} particles. This is the Avogadro constant (N_A). For one mole of gas we can write

$$pV = \tfrac{1}{3}N_A m\langle c^2\rangle$$

where m is the mass of one molecule of the gas. But

$$pV = RT$$

for one mole, therefore

$$\tfrac{1}{3}N_A m\langle c^2\rangle = RT$$

or

$$\tfrac{1}{3}m\langle c^2\rangle = \frac{RT}{N_A}$$

This can be written as

$$\tfrac{1}{3}m\langle c^2\rangle = kT$$

where k is known as the **Boltzmann constant** and is equal to R/N_A.

Dividing both sides by two and multiplying by three makes the left-hand side equal to the average translational kinetic energy of the gas particles:

$$\tfrac{1}{2}m\langle c^2\rangle = \tfrac{3}{2}kT$$

Therefore the mean kinetic energy of the gas particles is proportional to the absolute temperature of the gas, measured in kelvin. If the temperature (in kelvin) of a sample of gas is doubled, for example, the mean kinetic energy of the gas particles also doubles.

Molecular speeds in gases

The molecules of a gas are in continuous, random motion and have a wide range of speeds. A statistical analysis shows that the speeds are distributed as shown in Figure 21.2.

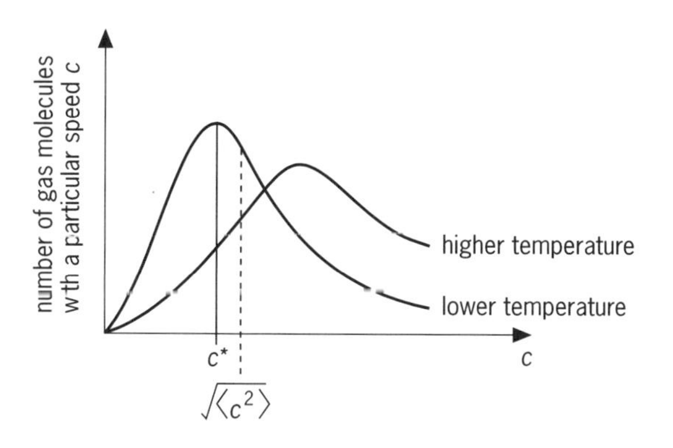

Figure 21.2 Molecular speed distribution in a gas

The key features of the graph are:

- the molecules can theoretically have speeds from zero to infinity;
- the most probable speed a molecule is likely to have at the lower temperature is c^*;
- the root mean square speed of the molecules at the lower temperature is $\sqrt{\langle c^2 \rangle}$;
- if the temperature of the gas is increased more molecules will have higher speeds.

Work done by a gas as it expands

Figure 21.3 shows a sample of gas in a cylinder sealed with a frictionless piston. If the gas expands it must push against a force due to the external pressure, which means the gas must do work.

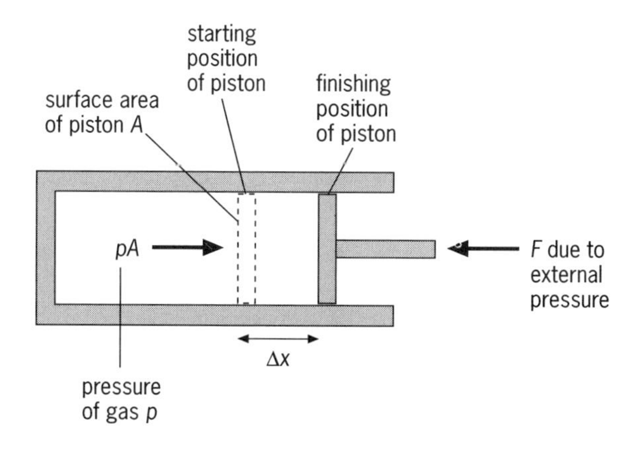

Figure 21.3 Expansion of a gas

The work done, ΔW, by the gas as it expands slightly from its equilibrium position is equal to the force, F, due to the external pressure, multiplied by the distance moved by the piston, Δx:

$$\Delta W = F\Delta x$$

But in equilibrium $F = pA$, the force on the piston (area A) due to the pressure of the gas, p.

$$\therefore \quad \Delta W = pA\Delta x$$

where p is assumed constant for this small change Δx.

But $A\Delta x = \Delta V$, the change in volume

So

$$\boxed{\Delta W = p\Delta V}$$

Therefore

work done by the gas $= p\,\Delta V$

providing the pressure in the cylinder does not change. Note that if the volume of the gas decreases, work is done *on* the gas.

Example 2

Calculate the work done by a gas at a pressure of 2×10^5 Pa when it increases its volume by $0.25\,\text{m}^3$. Assume that the pressure of the gas in the cylinder remains constant.

work done by the gas $= p\Delta V$
$= 2 \times 10^5\,\text{Pa} \times 0.25\,\text{m}^3$
$= 0.5 \times 10^5\,\text{J}$ or $5 \times 10^4\,\text{J}$

If a graph is drawn of pressure against volume, the area under the graph represents the total work done by the gas as it expands. Figure 21.4 shows two examples.

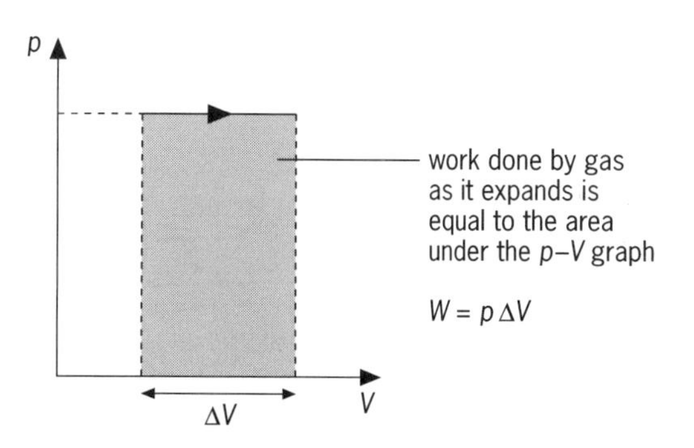

a Gas expanding at constant pressure

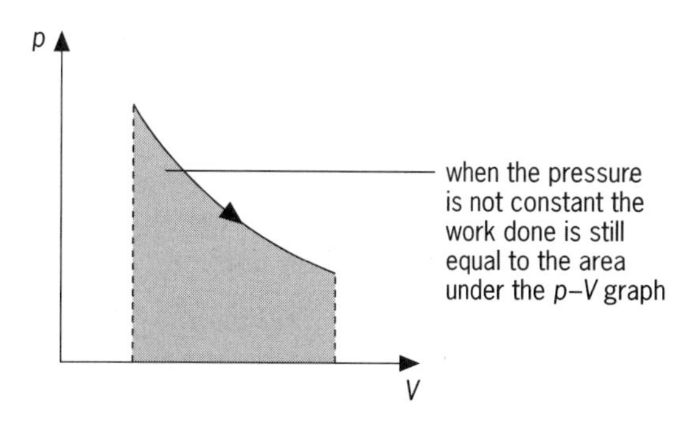

b Gas expanding when the pressure is not constant

Figure 21.4

Boltzmann constant, $k = 1.4 \times 10^{-23}\,\mathrm{J\,K^{-1}}$
Avogadro constant, $N_A = 6.0 \times 10^{23}\,\mathrm{mol^{-1}}$
Molar mass of hydrogen $= 0.002\,\mathrm{kg\,mol^{-1}}$
Molar mass of helium $= 0.004\,\mathrm{kg\,mol^{-1}}$
Molar mass of nitrogen $= 0.028\,\mathrm{kg\,mol^{-1}}$
Molar mass of oxygen $= 0.032\,\mathrm{kg\,mol^{-1}}$
Pressure $p = \frac{1}{3}\rho\langle c^2\rangle$

Level 1

1 A group of n particles have speeds $c_1, c_2, \ldots, c_n$. Write expressions for:
- **a** the mean square speed, $\langle c^2\rangle$, and state its units.
- **b** the root mean square speed, c_{rms}, and state its units.

2 a How do you find the mass of one molecule given the molar mass of a substance?
- **b** Express the density of a gas in terms of the molecular quantities:

  N, the number of molecules
  m, the mass of one molecule
  V, the volume occupied by the gas.
- **c** If the average kinetic energy of one *molecule* of a gas is E, what is the total kinetic energy of 3 *moles* of the gas?
- **d** Explain the relationship between the Boltzmann constant k and the molar gas constant R.

3 Calculate the root mean square speed of five particles with speeds, c_1 to c_5: 350, 470, 380, 430 and $410\,\mathrm{m\,s^{-1}}$.

4 Calculate the pressure exerted by argon of density $1.6\,\mathrm{kg\,m^{-3}}$ when the root mean square speed of its molecules is $420\,\mathrm{m\,s^{-1}}$.

5 a Does the average kinetic energy of a gas at a certain temperature depend on the mass of a gas molecule?
- **b** Does the mean square speed at a certain temperature of the molecules depend on the mass of a gas molecule?

Level 2

6 a *Estimate* the average kinetic energy, in joules, of 1 mole of gas at room temperature.
- **b** Now calculate an accurate value.

7 How does the assumption of no intermolecular forces affect your ideas about the internal energy of an ideal gas?

8 What is the average kinetic energy of a molecule of nitrogen, with relative molecular mass $M_r = 28$, at 300 K?

9 a Calculate $\langle c^2\rangle$ (mean square speed) for nitrogen with a density of $1.25\,\mathrm{kg\,m^{-3}}$ at a pressure of 100 kPa.
- **b** What is the root mean square speed under the same conditions?

10

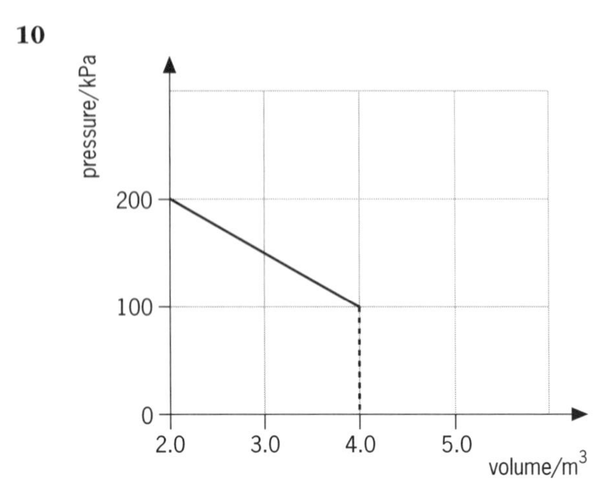

Figure 21.5

A p–V graph for an expanding gas is shown in Figure 21.5. How much work does the gas do as it expands?

Level 3

11 The root mean square speed of a molecule of molar mass M, at absolute temperature T, is given by the equation $c_{rms} = \sqrt{3RT/M}$.
- **a** Sketch a graph of c_{rms} against T, from $T = 0\,\mathrm{K}$ to 400 K, for helium (molar mass 0.004 kg), giving calculated values for at least two points, for instance at $T = 200\,\mathrm{K}$ and $T = 400\,\mathrm{K}$.
- **b** Sketch a graph of c_{rms} against M for the four gases with molar masses listed at the beginning of these questions. Include your calculated values. Take T as 300 K.

12 What is the temperature if the root mean square speed, c_{rms}, of nitrogen (molar mass 0.028 kg) is $410\,\mathrm{m\,s^{-1}}$?

Hint: Consider the kinetic energy of 1 mole.

13 The mass of a bromine molecule is four times that of an argon molecule. They are mixed in a container in thermal equilibrium at 300 K.
- **a** How does the root mean square speed of the bromine molecules compare with that of the argon molecules?
- **b** How do the root mean square speeds change if the container is heated to 400 K?

14 a By what factor must the root mean square speed of an ideal gas molecule increase to produce double the pressure at the same density of gas?
- **b** What happens to the root mean square speed of an ideal gas molecule if the absolute temperature is doubled?

15 a Calculate the mean square speed $\langle c^2 \rangle$ for hydrogen at 300 K, given that its molar mass is 0.002 kg.

b Use your answer to calculate the root mean square value of the speed.

c Explain why there is so little hydrogen in Earth's atmosphere.

Hint: The escape velocity from Earth is 11 000 m s^{-1}.

16 a What is the temperature if a molecule of oxygen has an average kinetic energy of 9.2×10^{-21} J?

b What is the molecule's root mean square speed?

17 A cylinder of gas has a frictionless piston with a cross-sectional area of 50 cm^2. When 200 J of energy is supplied, the piston is pushed 22 cm along the cylinder (Figure 21.6) against the pressure of the atmosphere (100 kPa).

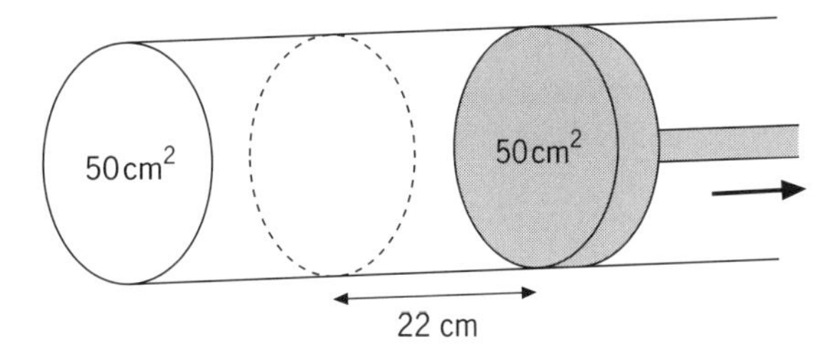

Figure 21.6

a Calculate the work done by the gas as it expands.

b Calculate the change in internal energy of the gas.

Level 1

1 a The mean square speed is the average (or mean) of the *squares* of the speeds,

$$\langle c^2\rangle = \frac{c_1^2 + c_2^2 + \ldots + c_n^2}{n}$$

The unit is that of speed squared, $\mathbf{m^2\,s^{-2}}$.

b The root mean square speed is the *square root* of the *mean* of the *squares* of the speeds,

$$c_{\text{rms}} = \sqrt{\frac{c_1^2 + c_2^2 + \ldots + c_n^2}{n}}$$

This is the square root of the answer to **a** and the units are $\mathbf{m\,s^{-1}}$.

2 a The molar mass of a substance, $M = m \times N_A$ where m is the mass of one molecule and N_A is the number of molecules in one mole (the Avogadro constant). You can rearrange this equation to give

$$m = \frac{M}{N_A}$$

So, to find the mass of one molecule divide the molar mass by the Avogadro constant.

b The mass of gas is equal to the number of molecules × the mass of one molecule:

$$\text{mass} = Nm$$

$$\text{density, } \rho = \frac{\text{mass}}{\text{volume}}$$

$$\therefore \quad \rho = \frac{Nm}{V}$$

c The number of molecules in each mole is equal to the Avogadro constant, N_A. If the average kinetic energy of each molecule is E then the total kinetic energy of three moles is $\mathbf{3N_AE}$.

d The Boltzmann constant, k, is used with reference to a single *molecule*, whereas the gas constant, R, is used with reference to one *mole*. R is equal to the number of molecules in one mole × k:

$$R = N_A k \qquad \text{where } N_A \text{ is the Avogadro constant}$$

3 Square each speed, find the average of the squares, then take the square root:

$$c_{\text{rms}} = \sqrt{\frac{350^2 + 470^2 + 380^2 + 430^2 + 410^2}{5}}$$

$$= \sqrt{\frac{840\,800}{5}} = \sqrt{168\,160} = \mathbf{410\,m\,s^{-1}}$$

4 The pressure is given by the equation

$$\begin{aligned} p &= \tfrac{1}{3}\rho\langle c^2\rangle \\ &= 0.33 \times 1.6\,\text{kg}\,\text{m}^{-3} \times 420^2\,\text{m}^2\,\text{s}^{-2} \\ &= 93 \times 10^3\,\text{kg}\,\text{m}^{-1}\,\text{s}^{-2} = \mathbf{93\,kPa} \end{aligned}$$

5 a No. $\langle E_k\rangle = \tfrac{3}{2}kT$ so the average kinetic energy depends *only* on the temperature.

b Yes. The mean square speed $\langle c^2\rangle = 3RT/M$. The mean square speed is inversely proportional to the mass of a single gas molecule (because $M = N_A m$).

Level 2

6 a Look at the answer to part **b**. Was your estimate anywhere near?

b The average kinetic energy of a single molecule of *any* gas, $\langle E_k\rangle = \tfrac{3}{2}kT$. So the kinetic energy of one mole of gas is

$$N_A \times \tfrac{3}{2}kT = \tfrac{3}{2}RT$$

Room temperature is about 300 K so the molar kinetic energy is

$$1.5 \times 8.3\,\text{J}\,\text{K}^{-1}\,\text{mol}^{-1} \times 300\,\text{K} = \mathbf{3700\,J\,mol^{-1}}$$

7 If there are no intermolecular forces there cannot be any internal potential energy so all the internal energy of an ideal gas must be kinetic. There is no work to be done against intermolecular forces so, for example, there will be no lowering of temperature when an ideal gas expands

8 The mass makes no difference. The average kinetic energy of one molecule of *any* gas at absolute temperature T is given by the equation

$$\langle E_k\rangle = \tfrac{1}{2}m\langle c^2\rangle = \tfrac{3}{2}kT$$

where k is the Boltzmann constant. In this case

$$\begin{aligned} \langle E_k\rangle &= 1.5 \times 1.4 \times 10^{-23}\,\text{J}\,\text{K}^{-1} \times 300\,\text{K} \\ &= \mathbf{6.3 \times 10^{-21}\,J} \end{aligned}$$

9 a The pressure of a gas is given by the equation $p = \tfrac{1}{3}\rho\langle c^2\rangle$. You can rearrange this to give

$$\begin{aligned} \langle c^2\rangle &= \frac{3p}{\rho} \\ &= \frac{3 \times 10^5\,\text{Pa}}{1.25\,\text{kg}\,\text{m}^{-3}} = \mathbf{2.4 \times 10^5\,m^2\,s^{-2}} \end{aligned}$$

$(\text{Pa} \equiv \text{N}\,\text{m}^{-2} \equiv \text{kg}\,\text{m}\,\text{s}^{-2}\,\text{m}^{-2} \equiv \text{kg}\,\text{m}^{-1}\,\text{s}^{-2})$

b The root mean square speed is just the square root of the mean square speed that you calculated in **a**:

$$c_{\text{rms}} = \sqrt{2.4 \times 10^5\,\text{m}^2\,\text{s}^{-2}} = \mathbf{490\,m\,s^{-1}}$$

10 The work done is equal to the area under a p–V graph. You can calculate the work done in a number of different ways.

The average pressure is 150 kPa and the volume changes from $2.0\,\text{m}^3$ to $4.0\,\text{m}^3$. So

$$\begin{aligned} W &= 150 \times 10^3\,\text{Pa} \times 2.0\,\text{m}^3 \\ &= 300\,\text{kPa}\,\text{m}^3 \\ &= 300\,\text{kN}\,\text{m}^{-2}\,\text{m}^3 = \mathbf{300\,kJ} \end{aligned}$$

Or you can consider the area as a trapezium, for which the area = base × average height. So

$$W = 2.0\,\text{m}^3 \times (200 + 100)/2\,\text{kPa} = \mathbf{300\,kJ}$$

Or you can consider the area as a triangle and a rectangle.

Whichever way you do it, you should get the same result. Doing it by two methods is a good way of checking that you have the correct answer.

Level 3

11 a

T/K	c/m s^{-1}
100	790
200	1120
300	1370
400	1580

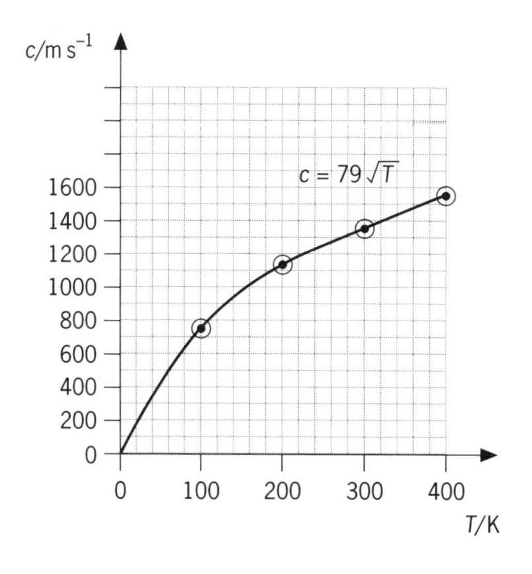

b

M/kg	c/m s^{-1}
0.002	1930
0.004	1370
0.028	516
0.032	480

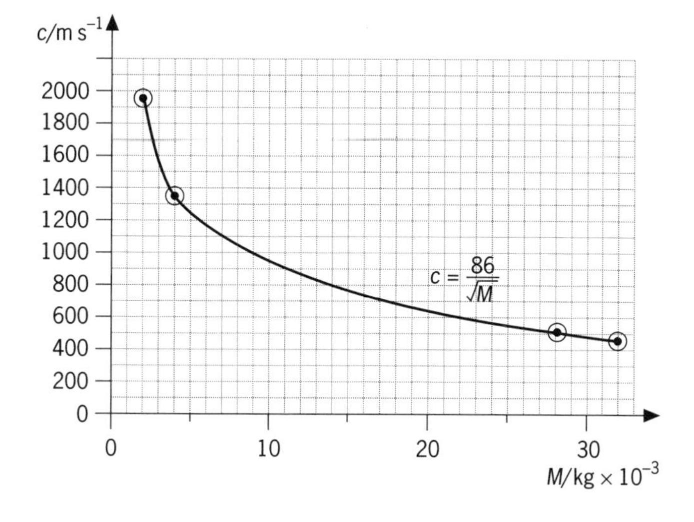

Figure 21.7

You should now have some feeling for how the root mean square speed of the molecules depends on temperature and on molar mass.

12 The kinetic energy of 1 mol of a gas is $E_k = \frac{1}{2}M\langle c^2\rangle = \frac{3}{2}RT$. You can rearrange this equation to give

$$T = \frac{M\langle c^2\rangle}{3R}$$

In this example,

$$T = \frac{0.028\,\text{kg mol}^{-1} \times 410^2\,\text{m}^2\,\text{s}^{-2}}{3 \times 8.3\,\text{J K}^{-1}\,\text{mol}^{-1}} = \mathbf{189\,K}$$

($J \equiv N/m \equiv kg\,m\,s^{-2}\,m \equiv kg\,m^2\,s^{-2}$)

13 a When mixtures of gases are in thermal equilibrium their average kinetic energies are the same.

$$\therefore \tfrac{1}{2}m_{bromine}\langle c^2_{bromine}\rangle = \tfrac{1}{2}m_{argon}\langle c^2_{argon}\rangle$$

which you can rearrange to give

$$\frac{\langle c^2_{bromine}\rangle}{\langle c^2_{argon}\rangle} = \frac{m_{argon}}{m_{bromine}}$$

Taking square roots on both sides of the equation gives

$$\frac{c_{bromine}}{c_{argon}} = \sqrt{\frac{m_{argon}}{m_{bromine}}}$$

where c is the rms speed. So

$$c_{bromine} = c_{argon} \times \sqrt{\tfrac{1}{4}} = c_{argon} \times \tfrac{1}{2}$$

On average, the bromine molecules move with **half** the speed of the argon molecules.

b The average kinetic energy is proportional to the absolute temperature:

$$E_K = \tfrac{1}{2}m\langle c^2\rangle = \tfrac{3}{2}kT$$

If the absolute temperature is increased from 300 K to 400 K the mean *square* speed of both gases will go up by the same factor, $\frac{4}{3}$, but the *root* mean square speed will increase by the *square root* of $\frac{4}{3}$.

$$\therefore c_{400} = c_{300}\sqrt{\tfrac{4}{3}} = \mathbf{c_{300} \times 1.15}$$

14 a The pressure is given by the equation $p = \frac{1}{3}\rho\langle c^2\rangle$ which you can rearrange to give

$$\langle c^2\rangle = \frac{3}{\rho} \times p$$

or, taking square roots on both sides of the equation,

$$c_{rms} = \sqrt{\frac{3}{\rho}} \times \sqrt{p}$$

If ρ is constant then c_{rms} is proportional to the square root of the pressure. If p doubles then the mean square speed must double and the rms speed must increase by a factor of $\sqrt{\mathbf{2}} = \mathbf{1.4}$.

b Similarly, $\langle c^2\rangle = 3RT/M$ so $c \propto \sqrt{T}$. If T doubles then the rms speed increases by a factor of $\sqrt{\mathbf{2}} = \mathbf{1.4}$.

15 a For 1 mole of a gas, $pV = RT = \frac{1}{3}M\langle c^2\rangle$. So, to make use of the information you have been given in the question, rearrange the equation to give

$$\langle c^2\rangle = \frac{3RT}{M}$$

$$= \frac{3 \times 8.3\,\mathrm{J\,K^{-1}\,mol^{-1}} \times 300\,\mathrm{K}}{0.002\,\mathrm{kg\,mol^{-1}}}$$

$$= \mathbf{3.7 \times 10^6\,m^2\,s^{-2}}$$

$J = kg\,m^2\,s^{-2}$, so the units work out OK. But always check!

b c_{rms} is the square root of $\langle c^2\rangle$.

$c_{rms} = \sqrt{\langle c^2\rangle} = \mathbf{1930\,m\,s^{-1}}$

c The value you have calculated is the root mean square speed of the hydrogen molecules. There is a wide range of speeds (see Figure 21.2) and some (*about 1 in 2000*) will go faster than the escape velocity. This means that the hydrogen will gradually leak out of the top of the atmosphere. The speeds of any slower hydrogen molecules that collide with other larger molecules will be increased, so even slow hydrogen molecules are likely to escape eventually.

16 a The average kinetic energy of a single molecule is given by the equation $E_K = \frac{3}{2}kT$. You can rearrange this equation to give

$$T = \frac{2E_K}{3k}$$

$$= \frac{2 \times 9.2 \times 10^{-21}\,\mathrm{J}}{3 \times 1.4 \times 10^{-23}\,\mathrm{J\,K^{-1}}} = \mathbf{438\,K}$$

b The kinetic energy of a single molecule is also given by $E_K = \frac{1}{2}m\langle c^2\rangle$ where m is the mass of the molecule. To find the mass of a single molecule of oxygen, divide the molar mass ($0.032\,\mathrm{kg\,mol^{-1}}$) by the Avogadro constant, N_A:

$$m = \frac{0.032\,\mathrm{kg\,mol^{-1}}}{6.0 \times 10^{23}\,\mathrm{mol^{-1}}}$$

You can rearrange the equation for kinetic energy to give

$$\langle c^2\rangle = \frac{2E_K}{m}$$

$$= \frac{2 \times 9.2 \times 10^{-21}\,\mathrm{J} \times 6.0 \times 10^{23}\,\mathrm{mol^{-1}}}{0.032\,\mathrm{kg\,mol^{-1}}}$$

$$= 345\,000\,\mathrm{m^2\,s^{-2}}$$

The root mean square speed is the square root of the mean square speed $= \mathbf{587\,m\,s^{-1}}$.

(You could have worked out the kinetic energy of 1 mole and then divided by half the molar mass, M, to get the value of $\langle c^2\rangle$ using the equivalent equation $E = \frac{1}{2}M\langle c^2\rangle$.)

17 a There are two ways of calculating the work done when the gas expands. You can consider the work done in moving the piston, using

$W = \text{force} \times \text{displacement}$

or you can consider the work done by the expanding gas, in which case, because the pressure is constant, you can use

$W = p\Delta V$

Considering the work done in moving the piston, the force on the piston is given by $F = pA$. The pressure p is 1 atmosphere = 100 kPa. The area of the piston A is $50\,\mathrm{cm^2}$, but you need to have it in $\mathrm{m^2}$.

$1\,m^2 = 100\,cm \times 100\,cm = 10^4\,cm^2$

$\therefore 1\,cm^2 = 10^{-4}\,m^2$

$50\,\mathrm{cm^2} = 50 \times 10^{-4}\,\mathrm{m^2}$

$\therefore F = pA = 100 \times 10^3\,\mathrm{Pa} \times 50 \times 10^{-4}\,\mathrm{m^2} = 500\,\mathrm{N}$

The displacement of the piston is 22 cm = 0.22 m

$\therefore$ work done, $W = 500\,\mathrm{N} \times 0.22\,\mathrm{m} = \mathbf{110\,J}$

You can also use the equation $W = p\Delta V$. The pressure is constant at 100 kPa. The change in volume is

$\Delta V = 0.22\,\mathrm{m} \times A\,\mathrm{m^2} = 0.22\,\mathrm{m} \times 50 \times 10^{-4}\,\mathrm{m^2}$

$\therefore$ work done, $W = 100\,\mathrm{kPa} \times 0.22\,\mathrm{m} \times 50 \times 10^{-4}\,\mathrm{m^2}$
$= \mathbf{110\,J}$

b 200 J was added and 110 J was used to do external work. The remaining 90 J has gone to raise the internal energy of the gas. The temperature will have risen.

22 Thermodynamics

Heat, internal energy and temperature difference

Heat is energy which is moving because of a temperature difference.

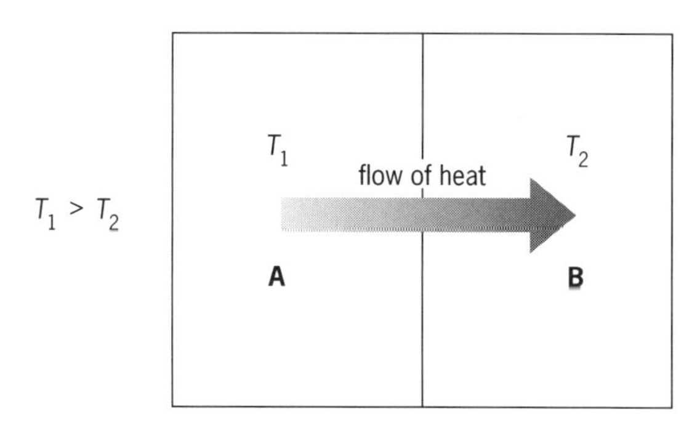

Figure 22.1 Heat flows from the object at a higher temperature, A, to the object at a lower temperature, B

If two objects are placed in 'thermal contact', heat will flow from the hotter object to the colder object (Figure 22.1). This flow of heat into an object may cause

- an increase in the internal energy of the object; or
- it may allow the object to do work; or
- it may cause both to occur.

The internal energy of an object is the sum of the kinetic and potential energies of the molecules within the object.

Thermal equilibrium

If two objects at the same temperature are in thermal contact there is no overall heat transfer. The objects are in **thermal equilibrium**.

Temperature is the major consideration when deciding if heat transfer will take place or if there is thermal equilibrium. This is summarised in the **Zeroth Law of Thermodynamics**, which states that **two bodies which are in thermal equilibrium with a third must be in equilibrium with each other**.

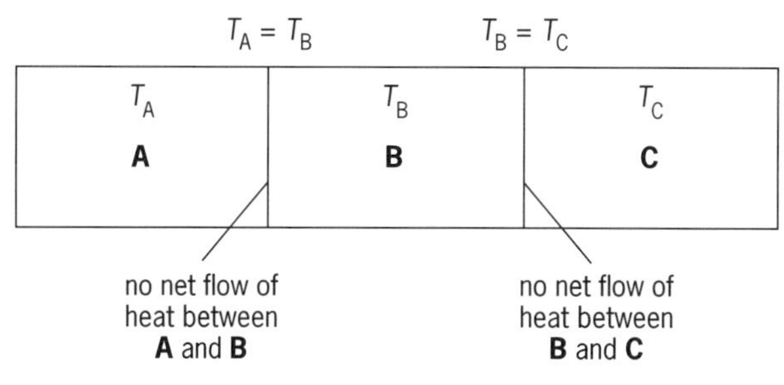

Figure 22.2 The Zeroth Law of Thermodynamics and the concept of temperature

First Law of Thermodynamics

The **First Law of Thermodynamics** describes the relationship between the heat which flows into a system, the work done and the consequent change in the internal energy of the system. It states that **the change in internal energy of a system** (ΔU) **is equal to the sum of the energy entering the system by heating** (ΔQ) **and the energy entering the system through work being done on it** (ΔW).

$$\Delta U = \Delta Q + \Delta W$$

We use here the convention that positive values for ΔQ and ΔW indicate that heat is being supplied *to* the system and work being done *on* it. Negative values for ΔQ and ΔW indicate that heat is being taken *from* the system and work being done *by* it.

Note that you may come across other conventions, such as positive ΔW = work done by the system; then $\Delta U = \Delta Q - \Delta W$.

Consider a sample of gas contained in a cylinder, one end of which consists of a frictionless piston, Figure 22.3. There are two ways in which we can alter the internal energy of the gas.

1. If the piston is moved downwards it will strike the gas molecules, causing the internal energy of the gas to increase (in the same way that striking a ball with a bat increases the ball's kinetic energy). Work has been done on the gas. If the piston is moved upwards work is done by the gas in expanding and the internal energy of the gas will decrease.
2. If the gas is heated, keeping the piston stationary so that no work is done on or by the gas, its molecules gain kinetic energy and its internal energy therefore increases. Cooling the gas under the same conditions will cause the internal energy to decrease.

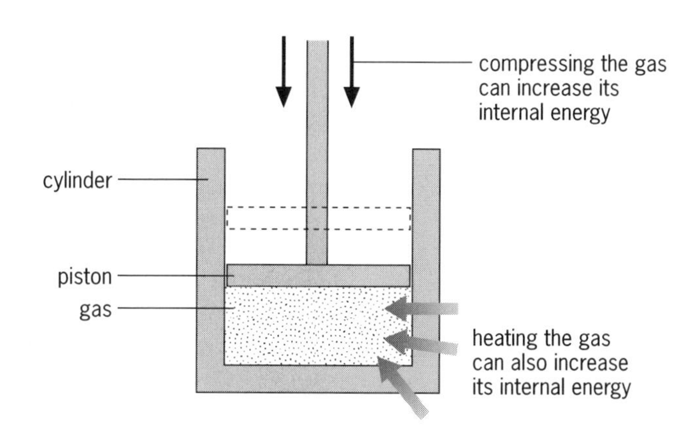

Figure 22.3 Changing the internal energy of a gas

If a gas is compressed and at the same time heated, from the Principle of Conservation of Energy we can describe the overall situation as

increase in internal energy = heat supplied to gas + work done on gas

or $$\Delta U = \Delta Q + \Delta W$$

Example 1

Calculate the increase in the internal energy of a gas which while being compressed has 50J of work done on it but loses 30J of heat to the surroundings.

$$\Delta U = \Delta Q + \Delta W$$

Heat is lost from the gas so ΔQ has a negative value.

$$\therefore \Delta U = -30\text{J} + 50\text{J} = 20\text{J}$$

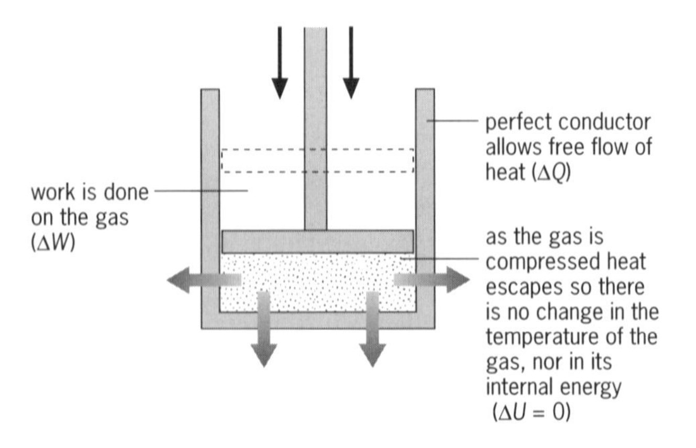

Figure 22.5 An isothermal change

Adiabatic and isothermal changes

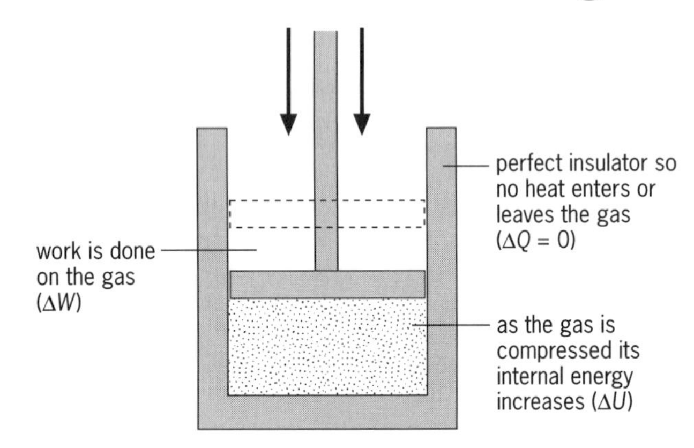

Figure 22.4 An adiabatic change

Consider a sample of gas trapped in a perfectly insulated cylinder, Figure 22.4. If the piston is pushed in and the gas compressed, work is done on the gas. If no heat enters or escapes, that is $\Delta Q = 0$, this is called an **adiabatic** compression.

For all changes

$$\Delta U = \Delta Q + \Delta W$$

but in this case $\Delta Q = 0$, so

$$\Delta U = \Delta W$$

The work done on the gas is equal to the increase in internal energy of the gas and the temperature of the gas will rise.

If the gas is allowed to *expand* adiabatically, work is done at the expense of the internal energy of the gas and the temperature of the gas will fall.

If the same sample of gas is placed in a cylinder which is a perfect conductor and is compressed slowly, Figure 22.5, any small increase in the temperature of the gas causes heat to flow out of the system until thermal equilibrium with the surroundings is once again established. The gas is compressed without an increase in temperature. This is called an **isothermal** compression.

Again

$$\Delta U = \Delta Q + \Delta W$$

but there is no change in the temperature of the gas $\Delta U = 0$ (assuming it is an ideal gas), so

$$\Delta W = -\Delta Q$$

The work done on the gas is equal to the heat escaping from the gas.

Similarly, if the gas is allowed to *expand* isothermally, heat energy will enter the gas as it does work and

$$-\Delta W = \Delta Q$$

In this case the work done by the gas is equal to the heat entering the gas.

The work done in the expansion of a gas is equal to the area under its p–V graph (see topic 21). From Figure 22.6 you can see that, for the same initial state and the same change in volume, more work is done in an isothermal expansion than in an adiabatic expansion.

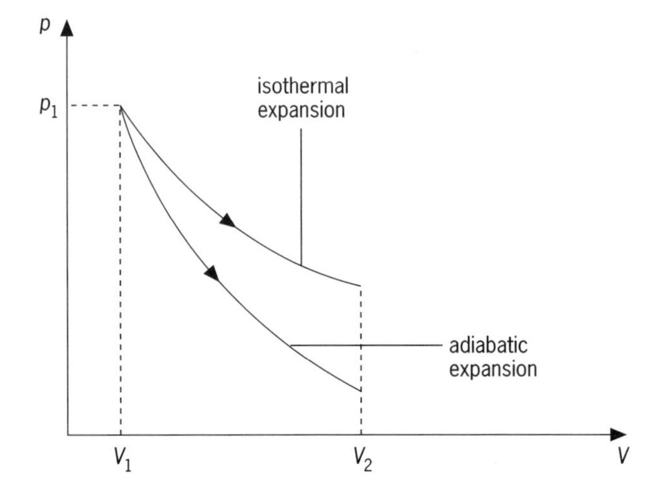

Figure 22.6 p–V curves for isothermal and adiabatic expansions

Heat engines

Heat engines convert heat into work. Heat is absorbed from a hot source, work is then done and some heat is then transferred to the surroundings or 'cold sink', Figure 22.7. For example, a small amount of fuel burnt in one of the cylinders of a car engine produces heat. This heat causes gases to expand and do work. The hot exhaust gases are then released into the surroundings. There is no way of avoiding the loss of energy when the exhaust gases are ejected. Heat engines can never be 100% efficient. This situation is described by the **Second Law of Thermodynamics**, which states that **no continually working heat engine can take in heat and completely convert it into work**.

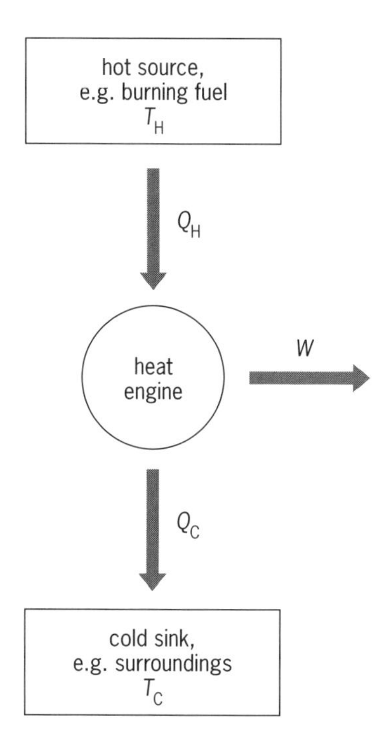

Figure 22.7 Energy transfers in a heat engine

The efficiency of a heat engine is defined as

$$\text{efficiency} = \frac{\text{work done by engine}}{\text{heat supplied to engine}} \times 100\%$$

Example 2

A heat engine absorbs 9 kJ of energy and does 3 kJ of useful work. Calculate the efficiency of the engine.

$$\begin{aligned}\text{efficiency} &= \frac{\text{work done by engine}}{\text{heat supplied to engine}} \times 100\% \\ &= \frac{3000\,\text{J}}{9000\,\text{J}} \times 100\% \\ &= 33\%\end{aligned}$$

Applying the Principle of Conservation of Energy to the heat engine, the difference between the heat supplied and the heat ejected to the surroundings must be equal to the work done by the engine:

$$W = Q_H - Q_C$$

$$\therefore \text{efficiency} = \frac{Q_H - Q_C}{Q_H} \times 100\%$$

Where Q_H is the heat taken from the hot source and Q_C is the energy ejected (Figure 22.7) to the cold surroundings.

Example 3

A heat engine absorbs 20 kJ of energy and ejects 5 kJ to the surroundings as it does work. Calculate the work done by the engine and its efficiency.

$$\begin{aligned}\text{work done} &= Q_H - Q_C \\ &= 20\,\text{kJ} - 5\,\text{kJ} \\ &= 15\,\text{kJ}\end{aligned}$$

$$\begin{aligned}\text{efficiency} &= \frac{Q_H - Q_C}{Q_H} \times 100\% \\ &= \frac{15\,\text{kJ}}{20\,\text{kJ}} \times 100\% \\ &= 75\%\end{aligned}$$

For an 'ideal' heat engine it can be shown that the heat supplied and ejected is related to the absolute thermodynamic temperature of the source (T_H) and the surroundings (T_C) by the equation

$$\frac{Q_C}{Q_H} = \frac{T_C}{T_H}$$

Substituting in the above expression for efficiency, we can write

$$\text{efficiency} = \frac{T_H - T_C}{T_H} \times 100\%$$

This is the theoretical maximum efficiency of a heat engine.

Example 4

An ideal heat engine operates between a source at 277 °C and the surroundings at 27 °C. Calculate the maximum efficiency of this engine.

$$\text{maximum efficiency} = \frac{T_H - T_C}{T_H} \times 100\%$$

The temperatures must be in kelvin.

$T_H = 277 + 273 = 550\,\text{K}$
$T_C = 27 + 273 = 300\,\text{K}$

$$\begin{aligned}\therefore \quad \text{maximum efficiency} &= \frac{550\,\text{K} - 300\,\text{K}}{550\,\text{K}} \times 100\% \\ &= 45\%\end{aligned}$$

Heat pumps

Heat pumps move heat from places of low temperature to places of high temperature. For example, a heat pump can be used to remove heat from the inside of a refrigerator and transfer it to the surroundings. Clearly, work will be needed to accomplish this transfer.

$$W = Q_H - Q_C$$

In fact the Second Law of Thermodynamics can also be expressed as: **heat cannot be transferred continually from an object to another object at a higher temperature unless external work is done**.

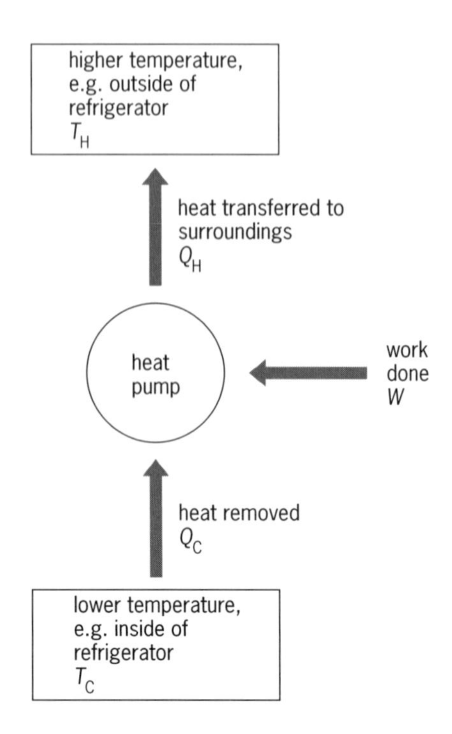

Figure 22.8 Energy transfers in a heat pump

Avogadro constant, $N_A = 6.0 \times 10^{23}\,\mathrm{mol^{-1}}$
Specific heat capacity of water, $c = 4200\,\mathrm{J\,kg^{-1}\,K^{-1}}$
Specific latent heat of vaporisation of water, $L_v = 2.26\,\mathrm{MJ\,kg^{-1}}$
Atmospheric pressure $= 1.0 \times 10^5\,\mathrm{Pa}$

Level 1

1 a Find the change in internal energy if 200 J of heat is supplied to a system and 120 J of work is done *on* the system.

b How much heat must be supplied for the internal energy of a system to rise by 350 J if the system also does 100 J of work *on its surroundings*?

c How much work must have been done *on* a system if 240 J of heat are supplied but the internal energy U rises by 760 J?

2 An engine burns fuel at a rate of 125 kW and produces 45 kW of useful power.

a What efficiency is this?

b 35% of the wasted energy goes to the cooling water. How many kW is this?

3 A cylinder of *ideal gas* fitted with a frictionless piston is heated isothermally, which means that the temperature stays constant.

a How does the heat supplied, ΔQ, relate to the change in internal energy, ΔU, and to the work done on or by the system, ΔW?

b Explain how the work occurs.

4 Show how the First Law of Thermodynamics applies to the following examples, explaining each energy change and distinguishing carefully between work and thermal energy transfer.

a A tub of ice cream is taken from the freezer into a warm room.

b The bit of a hand drill gets hot as it is used to drill a hole.

c A ball is dropped out of a window and bounces several times before coming to rest.

5 a What is a heat engine? Give an example in everyday use.

b What is a heat pump? Give an example in everyday use.

Illustrate each answer with a flow diagram of energy inputs and outputs.

Level 2

6 In the course of your studies you may see the First Law of Thermodynamics quoted as $\Delta U = \Delta Q + \Delta W$ in some texts, and as $\Delta Q = \Delta U + \Delta W$ in others. Your knowledge of algebra tells you that they are not the same equation. What is going on?

7 The pressure and volume of an *ideal gas* in a cylinder with a frictionless piston are altered. Comment on the work done (on or by the system), on any thermal transfer of energy and on any change in internal energy when the changes **a**, **b**, **c**, **d** and **e** in Figure 22.9 are carried out.

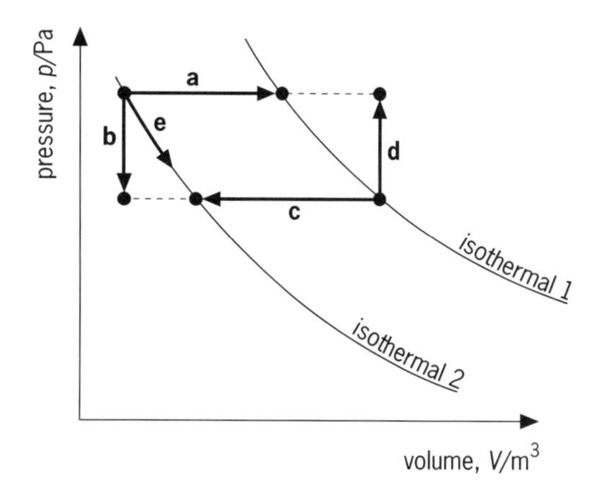

Figure 22.9

8 A system starts in state A and ends in state B, Figure 22.10.

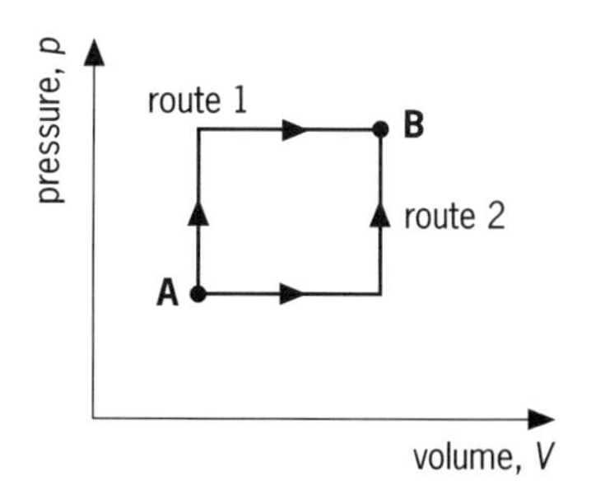

Figure 22.10

Routes 1 and 2 both take the system from state A to state B.

a Which one of the changes, ΔU, ΔQ or ΔW, is the same for both routes?

b Compare the amounts of work done on each route.

c What can you say about the thermal energy transfer on each route?

9 A steam engine (using steam at 100 °C) is operating on a wintry day with an outside temperature of −3 °C. What is the maximum possible efficiency?

Level 3

10 1.0 kg of water at 100 °C has a volume of 1.0×10^{-3} m^3. After changing into steam at the same temperature, and at normal atmospheric pressure, the volume becomes 1.7 m^3.

a Apply the First Law of Thermodynamics to find the change in internal energy, ΔU.

b The molar mass of water is 0.018 kg. Find the average binding energy per molecule at 100 °C.

11 A fixed mass of *ideal gas* with volume V_0 at 280 K and pressure 1.3×10^5 Pa is taken through the following cycle. Complete the p–V diagram started in Figure 22.11 to show the changes.

First an isothermal expansion to $2V_0$ this change to be labelled α).

a Calculate the pressure at the end of α.
Then, from the end-point of α, an expansion at constant pressure to $3V_0$ (label this β).

b Calculate the temperature at the end of β.
From the end-point of β an isothermal compression (label this change γ).

c Calculate the volume at the end of γ.
From the end-point of γ, a compression at constant pressure back to the initial state (label this change δ).

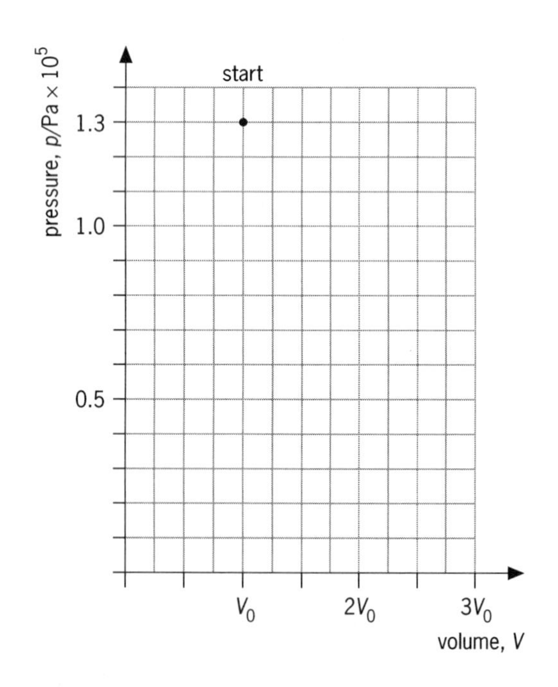

Figure 22.11

12 A power station boiler produces high pressure steam at 540 °C which turns a turbine to produce 560 MW of electricity from the generator. The steam is then cooled in a heat exchanger by 15 tonnes of water per second. The water is initially at 273 K but its temperature has risen 15 K by the time it leaves the cooler.

a What power does the water remove?

b What is the efficiency of the process?

c What is the theoretical maximum efficiency?

13 The cycle shown in Figure 22.12 is for a refrigerator. The two curves **B** and **D** are adiabatic processes, in which no energy is transferred thermally.

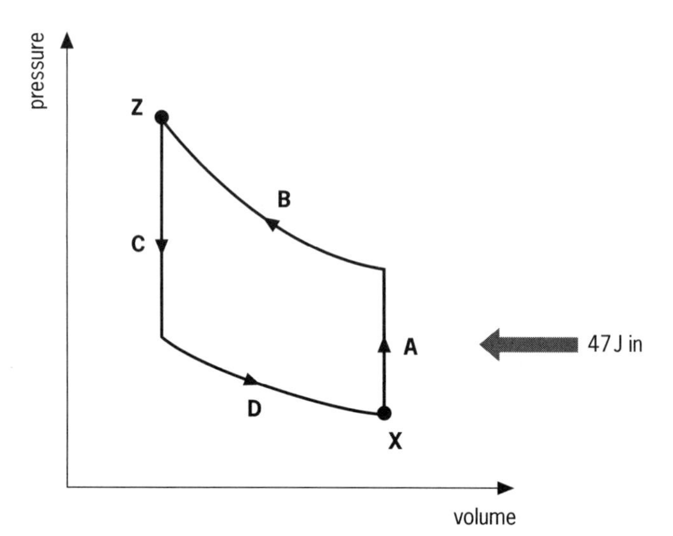

Figure 22.12

The area between **B** and the V-axis is 42 J and between **D** and the V-axis is 26 J. 47 J of energy are taken from the food in stage **A**.

	ΔU/J inc or dec?	ΔQ/J in or out?	ΔW/J on or by?
Change labelled **A**		47 into system	
Change labelled **B**			42
Change labelled **C**			
Change labelled **D**			26

a Copy and complete the table of values.

b Calculate the net work done by the fridge.

c Calculate the total change in thermal energy, ΔU, between states X and Z in Figure 22.12.

Level 1

1 a The change in internal energy is the sum of the heat supplied and the work done **on** the system. Taking the work done **on** the system to be positive,

$$\Delta U = \Delta Q + \Delta W = 200\,\text{J} + 120\,\text{J} = \mathbf{320\,J}$$

b The work ΔW, is taken to be negative because it is done **by** the system; so, from $\Delta U = \Delta Q + \Delta W$,

$$350\,\text{J} = \Delta Q - 100\,\text{J}$$

$$\therefore\ \Delta Q = 350\,\text{J} + 100\,\text{J} = \mathbf{450\,J}$$

c $$\Delta U = \Delta Q + \Delta W$$

$$760\,\text{J} = 240\,\text{J} + \Delta W$$

$$\therefore\ \Delta W = 760\,\text{J} - 240\,\text{J} = \mathbf{520\,J}$$

2 a $$\text{Efficiency} = \frac{\text{useful power}}{\text{input power}} \times 100\%$$

$$= \frac{45}{125} \times 100\% = \mathbf{36\%}$$

b Energy is wasted at a rate of $125\,\text{kW} - 45\,\text{kW} = 80\,\text{kW}$. The energy used for cooling water is 35% of 80 kW

$$= 0.35 \times 80\,\text{kW} = \mathbf{28\,kW}$$

3 a For an ideal gas, internal energy depends only on the temperature. In this case, the temperature stays constant, so there is no change in internal energy: $\Delta U = 0$.

$$\Delta U = \Delta Q + \Delta W = 0$$
$$\therefore\ \Delta Q = -\Delta W$$

where ΔW is negative because in this case work is done **by** the gas. So the heat gained is equal in magnitude to the work done **by** the gas.

b Work is done when the piston moves, against the atmosphere, to increase the volume of the cylinder.

4 a The ice cream gains energy from the room by thermal transfer. This is $+\Delta Q$. The internal energy of the ice cream goes up by ΔQ, while the internal energy of the room goes down by ΔQ. No mechanical work is done.

$\Delta W = 0$
$\Delta U = \Delta Q$ for the ice cream
$\Delta U = -\Delta Q$ for the room

b Consider the drill bit as 'the system' and everything else as its surroundings. The source of energy is the chemical energy that is changed into movement by your muscles. You do work $+\Delta W$ on the drill bit because there is force moving through a distance. The internal energy of the drill bit rises by $+\Delta U$ because there is friction, but there is no transfer of energy to the bit solely because of temperature difference, so

$\Delta Q = 0$
$\Delta U = \Delta W$

c The ball starts with gravitational potential energy. This is changed into internal energy of the ball and the ground, $+\Delta U$, because the bouncing jolts the molecules and increases their kinetic energy at each bounce. The gravitational field does work *on* the ball, $+\Delta W$. There is no transfer of energy solely because of temperature difference, so

$\Delta Q = 0$
$\Delta W = \Delta U$ for the ball and the ground

5 a A heat engine changes heat into work. See Figure 22.13a. The steam engine was the first type, in which burning fuel heats water to make steam which is used in the cylinder to produce movement. Since then, we have developed the **internal combustion engine** where the fuel is burnt inside the cylinder to produce movement.

b A heat pump uses work to transfer energy from a cooler place to a hotter place. It is a heat engine used in reverse. See Figure 22.13b. The **refrigerator** is one example. Heat pumps can be also used to heat houses or swimming pools by extracting energy from a source, for instance the ground or a nearby river.

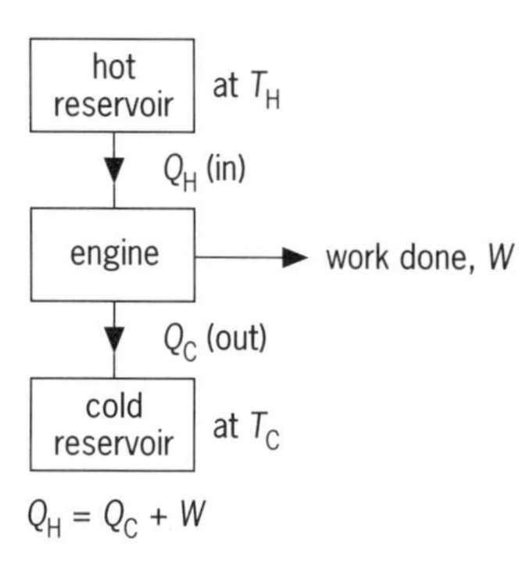

a Heat engine

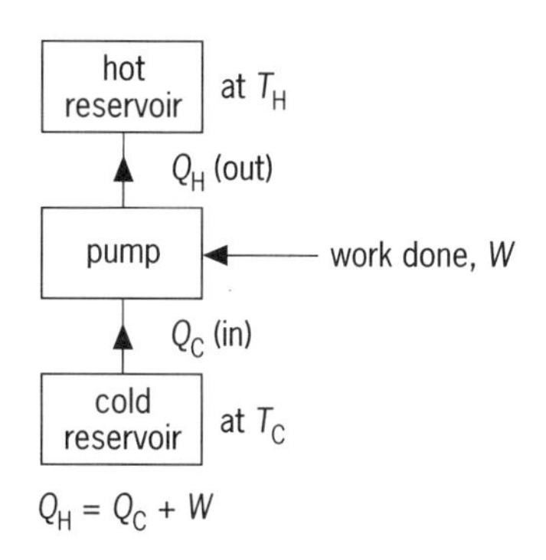

b Heat pump

Figure 22.13

Level 2

6 The difference arises because the first equation considers **work done on the system** to be **positive**. This equation tells you that the change in internal energy is equal to the sum of the energy transferred **to** the system by heating and the energy transferred **to** the system by work being done **on** it.

The second equation considers **work done by the system** to be **positive** and work done **on** the system to be negative. This equation tells you that energy transferred to the system thermally can be used to raise the internal energy and/or to **do** work. This is work done **by** the system, and would be negative in the first equation.

Check your syllabus to see which convention you are to use.

7 a p is constant and V increases. Work $= p\Delta V$ is done **by** the system and so the change to the system, ΔW, is taken to be negative by our convention. The final state (see Figure 22.9) is on a higher temperature isothermal, which means that there is an increase in internal energy, ΔU. So, there must be an input of thermal energy, ΔQ, to provide the increase in internal energy, ΔU, and the energy for the work done by the system, $-\Delta W$.

$$\Delta U = \Delta Q + \Delta W$$

where ΔW is negative.

b V is constant so $\Delta V = 0$, so $p\Delta V = \Delta W = 0$ and no work is done. p decreases, so $pV\,(=nRT)$ also decreases. This means that both T and the internal energy of the system decrease: ΔU is negative. The molecules must move more slowly. This can only happen if thermal energy is removed from the system: ΔQ is negative.

$$\Delta U = \Delta Q$$

where both ΔU and ΔQ are negative.

c p is constant and V decreases. Work $= p\Delta V$ is done **on** the system, which means we take ΔW to be positive. This tends to increase the internal energy, but the final state is on a lower isothermal, so the final temperature is lower than the initial temperature and overall ΔU is negative. This means that thermal energy must be transferred out of the system: ΔQ is negative.

$$\Delta U = \Delta Q + \Delta W$$

where ΔQ is negative and is 'bigger' than ΔW, so ΔU is negative.

d This is the reverse of **b**. Because there is no volume change, no work is done: $\Delta W = 0$. But the internal energy increases; ΔU is positive. Therefore thermal energy must be added: ΔQ is positive.

$$\Delta U = \Delta Q$$

where both ΔU and ΔQ are positive.

e This represents isothermal expansion. The temperature stays the same, so $\Delta U = 0$ because the internal energy of an ideal gas depends only on the temperature. The volume increases so work is done **by** the system: ΔW is negative. There must be an input of thermal energy to cover this: ΔQ is positive.

$$\Delta Q + \Delta W = 0$$

ΔQ and ΔW are equal in magnitude but opposite in sign (ΔW is negative).

8 a The change in internal energy, ΔU, is the same whatever route is taken because ΔU depends only on the initial and final states.

b On route 2 the expansion takes place at a lower pressure, so the work done by the system, $p\Delta V$, is less than on route 1.

The work done is the area under the line for a p–V graph.

c Because ΔU is the same on each route, ΔQ and ΔW must combine to produce the same change, no matter which route is taken. On route 2 less work is done by the system so less energy must be transferred thermally into the system to achieve the same change in internal energy.

9 You need to change all the temperatures to absolute temperatures in K. The engine is working between

$$(100 + 273) = 373\,\text{K} \quad \text{and} \quad (-3 + 273) = 270\,\text{K}$$

The maximum possible efficiency is given by

$$\frac{T_H - T_C}{T_H}$$

where T_H and T_C are the temperatures of the hot and cold reservoirs used by the engine. So the maximum possible efficiency is

$$\frac{373\,\text{K} - 270\,\text{K}}{373\,\text{K}} = 0.28 \text{ or } \mathbf{28\%}$$

Level 3

10 a You can find ΔQ from the thermal energy supplied to vaporise 1.0 kg of water: $\Delta Q = L_v = 2.26$ MJ per kg. ΔQ is positive because the energy goes into the system.

The work done in expanding against the atmosphere $\Delta W = p\Delta V$. You can take the change in volume, ΔV, to be $1.7\,\text{m}^3$ because the volume of water ($0.001\,\text{m}^3$) is negligible compared with the volume of steam. So

$$\Delta W = p\Delta V = 1.0 \times 10^5\,\text{Pa} \times 1.7\,\text{m}^3 = 1.7 \times 10^5\,\text{J}$$

In the equation $\Delta U = \Delta Q + \Delta W$ the work (ΔW) is taken to be negative because it is work done **by** the system,

$$\therefore \Delta U = 2.26 \times 10^6\,\text{J} - 1.7 \times 10^5\,\text{J} = \mathbf{2.09 \times 10^6\,J}$$

b When water turns into steam at 100 °C there is no change in temperature so the increase in internal energy is an increase in potential energy, the result of the breaking of intermolecular bonds. To find the 'strength' of individual bonds you need to know the total energy needed to break all the molecular bonds, and the number of molecules. You already know the total energy required – assuming you got the answer to part **a** correct – so now you need to calculate the number of molecules in a kilogram of water.

1 mole of water has mass 0.018 kg.

$\therefore$ 1.0 kg contains n moles where n (mol) $\times$ 0.018 (kg mol^{-1}) = 1.0 kg.

$$n = \frac{1.0}{0.018} = 55.6\,\text{mol}$$

The number of *molecules*, N, in n moles of a substance is given by the equation

$$N = n \times N_A$$

where N_A is the Avogadro constant. So the number of molecules in 1.0 kg of water,

$$N = 55.6\,\text{mol} \times 6.0 \times 10^{23}\,\text{mol}^{-1} = 3.3 \times 10^{25}\text{ molecules}$$

$\therefore$ the binding energy per molecule

$$= \frac{\Delta U}{N} = \frac{2.09 \times 10^6\,\text{J}}{3.3 \times 10^{25}} = \mathbf{6.3 \times 10^{-20}\,J}$$

11 The equation that you need to use in this question is

$$\frac{p_1V_1}{T_1} = \frac{p_2V_2}{T_2}$$

where subscript 1 refers to the conditions at the start of a change and subscript 2 refers to the conditions at the end of the change.

a Throughout the change labelled α, the temperature T is constant, so $p_1V_1 = p_2V_2$, which you can rearrange to give

$$p_2 = \frac{V_1}{V_2}p_1$$

But $V_1 = V_0$ and $V_2 = 2V_0$,

$$\therefore \quad p_2 = \frac{V_0}{2V_0} \times p_1 = \frac{p_1}{2} = \mathbf{0.65 \times 10^5\,Pa}$$

b Throughout the change labelled β, the pressure p is constant but the volume changes from $2V_0$ to $3V_0$. So, substituting the initial and final values of the volume for V_1 and V_2, and cancelling the p's, because $p_1 = p_2$,

$$\frac{p_1V_1}{T_1} = \frac{p_2V_2}{T_2} \quad \text{becomes} \quad \frac{2V_0}{T_1} = \frac{3V_0}{T_2}$$

which you can rearrange to give

$$T_2 = \frac{3V_0}{2V_0} \times T_1 = \frac{3}{2} \times T_1 = \frac{3}{2} \times 280\,\text{K} = \mathbf{420\,K}$$

c During the change labelled γ, the temperature T is constant so the general equation reduces to $p_1V_1 = p_2V_2$, which you can rearrange to give

$$V_2 = \frac{p_1V_1}{p_2}$$

But $V_1 = 3V_0$, from the answer to **b**, and the pressure doubles: $p_2 = 2p_1$. So

$$V_2 = \frac{p_1V_1}{p_2} = \frac{0.5p_2 3V_0}{p_2} = \mathbf{1.5\,V_0}$$

See Figure 22.14.

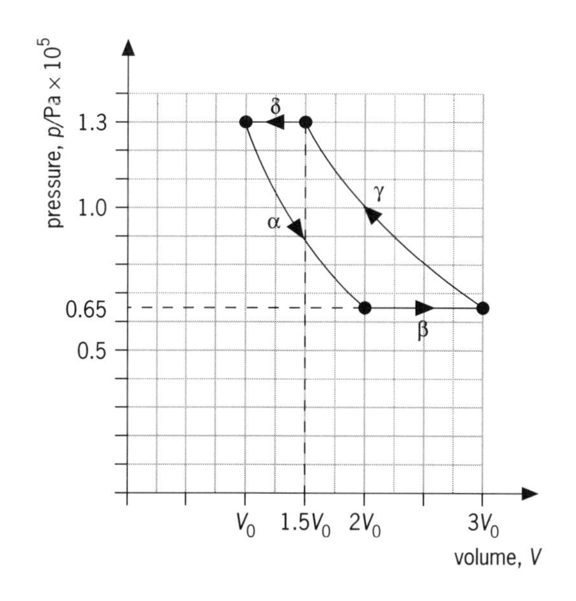

Figure 22.14

12 a You can calculate the waste power from the rise in temperature of the cooling water. 15 tonnes ($=15 \times 10^3$ kg) of water has a temperature rise of 15 K every second.

$$\text{power} = \frac{\Delta Q}{t} = \frac{m}{t}c\Delta\theta$$

$$= 15 \times 10^3\,\text{kg s}^{-1} \times 4200\,\text{J kg}^{-1}\text{K}^{-1} \times 15\,\text{K}$$

$$= \mathbf{945\,MW}$$

b Total power $=$ useful power $+$ power wasted $= 560\,\text{MW} + 945\,\text{MW} = 1505\,\text{MW}$

$$\text{efficiency} = \frac{\text{useful power}}{\text{total power}} \times 100\%$$

$$= \frac{560}{1505} \times 100\% = \mathbf{37\%}$$

Remember that temperatures must be absolute temperatures.

c $540\,°\text{C} = (540 + 273)\,\text{K} = 813\,\text{K}$
$15\,°\text{C} = (15 + 273)\,\text{K} = 288\,\text{K}$

The theoretical maximum efficiency is given by

$$\frac{T_H - T_C}{T_H} \times 100\% = \frac{813\,\text{K} - 288\,\text{K}}{813\,\text{K}} \times 100\% = \mathbf{64\%}$$

13 a

	Δ*U*/J inc or dec?	**Δ*Q*/J in or out?**	**Δ*W*/J on or by?**
Change labelled **A**	47 increase	47 into system	0
Change labelled **B**	42 increase	0	42 on system
Change labelled **C**	63 decrease	63 out of system	0
Change labelled **D**	26 decrease	0	26 by system

$\Delta Q = 0$ for the adiabatic processes, **B** and **D**.
$\Delta W = 0$ in the constant volume stages, **A** and **C**, because $\Delta V = 0$.
ΔU equals whichever of these two is not zero.
Work *on the system* increases *U*, work *by the system* decreases *U*.
ΔU must be zero at the end of the cycle because the system is back at its starting point.
The heat ejected during **C** is found from the total ΔU at Z, minus the work done by the system in stage **D**: $89\,\text{J} - 26\,\text{J} = 63\,\text{J}$.

b The net work is equal to the area inside the loop and is equal to the difference between the 42 J of work done *on the system* and the 26 J done *by the system*:

$42\,\text{J} - 26\,\text{J} = \mathbf{16\,J}$

(This is the energy for which you must pay.)

c Starting at X the system receives 47 J thermally in stage **A** and 42 J from work in **B** so the total change in internal energy when it reaches Z is

$47\,\text{J} + 42\,\text{J} = \mathbf{89\,J}$

23 The nuclear atom and radioactivity

Atomic size

An approximate value for the size of an atom may be obtained from the density and atomic mass of a substance, for example a metal. Knowing the number of atoms present in one mole (6.02×10^{23}) and the volume that one mole occupies, the volume of one atom and hence its diameter may be calculated. This gives a value of about 10^{-10} m.

Evidence for a nuclear atom

In 1911 a team of scientists carried out investigations to determine the structure of an atom. It was known that there was a positively charged part to atoms, associated with the mass of the atom, and that all atoms contained light negatively charged electrons. It was not known how they were arranged.

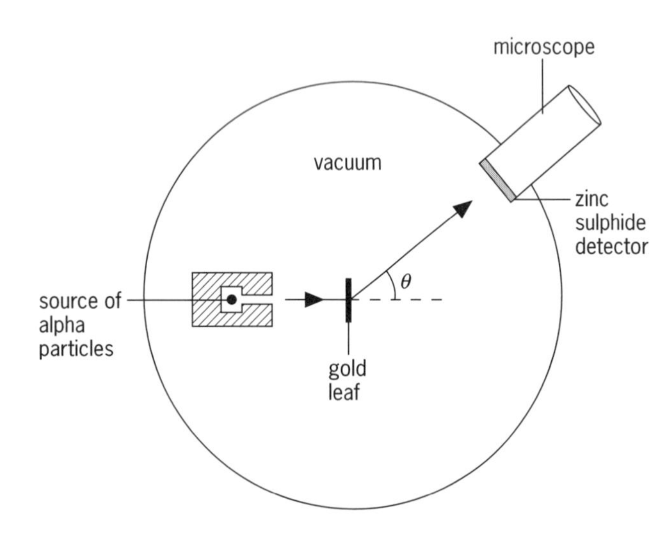

Figure 23.1 Investigating atomic structure

A thin piece of gold leaf was bombarded with alpha particles, Figure 23.1. Most of these particles passed undeviated through the gold leaf, a few were deflected through a small angle θ and a very small number of them were deflected through very large angles, occasionally greater than 90°, Figure 23.2. These results suggested that an atom is mainly empty space with a very small positively charged nucleus approximately 10^{-15} m across, which contains most of the mass of the atom.

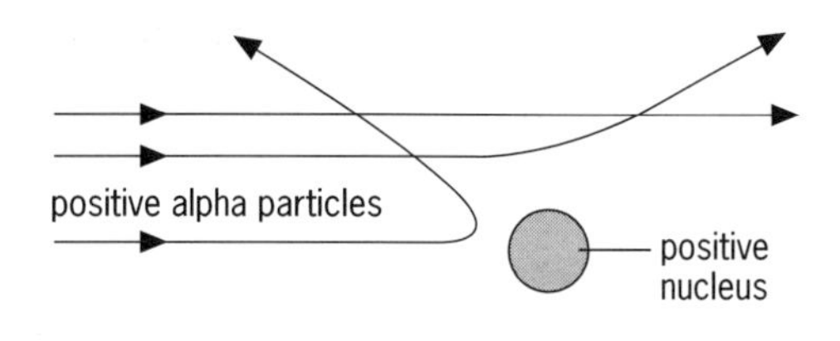

Figure 23.2 Scattering of alpha particles by an atomic nucleus

The nuclear model of the atom

Further experiments led scientists to more detailed theories about the structure.

- The nucleus of an atom contains two kinds of small particles called protons and neutrons (collectively called nucleons). A proton has an atomic mass of 1 u and carries a charge of $+e$. A neutron has an atomic mass of 1 u but carries no charge.
- Particles called electrons orbit the nucleus. They have a very small mass (approximately 1/1800th the mass of a proton) and carry a charge of $-e$. In an uncharged atom there are equal numbers of protons and electrons.

The **atomic mass unit** (u) is used for measuring the masses of atomic particles. The atomic mass of both a proton and a neutron is approximately 1 u.

Using these ideas and information from the periodic table it is possible to work out the atomic structures of different elements. See Figure 23.3. A nucleus with a particular structure of protons and neutrons is referred to as a 'nuclide'.

${}^{1}_{1}\mathrm{H}$ A hydrogen atom consists of one proton in the nucleus, no neutrons and one orbiting electron

${}^{4}_{2}\mathrm{He}$ A helium atom consists of two protons and two neutrons in the nucleus and two orbiting electron

${}^{7}_{3}\mathrm{Li}$ A lithium atom consists of three protons and four neutrons in the nucleus and three orbiting electrons

${}^{235}_{92}\mathrm{U}$ A uranium-235 atom consists of 92 protons and 143 neutrons in the nucleus and 92 orbiting electrons

Figure 23.3

The **proton number**, Z, of a nuclide is the number of protons that a particular element has in the nucleus of its atom. It is also the number of the element in the periodic table and used to be known as the atomic number.

The **nucleon number**, A, is the total number of protons and neutrons in the nucleus. It is also known as the mass number.

The number of neutrons in a nucleus, or its **neutron number**, is denoted by N.

$$A = N + Z$$

Isotopes

Some elements have more than one type of nucleus. Chlorine, for example, has two different nuclides, one containing 17 protons and 18 neutrons, and the other containing 17 protons and 20 neutrons, Figure 23.4. These two different nuclides are called **isotopes**. The presence of isotopes gives rise to non-integer relative atomic masses because the average mass is used.

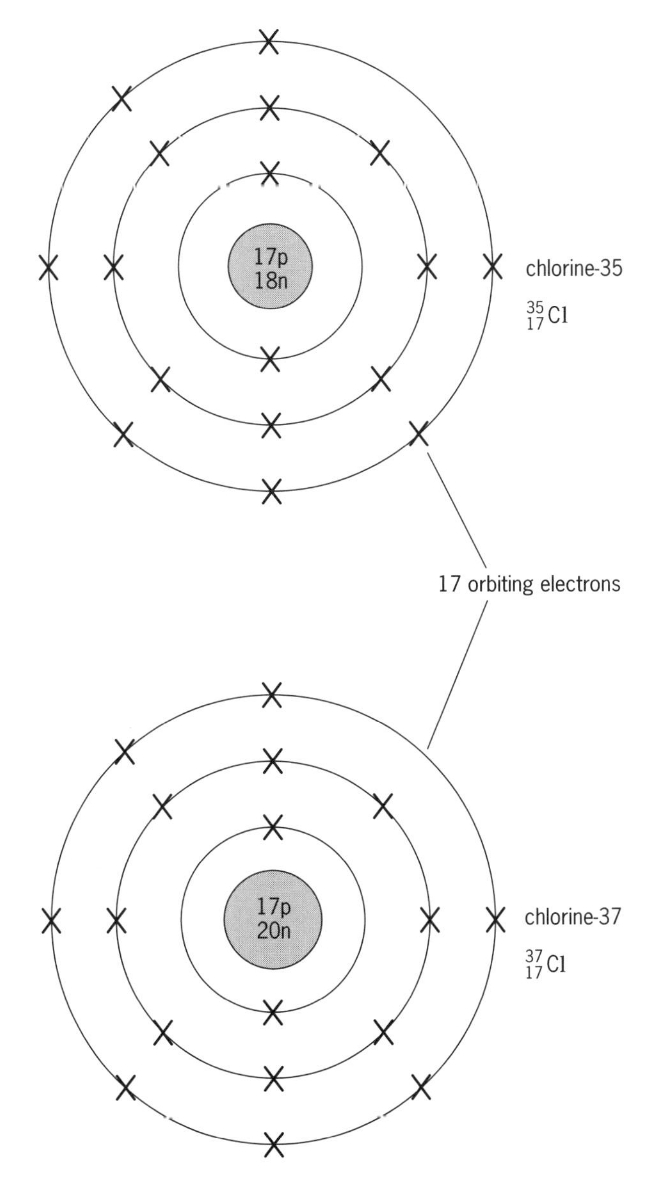

Figure 23.4 The two isotopes of chlorine

Stability of a nucleus

All protons are positively charged. There must therefore be repulsive electrostatic force between protons within a nucleus. For a nucleus to be stable there must therefore be a second, stronger force which binds all the particles together. This attractive, short-range force is called the 'strong force'. The relative sizes of these two forces depends upon how many protons and neutrons there are in a nucleus.

For lighter elements such as carbon and oxygen the number of protons in the nucleus is equal to the number of neutrons and the attractive forces are large enough to hold the nucleus together. For heavier elements it is necessary to have more neutrons than protons in the nucleus if it is to be stable. The graph in Figure 23.5 shows this pattern.

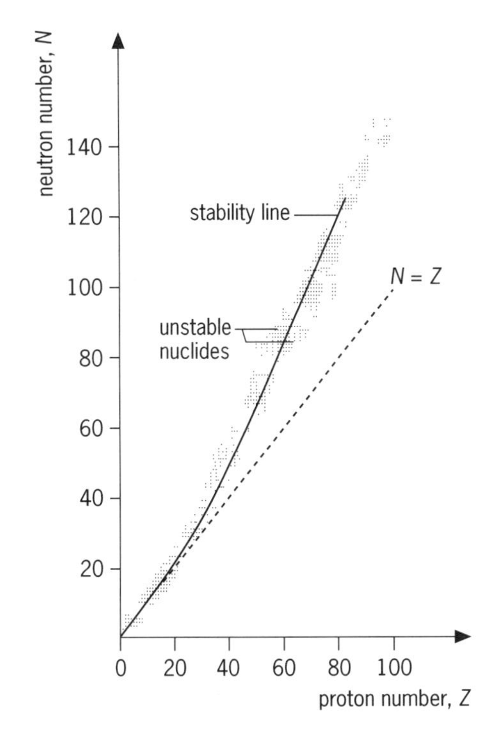

Figure 23.5 Stability of nuclides

Radioactive decay

Nuclides which do not lie on the 'stability line' of Figure 23.5 are unstable. They become more stable by **radioactive decay**, that is by emitting alpha, beta or gamma radiation. The final state of the nucleus is at a lower energy level. Some mass is converted to energy (see topic 25) which is carried away by the particles that are emitted.

Alpha decay

An alpha particle consists of two protons and two neutrons ($^4_2\alpha$). It is in fact a helium nucleus and carries a charge of $+2e$. A nuclide that emits an alpha particle will decrease its nucleon number by 4 and its proton number by 2.

To describe the emission of an alpha particle by a uranium-238 nucleus we can write

$$^{238}_{92}\mathrm{U} \rightarrow {}^{234}_{90}\mathrm{Th} + {}^4_2\alpha$$

The atom of uranium-238 is said to have decayed or 'transmuted' into an atom of thorium-234 by alpha emission. Proton number and nucleon number are conserved in such reactions.

Alpha decay tends to occur in heavy nuclides which lie below the stability line. The **daughter nuclide** produced lies closer to this line.

Beta decay

The most common form of beta decay is β^- decay, when the emitted particle is a fast-moving electron. The beta particle is created in the nucleus when a neutron is changed into a proton and an electron:

$$^1_0\mathrm{n} \rightarrow {}^1_{+1}\mathrm{p} + {}^0_{-1}\mathrm{e} + \bar{\nu}$$

The proton remains in the nucleus and the electron is emitted as a beta particle. A light, neutral particle called an antineutrino, $\bar{\nu}$, is also emitted.

A nuclide which emits a beta particle will increase its proton number by 1. Its nucleon number remains unchanged. For example,

$$^{14}_6\mathrm{C} \rightarrow {}^{14}_7\mathrm{N} + {}^0_{-1}\beta + \bar{\nu}$$

The atom of carbon-14 has transmuted into an atom of nitrogen-14 by beta emission.

Beta decay tends to occur in nuclides which lie above the stability line. The daughter nuclide produced lies closer to this line.

Some artificially produced radioactive nuclides decay by emitting a positively charged beta particle, β^+:

$$^{22}_{11}\mathrm{Na} \rightarrow {}^{22}_{10}\mathrm{Ne} + {}^0_{+1}\beta + \nu$$

The β^+ particle is a positive electron, $^0_{+1}\mathrm{e}$, called a **positron**. It is an antimatter particle and if it meets an electron they are both annihilated, leaving pure electromagnetic energy. A neutrino, ν, is emitted with the β^+ particle. Like its antimatter counterpart, the antineutrino, it is light (effectively zero mass) and has zero charge.

Gamma decay

Sometimes when an alpha particle or a beta particle has been emitted the daughter nuclide created has excess or residual energy. This energy is removed from the nucleus by the emission of a photon of electromagnetic radiation called gamma radiation. The photon has no charge and no mass and so causes no change to the proton or nucleon number of the nucleus.

For example, when an atom of uranium-238 emits an alpha particle the resulting thorium-234 nucleus contains too much energy and so emits gamma (γ) radiation:

$$^{234}_{90}\mathrm{Th} \rightarrow {}^{234}_{90}\mathrm{Th} + {}^0_0\gamma$$

Table 23.1 summarises the properties of alpha, beta and gamma radiations. One distinguishing feature is how they are affected by an electric or magnetic field; see Figure 23.6.

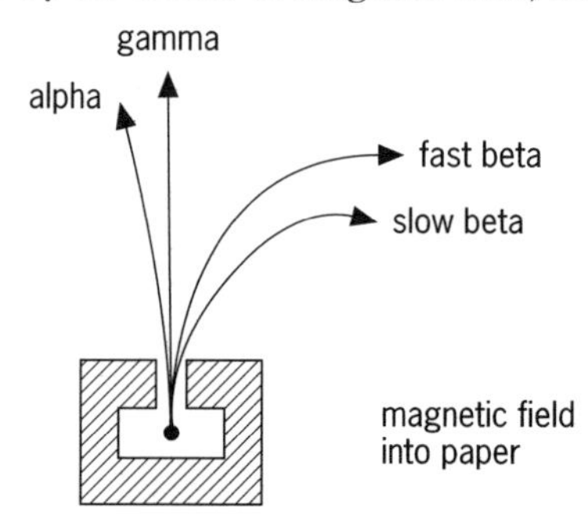

Figure 23.6 Deflection of alpha, beta and gamma radiation by a magnetic field (not to scale)

Table 23.1 Properties of alpha, beta and gamma radiations

Type of radiation	alpha	beta	gamma
Nature	particle consisting of 2 protons and 2 neutrons; identical to a helium nucleus	fast-moving electron or positron	electromagnetic wave
Charge	$+2e$	$-1e$ or $+1e$	no charge
Mass	4u	1/1836u	no mass
Speed	5–10% speed of light	90% speed of light	speed of light
Relative ionising power	10000	1000	1
Penetration	a few centimetres in air or one or two sheets of paper	much more penetrating than alpha; stopped by several mm of aluminium or 50–100cm of air	most penetrating of all three types of radiation; able to travel through several cm of lead
Deflection by magnetic and electric fields	small deviation	large deviation in opposite direction to alpha	no deviation
Detection	all three types of radiation can be detected using photographic plates, Geiger–Müller tubes and solid state detectors		

Measuring levels of radioactivity using a Geiger–Müller tube and counter

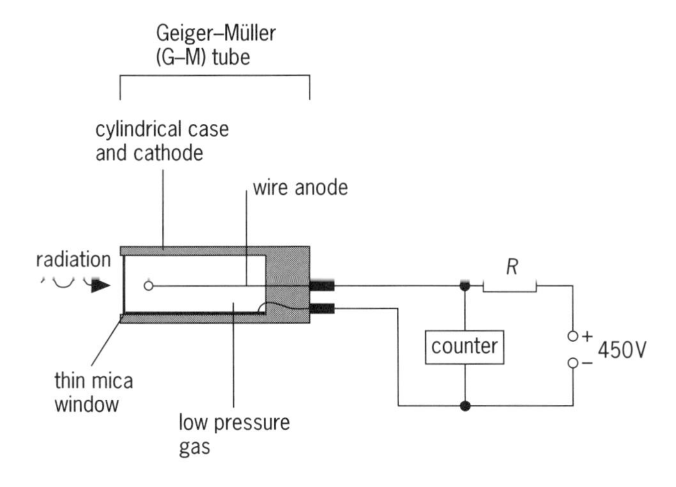

Figure 23.7 When alpha, beta or gamma radiation enters the G–M tube a pulse is created which triggers a counting circuit

When measuring the radiation emitted from a source it is important to take into account the level of the **background radiation**, that is any radiation present which has not been emitted by the source being investigated. True measurements are found by subtracting the 'background count' from measurements taken with the G–M tube.

Fall-off in intensity of radiation

Gamma radiation is not noticeably absorbed by air, which means its intensity follows an inverse square law: the intensity falls off as the square of the distance from the source.

$$\frac{I_1}{I_2} = \frac{d_2^2}{d_1^2}$$

In a vacuum alpha and beta radiations also follow this rule. In air or other materials they are absorbed so their intensity falls off more rapidly. They lose energy by ionising the atoms and molecules in their paths; that is, by stripping electrons from them. The higher the ionising power of the radiation (see Table 23.1) the greater the absorption and so the lower the penetration into the medium.

Danger of nuclear radiation

The ionising power of alpha, beta and gamma radiations makes them a hazard. The higher the ionising power, the more dangerous the radiation if it enters living cells.

Avogadro constant, $N_A = 6.0 \times 10^{23}\,mol^{-1}$
$1\,eV = 1.6 \times 10^{-19}\,J$
$e = 1.6 \times 10^{-19}\,C$

Level 1

1 $^{11}_{6}C$, $^{12}_{6}C$, $^{14}_{6}C$ are isotopes of carbon. The number at the top tells you how many ____________ there are in the nucleus and the number at the bottom tells you how many ____________ there are.

a Write down the missing words.

b What does the difference between the two numbers tell you?

c What is the

i atomic number
ii mass number
iii nucleon number
iv proton number
v neutron number

of each isotope given?

d How many electrons would each neutral atom have?

2 a The molar mass of iron is $56\,g\,mol^{-1}$. What is the mass of one iron atom?

b The molar mass of silver is $108\,g\,mol^{-1}$. How many atoms are there in 5.4 g of silver?

3 $1.0\,m^3$ of aluminium contains about 10^5 moles.

a How many atoms is this?

b If the atoms are arranged regularly within a cube how many will there be along each side?

c What is a rough figure for the diameter of an aluminium atom?

4 Imagine that your question sheet is badly copied and that you can't read the powers of ten for the radius of the nucleus. It says $6 \times 10^{-}$ m.

a What power of 10 should be there?

b How many nuclei would fit across the atom in question 3?

5 A source of alpha particles is often used in a smoke detector, as they are easily stopped by smoke. The alpha particles have energy 1.3 pJ and a range, in air, of 29 mm. By collision with atoms in air each alpha particle can create 5.2×10^5 ion-pairs per cm of air travelled through. How much energy is needed to create each ion-pair?

6 Describe the following nuclear reactions in words.

a $^{234}_{92}U \rightarrow ^{230}_{90}Th + ^{4}_{2}He$

b $^{238}_{92}U + ^{1}_{0}n \rightarrow ^{239}_{92}U + \gamma$

c $^{40}_{19}K \rightarrow ^{40}_{20}Ca + ^{0}_{-1}e + \overline{\nu}$

Level 2

7 The hydrogen atom has one proton and one electron. The helium atom has two protons, two neutrons and two electrons. Can you suggest why the helium atom is considerably smaller than the hydrogen atom?

8 Copy and complete the following table.

Mode of decay	Effect on parent nuclide Z	N	A
α emission			
β^- emission			
β^+ (positron) emission			
electron capture by nucleus			
emission of proton			
emission of neutron			

Level 3

9 A proton, an alpha particle and an electron are each accelerated through 5.0 kV.

a What will be the ratios of the kinetic energies that they acquire?

b How does this compare with the ratios of their velocities?

Take the proton velocity as standard.

10 When a stationary polonium-216 nucleus decays by alpha emission, the alpha particle that is ejected has a velocity of $1.8 \times 10^7\,m\,s^{-1}$.

a Calculate the recoil velocity of the remaining nucleus.

b Comment on the kinetic energy of the system before and after the decay.

11 $^{11}_{6}C$ is an artificial radioisotope which decays by emitting a positron.

a Write an equation for this reaction.

Use X if you don't know the symbol for an element.

b Describe what will happen to the positron after it is emitted.

c This event will give rise to electromagnetic radiation which can be detected and which identifies the location of the positron. What type of radiation is this and how is this phenomenon used in medicine?

12 a The intensity of some types of radiation falls off as the inverse square of the distance from the source. The intensity of others decreases exponentially. What factor makes the difference?

b Which of the following would follow the inverse square law?

i radio waves in space
ii gamma rays in concrete
iii alpha radiation in air
iv beta radiation in a vacuum
v light in water
vi X-rays in tissue
vii gamma rays in air.

c A cobalt-60 source gives a gamma dose rate of $120\,\mu Sv\,h^{-1}$ at a distance of 1.5 m in air. At what distance will the dose rate be $30\,\mu Sv\,h^{-1}$?

You don't need to worry about exactly what these units mean.

13 Explain the following statements.

a Small doses of gamma radiation are less harmful than small doses of beta radiation, even though gamma radiation is more penetrating.

b Alpha radiation is not generally harmful, but if an alpha emitter, for instance radon, is inhaled, it is very dangerous.

14 a Copy the stability line in Figure 23.5 and shade areas to show nuclides that decay

i by β^- emission
ii by β^+ emission
iii by α emission.

b Explain in general terms why certain decays occur in certain areas of the chart.

Level 1

1 a The number at the top tells you how many **nucleons** there are in the nucleus and the number at the bottom tells you how many **protons** there are.

b The difference between these two numbers is equal to the number of **neutrons**.

c i, iv The atomic number is the same as the proton number, and is the same for all isotopes of a particular element. In this case the atomic number for all of them is **6**.

ii, iii The mass numbers are the same as the nucleon numbers, and are **11, 12, 14**.

v The neutron numbers are **5, 6, 8**.

d The neutral atoms will each have **six electrons**, the same as the number of protons, so that the positive and negative charges cancel.

2 a The molar mass of $56\,\mathrm{g\,mol^{-1}}$ tells us that 1 mole of iron has a mass of 56 g. 1 mole contains the Avogadro constant, 6.0×10^{23}, 'particles'. So the mass of 6.0×10^{23} 'particles' (atoms) = 56 g.

$$\text{mass of 1 atom} = \frac{56\,\mathrm{g\,mol^{-1}}}{6.0\times10^{23}\,\mathrm{mol^{-1}}}$$

$$= \mathbf{9.3\times10^{-23}\,g} \text{ or } \mathbf{9.3\times10^{-26}\,kg}$$

b If you spotted that 5.4 is 1/20 of 108, then you would divide the Avogadro constant, N_A, by 20 – which gives 3.0×10^{22} atoms.

If not, solve it by proportion. If x is the number of atoms in 5.4 g,

$$\frac{x}{N_A} = \frac{5.4\,\mathrm{g}}{108\,\mathrm{g}}$$

$$\therefore \quad x = \frac{5.4N_A}{108} = \mathbf{3.0\times10^{22}} \text{ atoms}$$

3 a 1 mole of aluminium contains 6.0×10^{23} atoms. So 10^5 moles of aluminium contain $\mathbf{6.0\times10^{28}}$ atoms.

b If there are n atoms along each side of the cube there will be n^3 altogether.

So you have to take the cube root of the total number to find how many on each side:

$$\sqrt[3]{6.0\times10^{28}} = \mathbf{3.9\times10^{9}} \text{ (atoms on each side)}$$

c If n is the number of atoms along each side and d is the diameter of one, then all n of them will cover a distance $l = n\times d$. You can rearrange this simple equation to give an approximate value for the diameter of one atom:

$$d = \frac{l}{n}$$

In this case $n = 3.9\times10^9$ and $l = 1.0\,\mathrm{m}$

$$\therefore d = \frac{1.0\,\mathrm{m}}{3.9\times10^9} = \mathbf{2.5\times10^{-10}\,m}$$

4 a The radius of a nucleus is of the order of femtometres (10^{-15} m) so it should say $\mathbf{6\times10^{-15}}$.

b If N is the number of nuclei and d is the diameter of one nucleus (about 1.2×10^{-14} m) then $N\times d = 2.5\times10^{-10}$ m.

$$\therefore N = \frac{2.5\times10^{-10}\,\mathrm{m}}{1.2\times10^{-14}\,\mathrm{m}} = \mathbf{2.1\times10^{4}}$$

21 000 nuclei would go across one atom diameter.

Try to visualise this. If a nucleus were the size of a tennis ball, an aluminium atom would be about 1350 m across – almost a mile.

5 You need to divide the initial energy of an alpha particle by the number of ion-pairs that it will produce. The alpha particle will travel 29 millimetres, creating 5.2×10^5 ion-pairs every centimetre.

Did you notice the mixture of units?

29 mm = 2.9 cm, so the alpha particle will produce

$$2.9\,\mathrm{cm}\times5.2\times10^5\,\text{ion-pairs cm}^{-1} = 1.5\times10^6\,\text{ion-pairs}$$

before it is stopped. The energy available is 1.3×10^{-12} J, so each pair must take

$$\frac{1.3\times10^{-12}}{1.5\times10^6} = \mathbf{8.7\times10^{-19}\,J}$$

To check this is sensible, change to eV: $8.7\times10^{-19}\,\mathrm{J} = 5.4\,\mathrm{eV}$.

6 a Uranium decays to thorium by the emission of an alpha particle.

b Uranium is bombarded by neutrons. One neutron enters the nucleus giving a different isotope of uranium. The new nucleus has surplus energy, which is emitted as a gamma ray.

c Potassium decays to calcium by β^- emission. A neutron has changed into a proton and an electron, which is emitted as a β^- particle. An antineutrino is also emitted.

Level 2

7 The helium atom has double the charge at the centre, compared with the hydrogen atom. The electrostatic force on each electron is doubled and they are pulled in more closely than in the hydrogen atom, so the overall diameter is smaller. It is not the number of nucleons in the nucleus that determines the size of the atom, but the electron orbits.

8

Mode of decay	Effect on parent nuclide Z	 N	 A
α emission	down 2	down 2	down 4
β^- emission	up 1	down 1	no change
β^+ (positron) emission	down 1	up 1	no change
electron capture by nucleus	down 1	up 1	no change
emission of proton	down 1	no change	down 1
emission of neutron	no change	down 1	down 1

Level 3

9 a When a charge, q, travels through a potential difference, ΔV, it acquires kinetic energy, $E_k = q \times \Delta V$. The energy depends on the charge and the potential difference, but not on the mass of the particle.

The charge of both proton and electron is e, so they each gain 5.0 keV.
The charge of the alpha particle is $2e$, so it gains 10 keV.

The ratios of the energies are therefore

proton : alpha : electron = **1 : 2 : 1**

b Let v_p, v_α, v_e be the velocities of the particles. Let m be the mass of the proton; the electron mass is then approximately $m/2000$ and the mass of the alpha particle is $4m$.

Comparing the proton and the electron, their kinetic energies are equal, so

$$\tfrac{1}{2} \times m \times v_p^2 = \tfrac{1}{2} \times \frac{m}{2000} \times v_e^2$$

which you can rearrange to give

$$v_e^2 = 2000\, v_p^2$$

$$\therefore \quad v_e = \sqrt{2000} \times v_p = \mathbf{45\, v_p}$$

Comparing the alpha particle and the proton, the alpha has twice the energy of the proton, so

$$\tfrac{1}{2} \times 4m \times v_\alpha^2 = 2 \times \tfrac{1}{2} \times m \times v_p^2$$

Cancelling gives

$$2v_\alpha^2 = v_p^2$$

Rearranging and taking square roots gives

$$v_\alpha = \frac{1}{\sqrt{2}} \times v_p = \mathbf{0.7\, v_p}$$

The ratios of the velocities are therefore

proton : alpha : electron = **1 : 0.71 : 45**

10 a Momentum must be conserved (see topic 9). The momentum before the decay is zero, so the momentum after decay must also be zero. This means that the particles must move away in opposite directions. It doesn't matter which direction you take as positive – you just have to be consistent. In Figure 23.8 we have taken movement to the right to be positive, so the alpha particle has a positive velocity and the remaining nucleus will have a negative velocity.

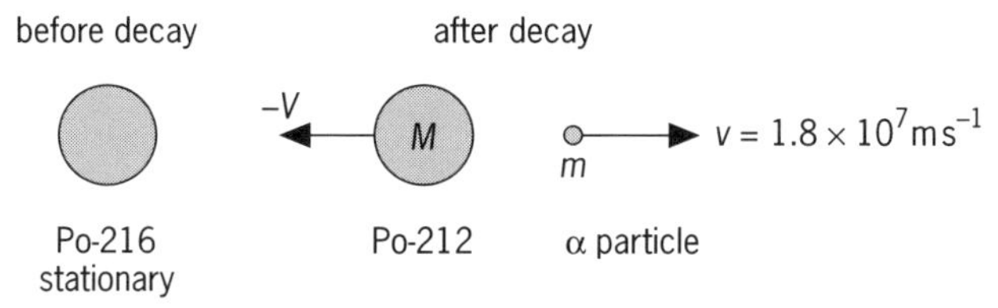

before decay | after decay

$-V$ | M | m | $v = 1.8 \times 10^7\,\text{m s}^{-1}$

Po-216 stationary | Po-212 | α particle

Figure 23.8

After the decay, total momentum $= mv + MV = 0$, where the small letters refer to the alpha particle and the large letters to the nucleus. You can rearrange this equation to give

$$V = \frac{-mv}{M}$$

$$m = 4\,\text{u}, \quad v = 1.8 \times 10^7\,\text{m s}^{-1}$$
$$M = 216\,\text{u} - m = 216\,\text{u} - 4\,\text{u} = 212\,\text{u}$$

(Remember that u is the unit of atomic mass.)

$$\therefore \quad V = \frac{-4\,\text{u} \times 1.8 \times 10^7\,\text{m s}^{-1}}{212\,\text{u}} = \mathbf{-3.4 \times 10^5\,m\,s^{-1}}$$

b The nucleus was stationary before the decay, so it had no kinetic energy, but both particles are moving after the decay. Energy is not a vector so the direction of their movement does not affect the total energy, which is just the sum of the two energies. Clearly energy has to come from somewhere. In fact it comes from a small change in mass of the remaining particles which has been converted to energy.

11 a ${}^{11}_{6}\text{C} \rightarrow {}^{11}_{5}\text{B} + {}^{0}_{+1}\text{e} + \nu$

b The positron will very soon meet an electron and they will annihilate each other.

c This gives rise to two gamma ray photons which go in opposite directions. See Figure 23.9.

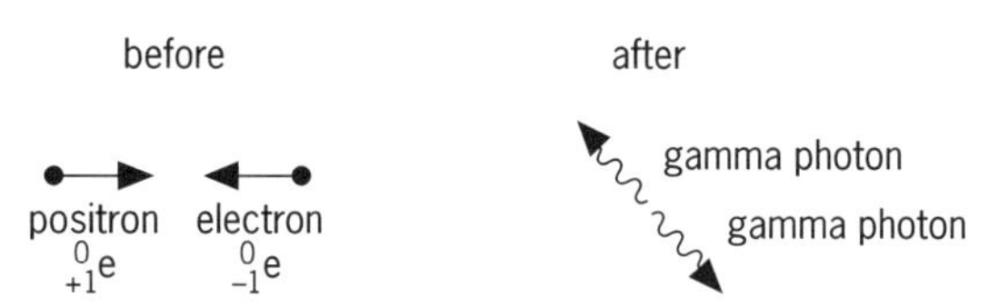

Figure 23.9 Positron–electron annihilation

The gamma photons can be detected by a scintillation counter that registers the coincidence of two signals. It traces them back to their origin, which is the position (to within a mm) of the radioactive carbon. Thus very precise mapping is possible. This is used in brain scans. The radioactive carbon is inhaled as carbon monoxide, CO. The process is called positron emission tomography or PET scan.

12 a The exponential law is followed if there is absorption of radiation by the material between the source and the detector. The inverse square law is really valid only in a vacuum, but it can be used where the absorption is small.

b **i**, **iv** and **vii** would follow the inverse square law; the other radiations are passing through relatively dense materials.

c There is little absorption of gamma rays by air so the inverse square law will be followed:

$$I = \frac{k}{d^2}$$

where k is a constant.

In this case the value of the second dose is one quarter of the dose at 1.5 m, so the distance away must be twice 1.5 m = **3.0 m**.

To solve the problem in more general cases, you can rearrange the equation to give

$$k = I \times d^2 = 120\,\mu Sv\,h^{-1} \times 1.5^2\,m^2 = 270\,\mu Sv\,h^{-1}\,m^2$$

Now you can find the dose rate at any distance, or the distance for any dose rate. So, for example, at a distance of 3.0 m the dose rate will be

$$\frac{270\,\mu Sv\,h^{-1}\,m^2}{9.0\,m^2} = 30\,\mu Sv\,h^{-1}$$

as expected.

13 a Gamma rays pass through the body with little interaction and do little damage. Beta radiation is more strongly ionising and will damage cells along its path until its energy is absorbed.

b Alpha radiation is usually stopped by the layer of dead skin on the outside of your body, but if it gets inside, in contact with soft tissue, it is highly ionising and damages cells.

14 a

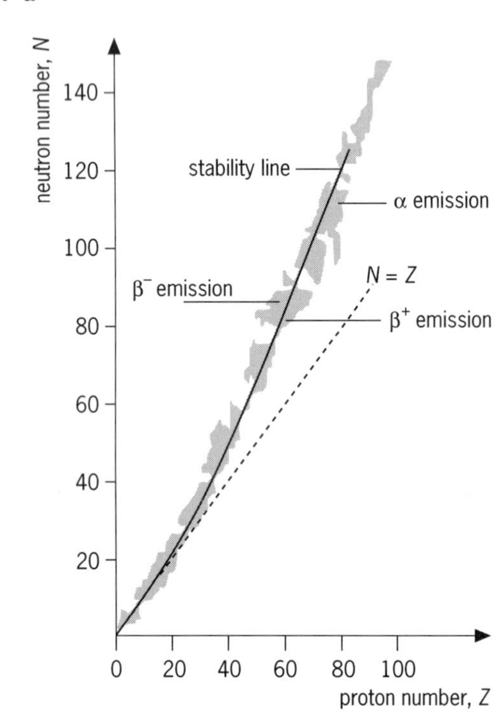

Figure 23.10

b The transitions that occur always bring the nuclide nearer to the stability line. Some transitions are shown in the following N–Z plots, Figure 23.11.

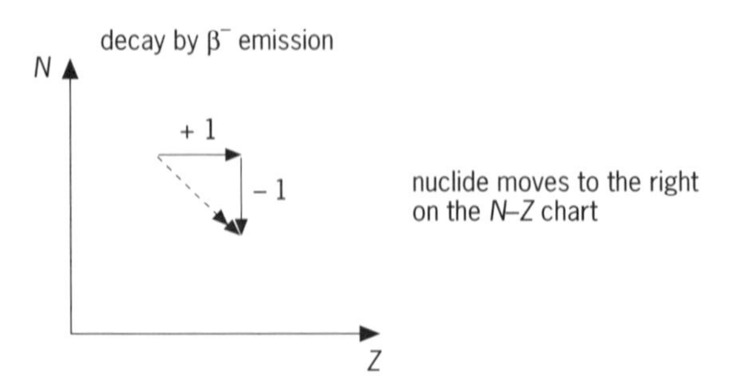

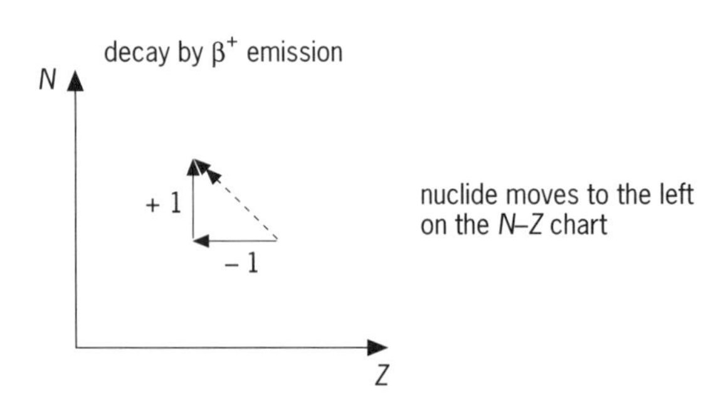

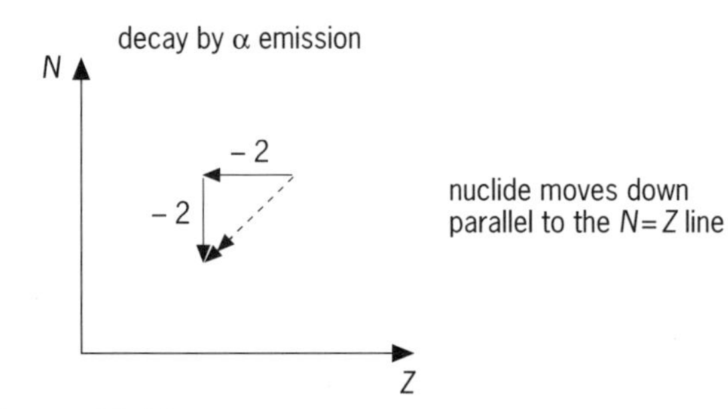

Figure 23.11

Compare these transitions with those given in the answer to question 8.

24 Decay series and rate of radioactive decay

Decay series

When an unstable nuclide decays by alpha or beta decay a new nuclide is created. This process is called 'transmutation'. The daughter nuclide created may also be unstable, in which case it, too, will decay. This process will continue until a stable nuclide is formed. The written set of transmutations which describe these decays is called a **decay series**. An example is drawn in Figure 24.1. The decay continues as long as it is resulting in lower energy states. The final mass is always less than the initial mass.

$${}^{238}_{92}\mathrm{U} \longrightarrow {}^{234}_{90}\mathrm{Th} + {}^{4}_{2}\alpha + \gamma$$

$${}^{234}_{90}\mathrm{Th} \longrightarrow {}^{234}_{91}\mathrm{Pa} + {}^{0}_{-1}\beta + \gamma$$

$${}^{234}_{91}\mathrm{Pa} \longrightarrow {}^{234}_{92}\mathrm{U} + {}^{0}_{-1}\beta + \gamma$$

$${}^{234}_{92}\mathrm{U} \longrightarrow {}^{230}_{90}\mathrm{Th} + {}^{4}_{2}\alpha + \gamma$$

$${}^{230}_{90}\mathrm{Th} \longrightarrow {}^{226}_{88}\mathrm{Ra} + {}^{4}_{2}\alpha + \gamma$$

Figure 24.1 Part of the decay series for uranium-238

Rate of radioactive decay

The precise moment that an unstable nucleus will decay cannot be predicted, nor can it be speeded up or slowed down by changes in pressure, temperature and so on. It is a totally random and spontaneous event. Some nuclei exist for thousands of years before they decay; others may exist for only a fraction of a second.

Decay curves and half-life

If a graph is plotted of the number of undecayed atoms in a sample against time, a curve similar to that shown in Figure 24.2 will be constructed.

Figure 24.2 Typical radioactive decay curve

A graph which has this shape is described as being **exponential**. All radioactive decay curves have this exponential shape.

Figure 24.2 shows that the time it takes for half the nuclides in a sample to decay is constant; that is, it is independent of the number of atoms in the sample and depends only on the type of nuclide. The time taken for half the atoms to decay is called the **half-life** of that particular nuclide.

Example 1

Calculate the fraction of the original atoms remaining in a sample of a radioactive isotope which has a half-life of 1.0 h after **a** 2.0 h, **b** 3.0 h and **c** 4.0 h.

a After two half-lives the fraction remaining is $\frac{1}{2} \times \frac{1}{2} = \frac{1}{4}$.

b After three half-lives the fraction remaining is $\frac{1}{2} \times \frac{1}{2} \times \frac{1}{2} = \frac{1}{8}$.

c After four half-lives the fraction remaining is $\frac{1}{2} \times \frac{1}{2} \times \frac{1}{2} \times \frac{1}{2} = \frac{1}{16}$.

The exponential law equation

The exponential shape of the decay curve means that the rate of emission of particles from a sample of radioactive material is proportional to the number of atoms in the sample; that is, if the number of undecayed atoms in the sample halves, the rate of emission also halves.

$$\frac{\Delta N}{\Delta t} \propto N$$

Using calculus notation,

$$\frac{dN}{dt} \propto N$$

or

$$\boxed{\frac{dN}{dt} = -\lambda N}$$

where N is the number of undecayed nuclei and λ is known as the **decay constant**. The minus sign indicates that the number of undecayed nuclei falls with time.

The number of particles emitted per second is numerically equal to dN/dt, which is known as the **activity**, A, of that sample and is measured in **becquerel** (Bq). 1 Bq is equal to 1 disintegration or 1 emission per second. Because A is a positive quantity,

$$A = -\frac{dN}{dt}$$

$$= \lambda N$$

Example 2

Calculate the activity of a sample of a radioactive isotope containing 1.0×10^{17} atoms and which has a decay constant of $2.0 \times 10^{-6}\,s^{-1}$.

$$A = -\frac{dN}{dt} = \lambda N$$

$$= -2.0 \times 10^{-6}\,s^{-1} \times 1.0 \times 10^{17}$$
$$= -2.0 \times 10^{11}\,Bq$$

The relationship which satisfies $dN/dt = -\lambda N$ is

$$\boxed{N_t = N_0 e^{-\lambda t}}$$

where N_0 is the number of undecayed atoms at time $t = 0$ and N_t is the number of undecayed atoms after time t. This is an exponential law equation.

Example 3

Calculate the number of undecayed atoms remaining after 1.0 minute in a sample which originally contained 5.0×10^{21} atoms of a radioactive isotope which has a decay constant, λ, of $8.0 \times 10^{-2}\,s^{-1}$.

$$N_t = N_0 e^{-\lambda t}$$
$$\lambda t = 8.0 \times 10^{-2}\,s^{-1} \times 60\,s = 4.8$$

Note that the exponent is always dimensionless.

$$\therefore N_t = 5.0 \times 10^{21} \times e^{-4.8}$$
$$= 5.0 \times 10^{21} \times 8.3 \times 10^{-3}$$
$$= 4.1 \times 10^{19} \text{ undecayed atoms}$$

The exponential law equation can also be written in terms of the activity of the sample:

$$\boxed{A_t = A_0 e^{-\lambda t}}$$

where A_0 is the activity of the sample at time $t = 0$ and A_t is the activity of the sample after time t.

Half-life and the decay constant

After one half-life ($t_{1/2}$) has passed the number of undecayed atoms remaining in a sample will be $N_0/2$. Substituting this into the exponential law equation for N_t and taking natural logs of both sides, we establish the relationship between the decay constant and the half-life of a radioactive atom:

$$\boxed{t_{1/2} = \frac{0.693}{\lambda}}$$

Example 4

Calculate the half-life of carbon-14 which has a decay constant of $4.1 \times 10^{-12}\,s^{-1}$.

$$t_{1/2} = \frac{0.693}{\lambda}$$

$$= \frac{0.693}{4.1 \times 10^{-12}\,s^{-1}}$$

$$= 1.7 \times 10^{11}\,s \text{ or } 5400 \text{ years}$$

$1\,\text{eV} = 1.6 \times 10^{-19}\,\text{J}$
e, the charge on an electron $= 1.6 \times 10^{-19}\,\text{C}$

Level 1

1 A nuclear battery using a beta source in an evacuated container can produce a current of $2.0\,\mu\text{A}$. What is the activity of the source?

2 Strontium-90 decays, emitting beta particles of maximum energy 0.54 MeV. At what rate is energy released by a piece of strontium-90 of activity 5.25×10^{15} Bq? Give your answer in watts.

3 a Radium-226 has a half-life of 51 Gs. What is the value of its decay constant, λ?
b About how long is a gigasecond?
c What is the half-life of a material with a decay constant of $8.8 \times 10^{-4}\,\text{s}^{-1}$?

4 Krypton has a half-life of 3.0 s. There are initially 1024×10^{24} undecayed atoms. About how many will remain undecayed after
a 3.0 s
b 9.0 s
c 15.0 s?

5 The decay of an isotope with a conveniently short half-life is being investigated. At first the uncorrected count rate is $164\,\text{s}^{-1}$ and the background count is $15\,\text{s}^{-1}$. After 7 minutes the uncorrected count has fallen to $53\,\text{s}^{-1}$. Give a good estimate for the half-life of the substance.

Level 2

6 An alpha source, of activity 150 Bq, is placed in an ionisation chamber. A current of 2.0×10^{-11} A is detected.
a How many singly charged ions per second are required to produce this current?
b How many ions does each alpha particle create, on average?
c How many pairs of ions is this?
d It takes 4.8×10^{-10} J of energy to create each pair. What is the power produced by the source?

7 The half-life of caesium-137 is 30 years.
a Calculate the decay constant, λ.
A sample of caesium has an activity of 2.0×10^5 Bq.
b How many undecayed atoms are there in the sample?

8 Radon-222 is a radioactive gas occurring naturally in rocks. It has a decay constant, λ, of $2.1 \times 10^{-6}\,\text{s}^{-1}$.
a What is the mass in grams of 1 mole of radon-222 atoms?
b How many radon-222 atoms are there in 1 g?
c Calculate the activity, A, of 1 g of radon-222.

9 ^{14}C emits β^- particles and has a half-life of 5600 years. The isotope is made by the interaction of cosmic rays with nitrogen in the upper atmosphere. It is then taken in by living organisms. When they die there is no further exchange of carbon so you can calculate the age of a specimen by the proportion of undecayed ^{14}C relative to the more abundant stable ^{12}C present in the specimen.
The ratio of ^{14}C to ^{12}C while a specimen is alive is $1:10^{10}$.
a How many atoms are there in 12 g of carbon?
b How many atoms of ^{14}C are in 12 g of carbon from a live specimen?
c Calculate the decay constant of ^{14}C.
d Show that the activity because of the ^{14}C isotope is 234 Bq.
e The activity measured from 12 g of carbon from a bone is 30 Bq. Roughly how old is the bone?

Level 3

10 The window of a gamma ray detector has an area of $4.0 \times 10^{-4}\,\text{m}^2$. It records an activity of 0.20 Bq when 3.0 m away (in air) from a point source of gamma rays. What is the activity of the source?

11 a Fill in the missing nucleon, proton and neutron numbers for the decays shown in Figure 24.3.

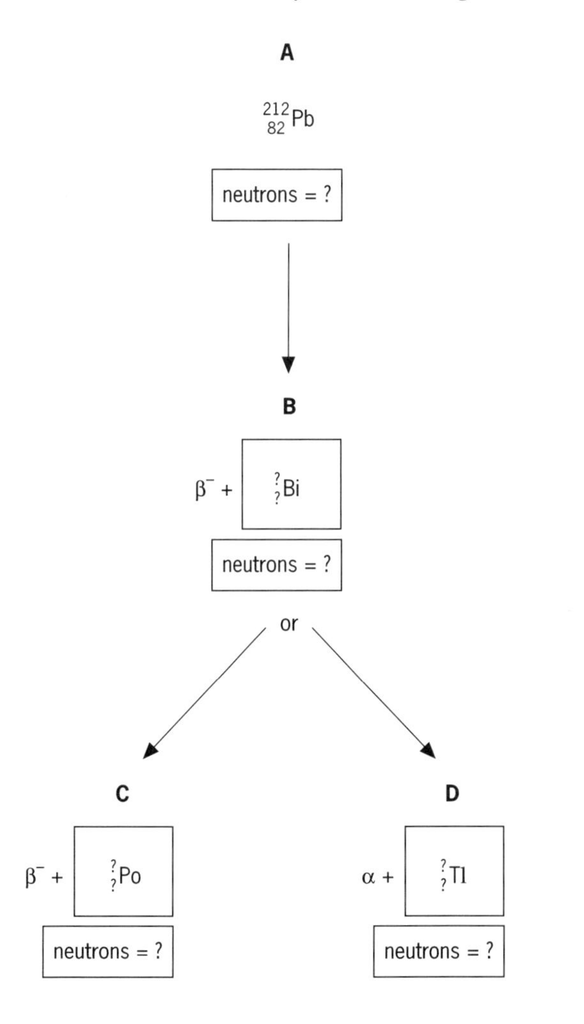

Figure 24.3

b Copy the chart of neutron number, N, versus proton number, Z, Figure 24.4, and put dots representing the nuclides at each stage **B**, **C**, **D** of the decay series in Figure 24.3. Then draw arrows to represent the changes. Label each nuclide and each decay.

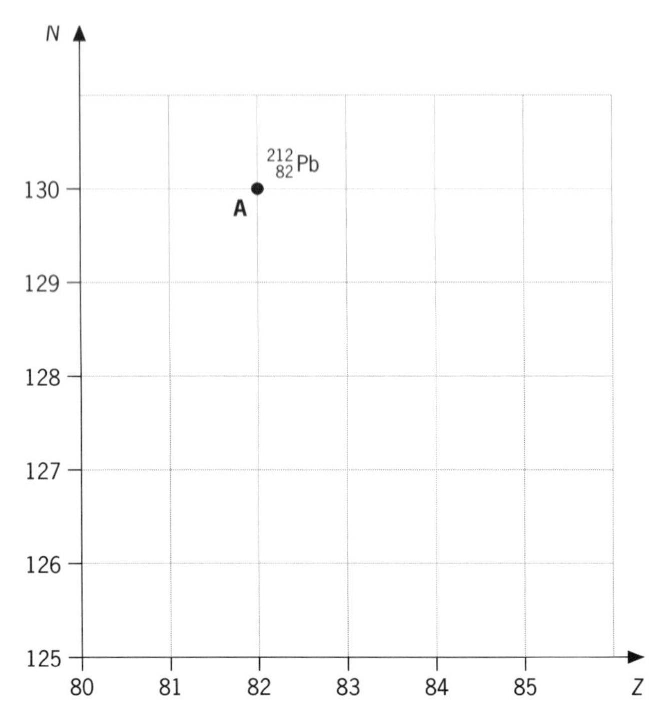

Figure 24.4

12 An ionisation chamber is a sealed box where all charge produced is detected. It contains a short-lived radioactive gas. The current created is monitored and recorded as the gas decays. A straight-line graph is drawn relating the current and the time, as shown in Figure 24.5.

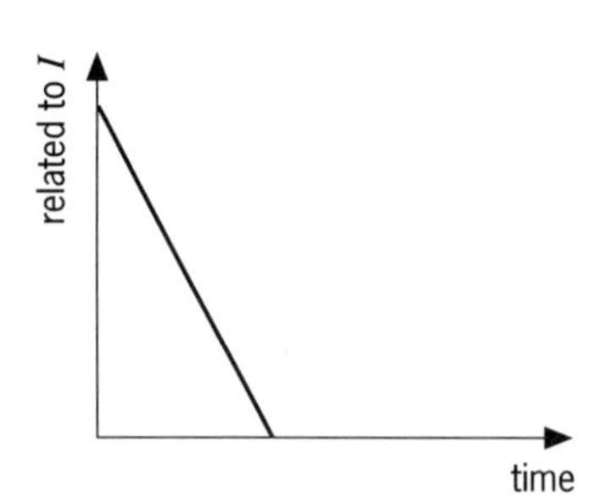

Figure 24.5

a From what you know about the way the activity of a radioactive material varies with time, write an equation describing the behaviour of the current with time.

b What must have been plotted on the vertical axis to produce this graph?

c What can you find out from the intercept on this axis?

d What can you find from the gradient?
Justify each of your answers.

13 A radioisotope of iron is used to measure wear in machines. ^{59}Fe, which has a half-life of 45 days, is mixed with non-radioactive iron.

A test component was produced from the mixture of radioactive and non-radioactive irons with a mass of 880 g. Its initial activity was 3.6×10^6 Bq. The component was fitted 10 days after it was made and the machine was then operated for 35 days. All the material worn from the test component collected in the oil which at the end had activity 60 Bq.

Calculate:

a The activity of the test component when it was installed.

b The activity that the test component would have had at the end of the test if there had been no wear.

c The mass removed from the test component during the time the machine operated.

Level 1

1 The current is generated by the beta particles (electrons) which are produced at a rate of one for each nuclear transformation. The activity of the source, in Bq (becquerel), is the rate (per second) at which beta particles are produced. In this case, it is also equal to the rate at which they flow around the circuit.

current, $2.0\,\mu A = 2.0 \times 10^{-6}\,A = Ne$

where N is the number of electrons flowing per second and e is the charge on an electron. So

$$N = \frac{2.0 \times 10^{-6}\,A}{1.6 \times 10^{-19}\,C} = 1.25 \times 10^{13}\,s^{-1}$$

Remember, the unit of charge, the coulomb (C), can be expressed as an amp second (A s).

$\therefore$ activity = $\mathbf{1.25 \times 10^{13}\,Bq}$

2 In this case each decay produces a beta particle of maximum energy $0.54 \times 10^6\,eV$. But $1\,eV = 1.6 \times 10^{-19}\,J$, so the maximum energy of each beta particle is

$$E_{max} = 0.54 \times 10^6\,eV \times 1.6 \times 10^{-19}\,J\,eV^{-1}$$

The activity, A, is the number of disintegrations per second. The total maximum energy released per second (=power) = $E_{max} \times A$

$$= 0.54 \times 10^6\,eV \times 1.6 \times 10^{-19}\,J\,eV^{-1} \times 5.25 \times 10^{15}\,s^{-1}$$
$$= 454\,J\,s^{-1} = \mathbf{454\,W}$$

3 a You can rearrange the equation for half-life,

$$t_{1/2} = \frac{0.693}{\lambda}$$

to give

$$\lambda = \frac{0.693}{t_{1/2}}$$

In this example

$$\lambda = \frac{0.693}{51 \times 10^9\,s^{-1}} = \mathbf{1.36 \times 10^{-11}\,s^{-1}}$$

b The prefix 'giga' means $\times 10^9$. So 1 gigasecond is about 30 years.

(Divide 1×10^9 s by $365 \times 24 \times 60 \times 60$ s year^{-1}.)

c
$$t_{1/2} = \frac{0.693}{\lambda}$$
$$= \frac{0.693}{8.8 \times 10^{-4}\,s^{-1}} = \mathbf{788\,s}$$

4 a After one half-life, the number of undecayed atoms will have decreased to half the original number, and for each further half-life the number will halve again. So, after 3.0 s (one half-life) there will be

$\frac{1}{2} \times 1024 \times 10^{24} = \mathbf{512 \times 10^{24}}$ undecayed atoms

b 9.0 s is three half-lives so the number will have halved three times. $(\frac{1}{2})^3 = \frac{1}{8}$, so the number left undecayed will be

$\frac{1}{8} \times 1024 \times 10^{24} = \mathbf{128 \times 10^{24}}$ atoms

c 15.0 s is five half-lives so the number will have halved five times. $(\frac{1}{2})^5 = \frac{1}{32}$, so the number left undecayed will be

$\frac{1}{32} \times 1024 \times 10^{24} = \mathbf{32 \times 10^{24}}$ atoms

When the time is equal to only a few half-lives you can write the numbers out, halving each time, but this gets very tedious as the number of half-lives increases, and it's impossible if the time isn't an exact number of half-lives.

5 You have to subtract the background count from all your measurements to get true readings.

The first corrected count rate is $164 - 15 = 149\,s^{-1}$
The second corrected count rate, after 7 minutes, is $53 - 15 = 38\,s^{-1}$.

38 is close to one-quarter of 149, so roughly 2 half-lives have elapsed. This gives a half-life of $\frac{1}{2} \times 7$ minutes = **3.5 minutes**.

Level 2

6 a The current of $2.0 \times 10^{-11}\,A$ ($= 2.0 \times 10^{-11}\,C\,s^{-1}$) is the result of the flow of a number of electrons or singly charged ions, N, per second, each with a charge, e.

current = $2.0 \times 10^{-11}\,C\,s^{-1} = 1.6 \times 10^{-19}\,C \times N$

which you can rearrange to give

$$N = \frac{2.0 \times 10^{-11}\,C\,s^{-1}}{1.6 \times 10^{-19}\,C}$$

$= \mathbf{1.25 \times 10^8}$ electrons or singly charged ions per second

b The source has an activity of 150 becquerel, which means that it will produce 150 alpha particles per second. These (150) alpha particles produce 1.25×10^8 electrons or singly charged ions (from the answer to **a**), so each alpha particle must produce, on average,

$\frac{1.25 \times 10^8}{150} = \mathbf{8.3 \times 10^5}$ electrons or singly charged ions

c There will be

$\frac{8.3 \times 10^5}{2} = \mathbf{4.15 \times 10^5}$ ion-pairs (per alpha particle)

d The power of the source is equal to the energy produced per second.

total energy per second = energy per pair $\times$ pairs per α $\times$ α's per second

$$= 4.8 \times 10^{-10}\,J\,pair^{-1} \times 4.15 \times 10^5\,pair\,\alpha^{-1} \times 150\,\alpha\,s^{-1}$$

$$= 30 \times 10^{-3}\,J\,s^{-1} = \mathbf{30 \times 10^{-3}\,W}$$

7 a 30 years = $30 \times 365 \times 24 \times 60 \times 60\,s = 9.5 \times 10^8\,s$

Rearrange the standard formula,

$$\lambda = \frac{0.693}{t_{1/2}} = \frac{0.693}{9.5 \times 10^8} = \mathbf{7.3 \times 10^{-10}\,s^{-1}}$$

b $A = \lambda N$

Remember that A is positive, although N is decreasing.

You can rearrange this equation to give

$$N = \frac{A}{\lambda}$$

$$= \frac{2.0 \times 10^5\,Bq}{7.3 \times 10^{-10}\,s^{-1}} = \mathbf{2.7 \times 10^{14}}\text{ atoms}$$

(1 becquerel = 1 decay per second)

8 a The relative mass is 222 so one mole of atoms will have mass **222 g**.

b One mole contains N_A atoms (N_A = the Avogadro constant = 6.0×10^{23}), and has a mass of 222 g. So the number of atoms in a 1 g sample is

$$\frac{N_A}{222} = \frac{6.0 \times 10^{23}\,mol^{-1}}{222\,mol^{-1}}$$

$$= \mathbf{2.7 \times 10^{21}\text{ atoms}}$$

c $A = \lambda N = 2.1 \times 10^{-6}\,s^{-1} \times 2.7 \times 10^{21} = \mathbf{5.7 \times 10^{15}\,Bq}$

9 a One mole of carbon-12 atoms has mass 12 g and contains 6.0×10^{23} atoms.

b If 1 in 10^{10} atoms are ^{14}C there will be 6.0×10^{13} 'active' atoms in 12 g.

This may seem a lot, but the contribution that they make to the mass of the one mole is only in the 10th significant figure – roughly 0.000 000 02.

c From the value of the half-life, the decay constant

$$\lambda = \frac{0.693}{t_{1/2}}$$

$$= \frac{0.693}{1.76 \times 10^{11}\,s} = \mathbf{3.9 \times 10^{-12}\,s^{-1}}$$

d Activity, $A = \lambda N$. In the one mole of carbon (12 plus 14), there are 6.0×10^{13} 'active' atoms with decay constant $\lambda = 3.9 \times 10^{-12}\,s^{-1}$, so

$$A = 3.9 \times 10^{-12}\,s^{-1} \times 6.0 \times 10^{13}\text{ atoms} = \mathbf{234\,Bq}$$

e 30 Bq is roughly $\frac{1}{8}$, or $(\frac{1}{2})^3$, of this activity, which indicates that three half-lives have passed. So the bone is about 3 × 5600 years = **16 800 years old**

Level 3

10 At a distance of 3.0 m from the source, the radiation is spread out over the surface of a sphere of radius 3.0 m. If the whole of this area could be covered with counters you would detect every event. But the radiation that you detect is only the part of the total activity that goes into the small area of the sphere that you cover with your detector.

$$\frac{\text{activity detected}}{\text{total activity}} = \frac{\text{area of detector}}{\text{total area}}$$

which you can rearrange to give

$$\text{total activity} = \frac{\text{activity detected} \times \text{total area}}{\text{area of detector}}$$

Surface area of sphere = $4\pi R^2$.

$$\therefore \text{total activity} = \frac{0.20\,Bq \times 4 \times \pi \times 3.0^2\,m^2}{4.0 \times 10^{-4}\,m^2}$$

$$= \mathbf{5.7 \times 10^4\,Bq}$$

11 a **A** neutrons = 130
B $^{212}_{83}Bi$; neutrons = 129
C $^{212}_{84}Po$; neutrons = 128
D $^{208}_{81}Tl$; neutrons = 127

b

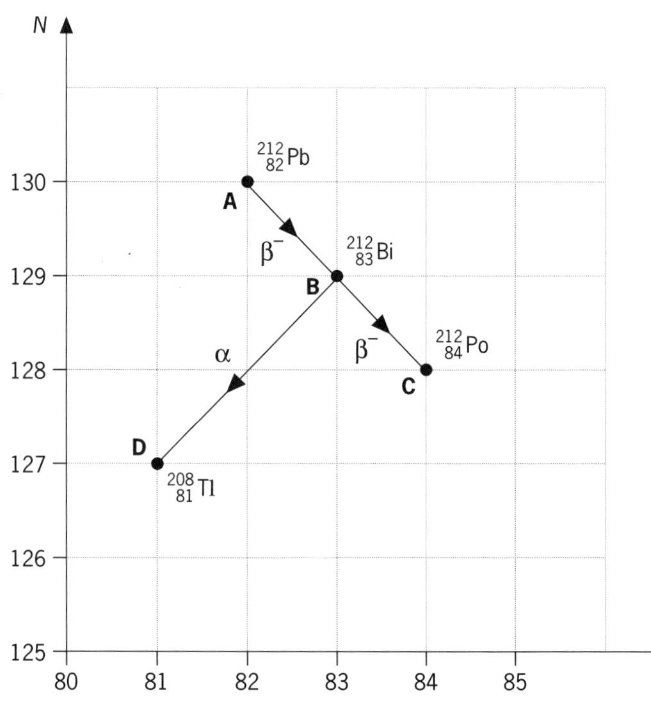

Figure 24.6

12 a The size of the current must be proportional to the activity of the radioactive gas. Because the activity of the gas decreases exponentially, the current must decrease exponentially too, so that

$$I_t = I_0 e^{-\lambda t}$$

where I_t is current at time t.

b A plot of I_t against time would give an exponential decay curve. So the vertical axis can't be just I_t. But, if you take the natural logarithms (logs to the base e) of both sides of the above equation,

$$\ln I_t = \ln I_0 + \ln e^{-\lambda t} = \ln I_0 - \lambda t$$

you can see that $\ln I_t$ is proportional to t. So, to get a straight-line graph as shown in Figure 24.5, you must plot the **natural logarithm of the current** against time.

You could use logs to the base 10 and still get a straight line, but it is less straightforward.

c The intercept when $t = 0$ is $\ln I_0$, so you can find the **initial value of the current**.

d The gradient is equal to $-\lambda$ where λ is the **decay constant**.

13 a $A_t = A_0 e^{-\lambda t}$

$$\lambda = \frac{0.693}{t_{1/2}}$$

At this point you might start converting the half-life and the other times in the question into seconds, but so long as they are all in the same units there is no need to do this. And, in fact, everything is in days.

$$\therefore \lambda = \frac{0.693}{45 \text{ days}}$$

When the test starts, after 10 days,

$$\lambda t = \frac{0.693}{45 \text{ days}} \times 10 \text{ days} = 0.154$$

So

$$A_{t=10} = A_0 e^{-0.154} = 3.6 \times 10^6 \,\text{Bq} \times 0.862$$
$$= \mathbf{3.1 \times 10^6\,Bq}$$

Make sure you know how to use the e^x button on your calculator.

b At the end of the test, the elapsed time $= 10 + 35 = 45$ days, which just happens to be equal to the half-life of the radioactive iron. So the activity will have halved from the original, to $\mathbf{1.8 \times 10^6\,Bq}$.

$(\lambda t = \dfrac{0.693}{45 \text{ days}} \times 45 \text{ days} = 0.693$

So $A_{t=45} = A_0 \times e^{-0.693} = A_0 \times 0.5)$

c The activity of the oil will be proportional to the mass of radioactive iron that has been collected. After 45 days the activity of 880 grams would be 1.8×10^6 Bq (from part **b**). So the activity per gram, after 45 days, is

$$\frac{1.8 \times 10^6 \,\text{Bq}}{880 \,\text{g}} = 2.045 \times 10^3 \,\text{Bq}\,\text{g}^{-1}$$

total activity = activity per gram × mass (in grams)

which you can rearrange to give

$$\text{mass (in grams)} = \frac{\text{total activity}}{\text{activity per gram}}$$

In this example,

$$\text{mass (in grams)} = \frac{60 \,\text{Bq}}{2.045 \times 10^3 \,\text{Bq}\,\text{g}^{-1}} = \mathbf{0.03\,g}$$

25 Fission, fusion and binding energy

Mass defect and binding energy

If we could accurately measure the masses of two protons and two neutrons and then add these four values together we should be able to determine the mass of a helium-4 nucleus. If we do so, however, we would find that the mass of the helium nucleus is slightly less than the sum of the masses of the two protons and two neutrons. See Figure 25.1.

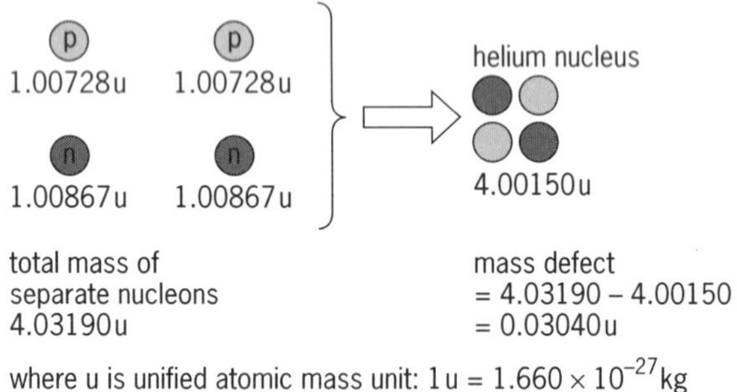

Figure 25.1 Combining nucleons into a nucleus involves a loss in mass

The difference between the masses is called the **mass defect**, Δm, of the nucleus. It arises because when the nucleons combine to form the nucleus some of their mass is released as energy. This is called the **binding energy** of the nucleus, defined as **the work done on the nucleus to completely separate all its nucleons.**

Mass–energy equivalence

Einstein showed, in his Special Theory of Relativity, that there is a relationship between mass and energy. That is, it is possible to interchange mass and energy. The relationship is given by the mass–energy equivalence equation:

$$E = mc^2$$

for the total energy equivalence of a mass, m, or

$$\Delta E = \Delta mc^2$$

for the energy released when there is a change in mass, where c is the speed of light in vacuo (2.998×10^8 or approximately $3.0 \times 10^8\,\mathrm{m\,s^{-1}}$).

In the example of Figure 25.1 the helium nucleus has a mass defect of 0.03040 u. The binding energy of the nucleus is therefore

$$0.03040 \times 1.660 \times 10^{-27}\,\mathrm{kg} \times (2.998 \times 10^8\,\mathrm{m\,s^{-1}})^2 = 4.5 \times 10^{-13}\,\mathrm{J}$$

If the binding energy of a nucleus is divided by the number of nucleons it contains, a value can be found for the binding energy per nucleon. This is a useful measure of the stability of a nucleus. The larger the binding energy per nucleon, the more stable the nucleus. For helium the binding energy is 4.5×10^{-13} J, so the binding energy per nucleon is

$$\frac{4.5 \times 10^{-13}}{4} = 1.1 \times 10^{-13}\,\mathrm{J}$$

Binding energies are more conveniently expressed in electronvolts rather than joules. The relationship between electronvolts and joules is $1\,\mathrm{eV} = 1.6 \times 10^{-19}\,\mathrm{J}$. So the binding energy per nucleon for helium is

$$\frac{1.1 \times 10^{-13}\,\mathrm{J}}{1.6 \times 10^{-19}\,\mathrm{J\ per\ eV}} = 7.08 \times 10^6\,\mathrm{eV\ or\ }7.1\,\mathrm{MeV}$$

Energy equivalence of the unified atomic mass unit

Using the relationship $E = mc^2$ we can calculate the energy equivalence of 1 u.

$$1\,\mathrm{u} = 1.660 \times 10^{-27}\,\mathrm{kg}$$

$$\therefore \quad E = 1.660 \times 10^{-27}\,\mathrm{kg} \times (2.998 \times 10^8\,\mathrm{m\,s^{-1}})^2$$
$$= 1.660 \times 10^{-27}\,\mathrm{kg} \times 8.988 \times 10^{16}\,\mathrm{m^2\,s^{-2}}$$
$$= 1.492 \times 10^{-10}\,\mathrm{J}$$

giving $1\,\mathrm{u} \equiv 931\,\mathrm{MeV}$

Variation of binding energy per nucleon with size of nucleus

The graph in Figure 25.2 shows how the binding energy per nucleon varies with the nucleon number of the nucleus.

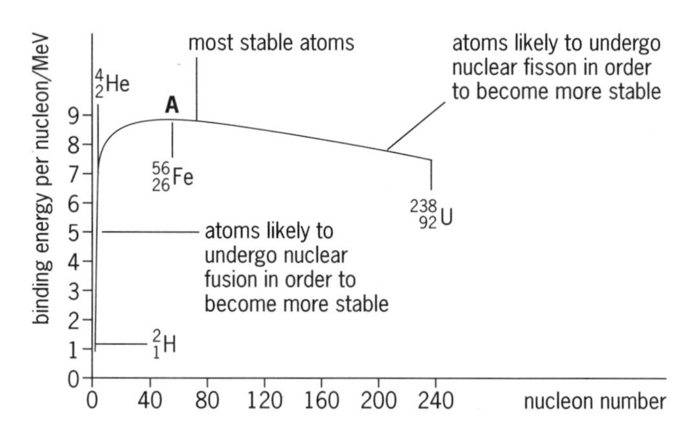

Figure 25.2 Binding energy per nucleon against nucleon number

From Figure 25.2 we can see that

- nuclei near the hump of the graph (A) have the largest binding energy per nucleon and so are the most stable;
- lighter atoms with small nuclei may become more stable by combining to form heavier nuclei – this process is called **nuclear fusion**;
- heavier nuclei may become more stable by splitting to form two smaller nuclei – this process is called **nuclear fission**.

Nuclear fission

Nuclear fission is the breaking up of a large heavy nucleus to form two smaller nuclei of approximately the same size. For example, if the nucleus of a uranium-235 atom captures a neutron it becomes a nucleus of uranium-236. Uranium-236 is unstable and undergoes fission. One possible reaction is:

$$^{235}_{92}U + ^{1}_{0}n \rightarrow ^{236}_{92}U \rightarrow ^{141}_{56}Ba + ^{92}_{36}Kr + 3^{1}_{0}n + \gamma + \text{energy}$$

The energy released by the fission of a single uranium-236 atom is approximately 200 MeV. The energy that would be released by 1 kg of uranium-236 if it were possible to cause all its atoms to undergo fission would be approximately 8×10^{13} J. This is one million times more than the energy released when 1 kg of coal is burnt!

The fission reaction above is self-sustaining; see Figure 25.3. It is a **chain reaction**.

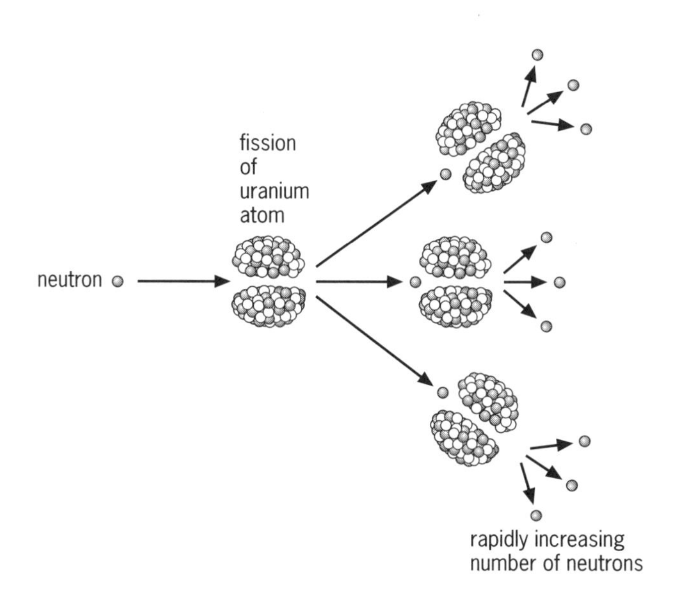

Figure 25.3 An uncontrolled chain reaction

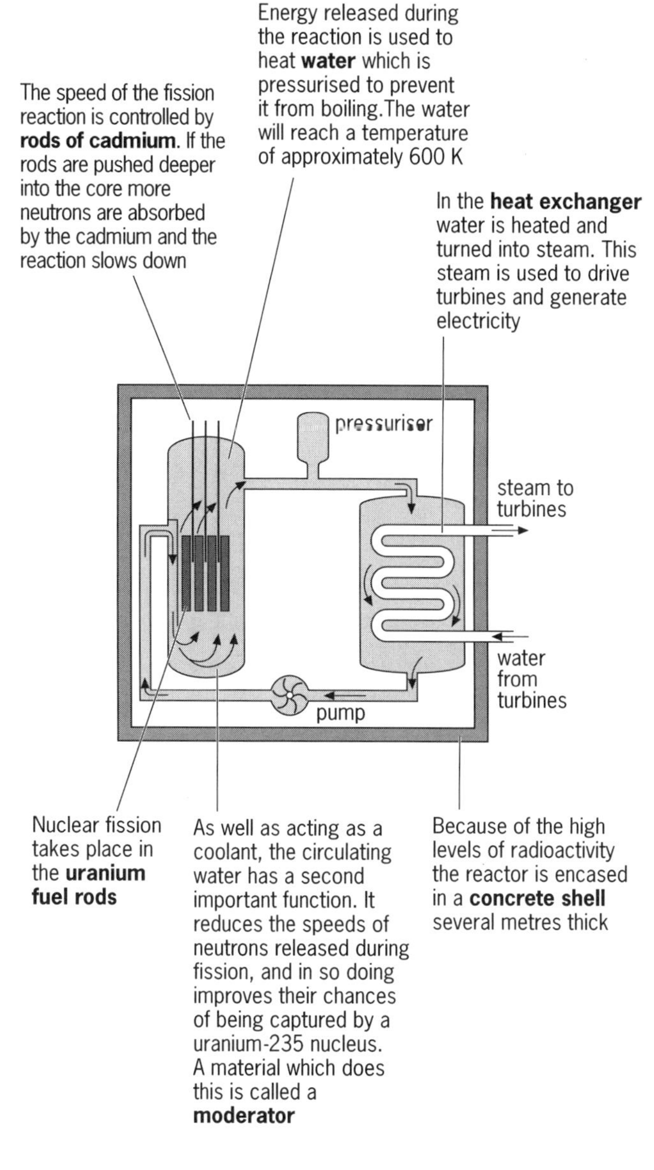

Figure 25.4 Pressurised water-cooled nuclear reactor

If uncontrolled, the rapid release of energy would lead to a nuclear explosion. If, however, the reaction is controlled; for example if only one-third of the emitted neutrons are allowed to continue the chain reaction, the rate of release of energy is restricted and this reaction can be used as the energy source for a nuclear reactor. Figure 25.4 shows the basic structure of the core of a nuclear reactor.

Nuclear fusion

Nuclear fusion is the joining together of two lighter nuclei to form a larger, heavier more stable nucleus. For example, two deuterium nuclei (hydrogen nuclei with a neutron as well as a proton) can fuse in the following reaction:

$$^{2}_{1}H + ^{2}_{1}H \rightarrow ^{3}_{2}He + ^{1}_{0}n + 3.27\,\text{MeV}$$

It is nuclear reactions like this which produce most of the energy that is emitted by the Sun. Controlled fusion reactions would be a far better source of energy than nuclear fission, because of the abundance of deuterium in water and because of the lack of radioactive waste. Unfortunately, for this reaction to take place the temperature of the fusing nuclei needs to be extremely high – of the order of 10^6 K.

Mass of proton = 1.008 u
Mass of neutron = 1.009 u
Mass of electron = 0.000549 u
$1\,\text{eV} = 1.6 \times 10^{-19}\,\text{J}$
$1\,\text{u} = 1.66 \times 10^{-27}\,\text{kg}$
Speed of light $= 3.0 \times 10^{8}\,\text{m s}^{-1}$
Avogadro constant, $N_A = 6.0 \times 10^{23}\,\text{mol}^{-1}$
Planck constant, $h = 6.6 \times 10^{-34}\,\text{J s}$
Density of water, $\rho = 1000\,\text{kg m}^{-3}$
Specific heat capacity of water, $c_w = 4200\,\text{J kg}^{-1}\,\text{K}^{-1}$

Level 1

1 Carbon-12 has its *atomic mass* defined as 12 u.
  a Calculate the mass of the electrons in atomic mass units (u).
  b What must be the mass of the nucleus in u?

2 Calculate the mass defect, in u, of $^{12}_{6}\text{C}$ using your answer to question 1b and the data supplied above.

3 In the Sun, fusion may take place between deuterium and tritium nuclei:

$$^{2}_{1}\text{H} + ^{3}_{1}\text{H} \rightarrow ^{4}_{2}\text{He} + ^{1}_{0}\text{n} + \text{energy}$$

The total final mass of the particles is 5.0113 u. The energy released is $2.8 \times 10^{-12}\,\text{J}$.
  a Find the mass equivalent, in kg, of the energy released.
  b Find the mass equivalent, in u, of the energy released.
  c Use the answer to **b** to estimate the total mass of the nuclei, in u, before fusion.

4 Use the data in Table 25.1 to answer the following questions.
  a Which is the most abundant isotope of boron?
  b How does $^{10}_{4}\text{Be}$ decay?
  c What is the energy of the particles emitted by $^{10}_{6}\text{C}$?
  d Which of the isotopes in the list has the longest half-life?

Level 2

5 Your younger sister refuses to believe that adding energy increases mass.
  a Estimate the mass increase when the temperature of 1.0 litre of water is raised from 20 °C to its boiling point.
  b Would you attempt to demonstrate the mass increase to her in the kitchen?

6 Uranium-235 absorbs a neutron and then undergoes fission. One mode of fission is as below.

$$^{235}_{92}\text{U} + ^{1}_{0}\text{n} \rightarrow ^{94}_{40}\text{Zr} + ^{140}_{56}\text{Ba} + 2\,^{1}_{0}\text{n} + 4\,^{0}_{-1}\text{e}$$

total mass: 236.053 u   235.836 u

  a Calculate the energy (in joules) released per fission.
  b If a fission bomb contained 10 kg of uranium, how much energy would it release if all the nuclei were split?

7 Show that the equation $\Delta E = \Delta mc^2$ is dimensionally correct.

Hint: Show that it has the same base units on each side.

8 The nuclides in Table 25.2 are all part of a decay sequence. Your friends say that because the binding energy gets less each time it should be energetically favourable to decay the other way, from lead to uranium.

Table 25.2

Nuclide	Binding energy/MeV	Binding energy per nucleon/MeV
$^{234}_{92}\text{U}$	1778.3	
$^{226}_{88}\text{Ra}$	1731.3	
$^{214}_{82}\text{Pb}$	1663.0	

  a Complete the third column and use it to convince them otherwise.
  b Sketch a graph to show the variation of binding energy per nucleon with atomic number. Comment on the general predictions this allows you to make.

Table 25.1 Extract from data book

Element Z	A	Abundance (%) or decay	Atomic mass/u	Energy of emitted particles/MeV	Half-life
4 Be	7	K	7.01693		53 day
	8	2α	8.00530	0.05	~3×10^{-16} s
	9	100	9.01218		
	10	β^-	10.01354	0.56	2.7×10^{6} a
	11	β^-	11.0216	11.5; 9.3	14 s
5 B	8	$\beta^+ + 2\alpha$	8.02461	β14	0.8 s
	9		9.01333		
	10	19.6	10.01293		
	11	80.4	11.009305		
	12	$\beta^-(+\alpha)$	12.01435	13.4	0.02 s
	13	β^-	13.01778		0.04 s
6 C	10	β^+	10.0168	2.1	19 s
	11	β^+	11.01143	0.96	20.5 min
	12	98.89	12		
	13	1.11	13.00335		
	14	β^-	14.003242	0.158	5570 a
	15	β^-	15.01060	9.8; 4.5	2.3 s

Level 3

9 Two almost stationary deuterium nuclei, $^{2}_{1}H$, can fuse to form a tritium nucleus, $^{3}_{1}H$, and a proton. This releases energy of 6.4×10^{-13} J.

a In what form is the energy released?
b Write an equation for this reaction.
c Calculate the ratio of the speed of the proton to that of the tritium nucleus.
d Calculate the ratio of the kinetic energy of the proton to that of the tritium nucleus.

10 A nuclear submarine needs 500 kW, which is provided by the fission of ^{235}U. The fuel is enriched uranium which may be 3% ^{235}U and the rest ^{238}U.

a Show that there are about 8×10^{22} atoms of ^{235}U in 1.0 kg of this fuel.
b The energy released at each fission is 3×10^{-11} J.
i How many fissions per second are needed to provide the required power?
ii How long will 1.0 kg of fuel last?

11 Cobalt-59 absorbs a neutron and then becomes a pure gamma source for medical use.

a Calculate the energy available to the emitted gamma ray photon, in J and eV.
b Calculate the wavelength of the gamma rays.

Use the data supplied at the beginning of the questions and:
mass of $^{59}Co = 58.9332$ u
mass of $^{60}Co = 59.9338$ u

12 Explain this statement seen in a physics textbook:

'1 u is equivalent to 931 MeV'.

13 Find out what is meant by

a 'thermal neutrons'
b a 'thermal reactor'.

Level 1

1 The neutral carbon-12 atom is the standard for atomic mass and by definition has a mass = 12.0000 u.

a There are six neutrons and six protons in the nucleus so there must be six electrons in a neutral atom.

$\therefore$ mass of the electrons $= 6 \times 0.000549\,\text{u}$
$= \mathbf{0.00329\,u}$

b The nuclear mass is the atomic mass minus the electron mass

$= 12.0000\,\text{u} - 0.00329\,\text{u} = \mathbf{11.9967\,u}$

2 Using the data supplied at the start of the questions, the total mass of six protons and six neutrons is 12.102 u. The nuclear mass, from question 1, is 11.9967 u.

$\therefore$ mass defect $= 12.102\,\text{u} - 11.997\,\text{u} = \mathbf{0.105\,u}$

3 a You can obtain an equation for the mass equivalent released during a reaction by rearranging the equation $\Delta E = \Delta mc^2$ to give

$$\Delta m = \frac{\Delta E}{c^2} = \frac{2.8 \times 10^{-12}\,\text{J}}{9.0 \times 10^{16}\,\text{m}^2\,\text{s}^{-2}} = \mathbf{3.1 \times 10^{-29}\,kg}$$

Check: $Jm^{-2}s^2 = Nm\,m^{-2}s^2 = kg\,m\,s^{-2}\,m\,m^{-2}s^2 = kg$

b $1\,\text{u} = 1.66 \times 10^{-27}\,\text{kg}$, so, to convert the mass equivalent in kg to u you must divide by $1.66 \times 10^{-27}\,\text{kg}\,\text{u}^{-1}$.

$$\therefore 3.1 \times 10^{-29}\,\text{kg} = \frac{3.1 \times 10^{-29}\,\text{kg}}{1.66 \times 10^{-27}\,\text{kg}\,\text{u}^{-1}} = \mathbf{0.0187\,u}$$

c Total initial mass = final mass + mass equivalent of energy released
$= 5.0113\,\text{u} + 0.0187\,\text{u} = \mathbf{5.0300\,u}$

4 a ^{11}**B**, which has an abundance of 80.4%.

b $^{10}_{4}$Be decays by **emission of β^-.**

c Energy = **2.1 MeV**.

This energy is shared between the positron and a neutrino.

d $\mathbf{^{10}_{4}Be}$ has the longest half-life, of $2.7 \times 10^6\,\text{a} = 2.7$ million years.

(a = years, from the Latin 'annus'.)

Level 2

5 a 1 litre of water has mass 1.0 kg and its specific heat is $c_w = 4200\,\text{J}\,\text{kg}^{-1}\,\text{K}^{-1}$. For an increase in temperature $\Delta\theta$, the energy increase is given by the equation

$$\Delta E = mc_w\Delta\theta$$

For 1 litre of water heated through 80 K,

$$\Delta E = 1.0\,\text{kg} \times 4200\,\text{J}\,\text{kg}^{-1}\,\text{K}^{-1} \times 80\,\text{K} = 336\,\text{kJ}$$

The mass increase is given by

$$\Delta m = \frac{\Delta E}{c^2} = \frac{336 \times 10^3\,\text{J}}{9.0 \times 10^{16}\,\text{m}^2\,\text{s}^{-2}} = \mathbf{3.7 \times 10^{-12}\,kg}$$

b This is thousandths of a microgram so it would not make an impressive demonstration. Electronic kitchen scales generally have a resolution of 2 grams!

6 a The mass defect (the difference between the starting mass and the finishing mass) is

$$\Delta m = 236.053\,\text{u} - 235.836\,\text{u} = 0.217\,\text{u}$$

In kg this is

$$0.217\,\text{u} \times 1.66 \times 10^{-27}\,\text{kg}\,\text{u}^{-1} = 3.6 \times 10^{-28}\,\text{kg}$$

The energy released is given by

$$\Delta E = \Delta mc^2 = 3.6 \times 10^{-28}\,\text{kg} \times 9.0 \times 10^{16}\,\text{m}^2\,\text{s}^{-2} = \mathbf{3.2 \times 10^{-11}\,J}$$

b One mole of uranium-235 has a mass of 235 g. So

$$10\,\text{kg} = \frac{10\,\text{kg}}{235 \times 10^{-3}\,\text{kg}\,\text{mol}^{-1}} = 42.55\text{ moles}$$

Each mole contains N_A atoms, so the total number in 10 kg is

$$42.55\,\text{mol} \times 6.0 \times 10^{23}\,\text{atoms}\,\text{mol}^{-1} = 2.55 \times 10^{25}\,\text{atoms}$$

If the nuclei of all these atoms follow this mode of fission, the total energy released will be given by

$$E = 2.55 \times 10^{25} \times 3.2 \times 10^{-11}\,\text{J} = \mathbf{8.2 \times 10^{14}\,J}$$

If this amount of energy is released in a short time the result is catastrophic.

7 On the left hand side of the equation,

E = energy = force × distance
= (mass × acceleration) × distance

Considering the base units, we have

$$\text{kg}\,\text{m}\,\text{s}^{-2}\,\text{m} = \text{kg}\,(\text{m}\,\text{s}^{-1})^2$$

which are the units of

mass × (velocity)^2

i.e. those of the right hand side of the equation.

8 a You obtain the values for the third column by dividing the total binding energy by the nucleon number. The way the *total* binding energy changes does not explain why the decay occurs. However, the way that the binding energy *per nucleon* rises from U to Pb shows that stability increases at each step in the decay. See Table 25.3.

b See Figure 25.2. The graph shows that it is energetically favourable for large nuclei to split into smaller ones and for small nuclei to fuse together to make larger ones. The maximum binding energy per nucleon is achieved in ^{56}Fe, which makes this the most stable nucleus.

Table 25.3

Nuclide	Binding energy/MeV	Binding energy per nucleon/MeV
$^{234}_{92}$U	1778.3	7.60
$^{226}_{88}$Ra	1731.3	7.66
$^{214}_{82}$Pb	1663.0	7.77

Level 3

9 a The energy from the fusion is released as the kinetic energy of the final particles.

b ${}^2_1H + {}^2_1H \rightarrow {}^3_1H + {}^1_1p$

c Using m for the mass of a nucleon, v for the velocity of the tritium and v' for the velocity of the proton; momentum is conserved, so

particle momentum before = particle momentum after

$$0 = 3mv + mv'$$

$$\therefore v' = -3v$$

The tritium leaves the site of the reaction with velocity v in one direction and the proton leaves, going three times as fast, in the opposite direction. The ratio of the speeds is

proton : tritium = **3 : 1**

d The mass of the tritium nucleus is three times the mass of the proton but the speed of the proton is three times the speed of the tritium nucleus. In the equation for kinetic energy speed is squared, so the proton has three times the kinetic energy of the tritium:

For the tritium,

$$E_k = \tfrac{1}{2} \times 3m \times v^2$$

For the proton,

$$E'_k = \tfrac{1}{2}m(v')^2 = \tfrac{1}{2} \times m \times (3v)^2 \quad = \tfrac{1}{2} \times 9m \times v^2$$

The ratio of the kinetic energies is therefore

proton : tritium = **3 : 1**

10 a If only 3% of the enriched uranium fuel is ${}^{235}U$, then in 1.0 kg of fuel there will be $0.03 \times 1\,kg = 30\,g$ of ${}^{235}U$. 1 mole of ${}^{235}U$ has a mass of 235 g and contains $N_A = 6.0 \times 10^{23}$ atoms. So, the number of atoms in 30 g is

$$\frac{6.0 \times 10^{23}\,\text{atoms mol}^{-1} \times 30\,\text{g}}{235\,\text{g mol}^{-1}} = \mathbf{7.7 \times 10^{22}\ atoms}$$

b i The power required, $500\,kW = 5 \times 10^5\,J\,s^{-1}$. Each fission releases $3 \times 10^{-11}\,J$, so the power required is equivalent to

$$\frac{5 \times 10^5\,\text{J s}^{-1}}{3 \times 10^{-11}\,\text{J (fission)}^{-1}} = \mathbf{1.67 \times 10^{16}\ fissions\ s^{-1}}$$

ii Initially there are 7.7×10^{22} atoms of ${}^{235}U$. If these are allowed to split at a rate of $1.67 \times 10^{16}\,s^{-1}$, they will all have split in

$$\frac{7.7 \times 10^{22}}{1.67 \times 10^{16}\,\text{s}^{-1}} = 4.6 \times 10^6\,\text{s} = \mathbf{53\ days}$$

11 a The reaction is

$${}^{59}Co + {}^{1}n \rightarrow {}^{60}Co + \gamma$$

Total mass before the reaction
$= 58.9332\,u + 1.0087\,u = 59.9419\,u$
Total mass after the reaction
$= 59.9338\,u$

So mass defect, $\Delta m = 59.9419\,u - 59.9338\,u$
$= 0.0081\,u$. In kg this is

$$\Delta m = 0.0081\,\text{u} \times 1.66 \times 10^{-27}\,\text{kg u}^{-1} = 1.3446 \times 10^{-29}\,\text{kg}$$

It doesn't matter whether you use atomic masses or nuclide masses to calculate the mass defect here, because the atoms of cobalt have the same number of electrons before and after.

This mass is converted into the energy of the gamma ray photon:

energy, $\Delta E = \Delta mc^2$
$= 1.3446 \times 10^{-29}\,kg \times 9.0 \times 10^{16}\,m^2\,s^{-2}$
$= \mathbf{1.2 \times 10^{-12}\,J}$

The energy of a gamma ray is more often given in eV:

$$\Delta E = \frac{1.2 \times 10^{-12}\,\text{J}}{1.6 \times 10^{-19}\,\text{J eV}^{-1}} = 7.6 \times 10^6\,\text{eV} = \mathbf{7.6\,MeV}$$

b The energy of any photon of radiation with wavelength λ is given by $E = h \times c/\lambda$, which you can rearrange to give

$$\lambda = \frac{hc}{E}$$

where h is the Planck constant and c is the speed of light.

In this example,

$$\lambda = \frac{6.6 \times 10^{-34}\,\text{J s} \times 3 \times 10^8\,\text{m s}^{-1}}{1.2 \times 10^{-12}\,\text{J}} = \mathbf{1.65 \times 10^{-13}\,m}$$

12 From Einstein's mass–energy relationship, $E = mc^2$, we can find how much energy is associated with any mass.

$1\,u = 1.66 \times 10^{-27}\,kg$ so the energy associated with 1 u is

$$1.66 \times 10^{-27}\,\text{kg} \times 9.0 \times 10^{16}\,\text{m}^2\,\text{s}^{-2} = 1.49 \times 10^{-10}\,\text{J}$$

But $1\,eV = 1.6 \times 10^{-19}\,J$ so the energy associated with 1 u is

$$\frac{1.49 \times 10^{-10}\,\text{J}}{1.6 \times 10^{-19}\,\text{J eV}^{-1}} = 9.31 \times 10^8\,\text{eV} = \mathbf{931\,MeV}$$

13 a When neutrons are slowed down so that they are in thermal equilibrium with their surroundings they are called 'thermal'. This is usually achieved in a fission reactor by repeated collisions with a 'moderator' until the neutron energy is about 0.025 eV. The neutrons then interact more readily with the uranium-235 nuclei.

b This type of reaction occurs in a 'thermal' reactor, as distinct from a 'fast' reactor where fission occurs by bombardment with fast neutrons.

26 Elasticity

If a force is applied to an object it may cause it to change shape. How much the object deforms depends upon

- its size and shape;
- the material it is made from; and
- the size and direction of the applied force.

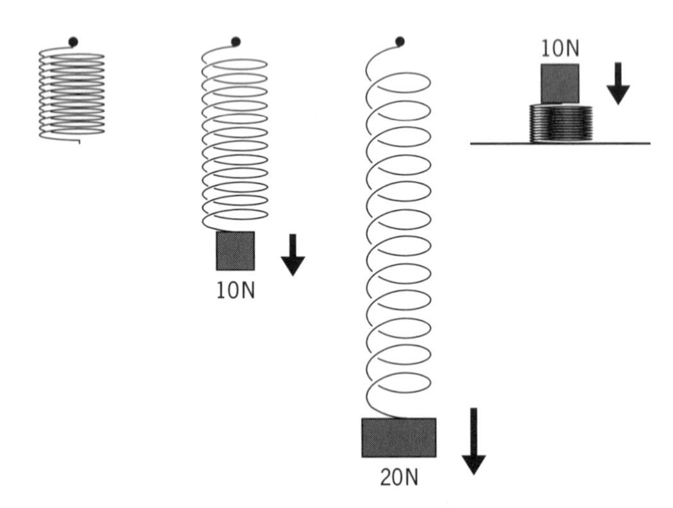

Figure 26.1 Springs are easily deformed by applied forces

Hooke's Law

The graph in Figure 26.2 shows how a spring or a metal wire deforms when subject to an increasing stretching force.

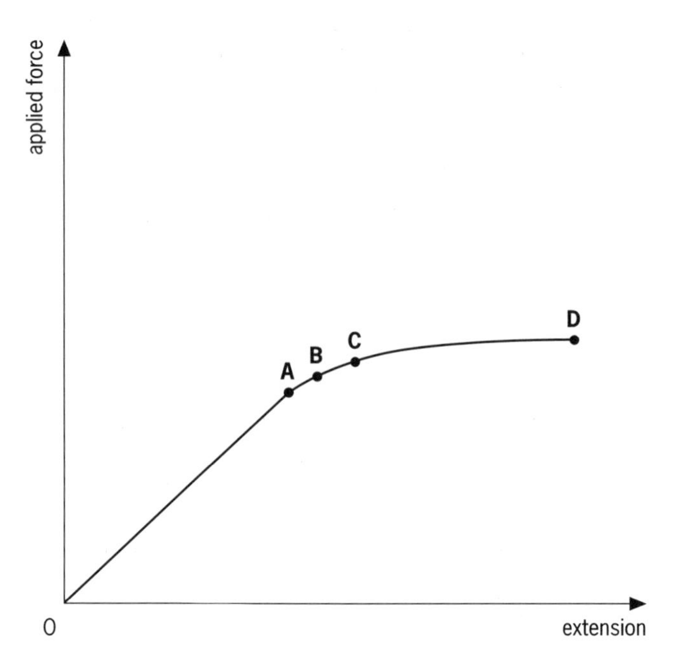

Figure 26.2

Between O and A the extension of the wire or spring is proportional to the applied force; if the applied force is doubled the extension of the wire or spring is doubled. We can express this relationship by the equation

$$F = k\Delta l$$

where Δl is the extension caused by a force F. k is the constant of proportionality known as the **force constant** for the wire or spring. The wire is said to be obeying **Hooke's Law**, which states that **the extension of an elastic body is directly proportional to the force that produces it**.

Points A, B, C and D on the force–extension curve correspond to changes in the physical behaviour of the wire or spring:

A Limit of proportionality

beyond this point the extension is no longer proportional to the applied force

B Elastic limit

beyond this point the applied force will cause permanent deformation

C Yield point

beyond this point small increases in the applied force cause large increases in extension

D Ultimate tensile strength

a force greater than this value will cause the wire to break

If an object undergoes **elastic** deformation it will return to its original shape when the applied force is removed. If an object undergoes **plastic** deformation it will not return to its original shape when the applied force is removed; some or all of the deformation will remain.

Stress and strain

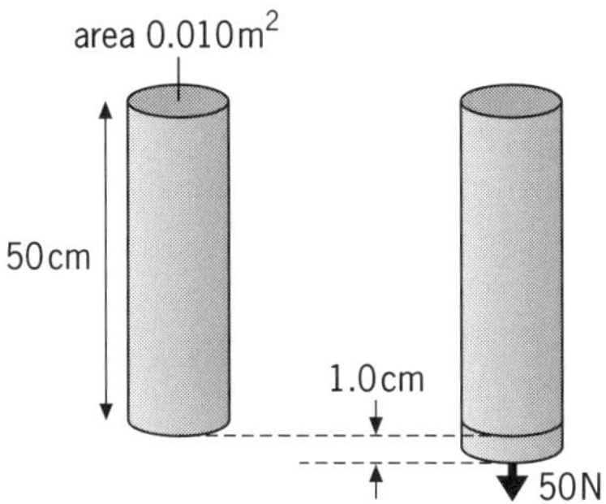

Figure 26.3

When a force is applied to an object, the object will experience a stress. **Stress** is defined as the **force per unit area** applied to the object and is measured in $\mathrm{N\,m^{-2}}$ or Pa.

$$\text{stress} = \frac{\text{applied force}}{\text{area of cross-section}} \quad \text{or} \quad \sigma = \frac{F}{A}$$

In the example shown in Figure 26.3 the applied stress the rod experiences is

$$\sigma = \frac{F}{A}$$
$$= \frac{50\,\mathrm{N}}{0.010\,\mathrm{m^2}}$$
$$= 5000\,\mathrm{Pa} \text{ or } 5.0\,\mathrm{kPa}$$

This applied stress causes the rod to change shape; that is, it is strained. **Strain** is defined as **change in length per unit length** and has no units.

$$\text{strain} = \frac{\text{change in length}}{\text{original length}} \quad \text{or} \quad \epsilon = \frac{\Delta l}{l}$$

In the example of Figure 26.3 the strain created in the rod by the applied stress is

$$\epsilon = \frac{\Delta l}{l}$$
$$= \frac{0.010\,\mathrm{m}}{0.50}\,\mathrm{m}$$
$$= 0.02$$

There are several different types of stress, including:

- **tensile** stresses – these are stresses which stretch an object, causing tensile strains (tension); and
- **compressive** stresses – these are stresses which squash or compress an object, causing compressive strains.

$$\text{tensile strain} = \frac{\text{increase in length of object}}{\text{original length of object}}$$

$$\text{compressive strain} = \frac{\text{decrease in length of object}}{\text{original length of object}}$$

Ductile and brittle materials

A **ductile** material is one which can be worked into shape or permanently deformed without fracturing (breaking). A **brittle** material cannot be permanently stretched; it fractures once the elastic limit has been exceeded. Figure 26.4 shows typical stress–strain graphs for ductile and brittle materials.

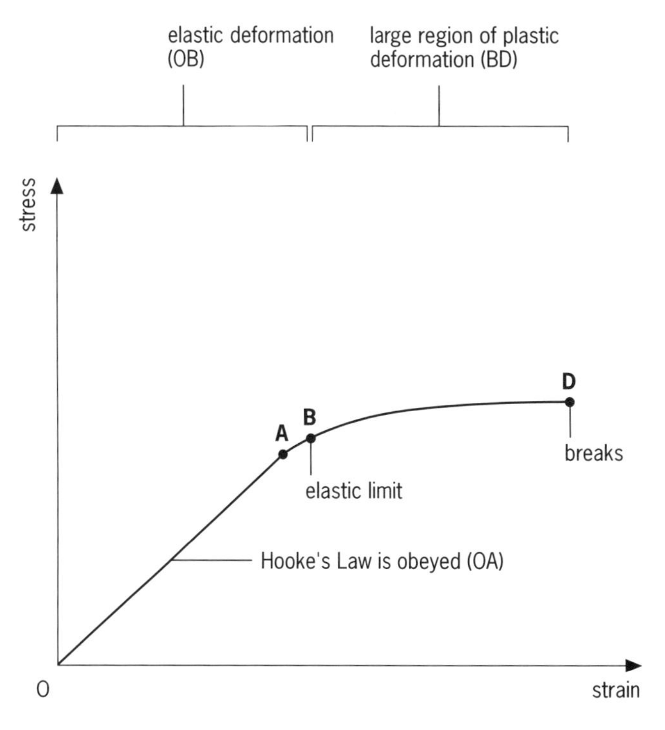

a Ductile material, for example copper

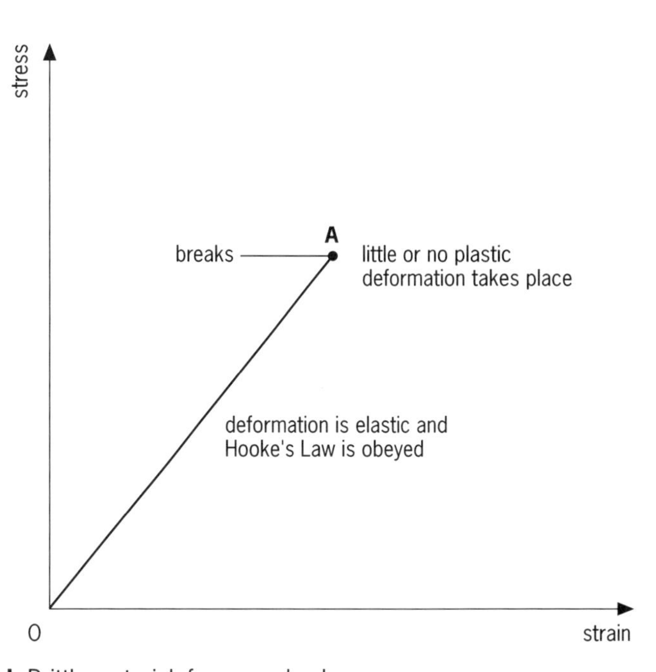

b Brittle material, for example glass

Figure 26.4

The Young modulus

When a stress is applied to an elastic material it produces a strain. If the stress does not exceed the limit of proportionality, the stress is proportional to the strain. This means that for a particular material the ratio stress/strain is a constant. This constant is known as the **Young modulus** (E) and has units Pa.

Do not confuse E for the Young modulus and E for energy.

$$E = \frac{\text{stress}}{\text{strain}}$$

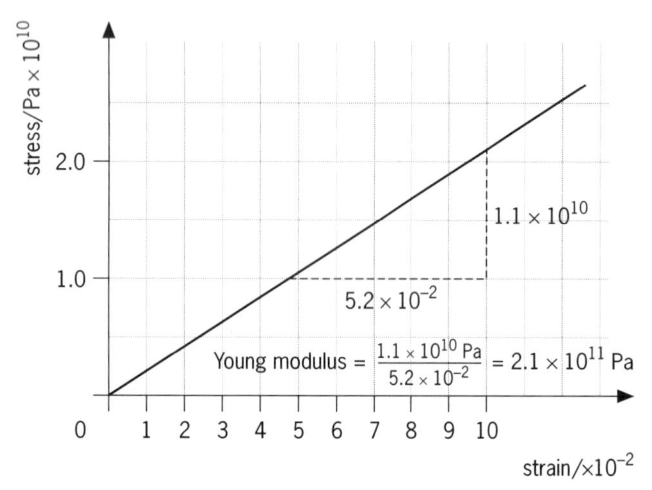

Figure 26.5 The Young modulus is equal to the gradient of the stress–strain graph

Examples of the Young modulus for different materials are given in Table 26.1.

Table 26.1 Values of the Young modulus

Material	The Young modulus/Pa
steel	2.1×10^{11}
copper	1.2×10^{11}
concrete	1.7×10^{10}
plastic	2.0×10^{9}
rubber	1.0×10^{6}

Calculating a value for the Young modulus

$$E = \frac{\text{stress}}{\text{strain}}$$

$$= \frac{F/A}{\Delta l/l}$$

$$= \frac{Fl}{A\Delta l}$$

Example 1

A steel wire 5.0 m long and with a cross-sectional area of 0.01 cm^2 extends by 2.5 mm when a mass of 10 kg is hung from one end. Calculate a value for the Young modulus for steel. (Take g as 10 m s^{-2}.)

$$E = \frac{Fl}{A\Delta l}$$

$$(1\,\text{cm}^2 = 10^{-4}\,\text{m}^2, \text{ so } 0.01\,\text{cm}^2 = 0.01 \times 10^{-4}\,\text{m}^2 = 1 \times 10^{-6}\,\text{m}^2)$$

$$\therefore E = \frac{10\,\text{kg} \times 10\,\text{m s}^{-2} \times 5.0\,\text{m}}{1 \times 10^{-6}\,\text{m}^2 \times 2.5 \times 10^{-3}\,\text{m}}$$

$$= 2 \times 10^{11}\,\text{Pa or } 200\,\text{GPa}$$

Experiment to determine the Young modulus for steel

The Young modulus for a material in the form of a wire can be found using the apparatus in Figure 26.6.

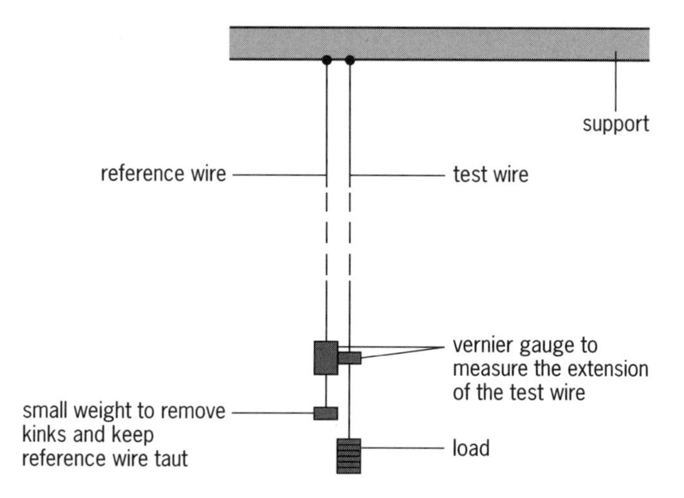

Figure 26.6 Experimental setup

Two identical wires are hung from a support. One of the wires, the reference wire, has a weight hung on it in order to remove any kinks. The test wire has a weight holder and is also loaded to remove any kinks. The two wires are connected by Vernier scales which allow measurement of the increase in length as the test wire is loaded. The diameter and original length of the test wire is also noted. A graph is then plotted of stress against strain, from which the Young modulus can be found.

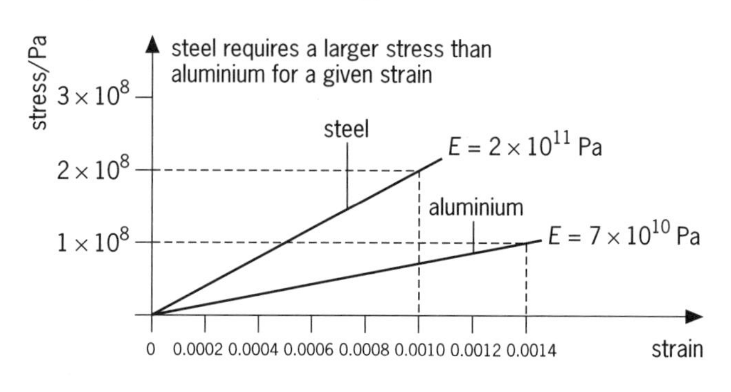

Figure 26.7 Typical results for steel and aluminium wires

Elastic energy stored in a stretched wire or spring

When a length of wire or spring is stretched by a force, work is done and energy is stored. Providing that the elastic limit has not been exceeded and no energy has been lost as heat, the energy stored as elastic potential energy is equal to the work done.

$$\text{work done in stretching} = \text{average force} \times \text{extension} = \tfrac{1}{2}F \times \Delta l$$

where F is the final force (see Figure 26.8). Therefore

$$\boxed{\text{energy stored} = \tfrac{1}{2}F\Delta l}$$

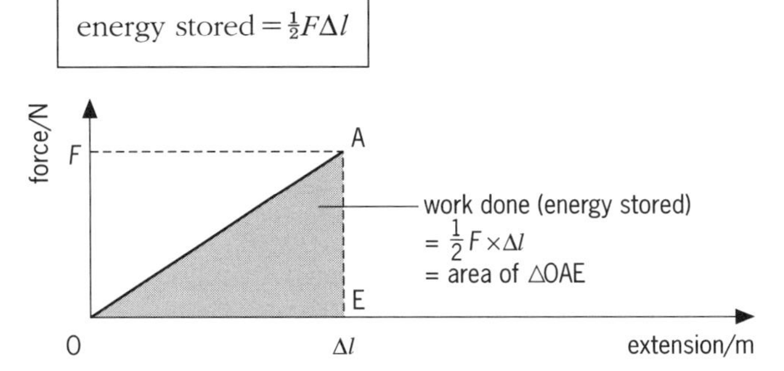

Figure 26.8 The energy stored in the wire or spring is equal to the area under the force–extension graph

Substituting for F from the Young modulus formula,

$$F = \frac{A\Delta lE}{l}$$

the above equation becomes

$$\boxed{\text{energy stored} = \frac{\tfrac{1}{2}EA\Delta l^2}{l}}$$

The volume of a wire is given by $A \times l$, therefore the energy stored *per unit volume* of a wire is

$$\frac{\frac{1}{2} \times \frac{EA\Delta l^2}{l}}{Al} = \frac{\tfrac{1}{2}E\Delta l^2}{l^2}$$

But $\Delta l/l$ is strain and $E\Delta l/l$ is stress, so this equation can be rewritten as

$$\boxed{\text{energy stored per unit volume} = \tfrac{1}{2}\,\text{stress} \times \text{strain}}$$

Example 2

Calculate the work done if a steel wire 2.0 m long and 0.010 cm² in cross-sectional area extends by 2.5 mm when a load is hung from its end. (E_{steel} = 200 GPa)

Using e for final extension,

work done on wire

$$= \frac{\tfrac{1}{2}EAe^2}{l}$$

$$= \frac{\tfrac{1}{2} \times 200 \times 10^9\,\text{Pa} \times 1.0 \times 10^{-6}\,\text{m}^2 \times (2.5 \times 10^{-3}\,\text{m})^2}{2.0\,\text{m}}$$

$$= 0.31\,\text{J}$$

Molecular explanation of plastic and elastic deformation

The particles of most solids and liquids are packed very closely together and are therefore not easily compressed. Between neighbouring particles there are strong intermolecular forces pulling the particles together. The closer the particles, the stronger these attractive forces. If, however, the separation between the particles becomes very small, repulsive forces try to push the particles apart. At a certain distance the attractive forces and the repulsive forces are equal. This distance is called the **equilibrium separation** and indicates the equilibrium position for particles in the solid.

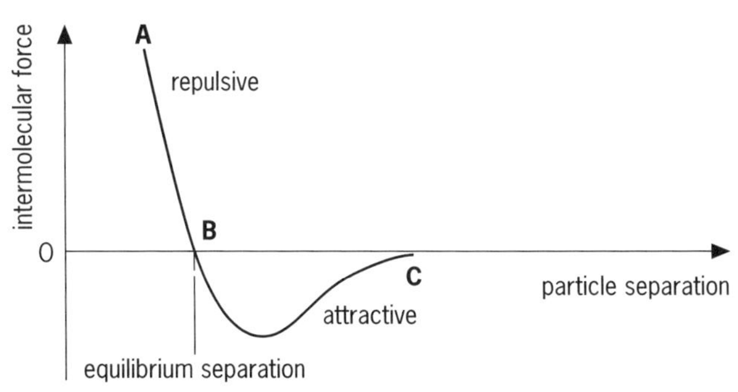

AB – the repulsive intermolecular forces are greater than the attractive forces
B – the attractive and repulsive intermolecular forces are equal
BC – the attractive intermolecular forces are greater than the repulsive forces

Figure 26.9 Variation of resultant intermolecular force with particle separation in a solid

If a small compressive force is applied to a solid its particles will be pushed closer together than the equilibrium position. The repulsive forces between the particles will increase. At some new smaller separation the increase in repulsive forces will balance the compressive force being applied. The solid has been compressed.

If the compressive force is removed the repulsive forces, being larger than the attractive forces, will increase the particle separation to the original equilibrium distance. The particle returns to its original size and shape.

Similarly, if a small tensile force is applied to a solid its particles will move further apart to new equilibrium positions – it will stretch. But when the tensile force is removed the particles will return to their original equilibrium distance and the solid to its original dimensions.

In both of the cases described the solid experienced elastic deformation: when the applied forces were removed the solid returned to its original shape. If larger forces are applied there is the possibility that neighbouring atoms will be made to slip over each other. If the applied forces are now removed the particles are unable to return to their original positions; the solid has retained some permanent deformation. It has undergone plastic deformation.

Take $g = 9.8\,\mathrm{m\,s^{-2}}$.

Level 1

1 Figure 26.10 shows the change in length, Δl, of a climbing rope as the force on it increases.

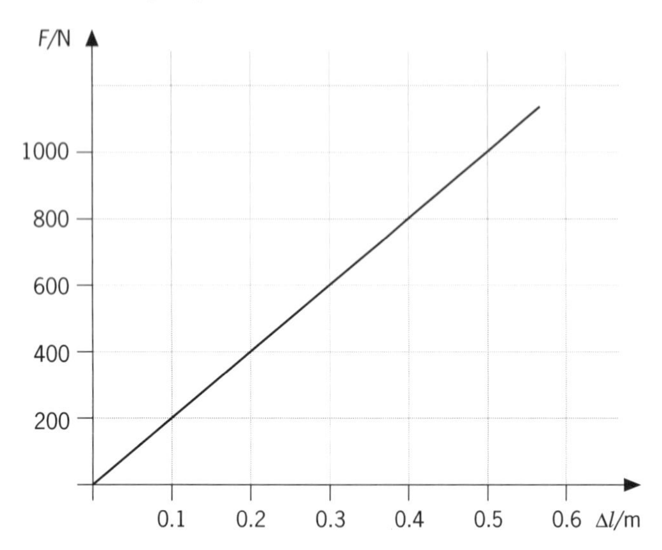

Figure 26.10

a What is the force constant, k, of the climbing rope?

b Provided that the elastic limit is not exceeded, by how much would you expect the rope to stretch if a force of
i 3000 N
ii 2.25 kN
is applied to it?

2 A spring has a force constant, k, of $800\,\mathrm{N\,m^{-1}}$.
a What tension, F, would stretch it by 40 mm?
b What *mass* would you have to hang on it (on Earth) to produce this tension?

3 A block of rubber with a force constant, k, of $50\,\mathrm{N\,m^{-1}}$ is compressed by a force of 1.0 N.
a What reduction in length is produced?
b How much energy is stored in the block?

4 A mass of 300 g is hung by a spring and stretches it by 6.0 cm.
a If *another two* identical springs are put in parallel to share the load (that is three springs in total, Figure 26.11), by how much will the springs stretch now?

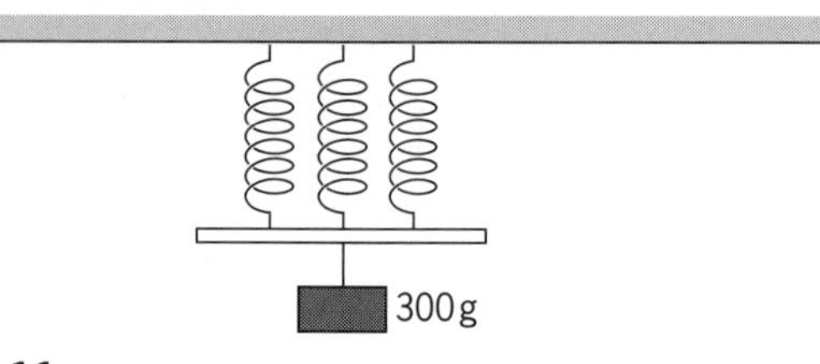

Figure 26.11

b By *what factor* has the force constant changed?

The springs are now all linked together in series, Figure 26.12.

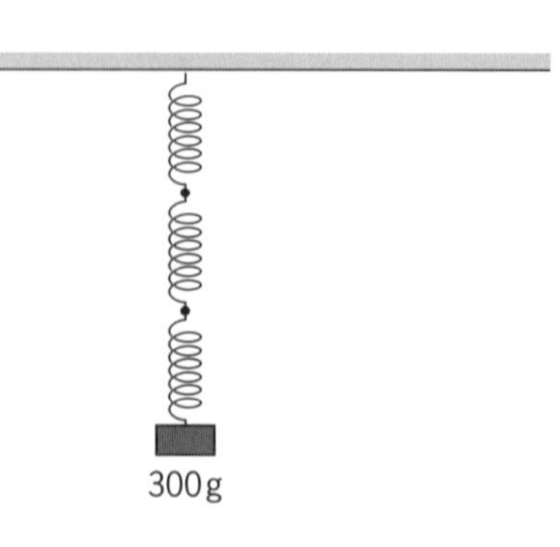

Figure 26.12

c By how much will each spring stretch and what is the total extension?
d By *what factor* has the force constant changed this time?

5 A human hair of cross-sectional area $2.5 \times 10^{-9}\,\mathrm{m^2}$ has a mass of 20 g hanging from it.
a Calculate the stress.
b Estimate the approximate diameter of the hair.

6 **a** A rope of length 80 m is stretched to give a strain of 10%. What is the new length of the rope?
b A rope of length 55 m is stretched to 57 m. What is the strain?
c A rope stretched to give a 15% strain is 47 m long. What was its original length?

7

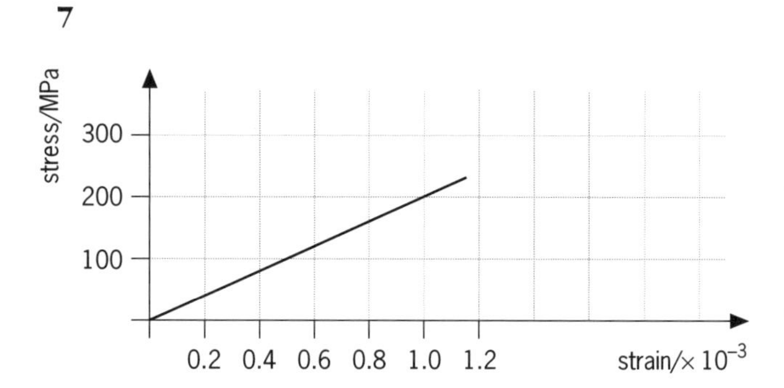

Figure 26.13

a What is the Young modulus of the material whose stress–strain graph is plotted in Figure 26.13?
b Show that the area under the graph is equal to the work done per unit volume of the material. (No calculation is needed.)

8 The Young modulus for copper is 130 GPa.
a What is the strain produced in a copper wire under a stress of 120 MPa?
b What stress would produce a strain of 0.0002?

9 **a** Find the change in length in question 8a if the length of the sample was 4.0 m.
b Find the area of cross-section and approximate diameter in question 8b if the force causing the stress is 22 N.

Level 2

10

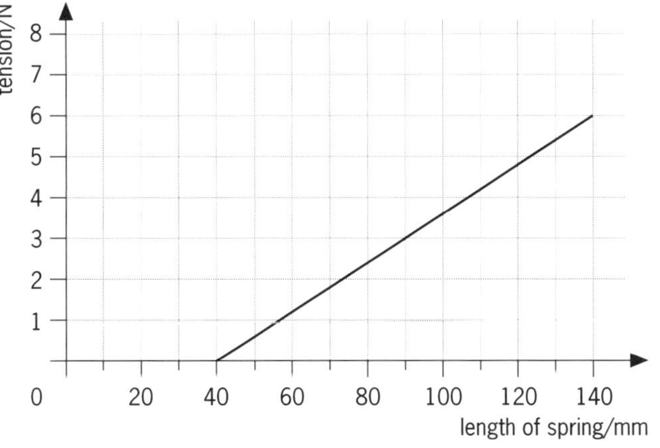

Figure 26.14

For the spring whose behaviour is shown in Figure 26.14 find

a the force constant

b the maximum strain shown (where strain = $\Delta l/l$)

c the energy stored at this maximum strain.

11 The scale of a newtonmeter, Figure 26.15, reads from 0 to 10 N and is 8.0 cm long.

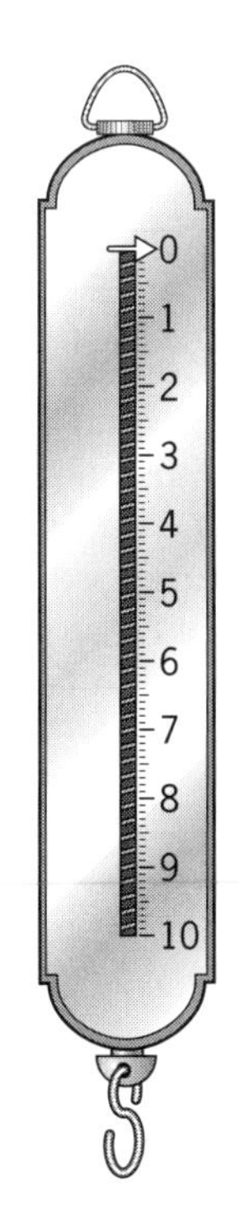

Figure 26.15 A newtonmeter

a Find the strain energy stored when it reads 3.0 N.

b How many times greater will the energy be at 6.0 N?

c Why don't you get *double* the energy when you double the tension?

12 A 5.1 kg mass is hanging from a wire whose diameter was measured by the micrometer screw gauge as shown in Figure 26.16.

a Read the diameter from the scale in the diagram and calculate the area of cross section of the wire.

b Hence, calculate the tensile stress in the wire.

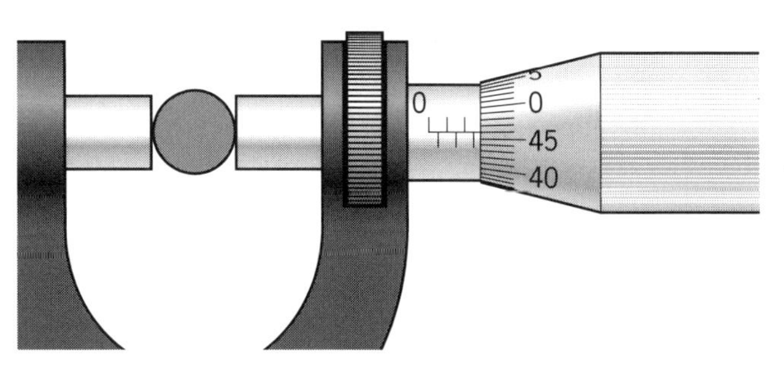

Figure 26.16 Measuring a wire's diameter with a micrometer screw gauge

13 In the body, tendons are under greater stress than muscles because they have a smaller area of cross section. See Figure 26.17. The cross-sectional area of a tendon is $\frac{1}{70}$ of the cross-sectional area of the muscle. What is the ratio of the stresses in the tendon and the muscle?

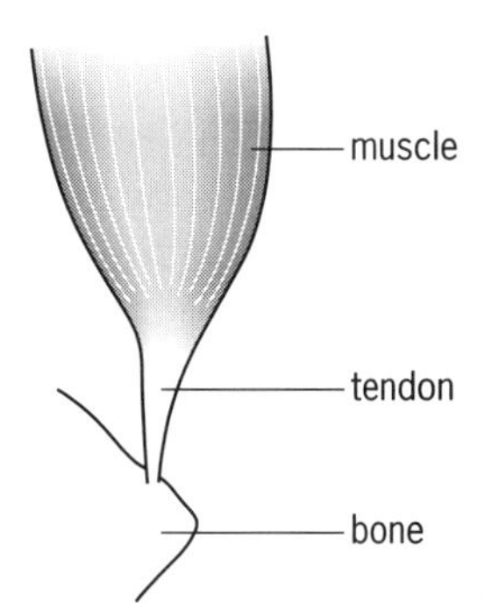

Figure 26.17

14 A tow-rope has diameter of 20 mm.

a **i** What is the tension in the rope when it is being used to accelerate a car of mass 900 kg at 1.2 m s^{-2}?

ii What would the tension be if it was hauling the car up a vertical cliff at the same rate?

b Calculate the stress in each case.

15 The headmaster wants a Foucault's pendulum in the foyer to demonstrate the rotation of the Earth. The stairwell allows a drop of 10.0 m. You have a large pendulum bob that weighs 20 N, and the caretaker has some steel wire with Young modulus, E = 200 GPa and diameter, d = 1.3 mm.

The caretaker wants to know if the extension due to the weight is going to make much difference to the length of wire he should cut. Can you advise him?

Level 3

16 A force of 2.0 N stretches a spring by 0.10 m. It is then displaced an extra 0.030 m and allowed to oscillate. What are the maximum and minimum tensions in the spring?

17 Figure 26.18 shows a force–extension graph for a rubber band stretched and then released.

a Estimate the energy stored in the rubber band after this process.

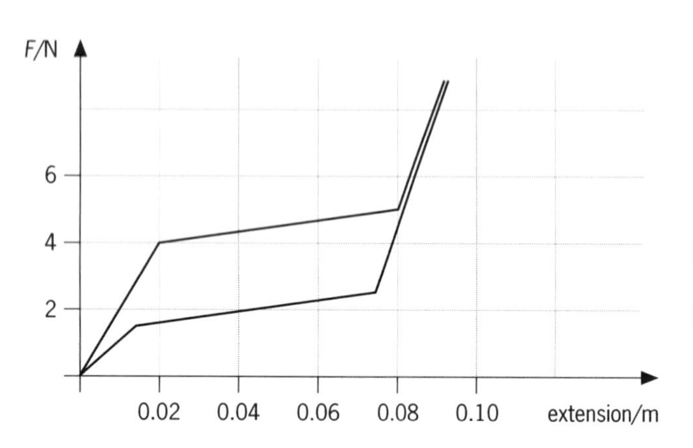

Figure 26.18

b How could you demonstrate that energy is stored?

18 A girl bungee jumping has a mass of 60 kg and the unstretched length of the rope is 40 m. She jumps from a height of 100 m and falls a total of 70 m before she is brought to rest for the first time.

a Describe the energy changes that take place during this fall.

b Calculate the total loss in gravitational potential energy.

c Use this to calculate the average force on her while she is being brought to rest.

d When does she feel the maximum force from the rope?

e What is its value?

f Three different ropes were tested on this jump, with force–extension characteristics shown in Figure 26.19. Explain what you would feel with each rope, and what aspect is equal for all three graphs.

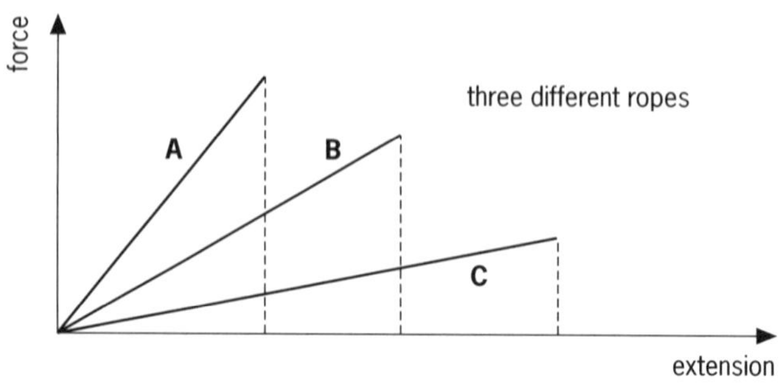

Figure 26.19

19 An elastic cord of cross-sectional area 4.0 mm^2 needs a force of 3.6 N to increase its length by 12%.

a Find the Young modulus of the material.

b The unstretched cord is 0.80 m long. Find the energy stored at this strain.

c Show that this is equal to $\frac{1}{2}$ stress × strain × volume of the cord.

20 The spacing of atoms in a solid may be taken as their 'diameter', estimated at about 2.0×10^{-10} m. The breaking strength of an interatomic bond is about 5.0×10^{-11} N.

Estimate the tension force that a 1 mm cube of material will withstand if the force is applied to opposite square faces before plastic deformation.

21 **a** Obtain an expression for the force needed to produce a certain strain, ϵ, in a material of Young modulus E and cross-sectional area A.

b Bone has a Young modulus of $E = 28$ GPa and your femur has a cross-sectional area of about 10^{-4} m^2. Calculate the force that will cause a compressive strain of 2%.

Level 1

1 a Extension, Δl, is proportional to the force, F, applied:

$$F = k \times \Delta l$$

where k is the force constant.

$$\therefore \text{ gradient of graph} = k = \frac{F}{\Delta l}$$

Taking values from the graph (Figure 26.10),

$$\frac{F}{\Delta l} = \frac{1000\,\text{N}}{0.5\,\text{m}} \quad \text{so} \quad k = \mathbf{2000\,N\,m^{-1}}$$

b i

$$\Delta l = \frac{F}{k} = \frac{3000\,\text{N}}{2000\,\text{N}\,\text{m}^{-1}} = \mathbf{1.5\,m}$$

Because the numbers are so easy you could probably 'see' the answer straight away, but it is worth writing it out properly so that you can do the same when the situation is more complicated.

ii Now we have $F = 2.25\,\text{kN} = 2250\,\text{N}$.

$$\Delta l = \frac{F}{k} = \frac{2250\,\text{N}}{2000\,\text{N}\,\text{m}^{-1}} = \mathbf{1.12\,m}$$

2 a First change the units to metres:

$$\Delta l = 40\,\text{mm} = 0.040\,\text{m}$$

$$F = k \times \Delta l = 800\,\text{N}\,\text{m}^{-1} \times 0.040\,\text{m} = \mathbf{32\,N}$$

b The force exerted by a mass, m, is $W = m \times g$. So

$$m = \frac{W}{g} = \frac{32\,\text{kg}\,\text{m}\,\text{s}^{-2}}{9.8\,\text{m}\,\text{s}^{-2}} = \mathbf{3.3\,kg}$$

3 a Hooke's Law, $F = k \times \Delta l$, works just as well in compression as in tension (stretching). In this example the force constant, $k = 50\,\text{N}\,\text{m}^{-1}$ and the applied force $= 1.0\,\text{N}$.

$$\therefore \Delta l = \frac{F}{k} = \frac{1.0\,\text{N}}{50\,\text{N}\,\text{m}^{-1}} = \mathbf{0.020\,m} \text{ or } \mathbf{20\,mm}$$

b Energy stored $= \frac{1}{2} \times F \times \Delta l$
$= 0.5 \times 1.0\,\text{N} \times 0.020\,\text{m} = 0.010\,\text{N}\,\text{m}$
$= \mathbf{0.010\,J}$ or $\mathbf{10\,mJ}$

4 a Each spring only supports $\frac{1}{3}$ of the load so it will stretch by only $\frac{1}{3}$ of 6.0 cm or **2.0 cm**.

b Force constant, $k = \dfrac{F}{\Delta l}$

Δl is reduced to $\frac{1}{3}$ of the value for a single spring so the force constant will be **3 times larger** than for a single spring.

c Each of the three springs 'feels' the whole load and so will stretch by **6.0 cm**, giving a total extension of **18.0 cm**.

d This arrangement has three times the extension for the same applied force, therefore the force constant, $k = F/\Delta l$, must be only $\frac{1}{3}$ **of the value** for a single spring.

5 a The force (weight) due to 20 g is

$F = mg = 0.020\,\text{kg} \times 9.8\,\text{m}\,\text{s}^{-2} = 0.20\,\text{N}$ (to 2 significant figures)

$$\text{stress} = \frac{\text{force}}{\text{area}} = \frac{F}{A} = \frac{0.20\,\text{N}}{2.5 \times 10^{-9}\,\text{m}^2}$$

$$= 8.0 \times 10^{7}\,\text{N}\,\text{m}^{-2} = \mathbf{80\,MPa}$$

b To find the approximate diameter, treat it as though it were square and take the square root of the area to find the length of side:

$$\sqrt{2.5 \times 10^{-9}\,\text{m}^2} = \mathbf{5.5 \times 10^{-5}\,m} \text{ or } \textbf{about } \mathbf{50\,\mu m}$$

6 a The rope is initially 80 m long. 10% of 80 m is 8.0 m so the new length is $l + \Delta l = 80\,\text{m} + 8\,\text{m} = \mathbf{88\,m}$.

More formally: strain, $\epsilon = \Delta l / l \quad \therefore \quad \Delta l = l \times \epsilon$

10% is 0.10 $\quad \therefore \quad \Delta l = 0.10 \times 80\,m = 8.0\,m$

b The change in length, $\Delta l = 57\,\text{m} - 55\,\text{m} = 2.0\,\text{m}$.

$$\text{strain} = \frac{\text{change in length}}{\text{original length}} = \frac{2.0\,\text{m}}{55\,\text{m}} = \mathbf{0.036}$$

You can multiply by 100 to give % strain = ***3.6%****.*

c When the rope is stretched, 47 m is 115% of the original length, l. To find 100% you divide 47 m by 115 and multiply by 100:

$$l = \frac{47\,\text{m}}{115} \times 100 = \mathbf{41\,m}$$

Check by adding 15% of 41 m to 41 m. You find 15% by multiplying by 0.15.

$$\Delta l = 41\,m \times 0.15 = 6.1\,m$$

$\therefore l + \Delta l = 41\,m + 6.1\,m = 47\,m$ *(to 2 significant figures)*

7 a The Young modulus, E = stress/strain. You can rearrange this equation to give stress = $E \times$ strain, which shows that E is equal to the gradient of the stress–strain graph.
From Figure 26.13,

$$\text{gradient} = \frac{200 \times 10^{6}\,\text{Pa}}{1.0 \times 10^{-3}}$$

$$= \mathbf{2.0 \times 10^{11}\,Pa} \text{ or } \mathbf{200\,GPa}$$

Note that strain has units of m/m so both the stress and the gradient are in Pa.

b The area under the graph is

$$\tfrac{1}{2}\,\text{stress} \times \text{strain} = \tfrac{1}{2} \times \frac{F}{A} \times \frac{\Delta l}{l}$$

But $\frac{1}{2}F \times \Delta l$ = work done, ΔW, and $A \times l$ = volume of the specimen, V. So the area under the graph is equal to $\Delta W / V$, the work done per unit volume.

8 a Rearrange E = stress/strain = σ/ϵ to give $\epsilon = \sigma/E$. In this case,

$$\epsilon = \frac{\sigma}{E} = \frac{120 \times 10^{6}\,\text{Pa}}{130 \times 10^{9}\,\text{Pa}} = \mathbf{9.2 \times 10^{-4}} \text{ or } \mathbf{0.092\%}$$

b Rearrange E = stress/strain = σ/ϵ to give $\sigma = \epsilon \times E$. In this case,

$$\sigma = \epsilon \times E = 0.0002 \times 130 \times 10^{9}\,\text{Pa} = \mathbf{26\,MPa}$$

9 a strain $= \dfrac{\text{change in length}}{\text{original length}}$ or $\epsilon = \dfrac{\Delta l}{l}$

which you can rearrange to give $\Delta l = \epsilon \times l$. In this example, from 8a above, $\epsilon = 9.2 \times 10^{-4}$.

$\therefore \Delta l = 9.2 \times 10^{-4} \times 4.0\,\text{m} = \mathbf{3.7 \times 10^{-3}\,m}$ or **3.7 mm**

b stress $= \dfrac{\text{force}}{\text{area}}$ or $\sigma = \dfrac{F}{A}$

which you can rearrange to give $A = F/\sigma$. In this example,

$$A = \frac{22\,\text{N}}{2.6 \times 10^{6}\,\text{Pa}} = 8.5 \times 10^{-7}\,\text{m}^2$$

approximate diameter $= \sqrt{A} = \sqrt{8.5 \times 10^{-7}\,\text{m}^2}$

$= \mathbf{9 \times 10^{-4}\,m}$ or **0.9 mm**

Level 2

10 a The maximum *extension*, Δl, of the spring is $140\,\text{mm} - 40\,\text{mm} = 100\,\text{mm} = 0.10\,\text{m}$.

$\therefore$ force constant, $k = \dfrac{F}{\Delta l} = \dfrac{6.0\,\text{N}}{0.10\,\text{m}} = \mathbf{60\,N\,m^{-1}}$

b The change in length is 100 mm; the original length was 40 mm.

$\therefore$ strain $= \dfrac{\Delta l}{l} = \dfrac{100}{40} = \mathbf{2.5}$ or **250%**

c Energy stored $\Delta W = \frac{1}{2}F\Delta l = 0.5 \times 6.0\,\text{N} \times 0.10\,\text{m}$
$= 0.30\,\text{N\,m} = \mathbf{0.30\,J}$

11 a The scale moves 0.80 cm for each newton so will be 2.4 cm (0.024 m) down for a load of 3.0 N.

energy stored in spring $= \frac{1}{2}F\Delta l$
$= 0.5 \times 3.0\,\text{N} \times 0.024\,\text{m} = \mathbf{0.036\,J}$

b If you consider the area under the graph, Figure 26.20, it is easy to see that the energy will be **4 times greater**.

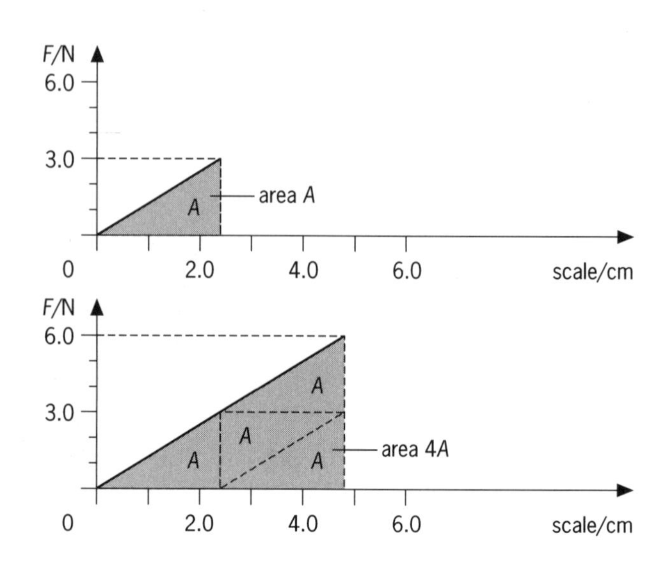

Figure 26.20

c When you double the tension **the extension doubles** as well.

In the equation

energy stored $= \frac{1}{2}F\Delta l$

you can substitute $\Delta l = F/k$, *to give*

energy stored $= \dfrac{1}{2}\dfrac{F^2}{k}$

where k is a constant, from which you can see that the energy is proportional to the ***square*** *of the tension.*

12 a The shaft of the screw gauge has two whole mm and one half showing $= 2.50\,\text{mm}$, then you add the $46 \times 10^{-2}\,\text{mm} = 0.46\,\text{mm}$ from the drum scale, giving a diameter, d, of $2.96\,\text{mm} = \mathbf{2.96 \times 10^{-3}\,m}$.

cross-sectional area, $A = \pi \times d^2/4 = \mathbf{6.88 \times 10^{-6}\,m^2}$

(Or you could use $A = \pi \times r^2$ *where* $r = d/2 = 1.48 \times 10^{-3}\,\text{m}$.*)*

b Force on wire $= mg = 5.1\,\text{kg} \times 9.8\,\text{m\,s}^{-2} = 50\,\text{N}$.

$$\text{Stress} = \frac{\text{force}}{\text{area}} = \frac{50\,\text{N}}{6.88 \times 10^{-6}\,\text{m}^2} = \mathbf{7.3\,MPa}$$

13 The tendon and the muscle support the same load (force). Rearrange stress = force/area, $\sigma = F/A$, to give $F = \sigma A$. Then

$$F = \sigma_{tendon}A_{tendon} = \sigma_{muscle}A_{muscle} \text{ or } \sigma_{tendon} = \frac{\sigma_{muscle} \times A_{muscle}}{A_{tendon}}$$

$A_{muscle}/A_{tendon} = 70$, so the stress in a tendon will be **70 times larger** than the stress in the corresponding muscle.

The ratio of the stresses is the inverse of the ratio of the areas.

14 a i Force = mass × acceleration, $F = ma$. So, to accelerate the car at $1.2\,\text{m\,s}^{-2}$ requires a force of $900\,\text{kg} \times 1.2\,\text{m\,s}^{-2} = \mathbf{1080\,N}$. This is the tension in the tow-rope.

ii If the rope is taking the weight as well as causing an acceleration, the total force must be greater by an amount equal to the weight of the car,

$$W = mg = 900\,\text{kg} \times 9.8\,\text{m\,s}^{-2} = 8820\,\text{N}$$

giving a total tension of $1080\,\text{N} + 8820\,\text{N} = \mathbf{9900\,N}$.

b The stress in the tow-rope is given by the equation stress = force/area, $\sigma = F/A$, where the area, A, is the cross-sectional area of the tow-rope $= \pi \times d^2/4$ or $\pi \times r^2$.

$r = 10\,\text{mm} = 10 \times 10^{-3}\,\text{m}$
$\therefore$ area $= \pi \times (10 \times 10^{-3}\,\text{m})^2 = 3.1 \times 10^{-4}\,\text{m}^2$

So, in the two cases:

i stress, $\sigma = \dfrac{1080\,\text{N}}{3.1 \times 10^{-4}\,\text{m}^2} = \mathbf{3.4\,MPa}$

ii stress, $\sigma = \dfrac{9900\,\text{N}}{3.1 \times 10^{-4}\,\text{m}^2} = \mathbf{32\,MPa}$

The stress pulling the car up the cliff is roughly ten times the stress when pulling along the road.

15 You need to calculate what the extension will be, given the diameter, length and Young modulus of the wire, and the weight of the pendulum bob that you are going to hang on the end of the wire.

You can expand the equation for the Young modulus, E = stress/strain, to give

$$E = \frac{\text{stress}}{\text{strain}} = \frac{\text{force/area}}{\Delta l/l} = \frac{\text{force} \times l}{\text{area}} \times \Delta l = \frac{Fl}{A\Delta l}$$

which you can rearrange to give

$$\Delta l = \frac{Fl}{AE}$$

This gives you the value that you need, Δl, in terms of things that you know, or that you can calculate from things that you know.

force = 20 N $\quad l = 10\,\text{m} \quad E = 200 \times 10^9\,\text{Pa}$
area $= \pi \times d^2/4 = \pi/4 \times (1.3 \times 10^{-3}\,\text{m})^2 = 1.3 \times 10^{-6}\,\text{m}^2$

So

$$\Delta l = \frac{Fl}{AE} = \frac{20\,\text{N} \times 10\,\text{m}}{1.3 \times 10^{-6}\,\text{m}^2 \times 200 \times 10^9\,\text{Pa}}$$

$$= \mathbf{7.7 \times 10^{-4}\,m} \approx 0.8\,\text{mm}$$

0.8 mm will be hardly noticeable, so the caretaker need not worry!

Level 3

16 Calculate the force constant:

$$k = \frac{F}{\Delta l} = \frac{2.0\,\text{N}}{0.10\,\text{m}} = 20\,\text{N}\,\text{m}^{-1}$$

The added tension when the end of the spring is displaced by an extra 0.030 m is then given by

$$F = k\Delta l = 20\,\text{N}\,\text{m}^{-1} \times 0.030\,\text{m} = 0.60\,\text{N}$$

This must be added to the equilibrium tension to give the maximum tension:

$2.0\,\text{N} + 0.6\,\text{N} = \mathbf{2.6\,N}$

and subtracted to give the minimum tension:

$2.0\,\text{N} - 0.6\,\text{N} = \mathbf{1.4\,N}$

17 a The energy stored is the area enclosed by the graph (Figure 26.18). This is close to a rectangle that is 2 N by 0.06 m, so it is about **0.12 J**.

b If you stretch and release a rubber band several times and then put it to your lips you can feel that it is warm. Rubber tyres get warm when vehicles move because the walls of the tyre are continually flexing as the wheel rotates.

18 a As she falls, going faster and faster, her gravitational potential energy decreases and her kinetic energy increases until the rope becomes taut. As the rope stretches it applies a gradually increasing upward force. Until this upward force balances the weight of the jumper, she will still be accelerating downwards, but at a gradually decreasing rate, so that her velocity and kinetic energy will still increase. An increasing proportion of the gravitational potential energy that she loses will become elastic strain energy in the rope and a decreasing proportion will be converted into kinetic energy.

Eventually the upward force will become greater than her weight, which means that she will experience a resultant upward force and will start to slow down and lose kinetic energy; but she is still going downwards and is still losing gravitational potential energy. Both these energy losses are converted into elastic strain energy in the rope. When all her kinetic energy has been converted into elastic strain energy, her velocity will be zero. At this point all the gravitational potential energy that she had at the start of the jump has been converted into elastic strain energy stored in the rope and she will have fallen as far as she is going to fall.

(We will leave her here, but remember that the rope has a lot of stored energy and she is about to head off upwards again. Will she ever stop?)

b The gravitational potential energy lost is given by

$$\Delta E_p = mg\Delta h$$
$$= 60\,\text{kg} \times 9.8\,\text{m}\,\text{s}^{-2} \times 70\,\text{m} = \mathbf{41\,kJ}$$

c The loss in gravitational potential energy (41 kJ) is equal to the energy stored in the rope, which is equal to

average force × distance over which she stopped

(This is the usual equation for work done or energy gained when a force does work.)

$$\therefore 41\,\text{kJ} = F_{av} \times 30\,\text{m}$$

and $F_{av} = \dfrac{41\,\text{kJ}}{30\,\text{m}} = \mathbf{1.4\,kN}$

d She feels the maximum force when she is stationary at the 'bottom' of her jump, because the rope is at maximum extension and maximum tension.

e The peak value of the force is twice the average, so $F_{max} = \mathbf{2.8\,kN}$.

f Rope A would give you a hard sharp jerk, stopping you in a short distance; rope B would give a medium deceleration and distance; rope C would be much more gentle, but there would be a very large extension. The areas under the graphs must all be the same and must equal the total energy absorbed.

19 a You are given the cross-sectional area of the elastic cord and the stress needed to produce a particular strain. You have to find the Young modulus, E.

$$E = \frac{\text{stress}}{\text{strain}} = \frac{\text{force/area}}{\text{strain}} = \frac{\text{force}}{\text{area} \times \text{strain}} = \frac{F}{A\epsilon}$$

so you have all you need.

Always check the units of the values you are given in a question. In this question the cross-sectional area is given in mm², and you need to convert this to m²:

$1\,\text{mm} = 1 \times 10^{-3}\,\text{m} \therefore 1\,\text{mm}^2 = 1 \times 10^{-6}\,\text{m}^2$

$$E = \frac{F}{A\epsilon} = \frac{3.6\,\text{N}}{4.0 \times 10^{-6}\,\text{m}^2 \times 0.12}$$

$= \mathbf{7.5 \times 10^6\,Pa}$ or $\mathbf{7.5\,MPa}$

b The energy stored in the cord is given by

$\Delta W = \frac{1}{2} F \times \Delta l$

$F = 3.6\,\text{N}$

$\Delta l = 12\%$ of $0.80\,\text{m} = 0.12 \times 0.80\,\text{m} = 0.096\,\text{m}$

$\therefore \quad \Delta W = \frac{1}{2} \times 3.6\,\text{N} \times 0.096\,\text{m} = \mathbf{0.17\,J}$

c stress $= \frac{\text{force}}{\text{area}}$, $\sigma = \frac{F}{A} = \frac{3.6\,\text{N}}{4.0 \times 10^{-6}\,\text{m}^2} = 9.0 \times 10^5\,\text{Pa}$

strain $= 0.12$

volume $= Al = 4.0 \times 10^{-6}\,\text{m}^2 \times 0.80\,\text{m} = 3.2 \times 10^{-6}\,\text{m}^3$

$\therefore \frac{1}{2}$ stress $\times$ strain $\times$ volume $=$
$\frac{1}{2} \times 9.0 \times 10^5\,\text{Pa} \times 0.12 \times 3.2 \times 10^{-6}\,\text{m}^3 = \mathbf{0.17\,J}$

20

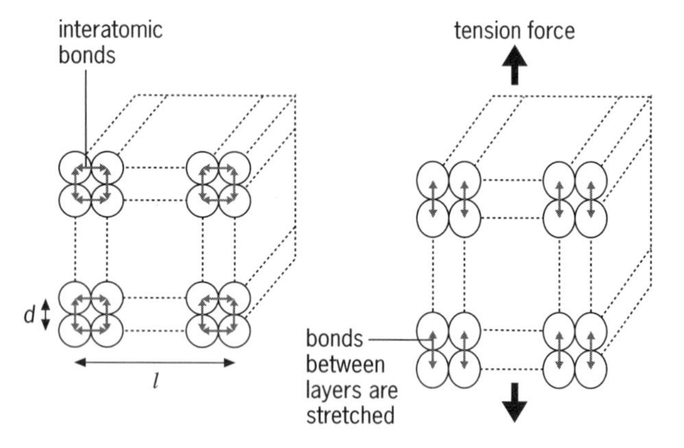

Figure 26.21

You have been told how strong the interatomic bonds are and how big the atoms are. You have to find out how many atoms there are in each 'layer' of atoms in the cube, and multiply by the individual interatomic bond-strength.

The number of atoms $\times$ their diameter $=$ the length of the side, $nd = l$.
So, the number of atoms $=$ length of the side/diameter of the atoms, $n = l/d$.
In this example,

$$n = \frac{1.0 \times 10^{-3}\,\text{m}}{2.0 \times 10^{-10}\,\text{m}} = 5.0 \times 10^6\,\text{atoms}$$

The number of atoms filling the square face is

$n^2 = (5.0 \times 10^6)^2 = 2.5 \times 10^{13}\,\text{atoms}$

Each atom has a bond to an atom in the next layer that will break when the interatomic bond strength is exceeded. The bonds are in parallel. So the total force required is

$2.5 \times 10^{13}\,\text{atoms} \times 5.0 \times 10^{-11}\,\text{N atom}^{-1} = \mathbf{1250\,N}$

21 a Young modulus $E = \frac{\text{stress}}{\text{strain}}$

$$= \frac{\text{force/area}}{\text{strain}} = \frac{\text{force}}{\text{area} \times \text{strain}} = \frac{F}{A\epsilon}$$

which you can rearrange to give

$F = E \times A \times \epsilon$

b $E = 28 \times 10^9\,\text{Pa}$; $A = 1 \times 10^{-4}\,\text{m}^2$; $\epsilon = 2\% = 0.02$

$\therefore F = E \times A \times \epsilon$
$= 28 \times 10^9\,\text{Pa} \times 1 \times 10^{-4}\,\text{m}^2 \times 0.02 = \mathbf{56\,kN}$

27 Motion in a circle

Angular displacement

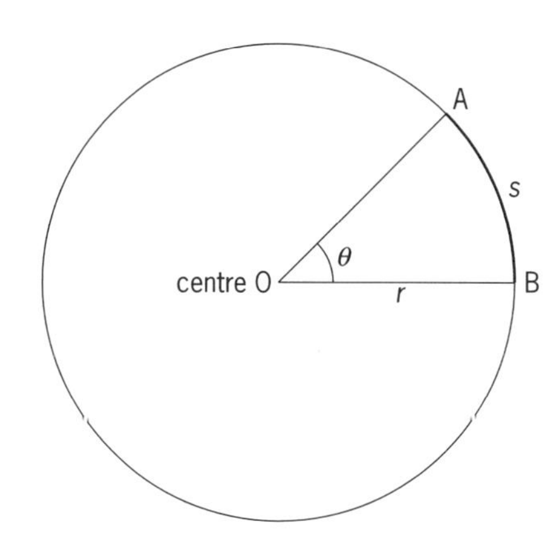

Figure 27.1

If an object moves from A to B along a circular path, Figure 27.1, it will pass through an **angular displacement** θ. If θ is measured in radians then

$$\theta = \frac{s}{r}$$

The angle in radians is defined as the ratio arc/radius.

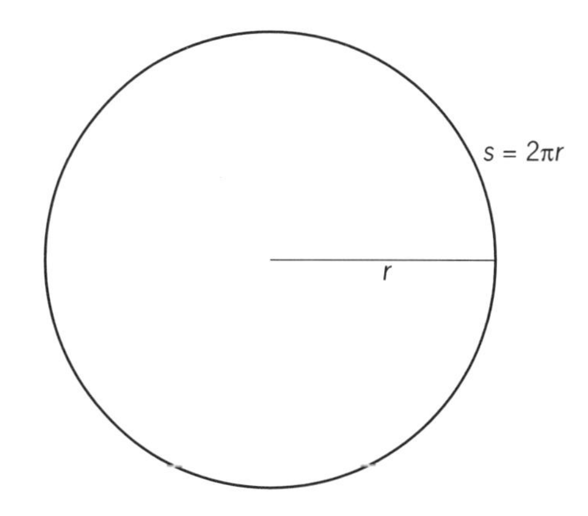

Figure 27.2

If the object travels around the circle once, Figure 27.2, then its angular displacement in radians is

$$\theta = \frac{s}{r} = \frac{2\pi r}{r} = 2\pi$$

The object will have travelled through 360°, so

$$2\pi \text{ radians} = 360°$$

and

$$1 \text{ radian} = \frac{360°}{2\pi} = 57.3°$$

Example 1

Calculate the angular displacement for an object which
a travels 20 cm along the arc of a circle of radius 10 cm,
b completes 5 revolutions.

a $\theta = \frac{s}{r}$

$= \frac{20}{10}$

$= 2$ radians (or 114.6°)

b An object travelling one complete revolution will have an angular displacement of 2π radians. Therefore after 5 revolutions its angular displacement will be 10π radians.

Angular velocity

If the object in Figure 27.1 travels steadily through θ radians in t seconds its **angular velocity**, ω, is given by

$$\omega = \frac{\theta}{t}$$

that is, angular velocity $= \dfrac{\text{angular displacement}}{\text{time taken}}$

Example 2

Calculate the angular velocity of an athlete who runs half-way around a circular running track in 30 s.

$\omega = \frac{\theta}{t}$

$= \frac{\pi \text{ rad}}{30 \text{ s}}$

$\therefore \omega = 0.10 \text{ rad s}^{-1}$ (or 6 degrees s^{-1})

(One revolution = 2π radians)

The relationship between v, r and ω

Consider an object moving around a circle with uniform speed, v, and angular velocity, ω.

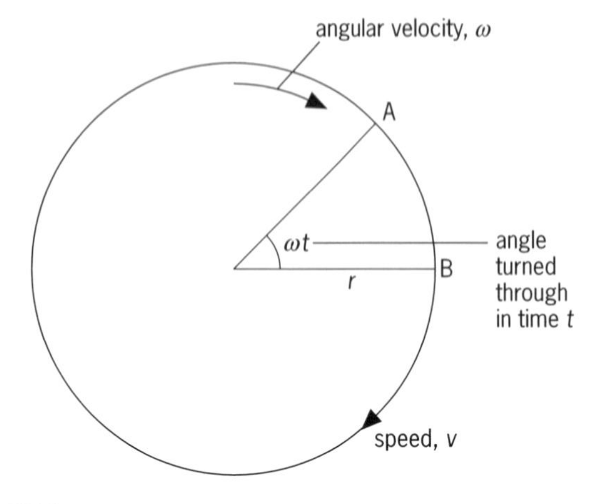

Figure 27.3

If the object moves along the arc AB in t seconds, Figure 27.3, then the angle turned through, $\theta = \omega t$

$$\therefore \qquad \frac{s}{r} = \omega t$$

Rearranging gives

$$\frac{s}{t} = r\omega$$

but

$$\frac{s}{t} = v$$

Therefore

$$\boxed{v = r\omega}$$

that is,

linear speed = radius × angular velocity

Example 3

Calculate the angular velocity of a racing car travelling at $50\,\mathrm{m\,s^{-1}}$ around a circular track of radius 500 m.

Rearranging

$$v = r\omega$$

gives

$$\omega = \frac{v}{r}$$

$$= \frac{50\,\mathrm{m\,s^{-1}}}{500\,\mathrm{m}}$$

$= 0.1\,\mathrm{rad\,s^{-1}}$ (or 5.7 degrees $\mathrm{s^{-1}}$)

Period and frequency

If T is the time it takes for an object to complete one revolution, called the **period**, then using

$$\text{time} = \frac{\text{distance}}{\text{speed}}$$

$$T = \frac{2\pi r}{v}$$

But $v = r\omega$, so, substituting for v and cancelling r,

$$\boxed{T = \frac{2\pi}{\omega}}$$

The period and frequency of rotation of an object moving in a circle are related by the equation $T = 1/f$. Therefore, from above,

$$\frac{1}{f} = \frac{2\pi}{\omega}$$

or

$$\boxed{\omega = 2\pi f}$$

ω can be called the **angular frequency**.

Example 4

Calculate the period and frequency of a satellite orbiting the Earth with an angular velocity of $0.07\,\mathrm{radians\,min^{-1}}$

$$T = \frac{2\pi}{\omega}$$

$$= \frac{2\pi}{0.07\,\mathrm{rad\,min^{-1}}}$$

(There is no need to change to seconds here.)

$$\therefore \qquad T = 90\,\mathrm{min}$$

$$f = \frac{1}{T}$$

$$= \frac{1}{90\,\mathrm{min}}$$

$$= 0.11\,\mathrm{rev\,min^{-1}}$$

Divide by 60 to find revs per second:

giving

$$f = 1.8 \times 10^{-3}\,\mathrm{Hz}$$

Acceleration in uniform circular motion

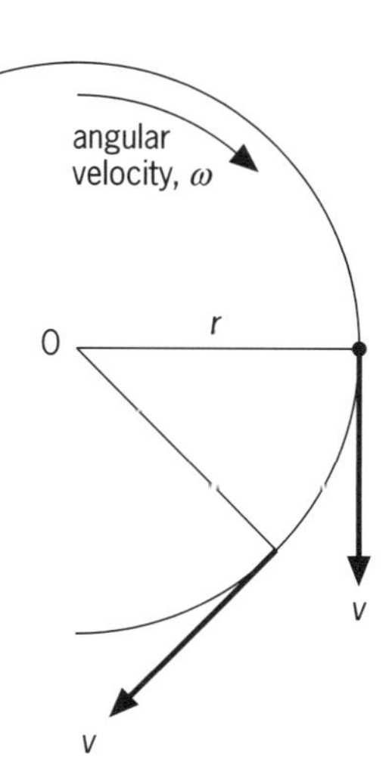

Figure 27.4

The object in Figure 27.4 is moving with a constant speed, v, but because it is continually changing direction its velocity (a vector quantity) is changing and so it is accelerating. The magnitude of the acceleration, a, is given by the formula

$$a = r\omega^2 \quad \text{or} \quad a = \frac{v^2}{r}$$

The direction of the acceleration (a vector quantity) is *towards the centre* of the circle around which the object is rotating. It is called the **centripetal acceleration**.

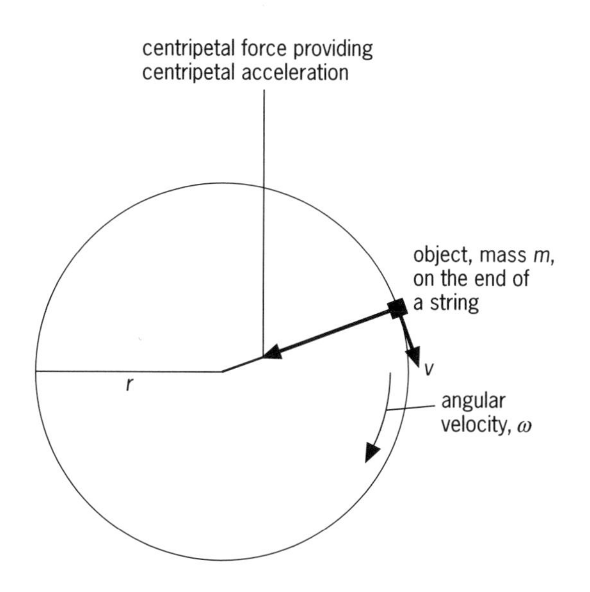

Figure 27.5 Circular motion on the end of a string

For an object to travel in a circle an unbalanced force must act on it to provide the centripetal acceleration. This force may be gravitational, electrostatic, mechanical or magnetic in origin. For the object in Figure 27.5 to travel in a circle there must be a force applied to it by the string. This force, called the **centripetal force**, acts towards the centre of the circle. If the mass of the object is m then, since $F = ma$, the magnitude of the centripetal force is given by the formula

$$F = mr\omega^2 \quad \text{or} \quad F = mv^2/r$$

If the force is removed, for example by cutting the string, the object will cease changing direction and will continue to move in a straight line along the tangent to the circle at that point.

Example 5

A metal ball of mass 2.5 kg is attached to the end of a piece of string 1.0 m long and made to travel in a horizontal circle. If the speed of the ball is $2.0\,\text{m}\,\text{s}^{-1}$, calculate **a** the acceleration of the ball towards the centre of the circle and **b** the tension in the string.

a The acceleration is given by

$$a = \frac{v^2}{r}$$

$$= \frac{(2.0\,\text{m}\,\text{s}^{-1})^2}{1.0\,\text{m}}$$

$$= 4.0\,\text{m}\,\text{s}^{-2}$$

b The tension force is given by

$$F = ma$$

$$= 2.5\,\text{kg} \times 4.0\,\text{ms}^{-2}$$

$$= 10.0\,\text{N}$$

Banking

If an aircraft in level flight wants to turn, it must create a centripetal force by banking (tilting). See Figure 27.6.

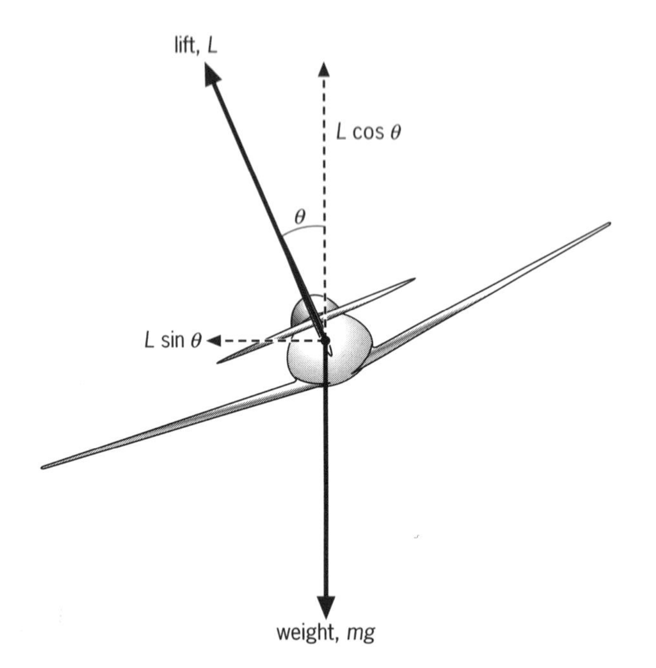

Figure 27.6 An aircraft banking

Resolving horizontally, we have an unbalanced component of the lift, $L \sin\theta$, which provides the centripetal force needed:

$$L \sin\theta = \frac{mv^2}{r} \qquad (1)$$

Resolving vertically, we have no unbalanced force as the component of the lift, $L \cos\theta$, is equal and opposite to the weight:

$$L \cos\theta = mg \qquad (2)$$

Dividing equation (1) by equation (2) gives

$$\tan\theta = \frac{v^2}{rg}$$

where θ is the angle the wing must make with the horizontal in order to follow a circle of radius r at speed v.

Similarly, if a car travels on a frictionless road that is banked at an angle θ, the centripetal force is provided by the unbalanced horizontal component of the reaction at the wheels, R. See Figure 27.7.

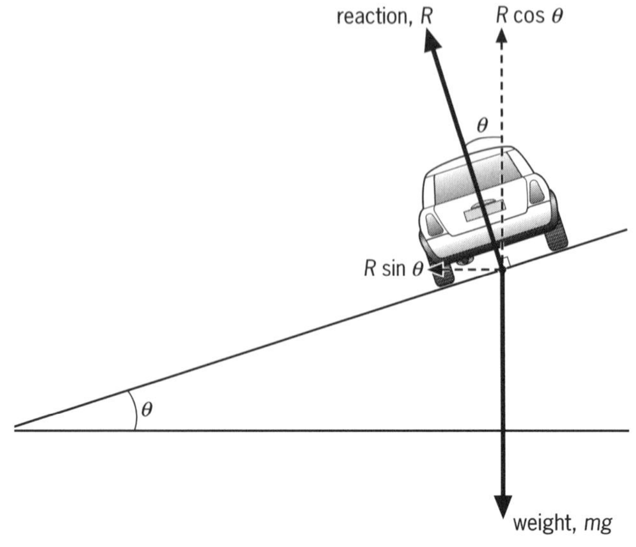

Figure 27.7 A car on a banked road

Resolving horizontally, we have an unbalanced force which provides the centripetal force:

$$R \sin\theta = \frac{mv^2}{r} \qquad (3)$$

There is no vertical motion, so resolving vertically we have balanced forces:

$$R \cos\theta = mg \qquad (4)$$

Dividing equation (3) by equation (4) gives the same result as before:

$$\tan\theta = \frac{v^2}{rg}$$

Example 6

Calculate the angle at which an aircraft must bank if it is to have a turning circle of 50 km while travelling horizontally with a speed of $500\,\mathrm{m\,s^{-1}}$. ($g = 10\,\mathrm{m\,s^{-2}}$)

$$\tan\theta = \frac{v^2}{rg}$$

$$= \frac{(500\,\mathrm{m\,s^{-1}})^2}{50 \times 10^3\,\mathrm{m} \times 10\,\mathrm{m\,s^{-2}}}$$

$$= 0.50$$

$$\therefore \quad \theta = \tan^{-1} 0.50 = 27°$$

Level 1

1 Draw three separate 'free body' diagrams of a satellite in circular orbit around the Earth, to show the direction of:

a the forces upon the satellite

b the velocity of the satellite

c the acceleration of the satellite relative to the centre of the circle.

Free body diagrams consider the forces on one object alone.

2 Linear velocity is the rate of change of displacement. Angular velocity is the rate of change of angular displacement or angle turned through in unit time. Angular velocity can be measured in *degrees per second* or *radians per second.*

Calculate the angular velocity in each of these units for:

a the second hand of a watch

Hint: How long does the hand take to go round once?

b a point on the edge (15 cm from the centre) of an old long-playing (LP) record rotating at $33\frac{1}{3}$ rpm.

3 When you walk round in a complete circle, you have covered a distance of $2\pi r$ and turned through an angle of 2π. Your linear speed, v, is r times the angular velocity, ω (in rad s^{-1}):

$$v = \omega r$$

Taking the examples from the previous question, **estimate** the linear speed of

a the tip of the second hand

b the point on the edge of the record.

4 The spreadsheet shown in Table 27.1 is to check that the two formulae for acceleration,

$a = \omega^2 r$ in column E and $a = v^2/r$ in column H

give the same result.

Enter the units needed to complete the calculations, then enter the formulae into a computer if possible (using *Excel*, for example). If not, complete rows 2, 3 and 5 by calculation.

Table 27.1

A	B	C	D	E	F	G	H
radius m	frequency Hz	ω	ω^2	$a = (\omega^2 r)$	v	v^2	$a = v^2/r$
0.1	1.0						
0.2	1.0						
0.3	1.0						
0.1	2.0						
0.2	2.0						
0.3	2.0						

5 At one point on an amusement park ride you go round an arc of a circle of radius 23 m at a velocity of 26 m s^{-1}. How many '*g*' do you experience?

Hint: Calculate the centripetal acceleration and compare with 9.8 m s^{-2}.

6 a A ball on a string is whirled at 22 m s^{-1} in a horizontal circle of radius 1.3 m. Find the time period and the frequency of the motion.

b A moon orbits its planet in 2.4×10^6 s at a distance of 3.8×10^8 m.

i What is its linear speed in m s^{-1}?

ii What is this in km h^{-1}?

c A disc spins at 18 Hz. Find the linear speed of a mark 0.050 m from the centre.

Level 2

7 A 'fast jet' flies at 400 m s^{-1} and is designed to undergo accelerations up to 12*g*. What, approximately, is the smallest radius turn it can execute at this speed?

8 You may be taught that there is no such thing as centrifugal force but that there can be an 'absence of centripetal force'. Use the example of water leaving clothes through the holes in the drum of a spin dryer to explain this statement.

In case you have never heard of centrifugal force, it was ***supposed*** *to act* ***outwards*** *from the centre of a circle.*

9 If there is little sideways friction to keep a vehicle moving in a circle then banking has to be used. A component of the normal reaction then provides the force towards the centre of the circle.

a Draw a diagram to illustrate this. Clearly distinguish the components from the actual force.

b A bobsleigh has mass 390 kg and the track is banked at 60° on a bend of radius 26.0 m.

i How fast is the bobsleigh able to go?

ii What angle of banking would be needed for the sleigh to be able to reach 130 km h^{-1}?

iii What would be the centripetal force required for the situation in **ii**?

Level 3

10 Standing on the equator you would be moving at $465\,\mathrm{m\,s^{-1}}$ due to the Earth's rotation. The centripetal acceleration of this motion is $0.034\,\mathrm{m\,s^{-2}}$.

a If you had a mass of 70 kg, what would be the centripetal force needed to keep you moving in this circle?

b What supplies this force?

c When you stand on bathroom scales the reading is the normal reaction pushing upwards. The reading is not your true weight because then there would be no unbalanced force acting towards the Earth's axis, to move you in a circle.

Draw a diagram showing the vector sum that applies here. Label each force and the resultant clearly.

d g at the equator is $9.78\,\mathrm{N\,kg^{-1}}$. Calculate your true weight at the equator and the reading you would see if you stood on bathroom scales.

e Imagine you are in orbit around the Earth. What is your acceleration towards the centre of the Earth? What is the size of the centripetal force needed?

f Use the previous example to explain what a set of bathroom scales would read if you stood on them while in orbit.

11 A helicopter, with supplies hanging from a rope, circles a lighthouse to deliver them. The load of 80 kg is on a circular path of radius 20 m at $3.3\,\mathrm{m\,s^{-1}}$.

a The rope cannot be vertical. Why not?

b By considering the tension in the rope and the centripetal force needed, establish the angle of the rope to the horizontal.

12 The force on a charged particle of mass m moving with velocity v in a magnetic field of strength B is given by $F = Bqv$ and acts at right angles to B and v. This means that the particle moves in a circular path.

a Derive an expression for the velocity of the particle.

b Show that the frequency of rotation in the circle is $Bq/2\pi m$. (This is known as the cyclotron frequency.)

13 For astronomical bodies such as a planet and its moon, the centripetal force is provided by the gravitational attraction between the two masses,

$$F = \frac{GM_1M_2}{r^2}$$

where G is the universal gravitational constant.

Show that Kepler's relation,

$$\frac{r^3}{T^2} = \text{constant}$$

is obeyed, where r is the radius of the orbit and T is the time period for one revolution.

Level 1

1

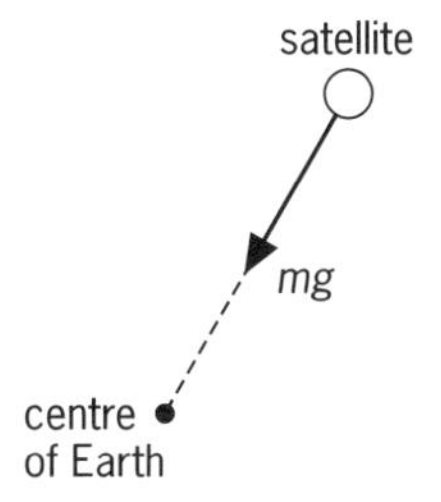

a There is only one force.

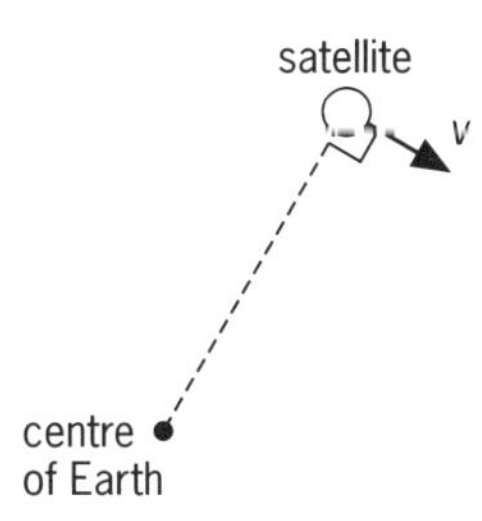

b The velocity is tangential to the circle.

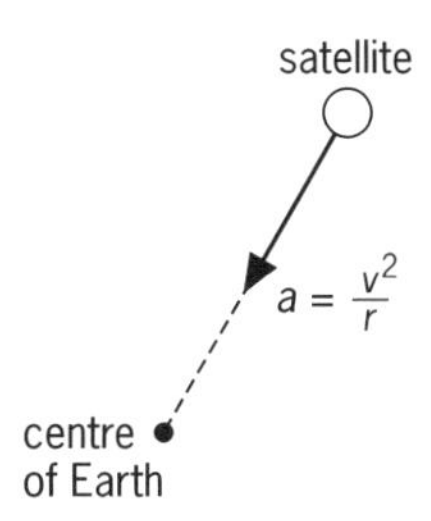

c The acceleration acts towards the centre. (That's the way the velocity changes.)

Figure 27.8

Remember that circular motion is not an equilibrium situation. There must be an unbalanced force for it to take place.

2 a It takes one minute for the second hand to go round once, so the angular velocity is

$$\omega = \frac{\text{angle}}{\text{time}} = \frac{\theta}{t} = \frac{360°}{60\,\text{s}} = \mathbf{6°\,s^{-1}}$$

or

$$\frac{2\pi\,\text{rad}}{60\,\text{s}} = \mathbf{0.10\,rad\,s^{-1}}$$

b A point on the edge of the LP sweeps out $33.33 \times 360°$ in one minute.

(rpm = revs per minute)

$$\therefore \quad \omega = \frac{33.33 \times 360°}{60\,\text{s}} = \mathbf{200°\,s^{-1}}$$

or

$$\frac{33.33 \times 2\pi\,\text{rad}}{60\,\text{s}} = \mathbf{3.5\,rad\,s^{-1}}$$

3 a From question 2a, $\omega = 0.10\,\text{rad}\,\text{s}^{-1}$, and if the second hand is about 1 cm long, or 0.01 m, then the linear speed is

$v = \omega r = 0.10\,\text{rad}\,\text{s}^{-1} \times 0.01\,\text{m} = \mathbf{1 \times 10^{-3}\,m\,s^{-1}}$

b From question 2b, $\omega = 3.5\,\text{rad}\,\text{s}^{-1}$ and $r = 15\,\text{cm} = 0.15\,\text{m}$, so

$v = \omega r = 3.5\,\text{rad}\,\text{s}^{-1} \times 0.15\,\text{m} = \mathbf{0.525\,m\,s^{-1}}$

4 See Table 27.2.

Table 27.2

A	B	C	D	E	F	G	H
radius m	frequency Hz	ω rad s^{-1}	ω^2 (rad $s^{-1})^2$	$a = \omega^2 r$ m s^{-2}	v m s^{-1}	v^2 (m $s^{-1})^2$	$a = v^2/r$ m s^{-2}
0.1	1.0	6.3	39.4	3.9	0.6	0.4	3.9
0.2	1.0	6.3	39.4	7.9	1.3	1.6	7.9
0.3	1.0	6.3	39.4	11.8	1.9	3.5	11.8
0.1	2.0	12.6	157.8	15.8	1.3	1.6	15.8
0.2	2.0	12.6	157.8	31.6	2.5	6.3	31.6
0.3	2.0	12.6	157.8	47.3	3.8	14.2	47.3

The formulae that have been entered are:

C2 = 6.28*B2 ($\omega = 2\pi f$)
D2 = C2*C2 (ω^2)
E2 = D2*A2 ($\omega^2 r$)
F2 = 6.28*A2*B2 ($v = 2\pi f r$)
G2 = F2*F2 (v^2)
H2 = G2/A2 (v^2/r)

Once you have set up the spreadsheet, you can use it to investigate the ways that changes in radius and frequency affect the velocity and acceleration.

5 Use the acceleration equation that involves the quantities you have been given. In this case you know r and v, so use

$$a = \frac{v^2}{r} = \frac{(26\,\text{m}\,\text{s}^{-1})^2}{23\,\text{m}} = 29\,\text{m}\,\text{s}^{-2} \text{ or } \textbf{about 3g}$$

This is about the limit of what it is sensible for untrained people to experience.

6 a You can rearrange the equation speed = distance/time, $v = s/t$, to give time = distance/speed, $t = s/v$. The ball goes around the circumference of a circle of radius 1.3 m at $22\,\text{m}\,\text{s}^{-1}$. Therefore the time it takes to go round once, the period, is

$$T = \frac{2\pi \times \text{radius}}{\text{speed}} = \frac{2\pi \times 1.3\,\text{m}}{22\,\text{m}\,\text{s}^{-1}} = \mathbf{0.37\,s}$$

The frequency, $f = 1/T = 1/0.37\,\text{s} = \mathbf{2.7\,Hz}$

b Taking the moon's orbit to be a circle, its speed is the circumference of the circle divided by the time for one revolution.

i $$v = \frac{s}{t} = \frac{2\pi r}{T} = \frac{2\pi \times 3.8 \times 10^8\,\text{m}}{2.4 \times 10^6\,\text{s}} = \mathbf{1.0 \times 10^3\,m\,s^{-1}}$$

ii In 1 hour, distance gone = speed × time,

$s = vt = 1.0\,\text{km}\,\text{s}^{-1} \times 60 \times 60\,\text{s} = 3600\,\text{km}$

so the speed is $\mathbf{3600\,km\,h^{-1}}$.

c You are given the frequency, f, and the radius, r, so, to find the velocity, you need to use the equation

$v = 2\pi f r = 2\pi \times 18\,\text{s}^{-1} \times 0.050\,\text{m} = \mathbf{5.7\,m\,s^{-1}}$

Level 2

7 The centripetal acceleration for motion in a circle is given by $a = v^2/r$, which you can rearrange to give $r = v^2/a$. In this example a_{max} can be $12g$, which is approximately $120\,m\,s^{-2}$. So the minimum value for the radius of the turn is

$$r_{min} = \frac{v^2}{a} = \frac{(400\,m\,s^{-1})^2}{120\,m\,s^{-2}} = \frac{160\,000\,m^2\,s^{-2}}{120\,m\,s^{-2}}$$

$$= \mathbf{1.3 \times 10^3\,m}\ (=1.3\,km)$$

8 To move the water in a circle there would have to be an inward force on it from the walls of the drum. Where there are holes in the side this force is not present and so the water cannot be kept in the circle but carries straight on, leaving the clothes and the drum.

9 a See Figure 27.7. The vertical component of the normal reaction force is equal and opposite to the weight:

$$R\cos\theta = mg$$

The horizontal component of the normal reaction force must provide the centripetal force towards the centre of the circle:

$$R\sin\theta = \frac{mv^2}{r}$$

b Dividing the second equation above by the first, $\tan\theta = v^2/rg$, which you can rearrange to give $v^2 = rg\tan\theta$ to calculate the velocity, or $r = v^2/g\tan\theta$ to calculate the radius of the circle.

i For the bobsleigh, the maximum speed around the curve is given by

$$v^2 = rg\tan\theta = 26.0\,m \times 9.8\,m\,s^{-2} \times \tan 60°$$
$$= 441.3\,m^2\,s^{-2}$$
$$\therefore\ \mathbf{v = 21\,m\,s^{-1}}\ (=76\,km\,h^{-1})$$

ii $130\,km\,h^{-1}$ is $36.1\,m\,s^{-1}$, so θ in this case is given by

$$\tan\theta = \frac{v^2}{rg} = \frac{(36.1\,m\,s^{-1})^2}{26.0\,m \times 9.8\,m\,s^{-2}} = 5.11$$

$$\therefore\ \theta = \tan^{-1}5.11 = \mathbf{79°}$$

iii The centripetal force is given by

$$F = \frac{mv^2}{r} = \frac{390\,kg \times (36.1\,m\,s^{-1})^2}{26\,m} = \mathbf{19.5\,kN}$$

(The centripetal acceleration is given by

$$a = \frac{v^2}{r} = \frac{(36.1\,m\,s^{-1})^2}{26.0\,m} = 50\,m\,s^{-2}$$

This is more than 5g: greater than the sensible limit for untrained people.)

Level 3

10 a The force needed to give a body of mass m an acceleration a is given by the equation $F = ma$. So the force needed to provide you with the necessary centripetal acceleration is

$$70\,kg \times 0.034\,m\,s^{-2} = \mathbf{2.4\,N}$$

b The force is supplied by the gravitational attraction of the Earth, which acts towards its centre.

c

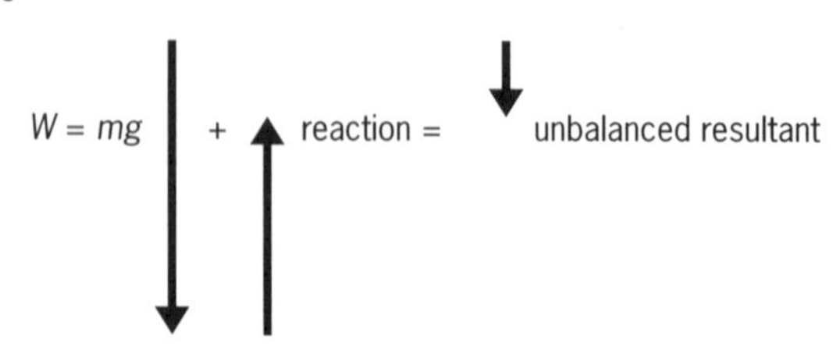

Figure 27.9

d At the equator, your true weight is

$$W = mg = 70\,kg \times 9.78\,m\,s^{-2} = \mathbf{684.6\,N}$$

You must subtract the centripetal force calculated in part a, 2.4 N, from this. What is left is what the scales will read, that is

$$684.6\,N - 2.4\,N = \mathbf{682.2\,N}$$

The reaction of the scales will balance most of the weight, but not the unbalanced force needed to keep your circular movement.

e When you are in orbit, you are in free fall and your acceleration is g at all times, so the necessary centripetal force must be your weight, mg (but note that g decreases with height above the Earth).

f All your weight is used to provide the centripetal force that you need to keep moving in a circle, and there is nothing 'left over'. If you 'stand' on the scales they will read zero. Not surprisingly, this situation is called 'weightlessness'.

11 a If the load were vertically below the helicopter there would not be an unbalanced horizontal force to move it in a circle. When the rope is at an angle, the horizontal component of the tension can provide the necessary centripetal force. See Figure 27.10.

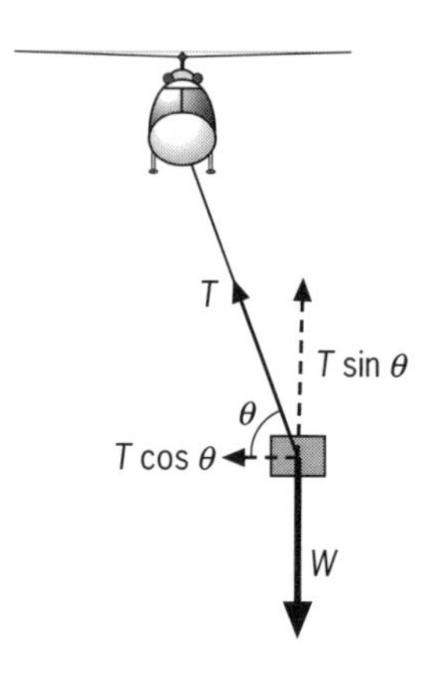

Figure 27.10

b From the diagram,

vertical component of the tension = weight of the supplies

$$T\sin\theta = 80\,\text{kg} \times 9.8\,\text{m s}^{-2} = 780\,\text{N}$$

The centripetal force needed to keep the supplies moving in a circle of radius r at a speed v is given by

$$F = \frac{mv^2}{r} = \frac{80\,\text{kg} \times (3.3\,\text{m s}^{-1})^2}{20\,\text{m}} = 44\,\text{N}$$

This force must be provided by the horizontal component of the tension, so

$$T\cos\theta = 44\,\text{N}$$

$$\therefore \frac{T\sin\theta}{T\cos\theta} = \tan\theta = \frac{780\,\text{N}}{44\,\text{N}} = 17.73$$

and $\theta = \tan^{-1} 17.73 = \mathbf{87°}$

12 The centripetal force is provided by the interaction between the magnetic field and the charged particle.

a The resulting force is $Bqv = mv^2/r$, which you can rearrange to give

$$v = \frac{rBq}{m}$$

b The velocity of a rotating object, $v = \omega r = 2\pi fr$, so $v = 2\pi fr = rBq/m$ which you can simplify to give

$$f = \frac{Bq}{2\pi m}$$

You will see that the frequency is independent of the velocity and radius of the circle.

13 From Newton's Law of Universal Gravitation (see topic 31), the attraction between two masses provides the centripetal force:

$$\frac{GM_1M_2}{r^2} = \frac{M_1v^2}{r}$$

where M_1 is the mass of the orbiting object and M_2 is the mass of the body around which M_1 is orbiting.

Cancelling M_1 and r,

$$\frac{GM_2}{r} = v^2 = \left(\frac{2\pi r}{T}\right)^2 = \frac{4\pi^2r^2}{T^2}$$

which you can simplify and rearrange to give

$$\frac{r^3}{T^2} = \frac{GM_2}{4\pi^2}$$

which is constant for *all* satellites orbiting a body with mass M_2.

28 Periodic motion: simple harmonic motion (shm)

A periodic motion is one which goes in cycles; that is, it continually repeats itself. One of the most common examples of this kind of behaviour is **simple harmonic motion (shm)**. Examples of shm include the swinging back and forth of a simple pendulum and the oscillation of a mass suspended on a spring.

The time taken for one complete oscillation of a vibrating system is known as the **period of oscillation**, T, measured in seconds. The number of oscillations the system performs in one second is known as the **frequency of oscillation**, f, measured in Hz or s^{-1}. The period and frequency of oscillation of a system are related by the equation:

$$T = \frac{1}{f} \quad \text{or} \quad f = \frac{1}{T}$$

The distance an oscillating object is from its equilibrium position is known as its **displacement**, x. The maximum displacement an oscillating object moves from its equilibrium position is known as the **amplitude** of vibration or oscillation, x_0.

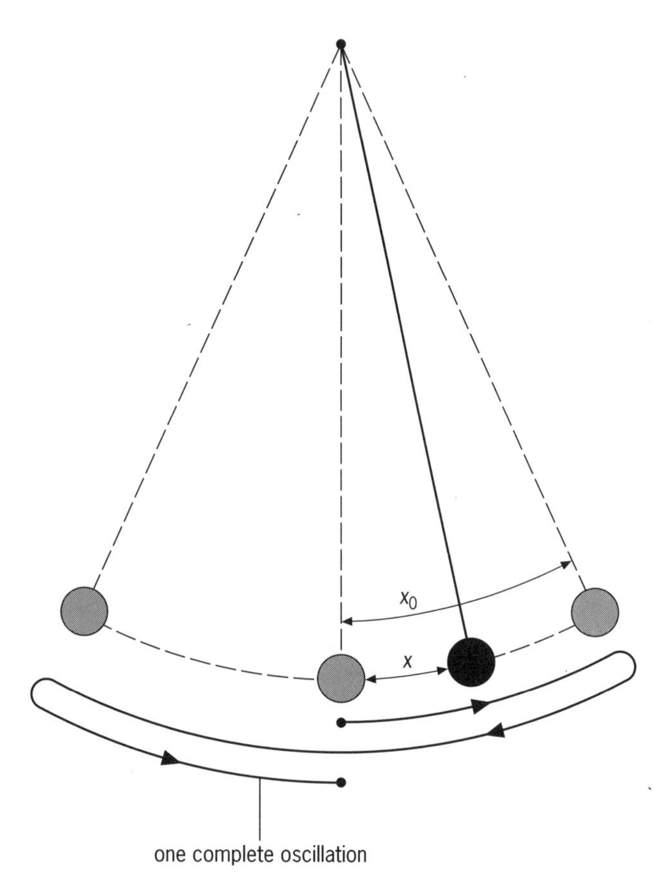

Figure 28.1 Periodic motion of a simple pendulum

Defining simple harmonic motion

Simple harmonic motion is a periodic motion such that the acceleration of the moving object is proportional to its displacement and is in the opposite direction to the displacement.

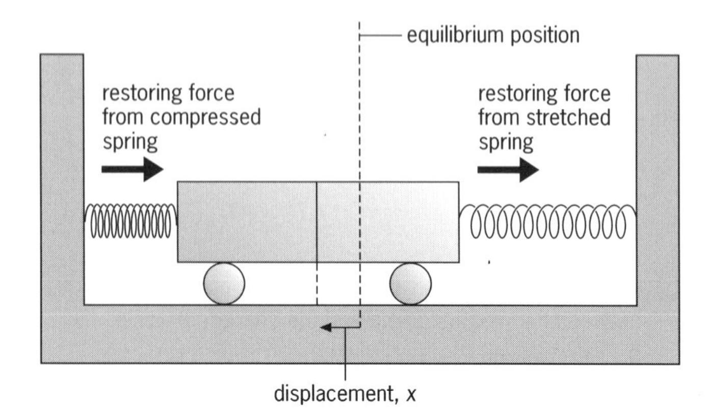

Figure 28.2

When the trolley in Figure 28.2 is moved to one side and released, the springs exert forces to return it to its original equilibrium position. The greater the displacement of the trolley the greater the restoring forces applied by the springs and so the greater the acceleration. The trolley undergoes simple harmonic motion.

From our definition we can describe shm by the equation

$$a = -\omega^2 x$$

where a is the acceleration of the object, x is its displacement and ω^2 is the constant of proportionality. The minus sign indicates that a is in the opposite direction to x.

The value of ω depends upon the nature of the shm being considered. It is related to the period of vibration by the equation

$$T = \frac{2\pi}{\omega}$$

Using the equation $T = \frac{1}{f}$ the above relationship may be written as

$$\omega = 2\pi f$$

ω is the angular velocity associated with the periodic motion and is measured in rad s^{-1}.

The equation defining shm can be written in terms of the frequency, f:

$$a = -(2\pi f)^2 x$$
$$= -4\pi^2 f^2 x$$

Example 1
An object moving with shm has an amplitude of vibration of 5.0 cm and a frequency of 10.0 Hz. Calculate **a** the period of vibration, T, and **b** the acceleration of the object at zero displacement and maximum displacement.

a $T = \frac{1}{f}$

$= \frac{1}{10.0\,\text{s}^{-1}}$

$= 0.10\,\text{s}$

b $a = -\omega^2 x$

At zero displacement $x = 0$ and therefore the acceleration of the object is zero.
At maximum displacement $x = 5.0$ cm, the acceleration of the object is given by

$a = -\omega^2 \times 0.050\,\text{m}$

$\omega = 2\pi f$ and $\omega^2 = 4\pi^2 f^2$

so

$a = -4\pi^2 \times (10.0\,\text{s}^{-1})^2 \times 0.050\,\text{m}$
$= -197\,\text{m}\,\text{s}^{-2}$

Displacement and time

From the equation $a = -\omega^2 x$ a second equation can be derived which describes how the displacement of a particle performing simple harmonic motion varies with time. This equation is

$$x = x_0 \sin \omega t \quad \text{or} \quad x = x_0 \sin 2\pi f t$$

if the displacement of the particle is zero when $t = 0$; or

$$x = x_0 \cos \omega t \quad \text{or} \quad x = x_0 \cos 2\pi f t$$

if the displacement is a maximum when $t = 0$.
x_0 is the maximum value of x; that is, the amplitude of vibration.

Example 2
Calculate the displacement of the object in the previous example at **a** $t = 0.010$ s and **b** $t = 0.025$ s, assuming that at time $t = 0$ the displacement $x = 0$.

a $x = x_0 \sin 2\pi f t$
$= 0.050\,\text{m} \times \sin(2\pi \times 10.0\,\text{s}^{-1} \times 0.010\,\text{s})$
$= 0.050\,\text{m} \times \sin(0.628\,\text{rad})$
$= 0.050 \times 0.588$
$= 0.029\,\text{m}$ or $2.9\,\text{cm}$

Use Radian Mode on your calculator, or change π radians into 180°.

$x = 0.050\,\text{m} \times \sin(2 \times 180° \times 10\,\text{s}^{-1} \times 0.010\,\text{s})$
$= 0.050\,\text{m} \times \sin 36°$
$= 0.050\,\text{m} \times 0.588$

b $x = x_0 \sin 2\pi f t$
$= 0.05\,\text{m} \times \sin(2\pi \times 10\,\text{s}^{-1} \times 0.025\,\text{s})$
$= 0.050\,\text{m} \times \sin(1.57\,\text{rad})$
$= 0.050\,\text{m} \times 1$
$= 0.050\,\text{m}$ or $5.0\,\text{cm}$

or

$x = 0.050\,\text{m} \times \sin(2 \times 180° \times 10 \times 0.025)$
$= 0.050 \times \sin 90°$
$= 0.050 \times 1$

If the displacement at time $t = 0$ is not zero or a maximum, the shm equation can be modified to

$$x = x_0 \sin(\omega t + \phi)$$

where ϕ is the initial **phase angle**; that is, the angular displacement of the particle at time $t = 0$.

Velocity and time

As a particle performs shm its velocity changes with time. Assuming that the displacement is zero at time $t = 0$,

$x = x_0 \sin \omega t$ or $x = x_0 \sin 2\pi f t$

and the velocity of the particle at time t is given by

$v = x_0 \omega \cos \omega t$ or $v = x_0 \omega \cos 2\pi f t$

The maximum velocity the particle can have is given by

$v_{max} = x_0 \omega$ or $v_{max} = x_0 \times 2\pi f$

when $\cos \omega t = 1$.

Example 3
A particle is performing shm with a period of 0.50 s and an amplitude of 4.0 cm. Assuming that the displacement of the particle was zero at time $t = 0$, calculate **a** the maximum velocity of the particle and **b** the velocity of the particle at time $t = 0.10$ s.

a $v_{max} = x_0 \omega = x_0 \times 2\pi f$

where $f = 1/T = 1/0.5\,\text{s}^{-1}$.

$\therefore\ v_{max} = 0.040\,\text{m} \times 2\pi \times \frac{1}{0.5}\,\text{s}^{-1}$

$= 0.50\,\text{m}\,\text{s}^{-1}$

b $v = x_0 \omega \cos \omega t$

From the previous answer $x_0 \omega = 0.50\,\text{m}\,\text{s}^{-1}$,

$\therefore\ v = (0.50\,\text{m}\,\text{s}^{-1}) \cos 2\pi f t$

Substituting the values and changing into degrees using $\pi\,\text{rad} = 180°$,

$v = (0.50\,\text{m}\,\text{s}^{-1}) \cos(2 \times 180° \times 2 \times 0.10)$
$= (0.50\,\text{m}\,\text{s}^{-1}) \cos 72°$
$= 0.15\,\text{m}\,\text{s}^{-1}$

Describing shm with graphs

The graphs in Figures 28.3a, b and c show how the displacement, acceleration and velocity of an object performing shm change with time, assuming that $x=0$ when $t=0$.

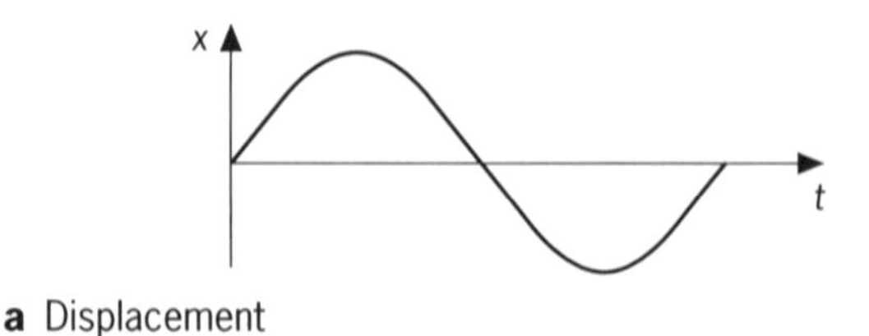

a Displacement

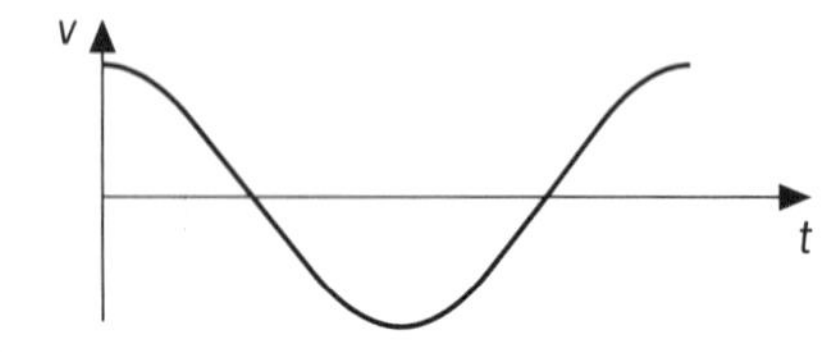

b Velocity

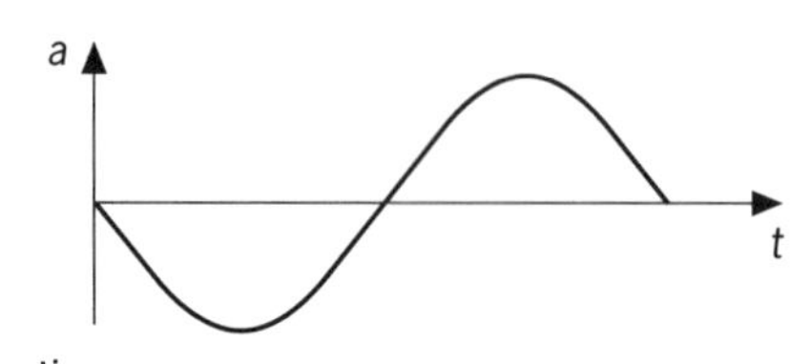

c Acceleration

Figure 28.3

Figure 28.4 shows how the acceleration of an object performing shm varies with the *displacement* of the object.

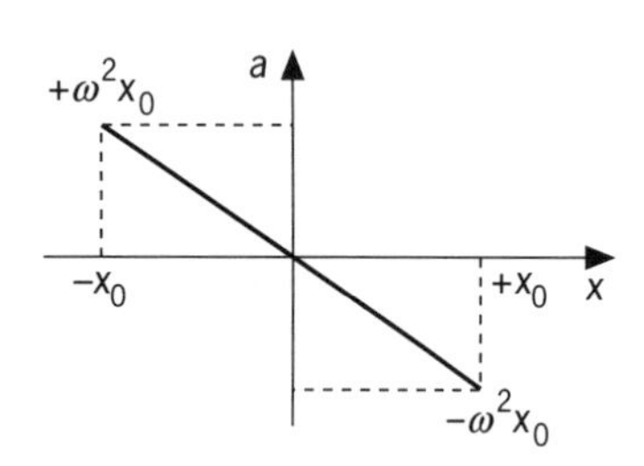

Figure 28.4 Acceleration against displacement

Figure 28.5 shows how the energy of an object performing shm changes from kinetic to potential during its oscillation; the total energy in ideal conditions (no resistive forces) remains constant.

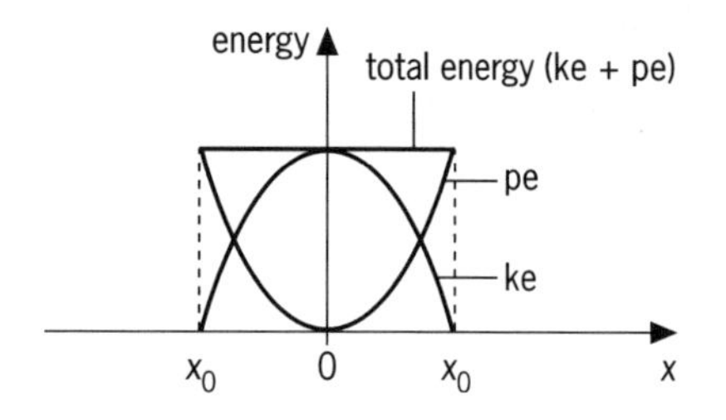

Figure 28.5 Change in kinetic energy (ke) and potential energy (pe) with displacement

Summary of shm

Table 28.1 summarises the values of the variables of shm at different points in the oscillation ABAC, illustrated in Figure 28.6 for the case of a simple pendulum.

Table 28.1 Summary of shm

Position	A	B	A	C
Displacement	zero	maximum	zero	maximum
Velocity	maximum	zero	maximum	zero
Acceleration	zero	maximum	zero	maximum
Energy	ke = max pe = min	ke = min pe = max	ke = max pe = min	ke = min pe = max

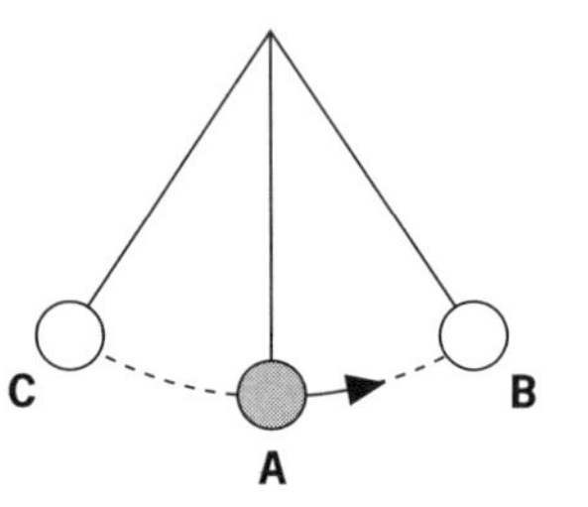

Figure 28.6

The simple pendulum

To show that a simple pendulum performs shm we must prove that the force acting on the bob during its oscillation is proportional to its displacement and in the opposite direction to it. Ideally the pendulum should consist of a point mass suspended on a light inelastic thread. In real experiments a small heavy bob is used.

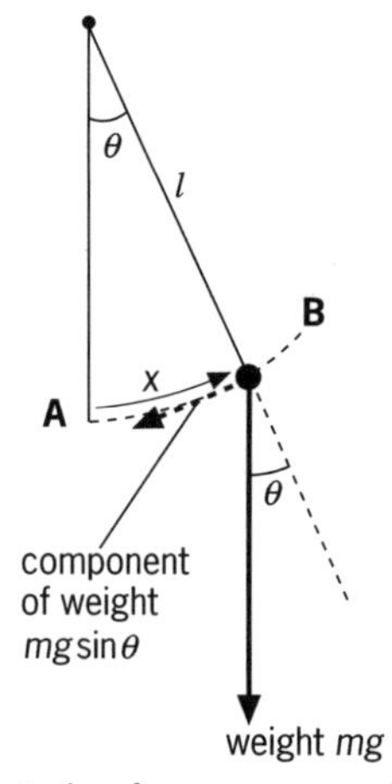

Figure 28.7 The restoring force on a pendulum bob

When the pendulum is displaced to one side, Figure 28.7, there is a component of its weight (mg) which acts as a restoring force. This force is at right angles to the pendulum and its value is given by

$$mg\cos(90° - \theta) = mg\sin\theta \qquad (1)$$

If the bob is displaced only slightly to one side of the equilibrium position, $\sin\theta$ is small. When $\sin\theta$ is very small it has the same value as θ measured in radians. The size of the angle, θ, in radians is given by

$$\theta = \frac{x}{l}$$

where x is the displacement of the bob along the arc AB. Substituting in equation (1), the restoring force can be written as

$$-mg \times \frac{x}{l}$$

The minus sign has been inserted because the force is in the opposite direction to the displacement x.

From Newton's Second Law, $F = ma$, therefore the acceleration a is given by

$$a = -g \times \frac{x}{l}$$

or $$a = -\frac{g}{l} \times x$$

This last equation confirms that the acceleration of the bob is proportional to its displacement and is in the opposite direction to the displacement; the pendulum is performing shm.

The standard equation which describes shm is

$$a = -\omega^2 x$$

Comparing this with the above-derived equation for a pendulum, it follows that

$$\omega = \sqrt{\frac{g}{l}}$$

The period of oscillation $T = 2\pi/\omega$, so for a pendulum

$$\boxed{T = 2\pi\sqrt{\frac{l}{g}}}$$

Note that

- the period of oscillation is not affected by the amplitude of vibration (this is true for all particles performing shm);
- the amplitude of oscillation of a pendulum must be small if it is to perform true shm.

Example 4

Calculate **a** the period, T, and **b** the frequency, f, of a simple pendulum of length 0.50 m. (Take g to be $= 10\,\text{m}\,\text{s}^{-2}$)

a $$T = 2\pi\sqrt{\frac{l}{g}}$$

$$= 2\pi\sqrt{\frac{0.50\,\text{m}}{10\,\text{m}\,\text{s}^{-2}}}$$

$$= 2\pi\sqrt{0.05\,\text{s}^2}$$

$$= 1.4\,\text{s}$$

b $$f = \frac{1}{T}$$

$$= \frac{1}{1.4\,\text{s}}$$

$$= 0.71\,\text{Hz}$$

Oscillations of a mass attached to a helical spring

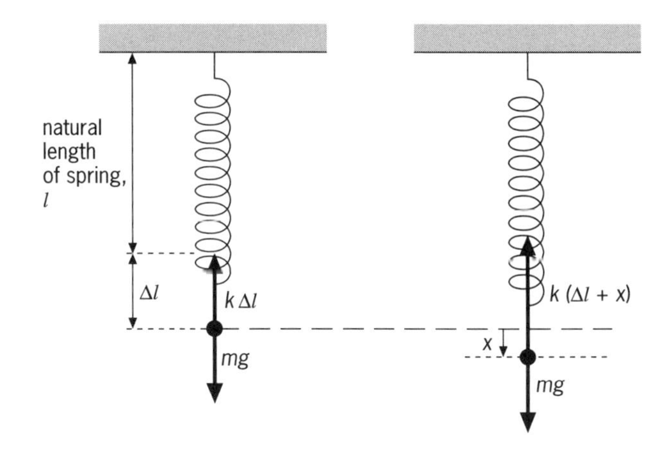

Figure 28.8

Consider a mass, m, suspended from a helical spring, as a result of which the spring is extended by an amount Δl, Figure 28.8. Assuming the spring obeys Hooke's Law, the force applied to it will be proportional to its extension:

$$mg = k\Delta l$$

where k is the force constant for the spring. k will have a larger value for a stiff spring and a lower value for a more flexible spring.

The mass m is then pulled down a further distance, x, and released. There is a resultant restoring force upwards, equal to

$$k(\Delta l + x) - mg$$

But $mg = k\Delta l$; therefore the restoring force is kx. Using Newton's Second Law,

$$ma = -kx$$

and $$a = -\left(\frac{k}{m}\right)x$$

The minus sign has been inserted because the acceleration is in the opposite direction to the displacement x.

This last equation shows that the acceleration of the mass is proportional to its displacement but is in the opposite direction to the displacement. The motion of the oscillating mass is therefore shm.

Comparing the equation with the standard shm equation, $a = -\omega^2 x$, it follows that $\omega^2 = k/m$ and, as $T = 2\pi/\omega$,

$$T = 2\pi\sqrt{\frac{m}{k}}$$

Example 5

A mass of 2.0 kg is hung from a spring whose force constant is $160\,\mathrm{N\,m^{-1}}$. The mass and spring are then made to oscillate vertically. Assuming the motion is simple harmonic, calculate the period, T, of the oscillations.

$$T = 2\pi\sqrt{\frac{m}{k}}$$

$$= 2\pi\sqrt{\frac{2.0\,\mathrm{kg}}{160\,\mathrm{N\,m^{-1}}}}$$

$$= 2\pi\sqrt{0.0125\,\mathrm{s^2}}$$

(Remember $\mathrm{N} = \mathrm{kg\,m\,s^{-2}}$.)

$\therefore\ T = 0.70\,\mathrm{s}$

Forced vibrations and resonance

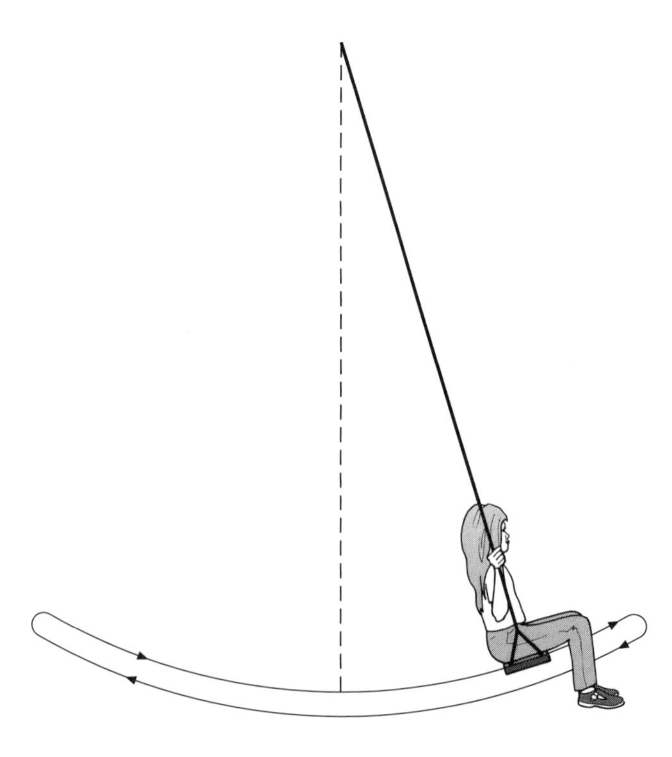

Figure 28.9 Free vibration

Any oscillating system, like the pendulum and the suspended mass just considered, has its own natural frequency. For example, a child's swing, pushed once, will oscillate back and forth at its own natural frequency.

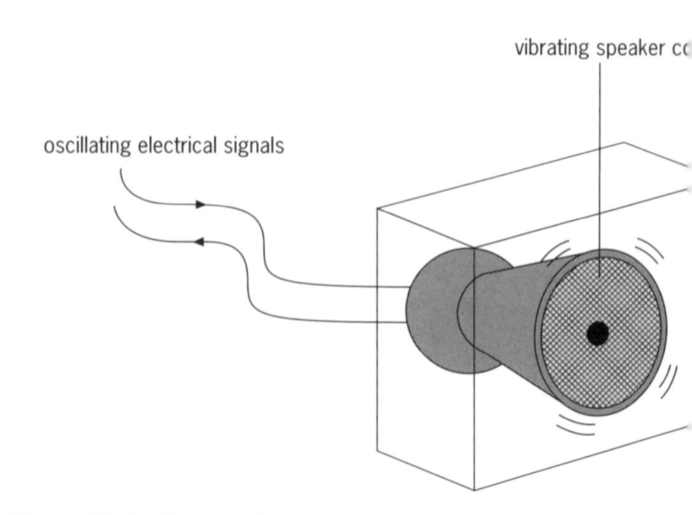

Figure 28.10 Forced vibration

Sometimes an object or a system can be forced to vibrate at a frequency other than its own natural frequency. For example, the cone of a loudspeaker is made to vibrate over a wide range of the frequencies in order to produce the sounds we hear.

If the driving frequency is the same as the natural frequency of the system being made to vibrate, the amplitude of vibration is a maximum. The system is said to be in **resonance**.

Barton's pendulums

The pendulum system illustrated in Figure 28.11 can be used to demonstrate resonance.

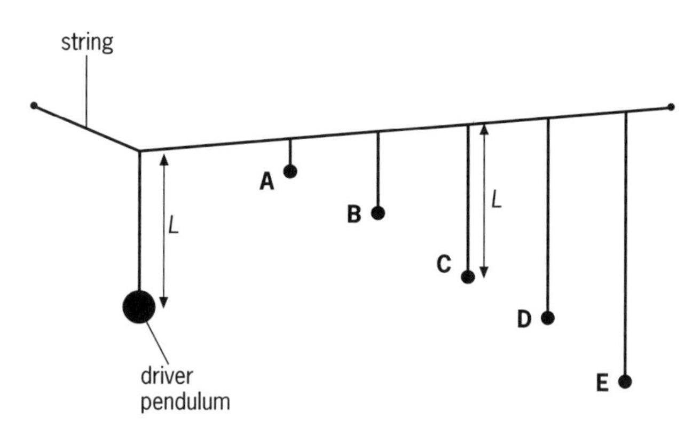

Figure 28.11 Barton's pendulums

The driver pendulum is pulled back and released. Its oscillating motion is transmitted along the supporting string to each of the other pendulums, causing them to oscillate. Pendulum C oscillates with a much larger amplitude than the other pendulums because it is the same length as the driver pendulum and therefore has the same natural frequency; pendulum C is in resonance.

If the driver pendulum is shortened, pendulum A or B can be made to resonate. If it is lengthened, pendulum D or E can be made to resonate.

Damping

Most oscillating systems have amplitudes which decrease with time. This decrease occurs because the system loses energy in overcoming resistive forces such as friction or air resistance. It is called **damping**. Damping does not affect the period of oscillation.

In some systems, such as pendulum clocks, damping can be a disadvantage. In others, such as the suspension system of a car, it can be very useful. With little or no damping a bump in a road could cause a car and its passengers to 'bounce' for too long. With too much damping the bumps would be hard and the ride uncomfortable. Figure 28.12 shows the result of different degrees of damping on the oscillations of a system.

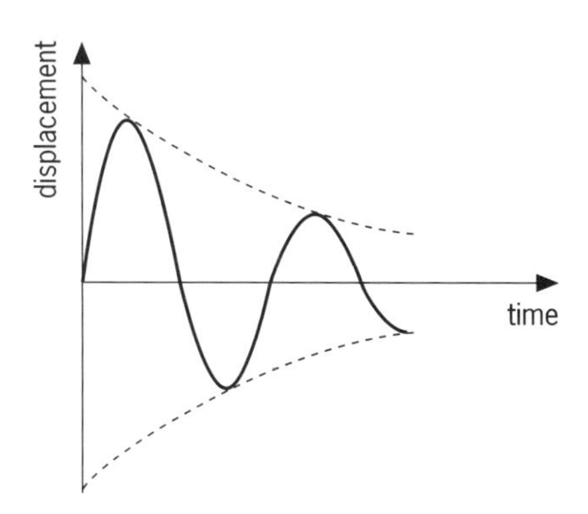

a Light damping

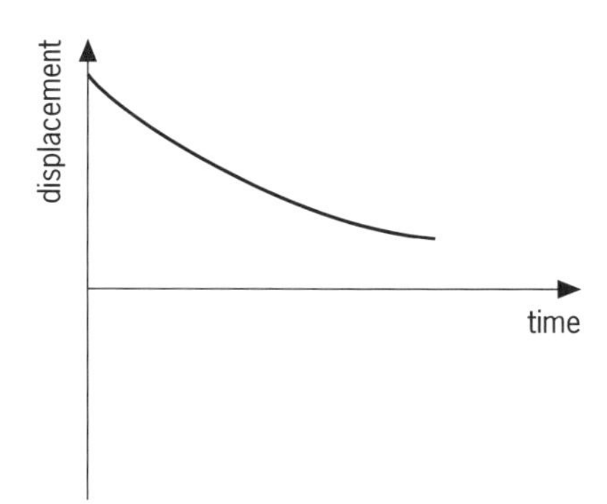

b Heavy damping

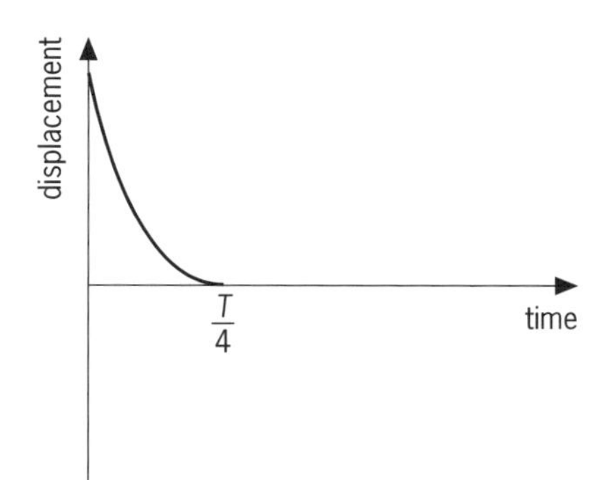

c Critical damping

Figure 28.12

Altering the degree of damping does not affect the natural frequency of a system; see Figure 28.13.

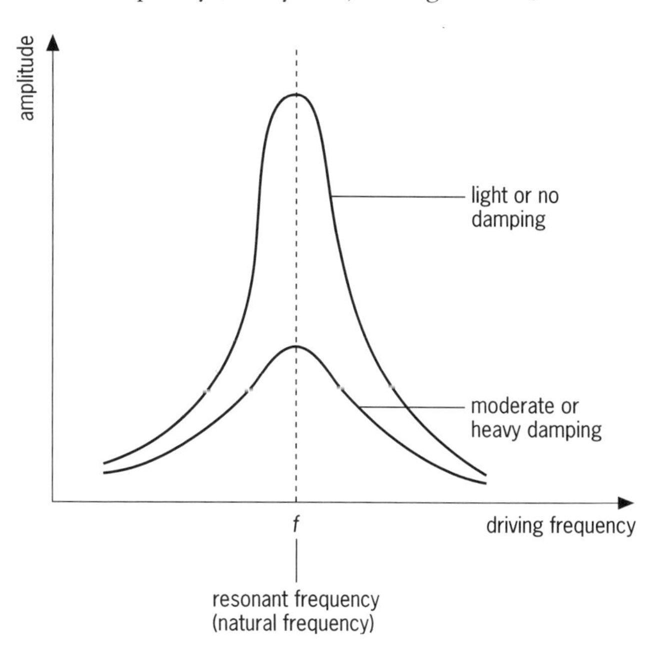

Figure 28.13

Level 1

1 Sketch two cycles of a displacement–time graph for an object performing shm, for which the displacement is zero at time zero.

- **a** Mark a position of maximum velocity, A.
- **b** Mark a position of zero velocity, B.
- **c** Mark a position of high acceleration, C, and of low acceleration, D.

2 Copy the axes in Figure 28.14 and sketch graphs of displacement, velocity and acceleration, all against time, for a pendulum bob. The first part of the displacement is shown. The period, T, and three instants of time t_1, t_2 and t_3 are shown. Choose suitable vertical scales.

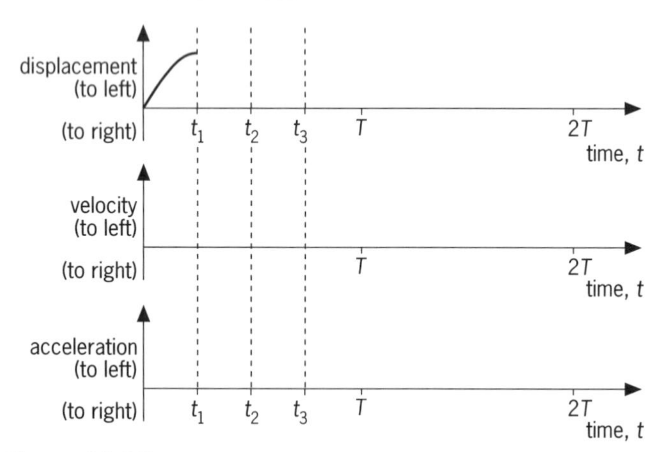

Figure 28.14

Comment on the value and direction of each quantity at the times t_1, t_2 and t_3.

3 A bungee jumper of mass 70 kg oscillates on the end of her bungee cord with a time period of 4.0 s. What is the force constant, k, of the cord?

4 When Galileo timed the incense holder swinging during a church service, what was the crucial observation that led to the development of the pendulum clock?

5 A mass of 30 g hanging from the spring of a newtonmeter oscillates with an amplitude of 0.040 m. The maximum force registered is 0.372 N.

- **a** Find the maximum acceleration.
- **b** Find the period, T.
- **c** Write the equation that describes the variation of displacement with time for this motion.

6 **a** In shm, the acceleration is proportional to the displacement. What must be the units of the constant of proportionality?

- **b** If one shm has a constant of proportionality 16 times larger than another, what does this tell you about the frequencies of the two motions?
- **c** Show that the expression for velocity

$$v = 2\pi f\sqrt{x_0^2 - x^2}$$

gives you the correct result when at the maximum ($x = x_0$) and minimum ($x = 0$) displacement.

- **d** Show that the maximum value of v is $x_0\omega$.

7

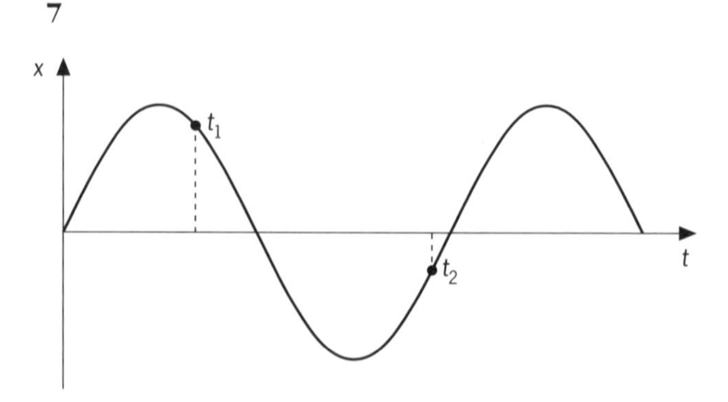

Figure 28.15

Figure 28.15 shows the displacement of an object performing shm, plotted against time.

- **a** At which time, t_1 or t_2 on the graph, does the object have the greater velocity?
- **b** At which of t_1 or t_2 does it have the greater acceleration?
- **c** Is the velocity at t_1 increasing?
- **d** Is the acceleration at t_1 increasing?

8 When the A-string (440 Hz) is plucked on a harp, the 880 Hz string vibrates as well. Explain this.

9 A standing wave on a string of length 1.5 m, which is fixed at both ends, has a frequency of 60 Hz and three 'loops' (antinodes).

- **a** Draw the wave and mark a wavelength on the diagram.
- **b** Calculate the wavelength.
- **c** Calculate the speed of the wave.
- **d** Will the string also resonate at 40 Hz and a standing wave be formed?
- **e** Will the string vibrate with eight 'loops' at 160 Hz?

10 The absorption curve in Figure 28.16 is obtained when radiation of different frequencies is incident on glass. It tells us something about the vibrations of the glass molecules.

- **a** Explain the shape of the curve.
- **b** In which region of the electromagnetic spectrum would you find the natural frequency of the molecules?

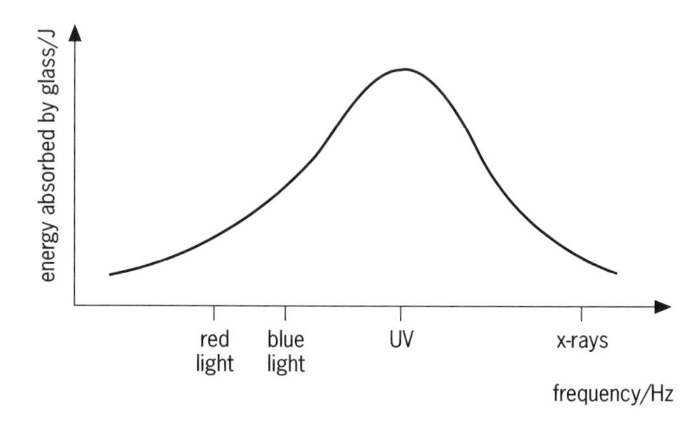

Figure 28.16

11 Some electrical circuits have a natural frequency and can resonate. Figure 28.17 shows the electrical oscillations dying away as energy is dissipated by the resistance of the circuit.

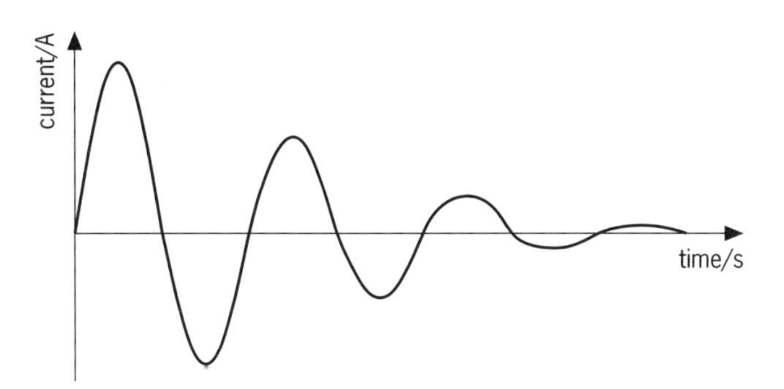

Figure 28.17

a What is this 'dying away' process called?

b Draw what you think would happen if a superconducting material could be used to make the wires of the circuit so that there was no resistance between the electrical components.

Level 2

12 Vibrations of the hydrogen chloride molecule, HCl, may be treated as a hydrogen atom oscillating at the end of a *fixed* spring, because the chlorine atom is so much heavier than the hydrogen atom. The force constant, k, of the bond joining the atoms is $540\,\mathrm{N\,m^{-1}}$ and the mass of the H atom is $1.66 \times 10^{-27}\,\mathrm{kg}$.

a Calculate the natural frequency of the system.

b Explain what you would expect to happen if radiation of exactly this frequency was incident on the molecule.

13 A mass of 0.25 kg oscillates horizontally at the free end of a spring with amplitude 0.20 m. It is on a frictionless surface. The force constant of the spring is $100\,\mathrm{N\,m^{-1}}$.

a Show that the maximum potential energy stored in the spring is 2.0 J.

b From kinetic energy considerations, find the maximum velocity of the mass.

c Show that this is the same value as that obtained by using $v_{max} = x_0\omega$.

14 The displacement of an oscillating object at time t is given by $x = 0.4\sin(6\pi t)$.

a What is the amplitude of the motion?

b What is the angular velocity?

c What is the frequency?

d At $t = 0.5$ s what is

i the displacement

ii the velocity

iii the acceleration?

Note: Use radians or change to degrees by π rad = 180°.

15 The graph in Figure 28.18 shows the way in which the square of the period, T, of a pendulum varies with its length, l, in metres. Use the data to calculate a value of g, the acceleration due to gravity.

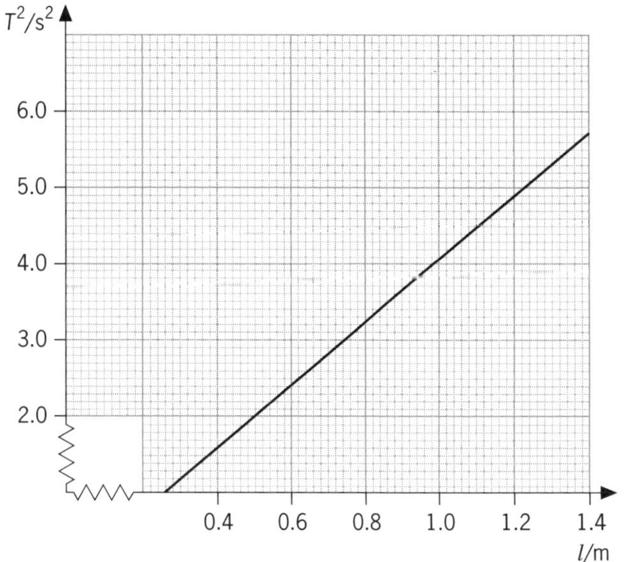

Figure 28.18

16

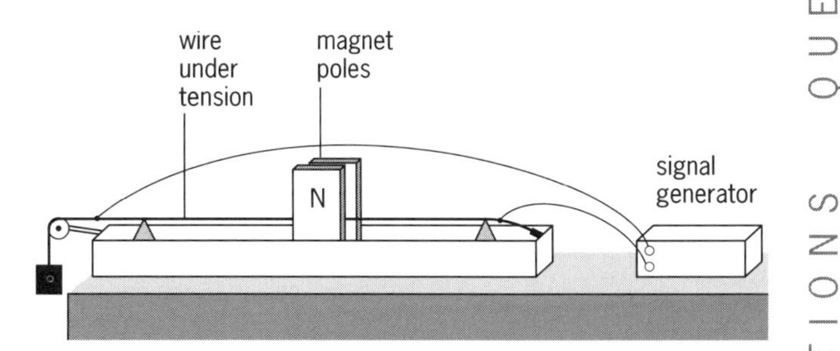

Figure 28.19

The apparatus shown in Figure 28.19 consists of a wire, in tension, with an alternating current of variable frequency flowing through it. A strong magnet is placed with its poles on either side of the centre of the wire.

a Describe what happens as the frequency of the current is increased, starting from a low value.

b The velocity of a wave on such a wire is given by the equation

$$v = \sqrt{\frac{\text{tension}}{\text{mass/unit length}}} = \sqrt{\frac{T}{m/l}}$$

Write an expression for the frequency of the fundamental mode of vibration for a wire of length l.

Hint: Use $v = f\lambda$.

17 An architect friend is telling you about his latest project and says he fitted 'dampers' to a tall chimney. What do these do?

18 The set of pendulums shown in Figure 28.20a is used to demonstrate resonance and damping. First the driver is set swinging and the motion observed, then the paper cones are slid down to the bobs and the process repeated. On axes like those in Figure 28.20b, plot the amplitude of vibration you would expect to see for each pendulum, with and without the cones.

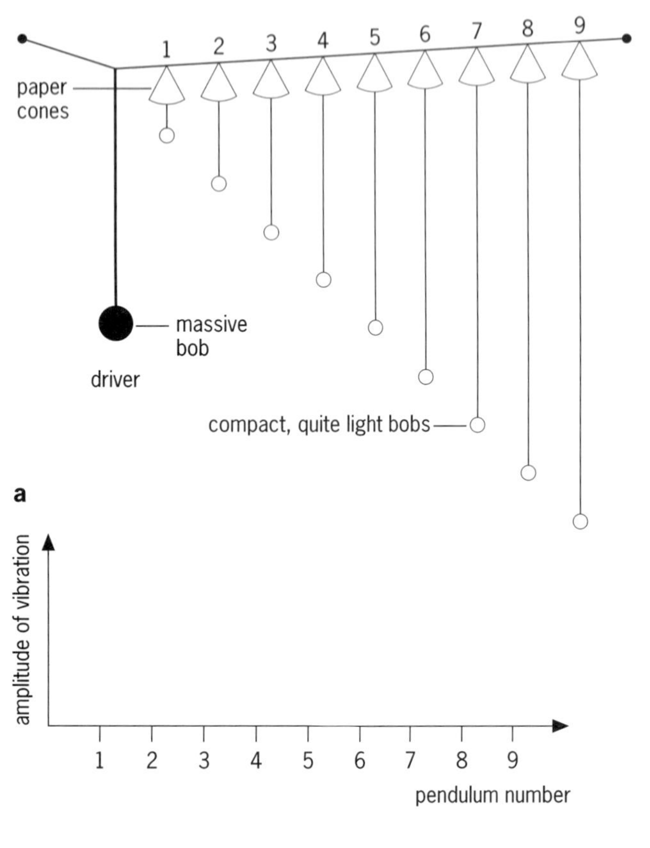

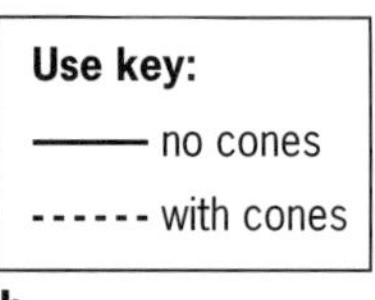

b

Figure 28.20

Level 3

19 Two identical masses are suspended using a total of four identical springs. One arrangement has two springs in series. A second arrangement has two springs in parallel. Comment on the force constant of each arrangement.

20

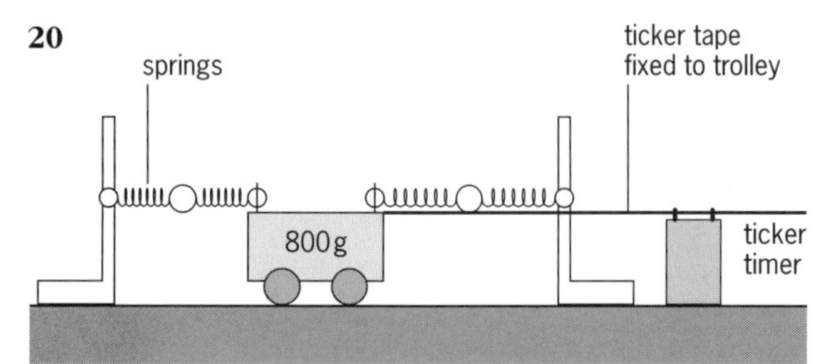

Figure 28.21

A trolley of mass 800 g oscillates between the stands as shown in Figure 28.21. The ticker tape is pulled through the timer on one half cycle, from right to left. The timer makes 50 dots per second.

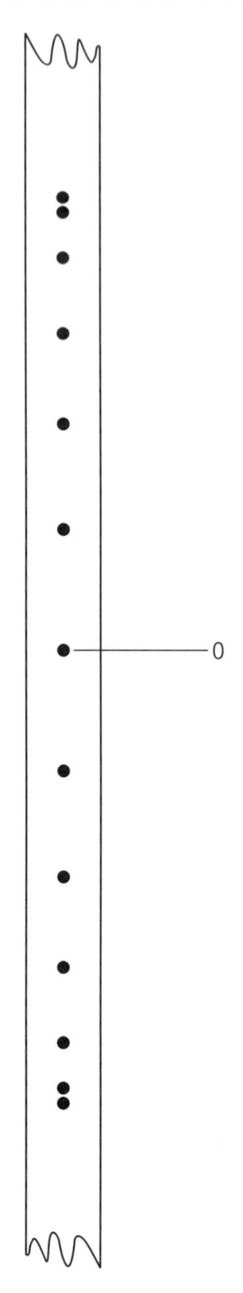

Figure 28.22

The tape in Figure 28.22 is reproduced **one-third real size**.

a What is the amplitude of the vibration?
b What is the frequency?
c What is the maximum velocity?
d Calculate the maximum acceleration and the maximum force experienced by the trolley.

21 The acceleration due to gravity on Mars is 0.37 times that on Earth.

a What would be the period, T, on Mars, of a pendulum that has a period of 0.48 s on Earth?
b Would there be a similar effect on the period of oscillations of a mass on a spring?

22 A Perspex tube 1.0 m long is clamped vertically with the lower end in a cylinder of water a little over 1.0 m long, Figure 28.23. The Perspex tube can be raised or lowered to change the length of the air column above the water.

A loudspeaker is fixed just above the top of the tube and is connected to a signal generator set at 500 Hz. The velocity of sound is $330\,m\,s^{-1}$.

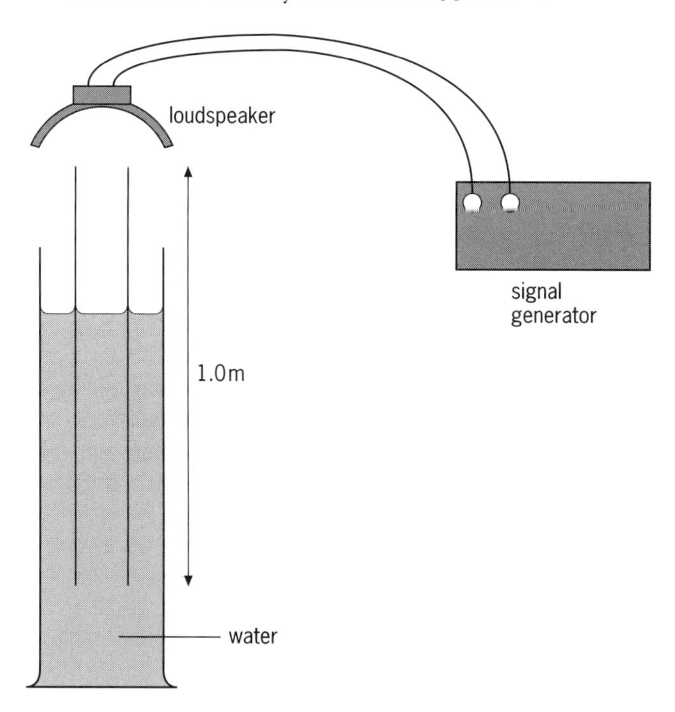

Figure 28.23

Describe, and explain with diagrams, what will be heard as the length of the air column is increased from the minimum.

23 Various parts of a washing machine, including the side panels, can flex and vibrate, and the whole machine can have its own natural frequency. Take this to be 75 rpm for yours. The drum of your washing machine is able to deal with a total load, including water, of 24 kg. It is supported by two large springs, Figure 28.24, but one of them has broken.

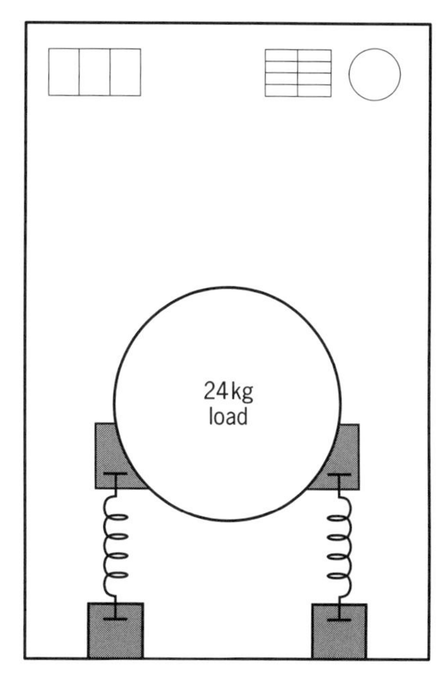

Figure 28.24

The service engineer has all sorts of different springs, labelled with their force constants, but doesn't know which one to use. What advice can you give him?

Level 1

1

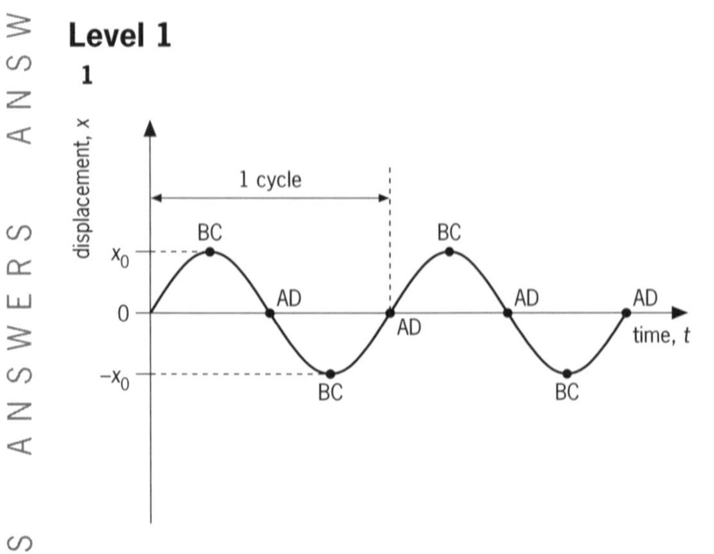

Figure 28.25

a The maximum velocity is at the steepest slope (A); this occurs when the object passes through the equilibrium position where the displacement, $x = 0$.

b The velocity is zero at the turning points of the graph (B); at these extremes of the motion, when $x = x_0$ (the amplitude), the object stops and turns round.

c Acceleration is proportional to displacement for shm, so at points C the acceleration will be high. This is when the displacement, x, is equal to x_0, the amplitude; here the object is stopping and turning round.

Low acceleration occurs at points D, when the displacement is zero; this is where the object goes through the equilibrium position. The object does not change speed or direction, so the acceleration must be zero.

*Note that the bottom of the curve is **not** the smallest displacement. It represents a large displacement 'in the other direction'.*

2

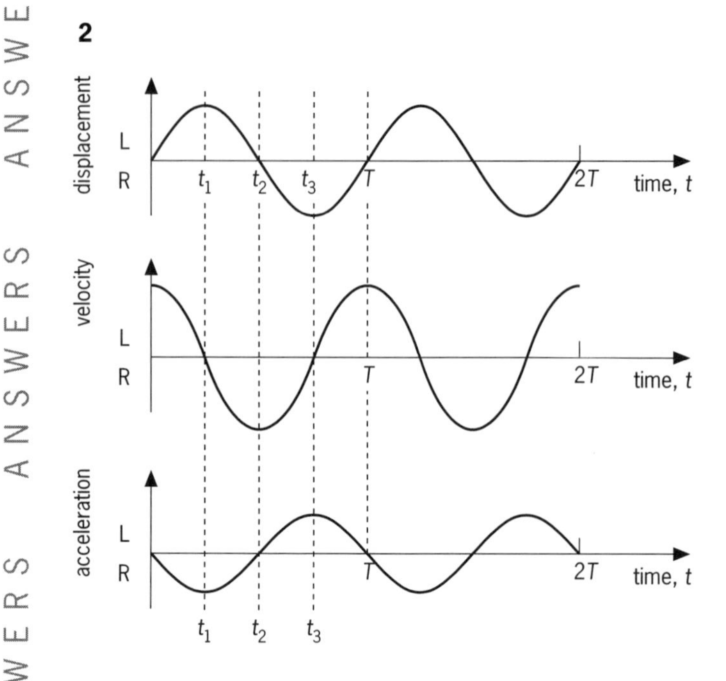

Figure 28.26

At time t_1
The displacement is a maximum to the left; it is equal to the amplitude.
The velocity is zero; the bob is turning round.
The acceleration is a maximum to the right, bringing the bob to a halt and starting to move it to the right.

At time t_2
The displacement is zero; the bob is at the central, equilibrium position.
The velocity is a maximum to the right, as the bob is swinging from left to right.
The acceleration is zero as the velocity is not changing.

At time t_3
The displacement is to the right but the bob is moving towards the left.
This means that the velocity is to the left and increasing.
The acceleration must be to the left to increase the velocity in that direction.

3 The period of oscillation, T, of a mass, m, suspended on a 'spring' is given by $T = 2\pi\sqrt{m/k}$, where k is the force constant. To derive an equation for the force constant, first square both sides of the equation to give $T^2 = 4\pi^2(m/k)$, then rearrange this equation to give

$$k = \frac{4\pi^2}{T^2}m$$

In this example the bungee jumper has a mass, $m = 70\,\text{kg}$ and the period, $T = 4.0\,\text{s}$, so

$$k = \frac{4\pi^2 \times 70\,\text{kg}}{(4.0\,\text{s})^2} = 173\,\text{kg}\,\text{s}^{-2} = \mathbf{173\,N\,m^{-1}}$$

(Remember that $N = kg\,m\,s^{-2}$ so $kg = N\,m^{-1}\,s^2$
$\therefore\ kg\,s^{-2} = N\,m^{-1}\,s^2\,s^{-2} = N\,m^{-1}$.)

4 Galileo saw that even though the amplitude of the swings got smaller the incense holder still took the same length of time to complete each oscillation. This meant that the speed must change continuously to compensate exactly for any change in amplitude! In a pendulum-controlled clock, during each oscillation a small amount of energy is transferred to the pendulum, from a spring or a weight, to compensate for any small losses that might otherwise eventually bring the pendulum, and the clock, to a stop. Small differences between the amplitudes of successive swings do not affect the accuracy of the clock – which is just what Galileo observed.

5 a As the mass oscillates, the reading of the newtonmeter will oscillate too. The force on the spring will be greatest when the mass is at the bottom of its oscillation, when the extension of the spring in the newtonmeter is greatest. The unbalanced restoring force and the acceleration of the mass are linked by the equation force = mass × acceleration or $F = ma$, which you can rearrange to give

$$a = \frac{F}{m}$$

This means that for a given mass, m, the acceleration will be a maximum when the applied force is a maximum. In this example $m = 30\,g = 0.030\,kg$, and the maximum restoring force, taking g as $10\,m\,s^{-2}$, is $F_{max} = 0.300 - 0.372 = -0.072\,N$. So

$$a_{max} = -\frac{0.072\,N}{0.030\,kg} = \mathbf{-2.4\,m\,s^{-2}}$$

($N = kg\,m\,s^{-2}$)

b For shm, the maximum acceleration,

$$a_{max} = -\omega^2 x_0 = -(2\pi f)^2 x_0$$

where x_0 is the amplitude and ω is the angular velocity of the oscillation which is equal to $2\pi f$.

Remember that the minus sign in this equation indicates that the acceleration is in the opposite direction to the displacement.

You can rearrange this equation to give

$$\omega^2 \quad \text{or} \quad (2\pi f)^2 = \frac{-a_{max}}{x_0}$$

In this example the amplitude, $x_0 = 0.040\,m$, and $a_{max} = 2.4\,m\,s^{-2}$ in the opposite direction, so if we take x as positive, we must take a as negative, or vice versa. We will take $x_0 = -0.040\,m$, so $a_{max} = +2.4\,m\,s^{-2}$.

$$\therefore \omega^2 = \frac{-a_{max}}{x_0} = \frac{-2.4\,m\,s^{-2}}{-0.040\,m} = +60\,s^{-2}$$

$$\omega = \sqrt{60\,s^{-2}} = 7.7\,s^{-1}$$

But $\omega = \dfrac{2\pi}{T}$

$$\therefore \quad T = \frac{2\pi}{\omega} = \frac{2\pi}{7.7\,s^{-1}} = \mathbf{0.81\,s}$$

c The displacement at time t is given by $x = x_0 \sin \omega t$, where x_0 is the amplitude. In this example

$$x = 0.04 \sin 7.7t$$

You could use $x = x_0 \cos \omega t$. It just means that the phase is different. At $t = 0$, $\cos \omega t$ is a maximum but $\sin \omega t$ is zero.

6 a $a = kx$ which you can rearrange to give $k = \dfrac{a}{x}$

Acceleration, a, has the unit $m\,s^{-2}$ and displacement, x, has the unit m. So

$$k = \frac{a(m\,s^{-2})}{x(m)} = \left(\frac{a}{x}\right) s^{-2}$$

k has the unit $\mathbf{s^{-2}}$

b The constant, $k = \omega^2 = (2\pi f)^2 = 4\pi^2 f^2$, so if k is 16 times larger then the frequency, f, must be 4 times larger since $k \propto f^2$ or $f \propto \sqrt{k}$.

c At maximum displacement, $x = x_0$ so $x_0^2 - x^2 = 0$ and

$$v = 2\pi f \sqrt{0} = 0$$

which is correct.

At minimum displacement (the equilibrium position), $x = 0$, so

$$v = 2\pi f \sqrt{x_0^2} = \pm 2\pi f x_0 = \pm \omega x_0$$

which is also correct.

d These are the largest values that v can have because x_0 is the largest value of x.

7 a At t_2 the displacement curve is steeper which means that the displacement is changing more quickly: the object has the **greater velocity at t_2**.

b At t_1 the displacement is greater so, as acceleration is proportional to displacement, the **acceleration is greater at t_1**.

c **At t_1 the velocity is increasing** because the slope is steeper to the right.

d The displacement is decreasing at t_1 so, because acceleration is proportional to the displacement, the **acceleration at t_1 is not increasing**.

8 When a string vibrates, it vibrates not only in the fundamental mode (antinode in the centre, nodes at each end) but also in several more complex modes, which give rise to **overtones**. The overtones are all multiples of the fundamental frequency, so 880 Hz vibrations will be present in the vibrations of the 440 Hz string and will cause the 880 Hz string to **resonate**.

9 a

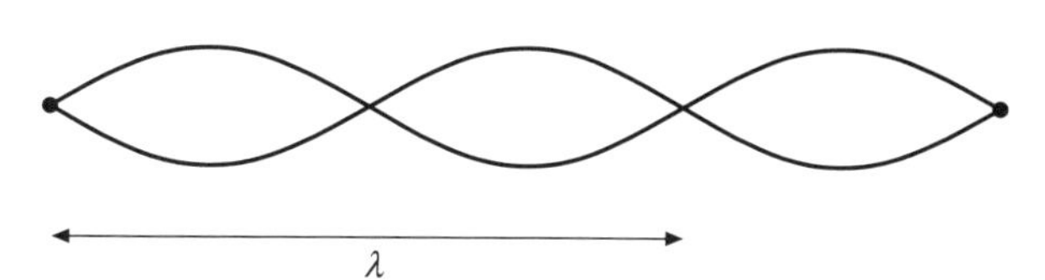

Figure 28.27

b The wavelength is equal to $\frac{2}{3}$ of the length of the string:

$$\lambda = \tfrac{2}{3} \times 1.5\,m = \mathbf{1.0\,m}$$

c The speed of the wave is given by

$$v = f\lambda = 60\,Hz \times 1.0\,m = \mathbf{60\,m\,s^{-1}}$$

d You need to see if the wavelength of the new frequency will 'fit' onto the string. The frequency has changed to 40 Hz but the velocity of the wave depends upon the string itself, which hasn't changed, so the velocity stays the same. The new wavelength is

$$\lambda = \frac{v}{f} = \frac{60\,m\,s^{-1}}{40\,s^{-1}} = 1.5\,m$$

so the string will oscillate as shown in Figure 28.28.

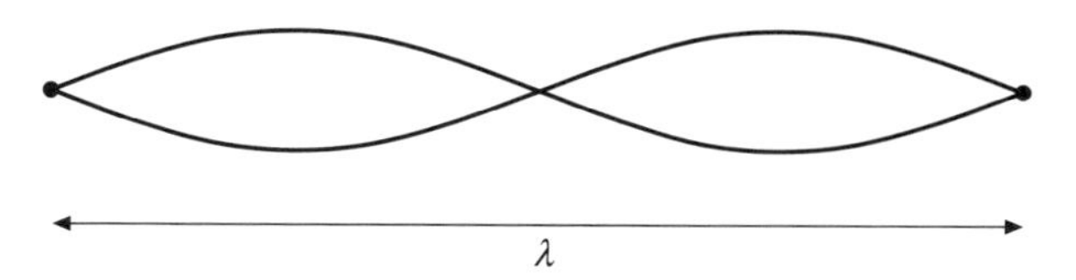

Figure 28.28

The string **will resonate at 40 Hz**; this is the first overtone.

e 160 Hz is 4 times the frequency in d.

$$f_1\lambda_1 = v = f_2\lambda_2$$

so the wavelength will be $\frac{1}{4}$ of the value in d and the string **will vibrate with 8 loops**.

10 a The curve is a typical resonance curve. (Compare it with Figure 28.13.) At frequencies far from the natural frequency of vibration of the molecules, very little motion is caused and there is little absorption of the radiation. The glass is transparent to radiation at these frequencies.

When the frequency of the incoming radiation is close to the natural frequency of the molecules, they are sent into large vibrations. The molecules absorb the energy from the incoming radiation.

b It can be seen from the curve that the glass molecules' natural frequency is in the **uv** (**ultraviolet**) region. (This absorption explains why you can't get a suntan through glass; but the plants in a greenhouse don't seem to mind.)

11 a This process is called **damping**.

b If the same electrical components were used, the frequency of oscillation of the circuit would be the same. If no energy were converted to heat by resistance, the amplitude would remain constant and the oscillations could go on for ever.

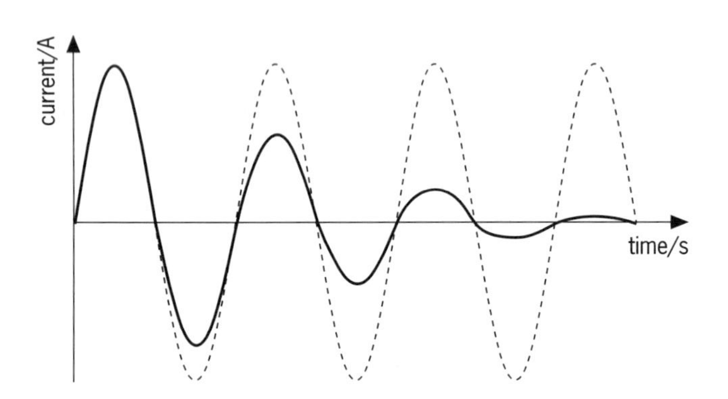

Figure 28.29

In fact, in the real world, there will always be some losses and nothing lasts for ever.

Level 2

12 a The period of oscillation of a mass, m, suspended on a 'spring' is given by $T = 2\pi\sqrt{m/k}$, where k is the force constant. The frequency of the oscillation is

$$f = \frac{1}{T} = \frac{1}{2\pi\sqrt{m/k}} = \frac{1}{2\pi}\sqrt{\frac{k}{m}}$$

In this example $k = 540\,\text{N m}^{-1}$ and $m = 1.66 \times 10^{-27}\,\text{kg}$, so

$$f = \frac{1}{2\pi}\sqrt{\frac{540\,\text{N m}^{-1}}{1.66 \times 10^{-27}\,\text{kg}}} = \mathbf{9.2 \times 10^{13}\,Hz}$$

$(\text{kg} = \text{N m}^{-1}\,\text{s}^2)$

b Because the frequency matches the natural frequency of the bond, you would expect the molecule to be set vibrating – to **resonate** – and thereby **absorb energy** from the radiation. This would show up as an absorption line in a spectrum, because that particular frequency of radiation would have been removed.

13 a The elastic potential energy, E, stored in a spring at displacement x is $E = \frac{1}{2}Fx$ (see topic 26) or, since $F = kx$, substituting for F, $E = \frac{1}{2}kx^2$.

The maximum elastic potential energy will be stored when $x = x_0$, the amplitude. The force needed to extend the spring by the amplitude is therefore kx_0. In this example $k = 100\,\text{N m}^{-1}$ and $x_0 = 0.20\,\text{m}$, so the maximum force $F_{max} = 100\,\text{N m}^{-1} \times 0.20\,\text{m} = 20\,\text{N}$.

So, according to the first expression,

$$E_{max} = \tfrac{1}{2} \times 20\,\text{N} \times 0.20\,\text{m} = \mathbf{2.0\,J}$$

or, according to the second expression,

$$E_{max} = \tfrac{1}{2} \times 100\,\text{N m}^{-1} \times (0.20\,\text{m})^2 = \mathbf{2.0\,J}$$

b In shm the total energy remains constant, but continuously changes from being all potential energy to being all kinetic energy, and back again (see Figure 28.5). This means that the value of the maximum kinetic energy, which will occur when the velocity is a maximum, is the same as the value of the maximum potential energy.

So, for the mass, m, on the end of the spring,

maximum kinetic energy, $\frac{1}{2}m(v_{max})^2 = E_{max}$

which you can rearrange to give

$$(v_{max})^2 = \frac{E_{max}}{\frac{1}{2}m}$$

But $m = 0.25\,\text{kg}$ and $E_{max} = 2.0\,\text{J}$ from part a, so

$$(v_{max})^2 = \frac{2.0\,\text{J}}{\frac{1}{2} \times 0.25\,\text{kg}} = 16\,\text{m}^2\,\text{s}^{-2}$$

$(\text{J} = \text{N m} = \text{kg m s}^{-2}\,\text{m} = \text{kg m}^2\,\text{s}^{-2})$

$$\therefore v_{max} = \sqrt{16\,\text{m}^2\,\text{s}^{-2}} = \mathbf{4.0\,m\,s^{-1}}$$

c The period of oscillation is

$$T = 2\pi\sqrt{\frac{m}{k}}$$

In this example

$$T = 2\pi\sqrt{\frac{0.25\,\text{kg}}{100\,\text{N m}^{-1}}} = 2\pi\sqrt{0.0025\,\text{s}^2}$$

$$= 0.31\,\text{s}$$

But

$$\omega = \frac{2\pi}{T} = \frac{2\pi}{0.31\,\text{s}}$$

so, according to the equation given,

$$v_{max} = x_0\omega = 0.20\,\text{m} \times \frac{2\pi}{0.31\,\text{s}} = \mathbf{4.0\,m\,s^{-1}}$$

14 A general form of the equation for shm is $x = x_0 \sin \omega t$. In this example $x = 0.4 \sin 6\pi t$.

a The amplitude, x_0, is **0.4** and should be in m.

b $\omega = \mathbf{6\pi\,s^{-1}}$. This is what the time, t, is multiplied by.

c You can rearrange the equation $\omega = 2\pi f$ to give $f = \dfrac{\omega}{2\pi}$. So, in this example, $f = 6\pi/2\pi = \mathbf{3\,Hz}$.

d When $t = 0.5\,\text{s}$:

i $x = 0.4\,\text{m} \times \sin(6\pi \times 0.5) = 0.4\,\text{m} \times \sin 3\pi = \mathbf{0}$

π radians = 180°. The sine of any whole-number multiple of π is zero.

ii $v = \omega x_0 \cos \omega t = 6\pi\,\text{s}^{-1} \times 0.4\,\text{m} \times \cos 3\pi$
But $\cos 3\pi = -1$

$\therefore v = \mathbf{-7.5\,m\,s^{-1}}$

In this case the minus sign tells us that the object is moving in the direction that will decrease the value of the displacement, which is actually zero at this moment in time.

iii Acceleration is proportional to displacement so the acceleration must also be **zero**. You should however remember that the equation for acceleration is

$$a = -\omega^2 x_0 \sin \omega t$$

This includes the $\sin \omega t$ term, which is zero when $\omega t = 3\pi$.

15 The period, T, of a simple pendulum is given by

$$T = 2\pi \sqrt{\frac{l}{g}}$$

If you square this equation you will get

$$T^2 = \frac{4\pi^2}{g} \times l$$

which tells you that if you plot a graph of T^2 against l you will get a straight line with a gradient of $4\pi^2/g$.
The gradient of the straight line in Figure 28.18 is

$$\frac{3.3\,\text{s}^2}{0.8\,\text{m}} = 4.125\,\text{m}^{-1}\,\text{s}^2$$

$5.3\,\text{s}^2 - 2.0\,\text{s}^2 = 3.3\,\text{s}^2$ on the T^2-axis; $1.3\,\text{m} - 0.5\,\text{m} = 0.8\,\text{m}$ on the l-axis.)

The gradient $= 4\pi^2/g$, which you can rearrange to give $g = 4\pi^2/\text{gradient}$. From the data in this example,

$$g = \frac{4\pi^2}{4.125} = \mathbf{9.6\,m\,s^{-2}}$$

(This answer is about 2% smaller than the value we normally use.)

16 a There will be a buzzing and the wire will vibrate each time it resonates with the applied frequency. There is an alternating force on the wire, at the frequency of the generator, because there is a force on a current-carrier in a magnetic field. Because this force is applied at the centre of the wire you will see an antinode there. Modes of vibration that have an antinode at the centre will resonate, as shown in Figure 28.30.

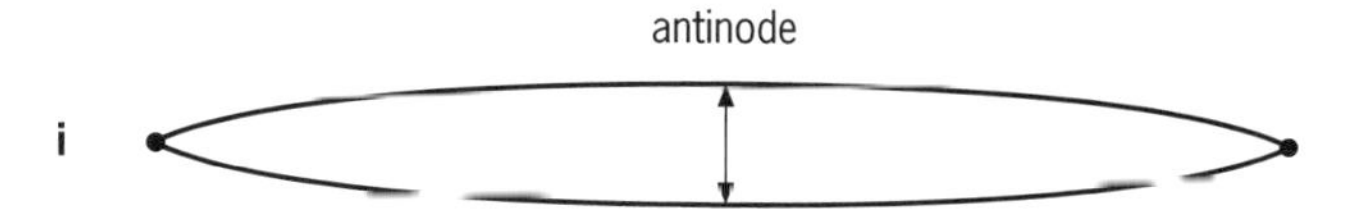

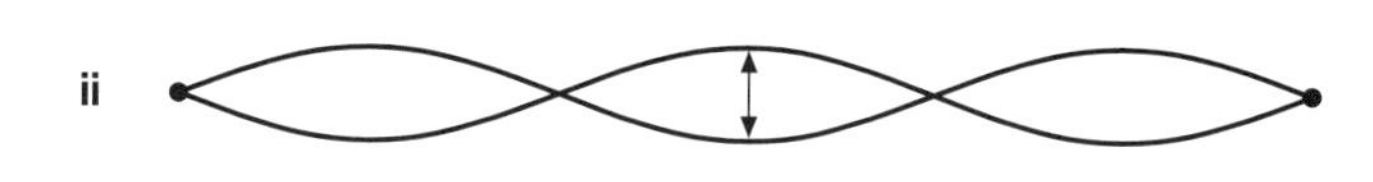

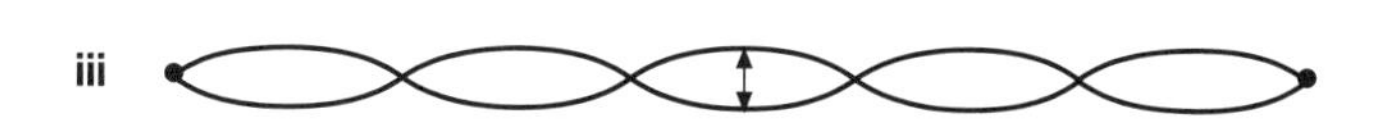

Figure 28.30

b For the fundamental mode, shown in Figure 28.30i, the wavelength is twice the length of the wire: $\lambda = 2l$.
The frequency is therefore

$$f = \frac{v}{\lambda} = \frac{1}{2l}\sqrt{\frac{T}{m/l}}$$

17 Dampers are devices that are used to prevent oscillations. They often convert the energy of vibration into heat using friction or the elastic properties of rubber.
(A tall chimney in a wind creates vortices which are shed alternately on either side and which push it to and fro at a frequency

$$f = 0.2 \times \frac{\text{wind velocity}}{\text{chimney diameter}}$$

Obviously the natural frequency of the chimney should be made to be well away from this value.)

18

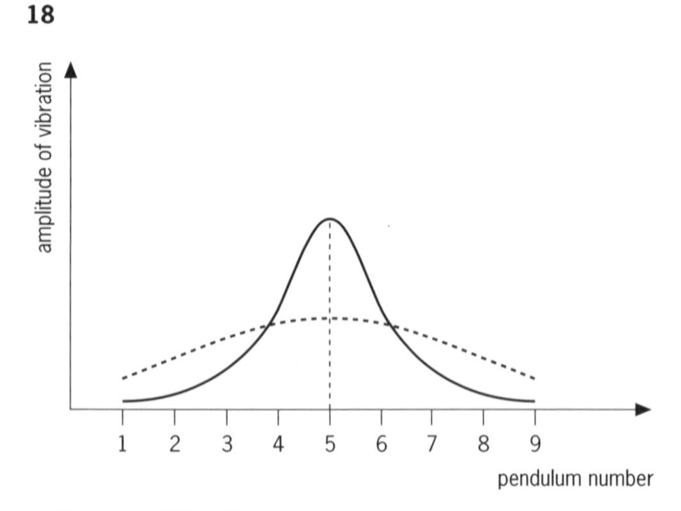

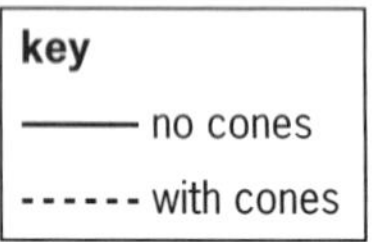

Figure 28.31

Without the cones there is little damping and you should see a sharp resonance. This means that only the pendulum that has the same length as the driver will have large amplitude.

With the cones, the damping is increased and the resonance is 'smeared out' (Figure 28.31), so that several lengths, centred on the resonant length, vibrate with medium amplitudes.

Level 3

19 Compare the performance of the two arrangements with the performance of a single spring supporting the same total mass.

With two springs in series the extension will be double, so the spring constant for this arrangement is half the constant for a single spring:

$$k_{\text{series}} = \frac{k_{\text{single}}}{2}$$

With two springs in parallel the extension will be halved, so the force constant is doubled:

$$k_{\text{parallel}} = 2 \times k_{\text{single}}$$

20 a From the centre of the tape to the furthest dot is 6.0 cm. Scaled up three times this gives an amplitude x_0 of 18 cm = **0.18 m**.

b The interval between dots is one-fiftieth of a second = 0.020 s. There are six gaps along half the tape, which represents one quarter of a cycle. So the time period of the oscillation is

$$T = 4 \times 6\,(\text{gaps}) \times 0.020\,(\text{s gap}^{-1}) = 0.48\,\text{s}$$

and the frequency is

$$f = \frac{1}{T} = \frac{1}{0.48\,\text{s}} = \mathbf{2.1\,Hz}$$

c From looking at the tape, you can see that the largest gaps, which correspond to the highest velocity, are on either side of the central dot and are 1.6 cm = 0.016 m wide. These measurements must be scaled up by a factor of three from the drawing to get the actual gaps, which are 4.8 cm = 0.048 m.

The dots are at time intervals of 0.02 seconds, so

$$v_{\max} = \frac{\text{max. displacement}}{\text{time}} = \frac{0.048\,\text{m}}{0.02\,\text{s}} = \mathbf{2.4\,m\,s^{-1}}$$

d The acceleration at any point is given by $a = -\omega^2 x$, where $\omega = 2\pi f$. The maximum acceleration occurs when $x = x_0$, the amplitude. From part b, $f = 2.1$ Hz. So

$$\begin{aligned} a_{\max} &= -\omega^2 x_0 \\ &= -(2\pi \times 2.1)^2\,\text{Hz}^2 \times 0.18\,\text{m} \\ &= \mathbf{-31.3\,m\,s^{-2}} \end{aligned}$$

The minus sign means the direction is opposite to that of the displacement.

The size of the maximum force is

$$ma_{\max} = 0.80\,\text{kg} \times 31.3\,\text{m\,s}^{-2} = \mathbf{25\,N}$$

The force is in the same direction as the acceleration.

21 a The period of oscillation of a pendulum of length l on Earth is

$$T_E = 2\pi\sqrt{\frac{l}{g}}$$

where g is the acceleration due to gravity on Earth.

The acceleration due to gravity on Mars, $g_M = 0.37 \times g$. So the period of the same pendulum on Mars is

$$T_M = 2\pi\sqrt{\frac{l}{0.37g}}$$

Comparing T_M with T_E the g's and l's cancel out, leaving

$$\frac{T_M}{T_E} = \sqrt{\frac{1}{0.37}} = 1.64$$

or

$$T_M = 1.64T_E = 1.64 \times 0.48\,\text{s} = \mathbf{0.79\,s}$$

b For a spring the time period is given by $T = 2\pi\sqrt{m/k}$. Neither the mass nor the force constant of the spring is affected by a change in the acceleration due to gravity, so the period of oscillation stays the same.

22 First, you need to find the wavelength of the sound coming from the speakers. This is

$$\lambda = \frac{v}{f} = \frac{330\,\text{m}\,\text{s}^{-1}}{500\,\text{s}^{-1}} = 0.66\,\text{m} = 66\,\text{cm}$$

When a standing wave of this wavelength can be set up in the tube there will be a louder sound because the air column resonates with the note from the speaker. There will always be a node at the lower end of the column of air, where the water surface closes the tube and stops the vibrations of the air molecules. There will always be an antinode at the top end.

The possible modes of vibration are shown in Figure 28.32. You would expect to hear the first loud sound at $\lambda/4$ (= 16.5 cm) and then as each half wavelength (= 33 cm) is added, so at 49.5 cm and 82.5 cm (or thereabouts, because the antinode is not exactly at the top of the tube).

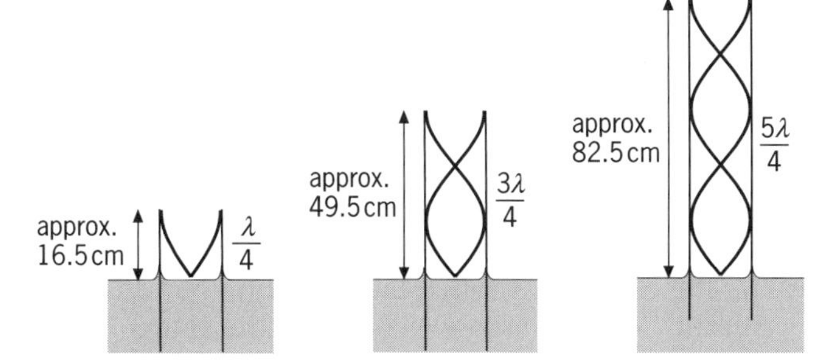

Figure 28.32

23 You need to avoid a spring which, with the load described, oscillates at the natural frequency of the machine. The natural frequency of a spring with force constant, *k*, and an attached mass, *m*, is given by the equation

$$f = \frac{1}{2\pi}\sqrt{\frac{k}{m}}$$

You can rearrange this to give the value of the force constant, *k*, that will give resonance for a particular combination of frequency and mass:

$$k = 4\pi^2 f^2 m$$

The resonant frequency of the machine is 75 rpm (revolutions per minute), equivalent to 1.25 Hz. Each spring supports about 12 kg. So, to avoid the natural frequency, you should choose a spring which doesn't have a force constant equal or near to

$$4\pi^2 \times (1.25\,\text{s}^{-1})^2 \times 12\,\text{kg} = 740\,\text{N}\,\text{m}^{-1}$$

29 Capacitors

A capacitor is a device which is used to store electric charge. It usually consists of two metal plates placed close together with an insulator between them, Figure 29.1a. The insulator may be air but is often a thin sheet of plastic. If we want to store a large amount of charge, a capacitor with large metal plates is needed. In order to keep these compact the plates are often 'rolled up' as shown in Figure 29.1b.

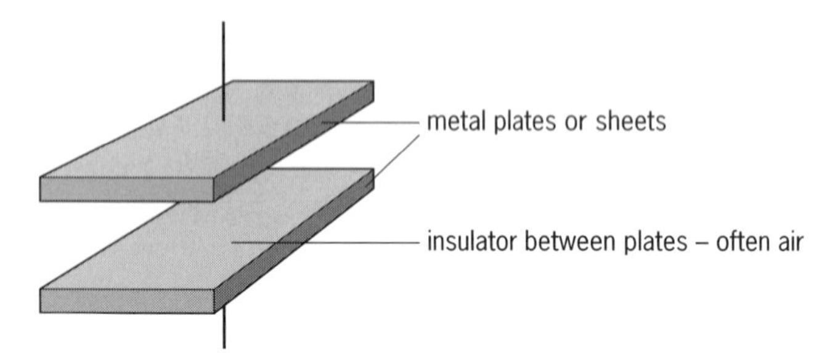

a Basic structure of a parallel plate capacitor

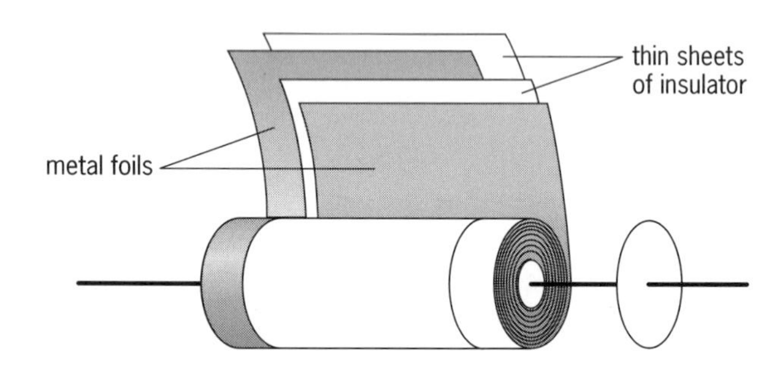

b Structure of a cylindrical capacitor

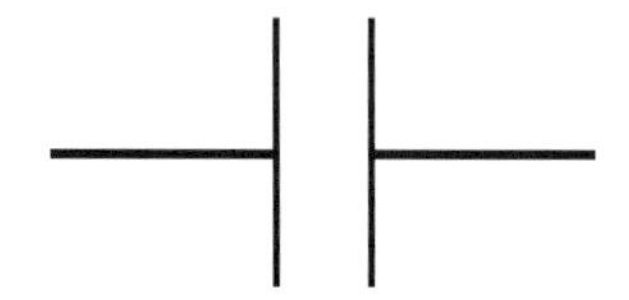

c Circuit symbol for a capacitor

Figure 29.1

If a battery is connected directly across the plates of a capacitor there is a momentary flow of current. Electrons flow from the negative side of the battery onto one of the plates, while at the same time an equal number of electrons flow off the second plate to the positive side of the battery. The overall effect of this flow of charge is that the plates are given equal and opposite charges. When fully charged the pd across the plates due to these charges is equal to the pd of the battery.

The measure of the extent to which a capacitor can store charge is called its capacitance C, defined as

$$C = \frac{Q}{V}$$

and measured in **farads** (F). If the pd across the plates of a capacitor increases by 1 volt when 1 coulomb of charge is added to it the capacitance of the capacitor is 1 farad.

Example 1

A charge of 200 C stored on a capacitor creates a pd of 100 V across its plates. Calculate the capacitance of this capacitor.

$$C = \frac{Q}{V}$$

$$= \frac{200\,\text{C}}{100\,\text{V}}$$

$$= 2\,\text{F}$$

Practical capacitors have capacitances much smaller than this. Usually they are measured in μF (10^{-6} F) or pF (10^{-12} F).

Most capacitors have marked on them their **maximum working voltage**. If too much charge is stored on them and the pd exceeds this maximum value, there is a danger that the insulation between the two plates will break down and allow the charge to leak across.

Capacitance and permittivity

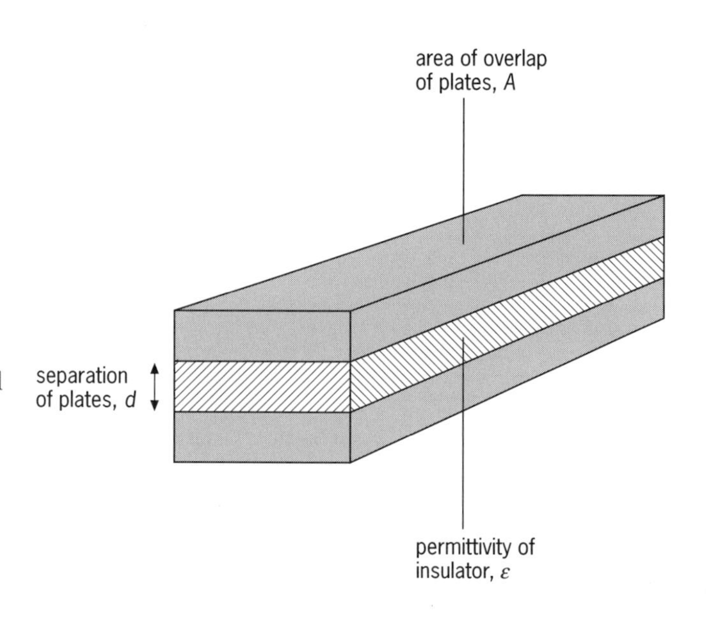

Figure 29.2

The capacitance of a parallel plate capacitor depends upon

- the area of overlap of the plates, A; the larger the overlap, the greater the capacitance.

  $C \propto A$

- the distance between the plates, d; the closer the plates, the greater the capacitance.

  $C \propto \frac{1}{d}$

- the type of insulator between the plates.

If the insulation is a vacuum we can write the equation relating all these factors as

$$C = \frac{\epsilon_0 A}{d}$$

where ϵ_0 is the constant of proportionality and is called the **permittivity of free space**.

Example 2

Calculate the capacitance of a parallel plate air capacitor consisting of two metal plates 1.0×10^{-4} m apart and of surface area $5.0 \times 10^{-3}\,m^2$. Assume that the permittivity of air has the same value as the permittivity of free space, $\epsilon_0 = 8.9 \times 10^{-12}\,F\,m^{-1}$.

$$C = \frac{\epsilon_0 A}{d}$$

$$= \frac{8.9 \times 10^{-12}\,F\,m^{-1} \times 5.0 \times 10^{-3}\,m^2}{1.0 \times 10^{-4}\,m}$$

$$= 450\,pF$$

If mica is placed between the plates of an air capacitor its capacitance would be seven times greater. We describe this situation by saying that mica has a **relative permittivity** (ϵ_r) or **dielectric constant** of 7.

For a capacitor with an insulating medium between its plates the above equation is then modified to

$$C = \frac{\epsilon_r \epsilon_0 A}{d}$$

Example 3

Calculate the capacitance of a parallel plate capacitor consisting of two metal plates of surface area $5.0 \times 10^{-3}\,m^2$ and separated by a piece of polythene 5.0×10^{-4} m thick. ($\epsilon_0 = 8.9 \times 10^{-12}\,F\,m^{-1}$, $\epsilon_r = 2.3$)

$$C = \frac{\epsilon_r \epsilon_0 A}{d}$$

$$= \frac{2.3 \times 8.9 \times 10^{-12}\,F\,m^{-1} \times 5.0 \times 10^{-3}\,m^2}{1.0 \times 10^{-4}\,m}$$

$$= 1000\,pF$$

Capacitors in parallel

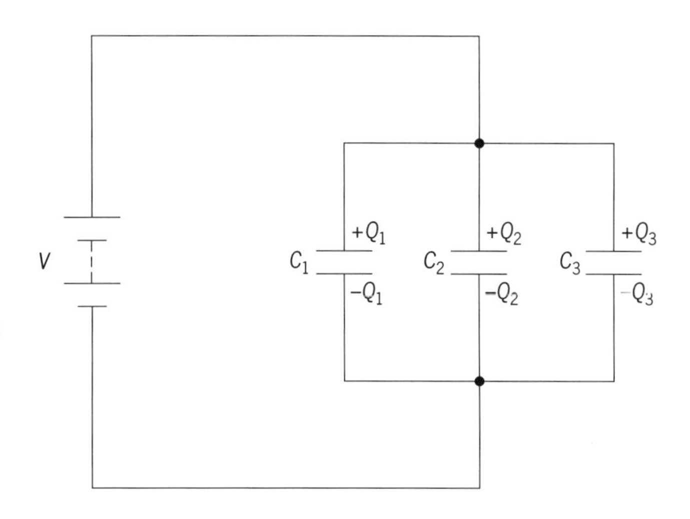

Figure 29.3 Capacitors in parallel

If several capacitors are connected in parallel as in Figure 29.3, the pd across each will be the same (V). It follows that the charge on each capacitor can be found from:

$$Q_1 = C_1 V \qquad Q_2 = C_2 V \qquad Q_3 = C_3 V$$

The total charge stored by this network of capacitors Q_{tot} is $Q_1 + Q_2 + Q_3$, or

$$Q_{tot} = C_1 V + C_2 V + C_3 V$$

The overall capacitance of the network $C_{tot} = Q_{tot}/V$,

$$\therefore C_{tot} = \frac{C_1 V + C_2 V + C_3 V}{V}$$

or, cancelling V,

$$C_{tot} = C_1 + C_2 + C_3$$

Example 4

Calculate the value of a single capacitor which could replace three capacitors of 2 μF, 3 μF and 6 μF connected in parallel

$$C_{tot} = C_1 + C_2 + C_3$$

$$= 2\,\mu F + 3\,\mu F + 6\,\mu F$$

$$= 11\,\mu F$$

Capacitors in series

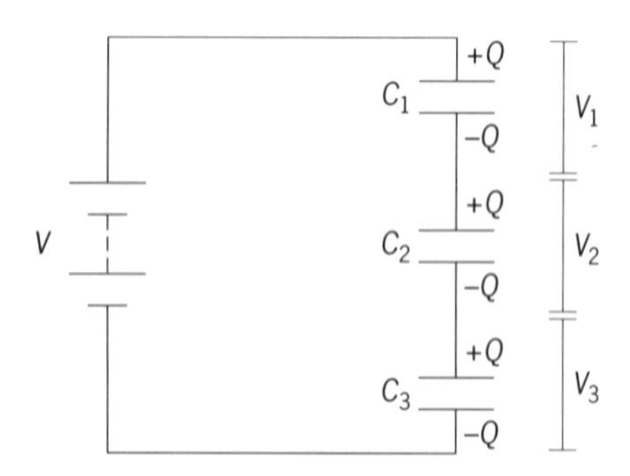

Figure 29.4 Capacitors in series

If several capacitors are connected in series as in Figure 29.4, the charge stored on each will be the same (Q). It follows that the pd across each capacitor can be found from:

$$V_1 = \frac{Q}{C_1} \qquad V_2 = \frac{Q}{C_2} \qquad V_3 = \frac{Q}{C_3}$$

The total pd across the network of capacitors V_{tot} is $V_1 + V_2 + V_3$, and

$$V_{tot} = \frac{Q}{C_{tot}}$$

$$\therefore \frac{Q}{C_{tot}} = \frac{Q}{C_1} + \frac{Q}{C_2} + \frac{Q}{C_3}$$

and so, cancelling Q,

$$\frac{1}{C_{tot}} = \frac{1}{C_1} + \frac{1}{C_2} + \frac{1}{C_3}$$

Example 5

Calculate the value of a single capacitor which could replace three capacitors of 2 μF, 3 μF and 6 μF connected in series.

$$\frac{1}{C_{tot}} = \frac{1}{C_1} + \frac{1}{C_2} + \frac{1}{C_3}$$

$$= \frac{1}{2\,\mu F} + \frac{1}{3\,\mu F} + \frac{1}{6\,\mu F}$$

$$= \frac{3}{6\,\mu F} + \frac{2}{6\,\mu F} + \frac{1}{6\,\mu F}$$

$$\frac{1}{C_{tot}} = \frac{6}{6\,\mu F} = \frac{1}{\mu F}$$

$$\therefore C_{tot} = 1\,\mu F$$

The combining of capacitors is thus the opposite to that of resistors (see topic 4).

Energy stored by a capacitor

When charge flows onto a capacitor, electrical potential energy is stored. The energy stored by a capacitor can be found using the equation

$$\text{energy stored} = \tfrac{1}{2}CV^2$$

Example 6

Calculate the energy stored by a 2 μF capacitor with a pd of 12 V across its plates.

$$\text{energy stored} = \tfrac{1}{2}CV^2$$

$$= \tfrac{1}{2} \times 2\,\mu F \times (12\,V)^2$$

(Note that the units $F \times V^2 = \frac{C}{V} \times V \times V = C \times V = J$.*)*

$$\therefore \text{energy stored} = 144\,\mu J$$

Using the equation $C = Q/V$ the energy equation can also be written as

$$\text{energy stored} = \tfrac{1}{2}QV \quad \text{or} \quad \tfrac{1}{2}\frac{Q^2}{C}$$

$\epsilon_0 = 8.9 \times 10^{-12}\,\text{F m}^{-1}$

Level 1

1 You transfer 24 nC to a capacitor and monitor the voltage across it. The voltage rises to 3.0 V. What size capacitor is it?

2 Charge runs into a capacitor at 2.0 mC per second (a current of 2.0 mA). After 40 s the potential difference across the capacitor has risen to 120 V.

a Calculate the charge transferred using data from the graph in Figure 29.5, or by any other method you choose.

b Calculate the capacitance of the capacitor.

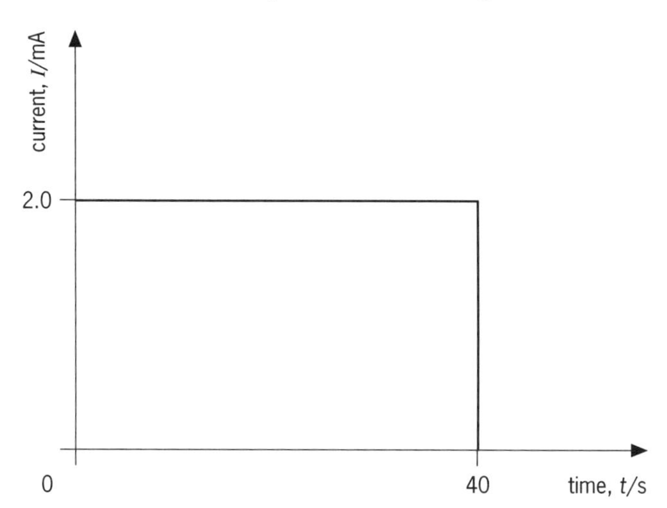

Figure 29.5

3 You can 'spoon' charge by touching a high voltage terminal with a metal object on an insulating handle. You then touch an isolated conductor with it. You can repeat this process to build up the charge on the conductor.

a If the capacity of the 'spoon' is 2 pF, how much charge is transferred each time from a 5 kV terminal?

b How much charge is transferred with a larger 'spoon' of capacity 16 pF?

c If you turn the voltage down to 3 kV, how much charge is transferred by each of the two 'spoons'?

4 When a potential difference is applied to a capacitor in series with a resistor, the current that flows initially is the same as if only the resistor were in the circuit. This is because there is initially no charge on the capacitor.

Calculate the current at $t = 0$ when 12 V is applied to a 5.0 kΩ resistor in series with a 6.0 μF capacitor.

5 You have two capacitors of 4.8 μF.

a How can you arrange them to give a capacitance of 9.6 μF?

b What other capacitance can be produced with a different arrangement of these capacitors?

6 A capacitor is charged as in Figure 29.6. Use the data in the graph to find

a the capacitance of the capacitor

b the energy stored when it is fully charged.

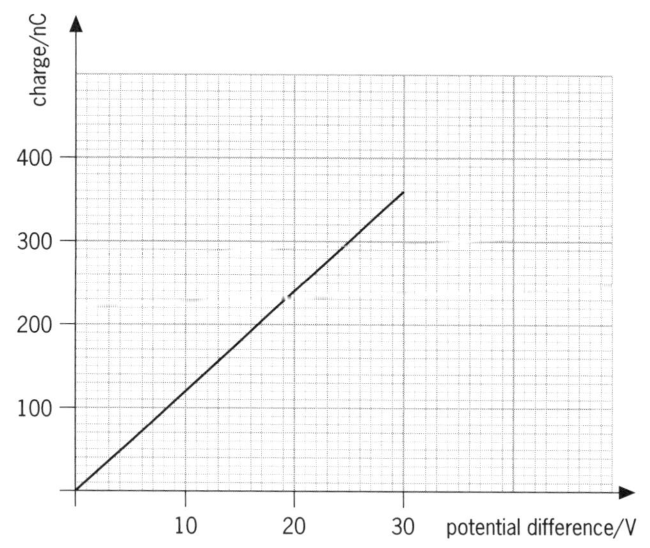

Figure 29.6

Level 2

7 The base of a thundercloud, Figure 29.7, forms a parallel plate capacitor with the surface of the Earth. The capacitance of the system is 200 nF and the potential difference between the two surfaces is 10^8 V.

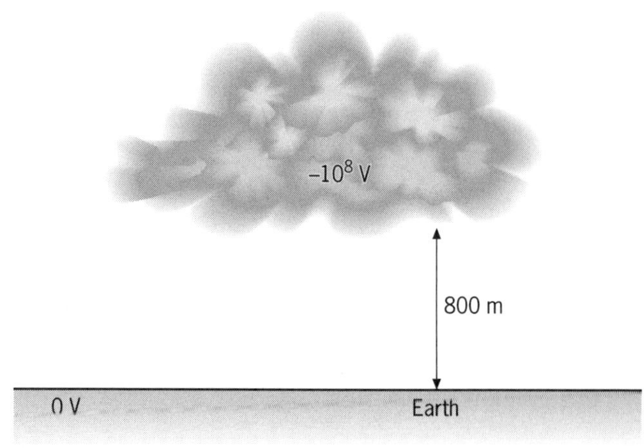

Figure 29.7

a How much electrical energy is stored in the system?

b Find out how this compares with the energy used in your home in a month.

8 The height above the ground, d, of the cloud in question 7 is 800 m.

a What area, A, has been estimated for the cloud base if it forms a capacitor of 200 nF, given that the capacitance of an air-filled parallel plate arrangement is $\epsilon_0 A/d$?

b If the cloud base were taken as square, what would be the length of one side?

9 In a grain silo, potentials of over 10 kV may be produced as the grain is handled. A spark of energy 10 mJ could ignite the dust and cause an explosion. Would a metal scoop of capacity 20 pF, left in contact with the grain, be a risk?

10 A parallel plate capacitor has plates that you can slide closer or further apart. It is charged by connection to a power supply. A high resistance voltmeter is connected across the plates and reads 20 V when they are first disconnected from the power supply.

What would the voltmeter read when the distance between the plates is

a halved

b doubled? (Assume no charge leaks away.)

What happened to the capacitance each time?

Level 3

11 Power capacitors for the National Grid are made from 'banks' or 'stacks' of about 40 capacitors in parallel, then 80 of these banks may be put in series.

a Explain the purpose of putting capacitors in parallel.

Hint: Calculate the capacity of 40 capacitors each of 20 μF.

b Explain the purpose of then arranging banks of them in series.

Hint: Calculate the working voltage if 80 capacitors each able to work at 200 V are linked.

12 A capacitor of 16 μF is required to operate at voltages not exceeding 1.1 kV. You have plenty of 4 μF capacitors that may be used up to 400 V. Work out a suitable arrangement using the principles in question 11. How many capacitors will you need to use in total?

13 An 8.0 μF capacitor is charged to 30 V and then removed from the battery. It is next connected across an uncharged 4.0 μF capacitor, Figure 29.8.

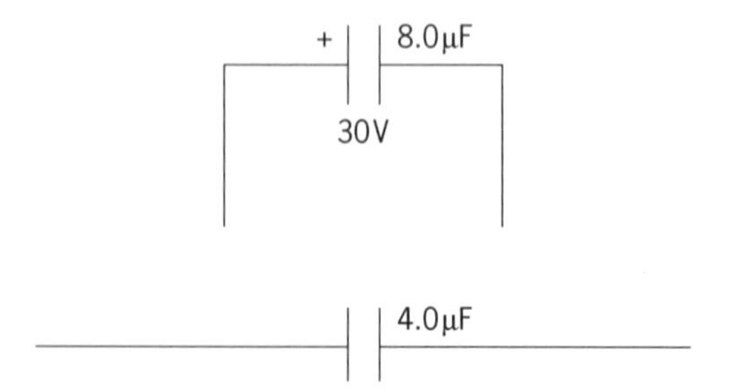

Figure 29.8

a Calculate the charge on the 8.0 μF capacitor at first.

b Calculate the capacitance of the combination of capacitors.

c What is the new potential difference across the 8.0 μF capacitor?

d How much charge is now on the 8.0 μF capacitor?

e How much charge has transferred to the 4.0 μF capacitor?

14 In the circuit of Figure 29.9, calculate the current flowing through, and potential difference across, *each* of the components,

a when the capacitor is first connected into the circuit

b when the capacitor is fully charged.

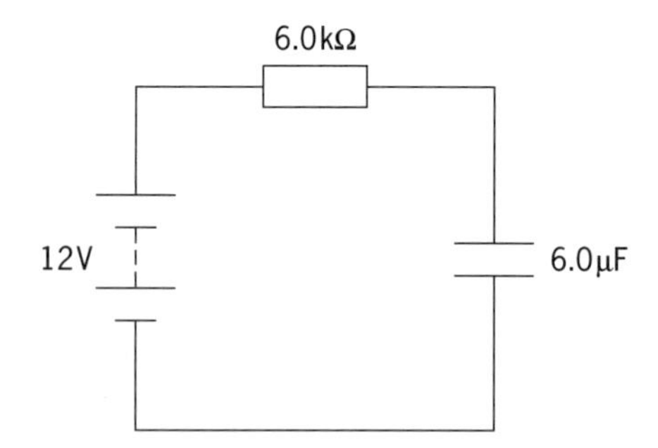

Figure 29.9

Level 1

1 The capacitance is the ratio of the charge to the final voltage produced:

$$C = \frac{Q}{V} = \frac{24 \times 10^{-9}\,C}{3.0\,V} = \mathbf{8.0\,nF}$$

2 a You can calculate the charge that has passed from

$$Q = I \times t = 2.0 \times 10^{-3}\,A \times 40\,s = \mathbf{0.08\,C}$$

This is the area under the graph.

b Capacitance, $C = \frac{Q}{V} = \frac{0.080\,C}{120\,V}$

$= \mathbf{6.7 \times 10^{-4}\,F}$ or $\mathbf{0.67\,mF}$ or $\mathbf{670\,\mu F}$

3 a $C = Q/V$ which you can rearrange to give $Q = CV$

$\therefore$ charge, $Q = 2 \times 10^{-12}\,F \times 5 \times 10^3\,V = \mathbf{10\,nC}$

b The capacitance of the large spoon is 8 times larger than that of the small spoon, so Q will be 8 times larger, or **80 nC**.

(Check by calculating: $16 \times 10^{-12}\,F \times 5 \times 10^3\,V = 80 \times 10^{-9}\,C$.)

c At 3 kV the charge per small spoon is $\frac{3}{5}$ of the original 10 nC, or **6 nC**.
Similarly, for the large spoon, $\frac{3}{5}$ of 80 nC is **48 nC**.

(Again you can check these by calculating $Q = C \times V$.)

4 Treating the circuit as if there were only the resistor (because, at $t = 0$, the capacitor does not affect the value of the current),

$$I = \frac{V}{R} = \frac{12\,V}{5.0 \times 10^3\,\Omega} = \mathbf{2.4\,mA}$$

5 a Capacitances add up if they are connected in parallel, Figure 29.10a.

$C_{tot} = C_1 + C_2 = 4.8\,\mu F + 4.8\,\mu F$
$= \mathbf{9.6\,\mu F}$ (or twice the value of one)

b If the two 4.8 μF capacitors are arranged in series, Figure 29.10b,

$$\frac{1}{C_{tot}} = \frac{1}{C_1} + \frac{1}{C_2} = \frac{2}{4.8 \times 10^{-6}\,F}$$

$\therefore C_{tot} = \mathbf{2.4\,\mu F}$ (or half the value of one)

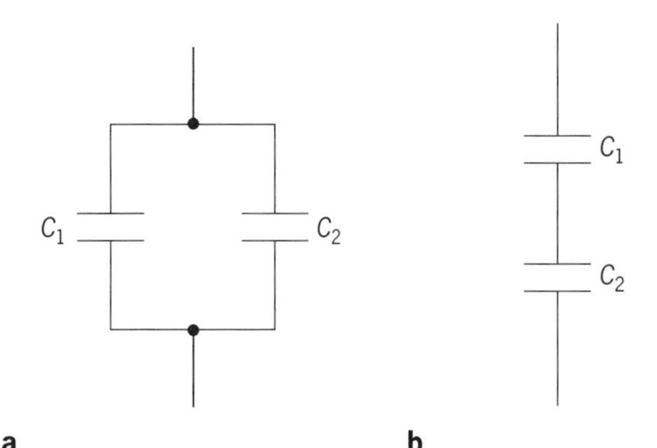

Figure 29.10

6 a The gradient of the straight-line graph is $C = Q/V$.

$$\therefore C = \frac{360\,nC}{30\,V} = \mathbf{12\,nF}$$

b The energy stored is given by $\frac{1}{2}QV$, which is the area under the graph:

$E = \frac{1}{2}V \times Q = 0.5 \times 30\,V \times 360\,nC$
$= \mathbf{5400\,nJ}$ or $\mathbf{5.4\,\mu J}$

Level 2

7 a The energy stored in a capacitor is given by $E = \frac{1}{2}CV^2$. In this example:

$$E = \tfrac{1}{2} \times 200 \times 10^{-9}\,F \times (10^8\,V)^2 = \mathbf{1 \times 10^9\,J}$$

b 1 kWh (or 3.6×10^6 J) is 1 'unit' of electricity. Look at your electricity and fuel bills. 1000 MJ might well be enough energy to last your household for a month!

8 a For the air-filled capacitor $C = \epsilon_0 A/d$, where e_0 is the permittivity of free space, A is the area and d the separation of the 'plates'. You can rearrange this equation to make the area of the 'plates' the subject:

$$A = \frac{Cd}{\epsilon_0}$$

$$= \frac{200 \times 10^{-9}\,F \times 800\,m}{8.9 \times 10^{-12}\,F\,m^{-1}} = \mathbf{1.8 \times 10^7\,m^2}$$

b The length of side of the square cloud is

$$\sqrt{1.8 \times 10^7\,m^2} = 4.3 \times 10^3\,m = \mathbf{4.3\,km}$$

9 The energy stored in a capacitor is given by $E = \frac{1}{2}CV^2$, so the energy available if the scoop is at 10 kV is

$$E = \tfrac{1}{2} \times 20 \times 10^{-12}\,F \times (10^4\,V)^2 = 1 \times 10^{-3}\,J \text{ or } 1\,mJ$$

There doesn't seem to be a serious risk, but if the voltage is only three times higher – the energy depends on V^2 – the energy will get much too close to the critical value. Flammable solvent vapour present could be ignited by energy as small as 1 mJ.

10 a If you halve the distance between the plates in a parallel plate capacitor you will **double the capacitance**. This will halve the potential difference, for a given charge, so the voltmeter will read **10 V**.

Looking at the equations, you can rearrange $C = Q/V$ to give $V = Q/C$. Q is constant as the plates have been disconnected, so V is inversely dependent on C, which is itself inversely dependent on d. This means that V is proportional to d for parallel plate capacitors.

b If the distance is doubled the **capacitance is halved**. The potential difference is doubled for the same charge, so the new reading will be **40 V**.

Note that you would have to do work against the attraction between the plates if you tried to pull them apart.

Level 3

11 a When capacitors are connected in parallel they all have the same potential difference across them and they each store the full amount of charge. This arrangement will hold more charge, but only at the normal working voltage of each capacitor. In this example, 40 capacitors of 20 μF at 1.0 V (say) will each hold 20 μC making a total of 800 μC. The effective capacity of this 'bank' is 800 μF ($C = Q/V$).

So, by connecting capacitors in parallel, a **larger capacity** has been achieved.

b When several capacitors (or banks of capacitors in parallel) are connected in series the total potential difference across them is shared. So if you connect 80 'banks', each able to work at 200 V, you can put 16 000 V across the lot. It is not simple to construct individual capacitors that will work at such high voltages.

So, by connecting capacitors in series, a **higher working voltage** has been achieved.

12 To operate safely at 1.1 kV it would be best to have at least three capacitors, or three banks of capacitors, in series. This would then be safe up to $3 \times 400\,\text{V} = 1.2\,\text{kV}$.

Three identical capacitors in series have effective capacitance one-third of the individual value. You want the total capacitance to be 16 μF, so you would have to connect three 'banks' of capacitors in series, each 'bank' having a capacitance of 48 μF.

You can make a 48 μF 'bank' by connecting 4 μF capacitors together. In parallel, capacitances add up, so we can use 12 of the 4 μF capacitors connected in parallel to make each 48 μF 'bank'. See Figure 29.11.

In total we need three 'banks' in series, each with 12 capacitors in parallel = **36 capacitors**.

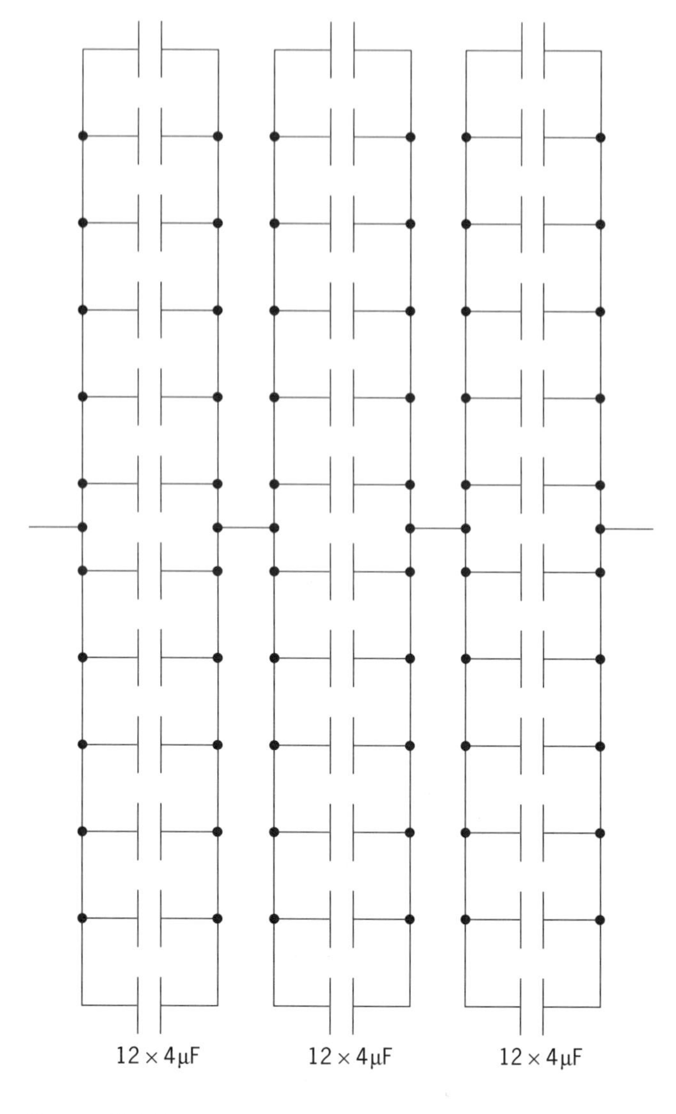

Figure 29.11

13 a The initial charge,

$$Q = CV = 8.0 \times 10^{-6}\,\text{F} \times 30\,\text{V} = \mathbf{240\,\mu C}$$

b When the two capacitors are connected in parallel, Figure 29.12, the total capacitance is

$$C_{tot} = C_1 + C_2$$

$$= 8.0\,\mu\text{F} + 4.0\,\mu\text{F} = \mathbf{12.0\,\mu F}$$

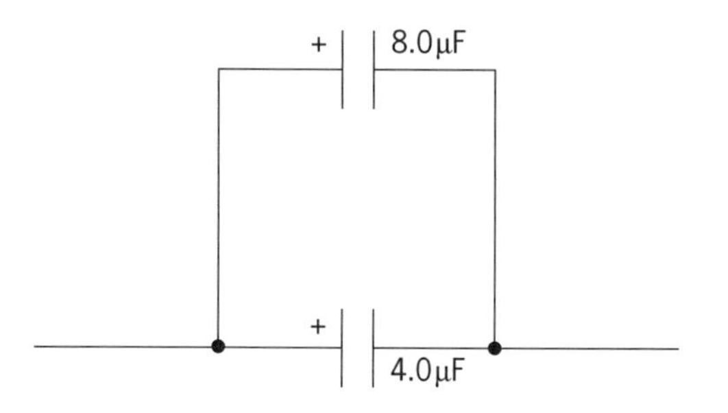

Figure 29.12

c Because no charge is lost, the potential difference across the combined capacitors after connection is given by

$$V = \frac{Q}{C_{tot}} = \frac{240 \times 10^{-6}\,\text{C}}{12 \times 10^{-6}\,\text{F}} = \mathbf{20\,V}$$

This is the same for each of the capacitors as they are in parallel.

d After connection, the potential across the 8.0 μF has dropped from 30 V to 20 V, so the new value of the charge is

$$Q = CV = 8.0\,\mu\text{F} \times 20\,\text{V} = \mathbf{160\,\mu C}$$

e The charge on the 4.0 μF capacitor is

$$Q = CV = 4.0\,\mu\text{F} \times 20\,\text{V} = \mathbf{80\,\mu C}$$

which is the charge transferred from the 8.0 μF capacitor.

Check that the final amounts of charge add up to the original: 160 μC + 80 μC = 240 μC.

14 a Initially the capacitor has no effect because it has zero charge and gives no opposition to the battery, so the initial current is given by

$$I = \frac{V_R}{R} = \frac{12\,\text{V}}{6.0 \times 10^3\,\Omega} = 2.0 \times 10^{-3}\,\text{A} = \mathbf{2.0\,mA}$$

The potential difference across the resistor is given by

$$V_R = IR = 2.0 \times 10^{-3}\,\text{A} \times 6.0 \times 10^3\,\Omega = \mathbf{12\,V}$$

There is no charge on the capacitor so

$$V_C = \frac{Q}{C} = \frac{0\,\text{V}}{6.0 \times 10^3\,\text{F}} = \mathbf{0\,V}$$

b When the capacitor is fully charged the potential difference across it is equal in magnitude but opposite in direction to the emf of the battery so no further current can flow.

$$V_R = IR = 0\,\text{A} \times 6.0 \times 10^3\,\Omega = \mathbf{0\,V}$$

and

$$V_C = \mathbf{12\,V}$$

30 Charging and discharging a capacitor

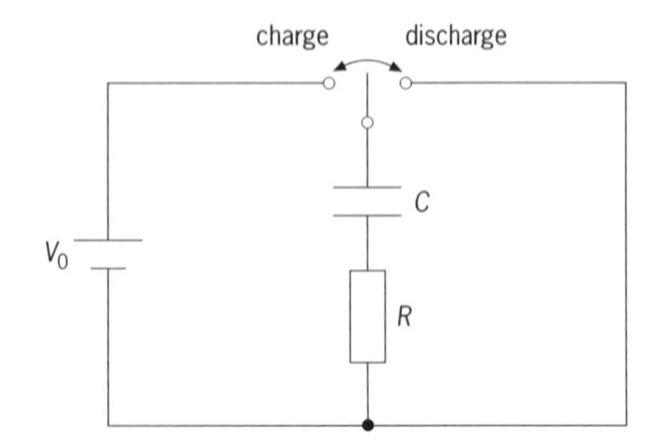

Figure 30.1 Charging and discharging circuit

When the two-way switch in the circuit of Figure 30.1 is moved to the left, charge flows onto the capacitor. Initially the rate of flow of charge is high but this decreases as the amount of charge stored on the plates increases. The three graphs in Figure 30.2 show how the charge on the capacitor, the current flowing in the circuit and the pd across the capacitor vary with time.

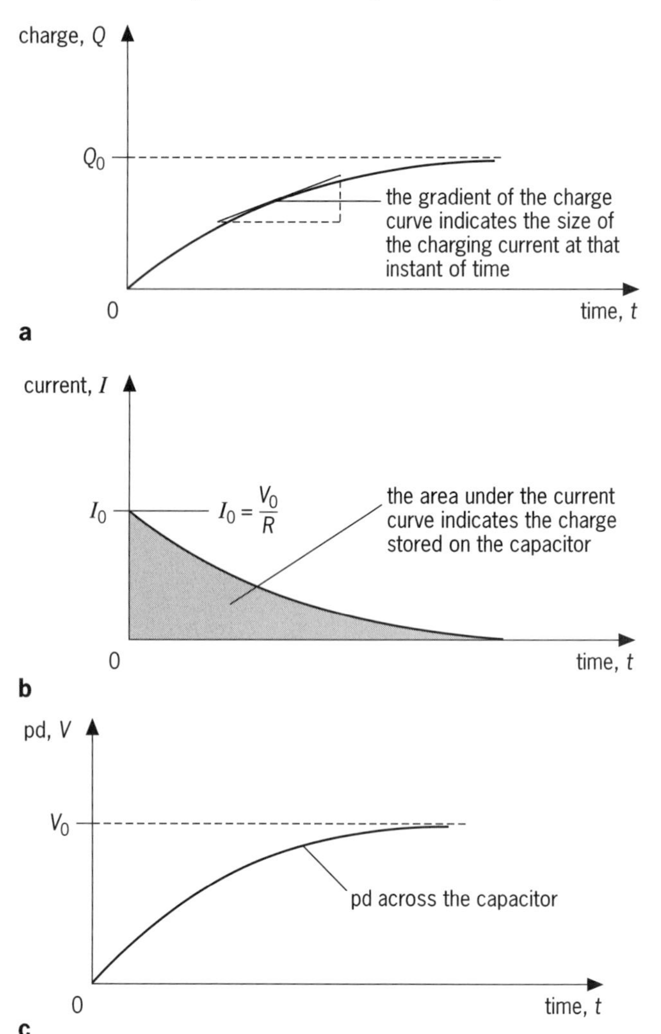

Figure 30.2 Charging a capacitor through a resistor: variation of charge, current and pd with time

When the switch is moved to the right, charge flows from the capacitor and through the resistor *R*. The three graphs in Figure 30.3 show how the charge on the capacitor, the current flowing in the circuit and the pd across the capacitor vary with time.

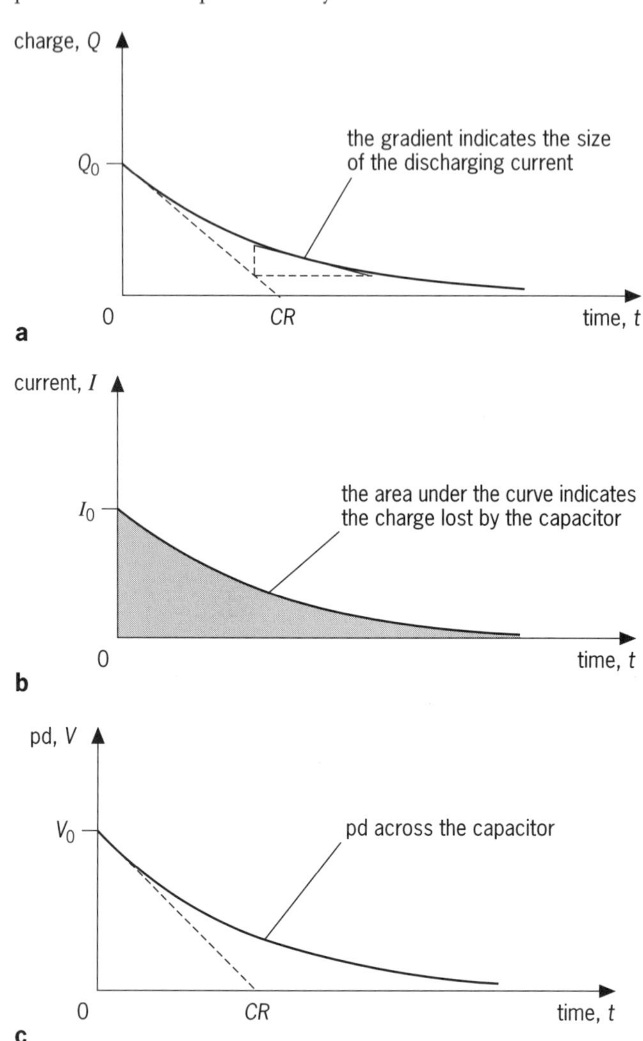

Figure 30.3 Discharging a capacitor through a resistor: variation of charge, current and pd with time

The discharging graphs are examples of **exponential decays** and are described by the equations

$$Q = Q_0 e^{-t/CR} \qquad I = I_0 e^{-t/CR} \qquad V = V_0 e^{-t/CR}$$

CR is called the **time constant** and is the time it would take the capacitor to completely discharge if its initial rate of discharge continued (see Figure 30.3). After a time $t = CR$ the charge remaining on the capacitor is $0.37 \times$ original charge.

These equations confirm that a capacitor discharges more slowly through a resistor of high value.

Example 1

A 100 μF capacitor is discharged through a 100 Ω resistor.

a Calculate the time constant for this arrangement.
b Calculate the half-life value for this arrangement.

a Time constant $= CR$
$= 100 \times 10\,F \times 100\,\Omega$
$= 10\,\text{s}$

b Half-life value $= 0.69\,CR$
$= 0.69 \times 10\text{s}$
$= 6.9\,\text{s}$

Using decay curves like those in Figure 30.3 it is possible to work out a **half-life** value for a charged capacitor, that is the time it takes to lose half its initial charge. Alternatively, a value can be calculated using the equation

$$\text{half-life} = 0.69\,CR$$

Level 1

1

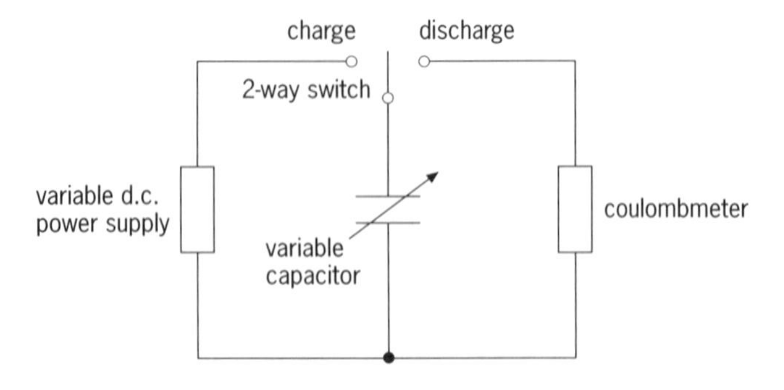

Figure 30.4

A capacitor is charged and then discharged through a coulombmeter using the circuit shown in Figure 30.4. A coulombmeter is a device for measuring charge. Two experiments are carried out.

a Keeping the capacitance constant, the voltage is increased, in steps, and the average of several measurements of the charge on the capacitor at each voltage is recorded. Sketch a graph showing how the charge would change with the voltage. Explain the physical significance of the gradient.

b Keeping the voltage constant, the capacitance is increased and the average of several readings of the charge at each capacitance value is recorded. Sketch a graph showing how the charge would change with the capacity. Explain the physical significance of the gradient.

2 In the circuit of Figure 30.5, both meters are moving-coil, centre-zero meters, so that the direction of the current can be seen easily.

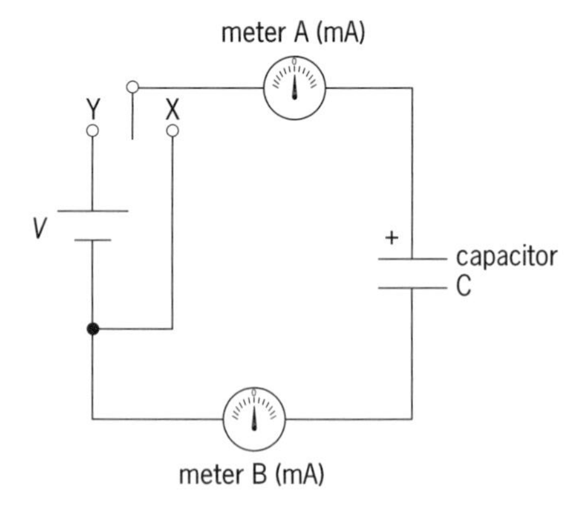

Figure 30.5

a When the switch is moved from X to Y the pointer of meter A swings clockwise by three divisions and then returns to the centre. What will meter B do?

b When the switch is moved back to X what will each meter do?

c Does this prove that charge passes through the capacitor?

3 A capacitor is charging. Sketch graphs to show:

a charge against time

b potential against time

c current against time.

Write a sentence to describe why each of your graphs has the shape you have shown.

4 A capacitor is discharged through a 5.0 kΩ resistor while a cathode ray oscilloscope (CRO) is connected across the resistor, Figure 30.6a. The CRO monitors the potential difference, which is proportional to the current flowing in the resistor.

The capacitor is charged up to 10 V; the voltage sensitivity of the CRO is set to 2.0 V per division and the time base is set to 0.10 s per division. The resulting trace is shown in Figure 30.6b.

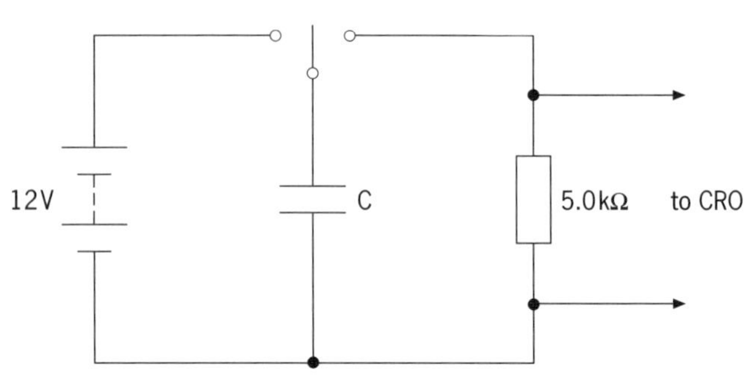

a

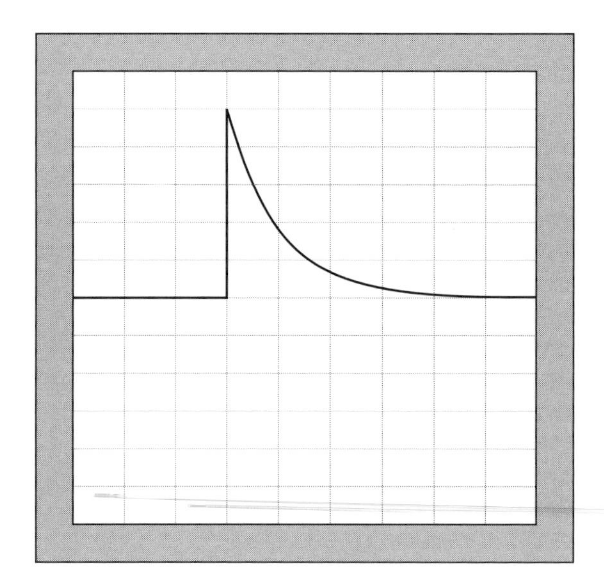

b

Figure 30.6

a What is the maximum discharge current?

b *Estimate* the quantity of charge that flows round the circuit during discharge.

c *Estimate* the capacitance of the capacitor.

d Sketch the traces that would be obtained if the capacitance is

i halved

ii doubled.

5 Considering the circuit in Figure 30.7, explain the following facts.

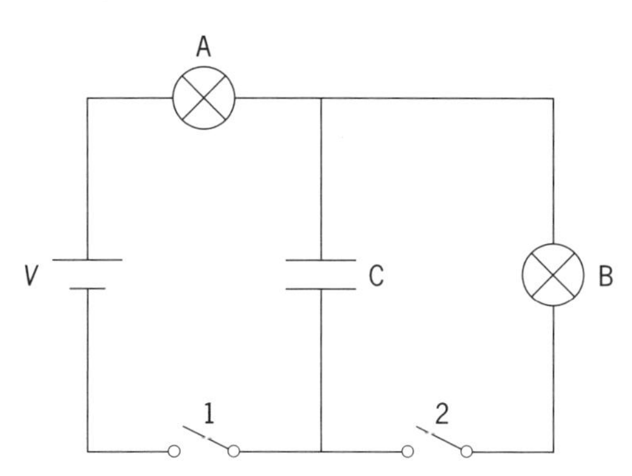

Figure 30.7

a Lamp A flashes only briefly when switch 1 is pressed and held down.
b If switch 1 is then released and switch 2 is pressed, lamp B flashes only briefly.

Level 2

6

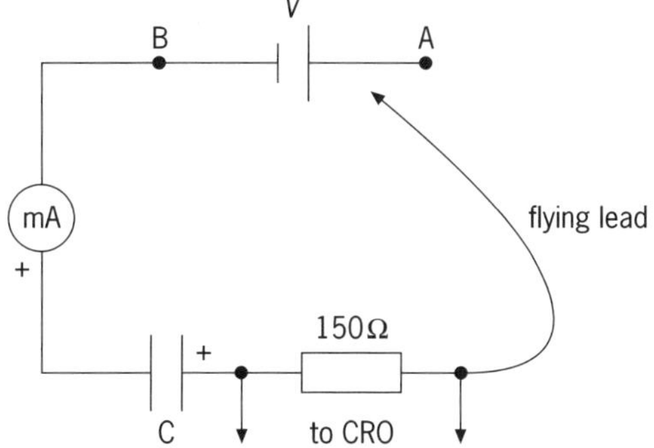

Figure 30.8

a Describe qualitatively the deflection seen on the meter in this circuit (Figure 30.8) if the flying lead is
i first connected to A
ii then connected to B
iii connected to A, then disconnected and then reconnected to A.
b When the cathode ray oscilloscope is connected as shown, sketch the traces seen for situations **i** and **ii** above.
c Sketch separate CRO traces to show what would be seen if
i a capacitor with twice the capacitance is used
ii a resistor with twice the resistance is used
iii the voltage is doubled.

7 A 220 μF capacitor is charged to 12 V and disconnected. During the next 30 s charge leaks from one plate across to the other and the potential difference falls to 10.5 V. What is the average leakage current between the plates?

Level 3

8

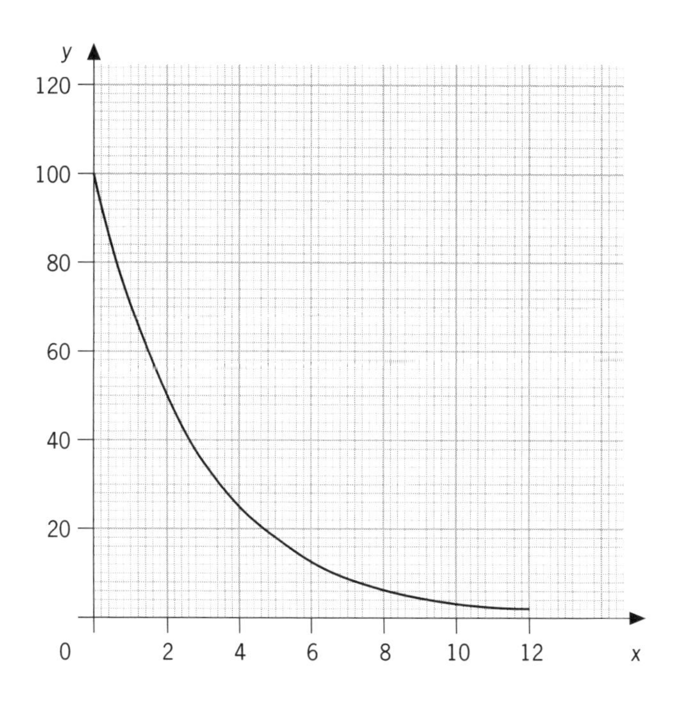

Figure 30.9 Exponential decay

Use data from the graph in Figure 30.9 to show that
a an exponential decay curve has a 'constant ratio' property

Hint: Compare the ratio of values of y for $x = 0$ and $x = 2$ with the ratio for $x = 2$ and $x = 4$, and with the ratio for $x = 4$ and $x = 6$.

b the rate of change of y with x, dy/dx, is proportional to y.

Hint: draw tangents to the curve at $y = 80$, $y = 40$ and $y = 20$. Calculate the gradients of the tangents.

9 A camera flash is operated by a 2.2 mF capacitor charged up to 40 V. It discharges in about 1.0 ms. What power is this?

10 When a capacitor discharges the time taken for both the charge and pd to halve is

$$T = 0.69 \times CR$$

A 3.0 pF capacitor is initially charged to 500 V, and then has its terminals linked by a 2.0 kΩ resistor.
a Calculate the 'time constant', CR, for this circuit.
b Show that the 'half-value' time – at which the pd will have fallen to 250 V – is 4.1 ns.
c Calculate the pd after 20.5 ns.

Hint: Use the half-value time from part b.

Level 1

1 a The average charge should increase steadily as the voltage is increased; see Figure 30.10a. The gradient of the graph is Q/V and is constant. This is the ratio we call **capacitance**:

$$C = \frac{Q}{V}$$

b The average charge should increase steadily as C increases; see Figure 30.10b. The gradient is Q/C and is constant. This ratio is the constant **voltage** that was used:

$$V = \frac{Q}{C}$$

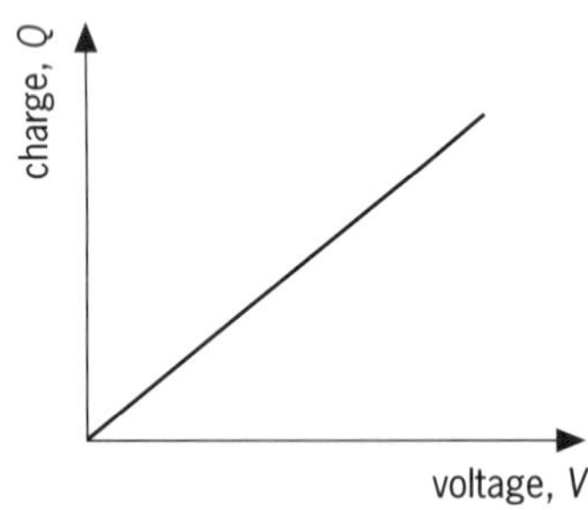

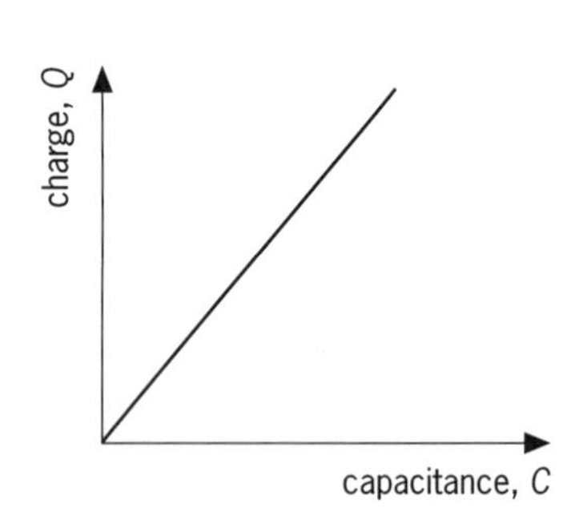

Figure 30.10

2 a The capacitor charges when the switch is at Y. Electrons leave one plate of the capacitor, which becomes positive, and they pile up on the negative plate. Meter B will show the **same deflection** in the **same direction** as A, while the charge flows.

b When the switch goes to X the capacitor can discharge and the electrons flow back again. Each meter shows a momentary deflection of **three divisions anticlockwise**.

c **No.** The meters just tell you what is moving round in the circuit. The charge is going to and fro in the wires of the circuit but not across the gap in the capacitor.

3 See the graphs in Figure 30.2.

a The charge rises rapidly at first and then more slowly until the maximum charge is reached.

b The potential difference (pd) across the plates rises rapidly at first then more slowly until the plates have a pd equal to that of the supply across them.

c The current is large at first but falls as the pd across the plates increases in opposition to the supply voltage.

4 a The maximum potential difference is represented by 5 vertical divisions on the trace. The sensitivity is set at 2.0 V per division, so

$$V_{max} = 5\,\text{div} \times 2.0\,\text{V}\,\text{div}^{-1} = 10\,\text{V}$$

The maximum current is therefore

$$I_{max} = \frac{V_{max}}{R} = \frac{10\,\text{V}}{5.0 \times 10^3\,\Omega} = 2.0 \times 10^{-3}\,\text{A} = \mathbf{2.0\,mA}$$

b The total charge flowing, Q, is given by the area under a graph of current against time. In this case you can take, to a good approximation, the area as a triangle. See Figure 30.11.

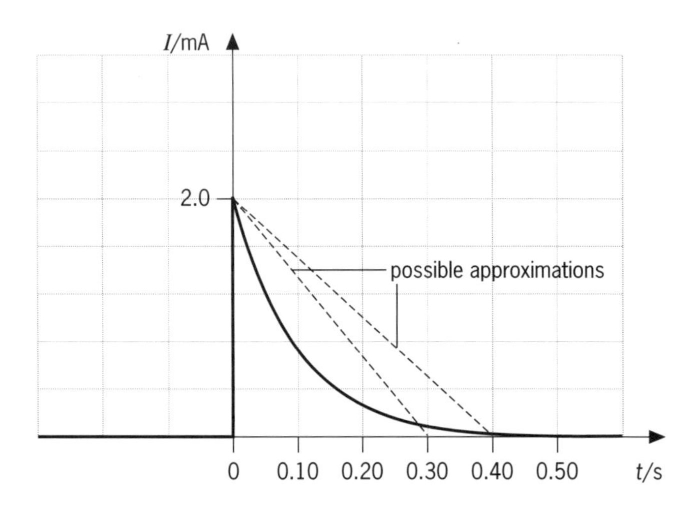

Figure 30.11

The time base is set to 0.10 s per division so the 'base' of the triangle can be taken as 0.40 s (or 0.30 s); the 'height' is 2.0 mA.

$$\therefore \quad Q = \tfrac{1}{2} \times 0.40\,\text{s} \times 2.0 \times 10^{-3}\,\text{A}$$
$$= 0.40 \times 10^{-3}\,\text{A s}$$
$$= \mathbf{0.40\,mC}$$

(If you took the base as 0.30 s, you should get 0.30 mC.)

c The capacitance is given by

$$C = \frac{Q}{V} = \frac{0.40 \times 10^{-3}\,\text{C}}{10\,\text{V}} = \mathbf{40\,\mu F}$$

(or 30 μF if $Q = 0.30$ mC)

d i If the capacitance is halved, the time taken for the current to fall, which is related to the time constant CR, will be halved. However, the maximum current is not affected. See Figure 30.12i.

ii If the capacitance is doubled, the time constant CR will be doubled. Again, the maximum current is not affected. See Figure 30.12ii.

i original trace

t

capacitance halved

$\frac{t}{2}$

ii original trace

t

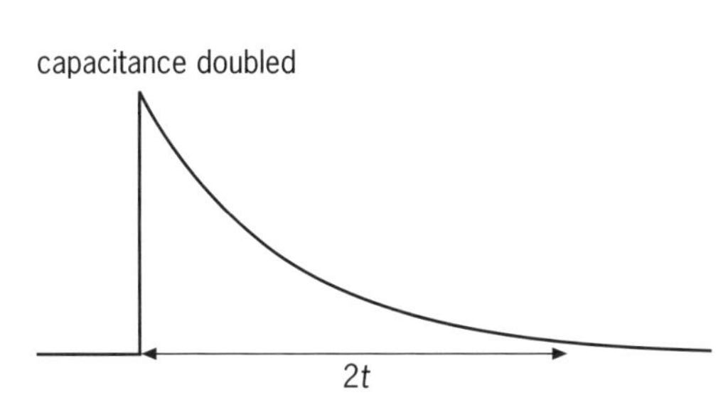

Figure 30.12

5 a Lamps only light when a large enough current flows through them. Current cannot pass through a capacitor, but charge builds up on it until the potential difference is equal and opposite to the emf of the battery. The current is large at first but gets smaller and smaller as the potential difference across the capacitor increases and opposes the battery. Lamp A only lights at the start when the full emf of the battery determines the current.

b When switch 1 is opened no more charge can flow from the battery even when the potential difference across C drops. When switch 2 is pressed, charge flows from C, through lamp B, rapidly at first, making a large enough current to light the lamp, but getting less and less as the capacitor empties (see the discharging graphs of charge and current in Figure 30.3).

For this ***exponential decay*** *the rate of flow of charge, that is the current, is proportional to the remaining charge.*

Level 2

6 a **i** The meter will deflect and then return to zero as the capacitor charges. The meter shows conventional current flowing from the + of the battery.
ii The meter will deflect the other way as the capacitor discharges.
iii There will be no deflection because the capacitor is already charged and has a potential difference equal to, but opposing that of the battery.

b

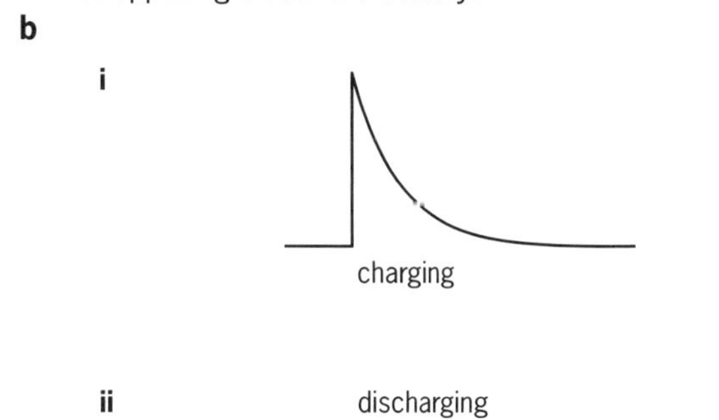

Figure 30.13

c **i** If C doubles, the time constant CR will be doubled, but V_{max} is not affected.

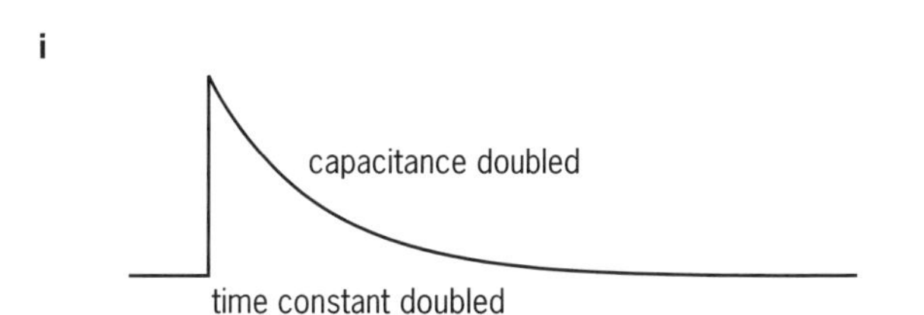

ii If R doubles, the time constant CR will be doubled. I_{max} will be halved but V_{max} ($= I_{max}R =$ battery V) is unchanged.

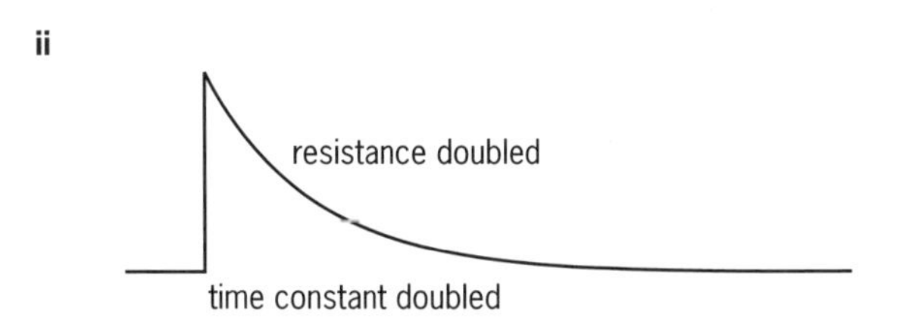

iii If the pd is doubled, the time constant will not change but I_{max} and V_{max} will be doubled.

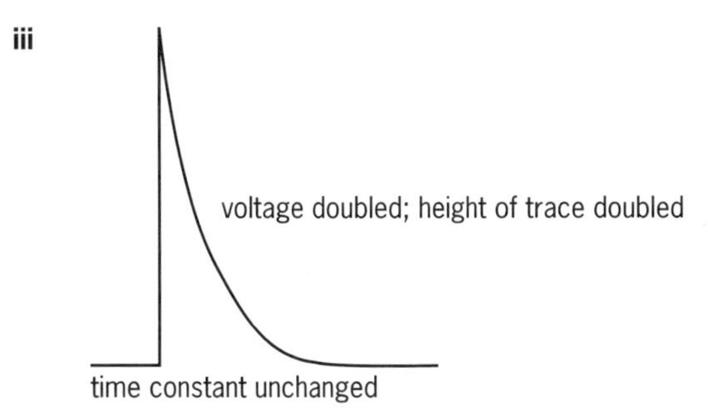

Figure 30.14

7 The *quickest* way to find the charge which has leaked away is to look at the fall in voltage, because $Q = CV$, so $\Delta Q = C \times \Delta V$.

$$\Delta V = 12\,\text{V} - 10.5\,\text{V} = 1.5\,\text{V}$$

So

$$\Delta Q = C\Delta V = 220 \times 10^{-6}\,\text{F} \times 1.5\,\text{V} = 330\,\mu\text{C}$$

If you didn't spot this 'trick' you could have worked out the separate charges:

$$Q_{12V} = CV = 220\,\mu F \times 12\,V = 2640\,\mu C$$
$$Q_{10.5V} = 220\,\mu F \times 10.5\,V = 2310\,\mu C$$

and subtracted one from the other to give 330 μC.

The average leakage current is then

$$I_{av} = \frac{\Delta Q}{t} = \frac{330 \times 10^{-6}\,\text{C}}{30\,\text{s}} = \mathbf{11\,\mu A}$$

Level 3

8 a The 'constant ratio' means that whatever interval you choose on the *x*-axis, the ratio of *y* at the start and finish of that interval is the same for all other equal intervals. For example, the values in Table 30.1 show that **equal increases in x give 'constant ratio' changes in y**.

Table 30.1

x_1 to x_2	Δx	y_1/y_2
0 to 2	2	100/50 = 2
2 to 4	2	50/25 = 2
4 to 6	2	25/12.5 = 2

b You have to compare the gradient with the value of *y* at the point the gradient is taken. For example, from the values in Table 30.2 you can see that the **gradient halves when y halves**. These values satisfy the equation

$$\frac{dy}{dx} = -0.35y$$

Table 30.2

y	dy/dx
80	−98/3.5 = −28
40	−77/5.5 = −14
20	−52/7.5 = −6.9

9 The energy stored in a capacitor is given by $E = \frac{1}{2}CV^2$ (see topic 29). In this example

$$\text{energy, } E = 0.5 \times 2.2 \times 10^{-3}\,\text{F} \times (40\,\text{V})^2 = 1.76\,\text{J}$$

Power = rate of change of energy, or

$$\frac{\text{energy}}{\text{time}} = \frac{1.76\,\text{J}}{1 \times 10^{-3}\,\text{s}} = \mathbf{1.76\,kW}$$

(Compare this with the most powerful light bulb that you have in the house!)

10 a $CR = 3 \times 10^{-12}\,\text{F} \times 2.0 \times 10^{3}\,\Omega = 6.0 \times 10^{-9}\,\text{s} = \mathbf{6.0\,ns}$

Check the units: $F\Omega = (CV^{-1})(VA^{-1})$. *But* $A^{-1} = (Cs^{-1})^{-1} = C^{-1}s$, *so*

$$F\Omega = CV^{-1}VC^{-1}s = s$$

b The 'half-life' time, $t_{1/2}$, is given by

$$t_{1/2} = 0.69 \times CR = 0.69 \times 6\,\text{ns} = \mathbf{4.1\,ns}$$

c The potential difference (pd) falls to half of its previous value as each half-life passes. 20.5 ns is 5 × 4.1 ns. After five half-lives the pd will be

$$0.5 \times 0.5 \times 0.5 \times 0.5 \times 0.5 = (0.5)^5 \text{ of its initial value} = (0.5)^5 \times 500\,\text{V} = \mathbf{15.6\,V}$$

On your calculator you may be able to do this by

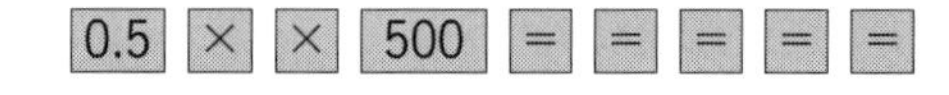

If not, use the sequence

31 Gravitational fields

A gravitational field is a region in space where gravitational forces act on a mass. Two of the most common types of field are radial and uniform (also see topics 32 and 33). These are represented by field lines; see Figure 31.1.

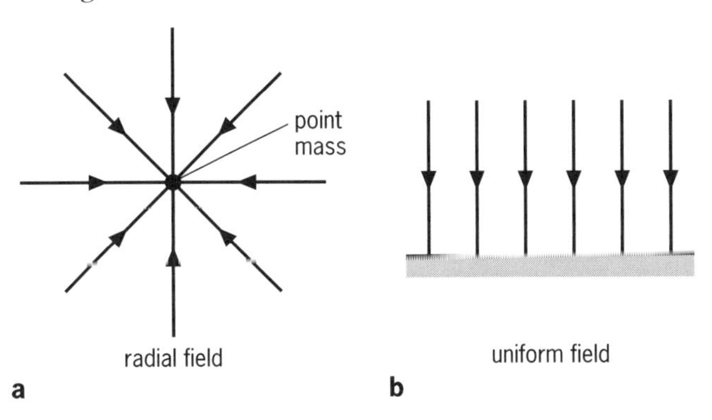

Figure 31.1 Representations of gravitational fields

Gravitational field strength

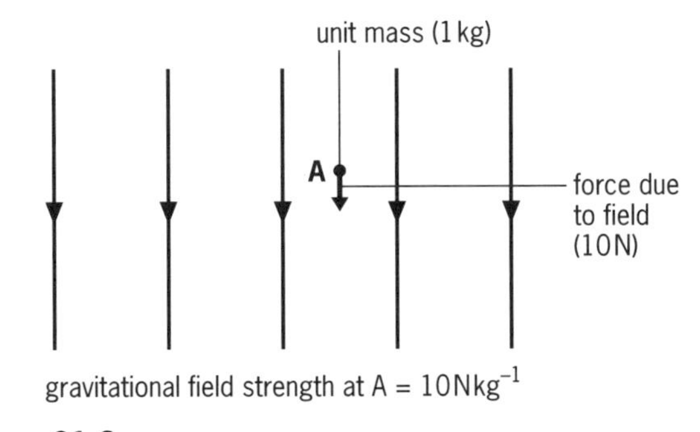

Figure 31.2

The **gravitational field strength**, g, at a point A, Figure 31.2, is defined as being numerically equal to the force in newtons exerted on unit mass (1 kg) placed at that point. Gravitational field strength has the unit N kg^{-1}. The direction of the field is the direction in which a point mass placed at A would move.

In general, if a mass m experiences a force F when placed in a gravitational field, the field strength, g, at that point is given by

$$g = \frac{F}{m}$$

Example 1

Calculate the gravitational field strength at a point if the force exerted on a mass of 5 kg placed here is 45 N.

$$g = \frac{F}{m} = \frac{45\,\text{N}}{5\,\text{kg}} = 9\,\text{N kg}^{-1}$$

Gravitational field due to a spherical mass

When calculating the effect of a gravitational field we are usually considering spherical masses such as the Earth, Figure 31.3.

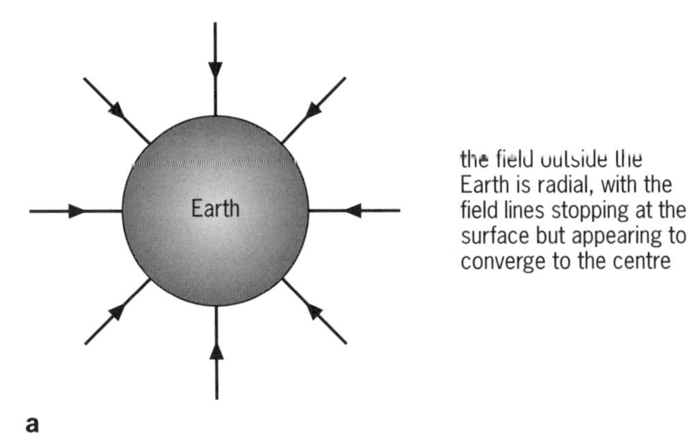

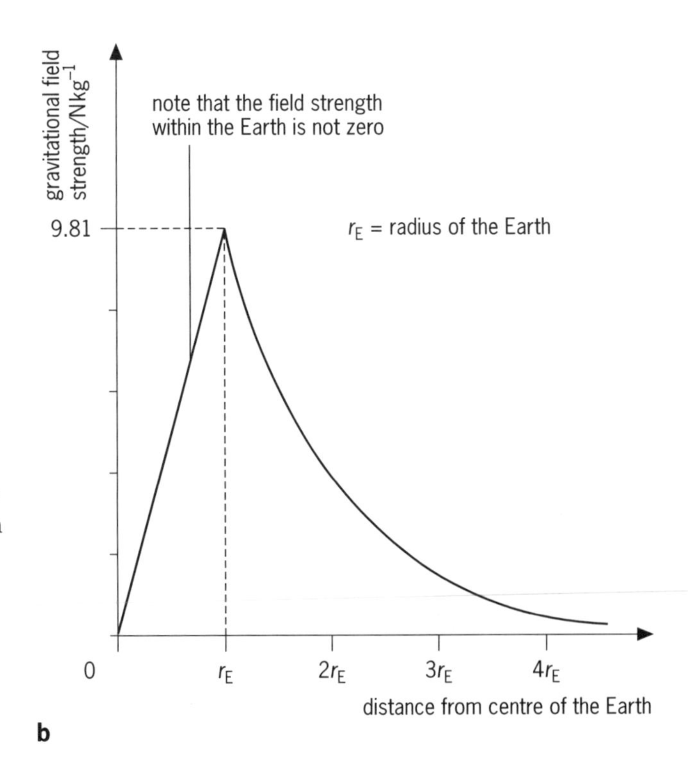

Figure 31.3 Gravitational field due to the Earth

The gravitational field at the surface of the Earth is approximately uniform (as Figure 31.1b). Its strength, g, is usually taken to be 9.81 N kg^{-1}. The force of attraction between a mass, m, and the Earth is commonly called the weight of the object and is equal to the mass times the gravitational field strength, mg. From Newton's Second Law it is clear that g is also the acceleration due to gravity.

Newton's Law of Universal Gravitation

In his **Law of Universal Gravitation** Newton stated that **every body attracts every other body with a force which is directly proportional to the product of their masses and inversely proportional to the square of the distance between them**.

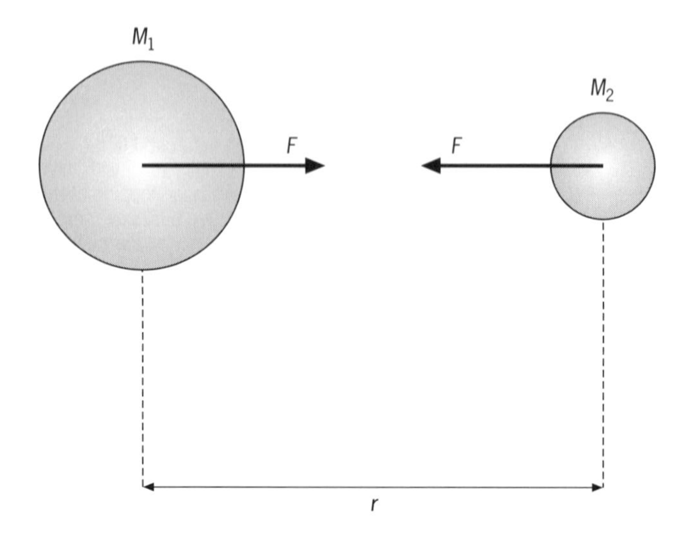

Figure 31.4

For uniform spheres, using the annotation of Figure 31.4, the law can be written as

$$F \propto \frac{M_1 M_2}{r^2}$$

or

$$\boxed{F = \frac{G M_1 M_2}{r^2}}$$

where G is the constant of proportionality, known as the **universal gravitational constant**, and has a value of $6.67 \times 10^{-11}\,\mathrm{N\,m^2\,kg^{-2}}$. The force is always attractive. Sometimes this force is written with a negative sign to indicate that it is attractive.

Note that for a uniform, spherical object it may be assumed that its mass is concentrated at its centre.

Example 2

Assuming that the Sun and the Earth are uniform spheres, calculate the force of attraction between them.

mass of Sun $= 2.0 \times 10^{30}\,\mathrm{kg}$
mass of Earth $= 6.0 \times 10^{24}\,\mathrm{kg}$
distance between Earth and Sun $= 1.5 \times 10^{11}\,\mathrm{m}$

$$F = \frac{G M_1 M_2}{r^2}$$

$$= \frac{6.67 \times 10^{-11}\,\mathrm{N\,m^2\,kg^{-2}} \times 2.0 \times 10^{30}\,\mathrm{kg} \times 6.0 \times 10^{24}\,\mathrm{kg}}{(1.5 \times 10^{11}\,\mathrm{m})^2}$$

$$= 3.56 \times 10^{22}\,\mathrm{N}$$

Calculating the gravitational field strength due to a spherical mass

Consider a small mass, m, on the surface of the Earth. From Newton's Law of Universal Gravitation the force exerted on this mass by the Earth would be

$$F = \frac{GMm}{r^2}$$

where M is the mass of the Earth and r here is the radius of the Earth.

This force of attraction is the weight of the object and has the value mg. Therefore

$$mg = \frac{GMm}{r^2}$$

or

$$\boxed{g = \frac{GM}{r^2}}$$

More generally, this equation describes the field strength, g, in $\mathrm{N\,kg^{-1}}$, a distance r from a point mass M; or a distance r from the centre of a uniform spherical mass M, where r is greater than or equal to the radius. It is shown by the curve in Figure 31.3b for distances greater than the Earth's radius.

g in the above equation is also the acceleration due to gravity, in $\mathrm{m\,s^{-2}}$, created at this point by the mass M. (Check that $\mathrm{N\,kg^{-1}} \equiv \mathrm{m\,s^{-2}}$.)

Example 3

Calculate the field strength on the surface of the planet Jupiter given the following information.

mass of Jupiter $= 1.90 \times 10^{27}\,\mathrm{kg}$
radius of Jupiter $= 7.15 \times 10^{7}\,\mathrm{m}$
universal gravitational constant, $G = 6.67 \times 10^{-11}\,\mathrm{N\,m^2\,kg^{-2}}$

$$g = \frac{GM}{r^2}$$

$$= \frac{6.67 \times 10^{-11}\,\mathrm{N\,m^2\,kg^{-2}} \times 1.90 \times 10^{27}\,\mathrm{kg}}{(7.15 \times 10^{7}\,\mathrm{m})^2}$$

$$= 24.8\,\mathrm{N\,kg^{-1}}$$

Gravitational potential

If an object of mass, m, on the surface of the Earth is lifted vertically through a distance, Δh, work is done on the object. The amount of work done can be found using the equation

work done = force × distance moved in the direction of the force

The force needed to lift it is equal to the weight of the object, mg.

$\therefore$ work done, $\Delta W = mg \times \Delta h$

The work done on the object is equal to the gravitational potential energy it gains (topic 10):

$$\text{gravitational potential energy gained, } \Delta E_p = \Delta W = mg\Delta h$$

If the same object is moved horizontally, that is at right angles to the field lines, no work is done and there is no potential energy gained or lost.

The **gravitational potential**, V, at a point A, Figure 31.5, is defined as being the work that must be done in moving a *unit mass* from infinity to that point.

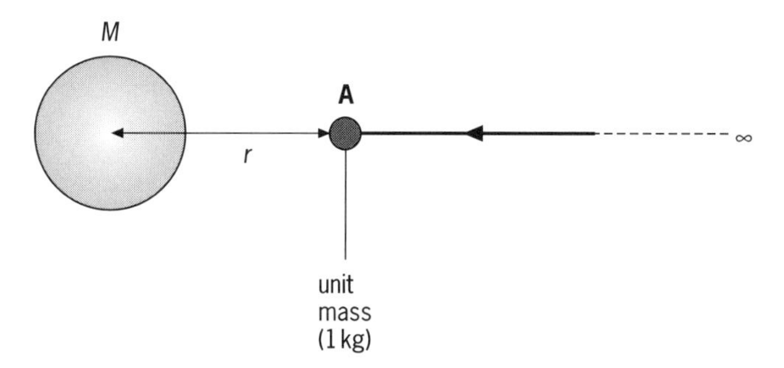

Figure 31.5

In a field due to a point mass or a uniform spherical mass, the gravitational potential of any point can be calculated using the equation

$$\boxed{V = \frac{-GM}{r}}$$

where M is the mass of the object creating the field and r is the distance of the point being considered from the centre of the mass. V has the unit $\mathrm{J\,kg^{-1}}$. The minus sign indicates that work must be done in order to move the unit mass *from* the point *to* infinity (because the field is attractive).

Example 4

Calculate the gravitational potential at the surface of the Earth.

radius of Earth $= 6.4 \times 10^6\,\mathrm{m}$
mass of Earth $= 6.0 \times 10^{24}\,\mathrm{kg}$

$$V = \frac{-GM}{r}$$

$$= \frac{-6.67 \times 10^{-11}\,\mathrm{N\,m^2\,kg^{-2}} \times 6.0 \times 10^{24}\,\mathrm{kg}}{6.4 \times 10^6\,\mathrm{m}}$$

$$= -6.3 \times 10^7\,\mathrm{J\,kg^{-1}} \quad \text{or} \quad -63\,\mathrm{MJ\,kg^{-1}}$$

The gravitational **potential difference**, ΔV, between two points is the work done in moving unit mass between the points. If a mass m at a point A in a gravitational field is moved a distance Δh in opposition to the field, Figure 31.6,

$$\text{work done per unit mass} = \frac{\Delta W}{m} = \Delta V$$

(change in potential)

$$\text{total work done} = \Delta W = m\Delta V$$

(change in potential energy)

which is equal to $mg\Delta h$.

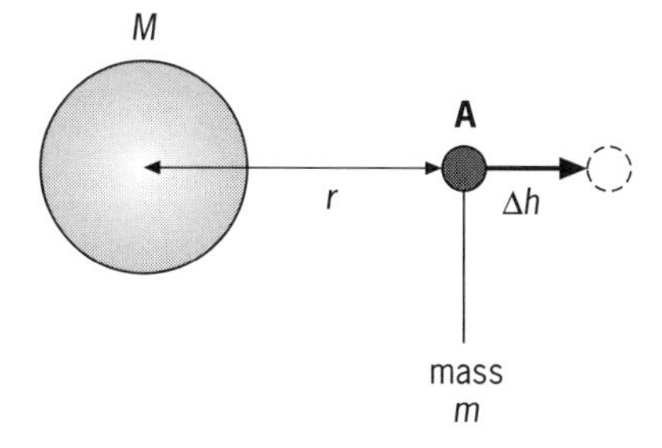

Figure 31.6

From the above it can be seen that

$$\Delta V = g\Delta h$$

and so

$$\frac{\Delta V}{\Delta h} = g$$

As the changes ΔV, Δh become infinitesimally small, this can be written in calculus notation:

$$\boxed{\frac{dV}{dh} = g}$$

The quantity dV/dh is called the **potential gradient**; the rate of change of gravitational potential with distance, in $\mathrm{J\,kg^{-1}}$ per m. It is equal to the field strength. (Check that $\mathrm{J\,kg^{-1}\,m^{-1}} \equiv \mathrm{N\,kg^{-1}}$.)

Universal gravitational constant, $G = 6.7 \times 10^{-11}\,\mathrm{N\,m^2\,kg^{-2}}$
Radius of Earth, $r_E = 6.4 \times 10^6\,\mathrm{m}$
Mass of Earth, $M_E = 6.0 \times 10^{24}\,\mathrm{kg}$

Level 1

1 a Is gravitational field strength, g, a vector or a scalar?
b $g = F/m$ means that, in addition to being gravitational field strength, g is also equal to another physical quantity. Which one is it?
c A mass of 3.2 kg has a weight of 8.2 N. What is the gravitational field strength?
d What is the force on a mass of 4.8 kg in a gravitational field with $g = 2.6\,\mathrm{N\,kg^{-1}}$?

2 A body of mass 2.0 kg is 3.0 m from the surface of the Earth, Figure 31.7.

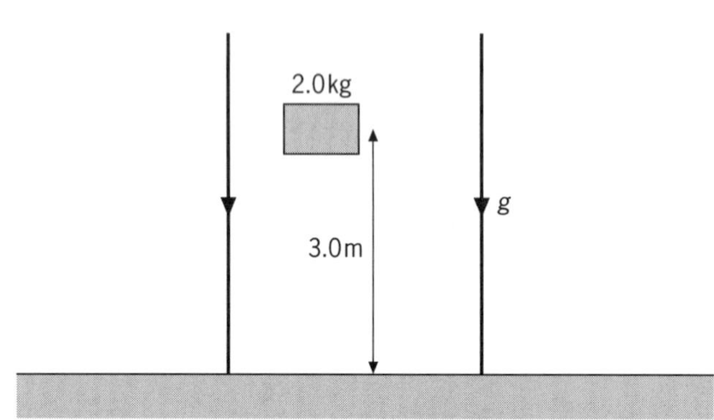

Figure 31.7

a What is the likely gravitational field strength, g?
b What is the force on the body?
c What is the work done by the field on the body if it moves to the Earth's surface?
d What type of energy does the body gain when it moves?
e What type of energy does the body lose when it moves?

3 Use the concept of gravitational potential to calculate the energy change when a 12 kg mass is moved from a point at potential $25\,\mathrm{J\,kg^{-1}}$ to one of potential $40\,\mathrm{J\,kg^{-1}}$.

4

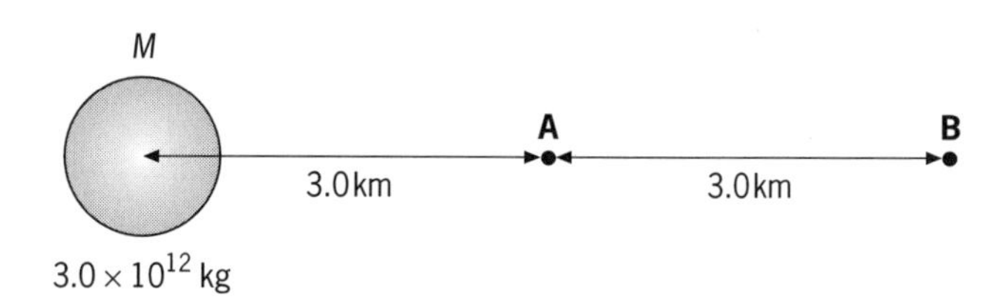

Figure 31.8

a Calculate the gravitational potential at A, 3.0 km away from the centre of a spherical asteroid of mass $M = 3.0 \times 10^{12}\,\mathrm{kg}$ (Figure 31.8). The asteroid is in space where the surrounding potential is taken as zero.
b Calculate the gravitational potential at B, 6.0 km away from the centre of the same asteroid.
c How much work would be done moving a 600 kg object from A to B?
d Which way would you be moving the object if you were doing work against the field?

Level 2

5 a The force on a 1.0 kg mass at the surface of the Earth is 9.8 N. What will it be when the mass is $12.8 \times 10^6\,\mathrm{m}$ above the surface?
b Find the height above the surface of the Earth at which a satellite's orbit would be synchronous with the Earth's rotation; it must have a period of 24 hours.

6 a Calculate the value of g at the surface of the Earth.
b Calculate the value of g at $1.1 \times r_E$ from the centre of the Earth.

7 Pluto has a moon that orbits the planet every 4.8 days at a radius that is estimated as $5.0 \times 10^7\,\mathrm{m}$. Calculate the mass of Pluto.

8 a What is the unit of gravitational potential?
b What gravitational potential gradient gives a gravitational field strength of $32\,\mathrm{N\,kg^{-1}}$?
c Define the gravitational potential difference between two points.

Level 3

9 What force constant, k, would a spring need to have if it had a 50 g mass hanging from it and a 0.1% change in g has to produce a change in extension of about 100 μm?

10 a From the basic law of gravitational attraction, show that the acceleration due to gravity at the Earth's surface is $g = GM_E/r_E^2$.
b Cavendish's work in measuring G led to him being called 'the man who weighed the Earth'. Use G, the radius of the Earth, r_E, and the acceleration due to gravity at the surface of the Earth, g, to find the Earth's mass, M_E.

11 Saturn's rings are particles in a circular orbit. The nearest are 70 Mm and the furthest are 140 Mm from the centre of the planet. The outermost particles travel with a speed of $17\,\mathrm{km\,s^{-1}}$.
a Show that the speed, v, of a particle in an orbit of radius r around a planet of mass M is given by $v = \sqrt{GM/r}$.
b What is the mass of Saturn?

Hint: Use the data given for one of the outer particles.

c How long does one orbit take for the outermost particles?
d How fast do the innermost particles go?

12 Figure 31.9 shows a region of space near the Earth's surface.

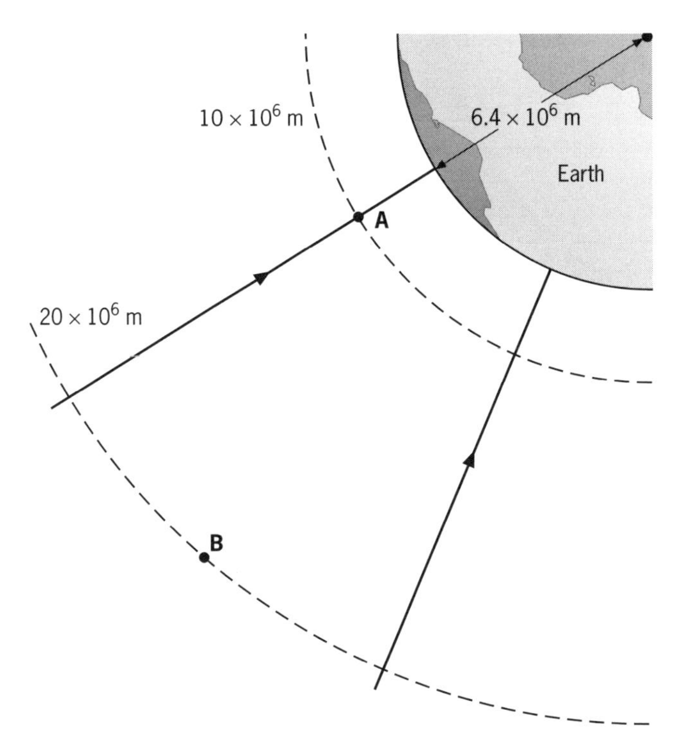

Figure 31.9

The gravitational potential at infinity is zero.

a Calculate the gravitational potential at A and at B.

b A spaceship of mass 2200 kg moves from A to B. What is the change in the potential energy of the spaceship?

c It later returns to Earth, passing through B at $4.0\,\text{km}\,\text{s}^{-1}$. How fast will it be going when it goes through A?

13 a What is the gravitational potential energy of an object of mass 500 kg at the Earth's surface? Take the gravitational potential at infinity to be zero.

b How much kinetic energy must it have at the surface to just escape from the Earth's gravitational field?

c What vertical velocity would it have to have, at the surface, to just escape?

d Does the velocity required depend on the mass of the body?

Level 1

1 a Gravitational field strength is a **vector**; it is force per unit mass (F/m), and force is a vector.

b Newton's Second Law of Motion says that

$$\text{force} = \text{mass} \times \text{acceleration} \quad \text{or} \quad F = ma$$

So

$$\frac{F}{m} = a$$

Therefore g is an acceleration; the 'acceleration due to gravity'.

c Gravitational field strength = force per unit mass

$$g = \frac{\text{weight}}{\text{mass}} = \frac{8.2\,\text{N}}{3.2\,\text{kg}}$$
$$= \mathbf{2.6\,N\,kg^{-1}}$$

d Force = mass × gravitational field strength

$$F = mg = 4.8\,\text{kg} \times 2.6\,\text{N}\,\text{kg}^{-1} = \mathbf{12\,N}$$

2 a The value of g is usually taken to be about **9.8 N kg^{-1}**.

b Force = mass × gravitational field strength

$$F = mg = 2.0\,\text{kg} \times 9.8\,\text{N}\,\text{kg}^{-1} = \mathbf{19.6\,N}$$

c Work = force × distance moved in the direction of the force

$$W = 19.6\,\text{N} \times 3.0\,\text{m} = \mathbf{58.8\,J}$$

d The body gains kinetic energy.

e The body loses gravitational potential energy.

3 Change in potential, $\Delta V = 40\,\text{J}\,\text{kg}^{-1} - 25\,\text{J}\,\text{kg}^{-1} = 15\,\text{J}\,\text{kg}^{-1}$
Total energy change $= m\Delta V = 12\,\text{kg} \times 15\,\text{J}\,\text{kg}^{-1} = \mathbf{180\,J}$

4 a The gravitational potential at distance r from a mass M is given by the equation

$$V = -\frac{GM}{r}$$

$$\therefore V_A = \frac{-6.7 \times 10^{-11}\,\text{N}\,\text{m}^2\,\text{kg}^{-2} \times 3.0 \times 10^{12}\,\text{kg}}{3.0 \times 10^3\,\text{m}}$$
$$= \mathbf{-6.7 \times 10^{-2}\,J\,kg^{-1}}$$

b The distance has doubled, so the potential has halved.

$$V_B = \mathbf{-3.35 \times 10^{-2}\,J\,kg^{-1}}$$

See Figure 31.10.

c The work done against the gravitational force, $\Delta W = m\Delta V$.

$$\Delta V = V_B - V_A$$
$$= -3.35 \times 10^{-2}\,\text{J}\,\text{kg}^{-1} - (-6.7 \times 10^{-2})\,\text{J}\,\text{kg}^{-1}$$
$$= 3.35 \times 10^{-2}\,\text{J}\,\text{kg}^{-1}$$

$$\therefore \Delta W = 600\,\text{kg} \times 3.35 \times 10^{-2}\,\text{J}\,\text{kg}^{-1} = \mathbf{20\,J}$$

d Gravitational force is always attractive, so in doing work against the field you are moving the smaller mass away, that is from A to B.

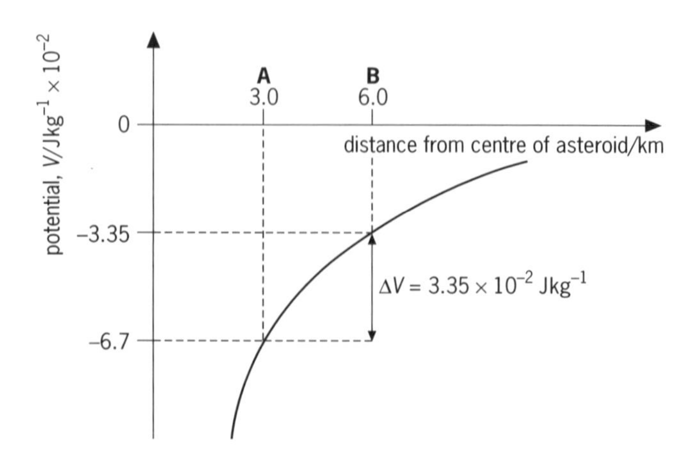

Figure 31.10 Variation in gravitational potential with distance from a spherical asteroid

Level 2

5

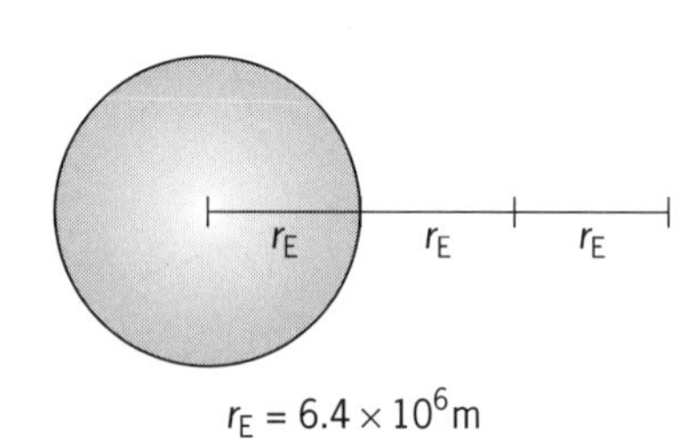

Figure 31.11

a $12.8 \times 10^6\,\text{m}$ is twice the radius of the Earth, r_E, so the total distance from the centre is $3 \times r_E$, Figure 31.11. The distance is trebled so the force goes down by $1/3^2 = 1/9$. So

$$F = \frac{9.8\,\text{N}}{9} = \mathbf{1.1\,N}$$

b The gravitational attraction provides the centripetal force (see topic 27) necessary to hold the satellite in circular orbit.

$$F = \frac{GmM_E}{r^2} = m\omega^2 r$$

where m is the mass of the satellite, $\omega = 2\pi/T$ and T is the time period of the orbit.

You can rearrange the equation and substitute for ω to give

$$r^3 = \frac{GM_E T^2}{4\pi^2}$$

The period must be 24 hours, so

$$T = 24 \times 60 \times 60\,\text{s}$$
$$= 86\,400\,\text{s}$$

$$\therefore r^3 = \frac{1}{4\pi^2} \times 6.7 \times 10^{-11}\,\text{N}\,\text{m}^2\,\text{kg}^{-2} \times 6.0 \times 10^{24}\,\text{kg} \times (86\,400\,\text{s})^2 = 7.6 \times 10^{22}\,\text{m}^3$$

$$r = 42.4 \times 10^6\,\text{m}$$

So the satellite's height above the Earth would be $(42.4 - 6.4) = \mathbf{36.0 \times 10^6\,m}$.

6 a The gravitational force on a mass m at the Earth's surface is

$$mg = \frac{GmM}{r_E^2}$$

so

$$g = \frac{GM}{r_E^2} = \frac{6.7\times10^{-11}\,\text{N m}^2\,\text{kg}^{-2}\times6.0\times10^{24}\,\text{kg}}{(6.4\times10^6\,\text{m})^2}$$

$$= \mathbf{9.81\,N\,kg^{-1}}$$

b You could just calculate it again at the new distance but it is better to look for patterns. Because $g \propto 1/r^2$,

$$\frac{g_2}{g_1} = \frac{r_1^2}{r_2^2}$$

which you can rearrange to give

$$g_2 = \frac{r_1^2}{r_2^2}\times g_1$$

So, for $r_1 = r_E$ and $r_2 = 1.1\times r_E$,

$$g_2 = \frac{r_E^2}{(1.1\times r_E)^2}\times9.81\,\text{N kg}^{-1}$$

Cancelling r_E^2,

$$g_2 = \frac{1}{1.21}\times9.81\,\text{N kg}^{-1} = \mathbf{8.1\,N\,kg^{-1}}$$

7 The equation for the orbit of Pluto's moon is

$$\frac{GmM}{r^2} = m\omega^2 r$$

where $\omega = 2\pi/T$, m is the mass of the moon, M is the mass of Pluto and r is the radius of the orbit.

T, the time period = 4.8 days = 414 720 s

You can rearrange the equation to give Pluto's mass:

$$M = \left(\frac{2\pi}{T}\right)^2\frac{r^3}{G}$$

Substituting the values given,

$$M = \frac{4\pi^2\times(5.0\times10^7\,\text{m})^3}{(414\,720\,\text{s})^2\times6.7\times10^{-11}\,\text{N m}^2\,\text{kg}^{-2}} = \mathbf{4.3\times10^{23}\,kg}$$

8 a Gravitational potential is the work done or energy gained per unit mass, so the unit is **joules per kilogram, J kg^{-1}**.

b The potential gradient would be **32 J kg^{-1} every m**.

c The gravitational potential difference between two points is **the work done or energy gained moving 1.0 kg between the two points**.

Level 3

9 In considering a mass on a spring (see topics 26 and 28) it is usual to use $g = 9.8\,\text{m s}^{-2}$ (unit of acceleration). A 0.1% change in g is

$$\frac{0.1}{100}\times9.8\,\text{m s}^{-2} = 9.8\times10^{-3}\,\text{m s}^{-2}$$

For a spring with force constant k, $F = kx$ where x is the extension. So an increase in the applied force, ΔF, will produce an increase in extension

$$\Delta x = \frac{\Delta F}{k}$$

which you can rearrange to give

$$k = \frac{\Delta F}{\Delta x}$$

The change in force applied by a mass m, due to a change in g, is $\Delta F = m\Delta g$. So, for an increase in g of Δg and a required extension Δx, the equation becomes

$$k = \frac{m\Delta g}{\Delta x}$$

$$= \frac{0.050\,\text{kg}\times9.8\times10^{-3}\,\text{m s}^{-2}}{100\times10^{-6}\,\text{m}} = \mathbf{4.9\,N\,m^{-1}}$$

10 a For a mass m at the surface of the Earth the force of attraction, according to the Law of Gravitation, is

$$F = \frac{GmM_E}{r_E^2}$$

This force is equal to the weight, $W = mg$. Therefore, cancelling m,

$$g = \frac{GM_E}{r_E^2}$$

b Rearranging the equation in part a,

$$M_E = \frac{r_E^2\times g}{G}$$

$$= \frac{(6.4\times10^6\,\text{m})^2\times9.8\,\text{m s}^{-2}}{6.7\times10^{-11}\,\text{N m}^2\,\text{kg}^{-2}} = \mathbf{6.0\times10^{24}\,kg}$$

(On the bottom line, $\text{N m}^2\,\text{kg}^{-2} = (\text{kg m s}^{-2})\text{m}^2\,\text{kg}^{-2} = \text{m}^3\,\text{s}^{-2}\,\text{kg}^{-1}$. This just leaves kg.)

11 a The centripetal force keeping a particle, of mass m, in orbit is provided by the gravitational pull of the planet mass M:

$$\frac{mv^2}{r} = \frac{GmM}{r^2}$$

You can rearrange this equation to give

$$v^2 = \frac{GM}{r} \quad \text{or} \quad v = \sqrt{\frac{GM}{r}}$$

b You can also rearrange the equation to give

$$M = \frac{rv^2}{G}$$

Using the data supplied,

$$M = \frac{140 \times 10^6\,\text{m} \times (17 \times 10^3\,\text{m s}^{-1})^2}{6.7 \times 10^{-11}\,\text{N m}^2\,\text{kg}^{-2}}$$

$$= \mathbf{6.0 \times 10^{26}\,kg}$$

c The particles go around a circle of radius 140 Mm at a speed of 17 km s^{-1}.

$$\text{time for one orbit} = \frac{\text{distance}}{\text{speed}}$$

$$= \frac{2\pi r}{v} = \frac{2\pi \times 140 \times 10^6\,\text{m}}{17 \times 10^3\,\text{m s}^{-1}}$$

$$= \mathbf{52 \times 10^3\,s} \quad \text{(This is about 14 h.)}$$

d Using the equation obtained in part a and the mass obtained in part b, find the speed when $r = 70 \times 10^6$ m:

$$v^2 = \frac{GM}{r} = \frac{6.7 \times 10^{-11}\,\text{N m}^2\,\text{kg}^2 \times 6.0 \times 10^{26}\,\text{kg}}{70 \times 10^6\,\text{m}}$$

$$= 5.74 \times 10^8\,\text{m}^2\,\text{s}^{-2}$$

So

$$\mathbf{v = 2.4 \times 10^4\,m\,s^{-1}}$$

12 a The gravitational potential at A, at 10×10^6 m from the centre of the Earth, is given by

$$V_A = \frac{-GM_E}{r}$$

$$= \frac{-6.7 \times 10^{-11}\,\text{N m}^2\,\text{kg}^{-2} \times 6.0 \times 10^{24}\,\text{kg}}{10 \times 10^6\,\text{m}}$$

$$= \mathbf{-40\,MJ\,kg^{-1}}$$

At B, which is twice as far away, it will be half this:

$$V_B = \mathbf{-20\,MJ\,kg^{-1}}$$

b The change in potential, ΔV, in moving from A to B is 20 MJ kg^{-1}, so the change in potential energy is

$$m\Delta V = 2200\,\text{kg} \times 20 \times 10^6\,\text{J kg}^{-1} = \mathbf{4.4 \times 10^{10}\,J}$$

c The kinetic energy of the spaceship at B is

$$\tfrac{1}{2}mv_B{}^2 = \tfrac{1}{2} \times 2200\,\text{kg} \times 16 \times 10^6\,\text{m}^2\,\text{s}^{-2} = 1.8 \times 10^{10}\,\text{J}$$

When it gets to A, gravitational potential energy has been lost and turned into kinetic energy, so the total kinetic energy of the ship will be

$$1.8 \times 10^{10}\,\text{J} + 4.4 \times 10^{10}\,\text{J} = 6.2 \times 10^{10}\,\text{J} = \tfrac{1}{2}mv_A{}^2$$

So $$v_A{}^2 = \frac{6.2 \times 10^{10}\,\text{J}}{\tfrac{1}{2} \times 2200\,\text{kg}} = 56 \times 10^6\,\text{m}^2\,\text{s}^{-2}$$

and $v_A = \mathbf{7.5\,km\,s^{-1}}$

13 a At the Earth's surface, $r = r_E$, so the gravitational potential is

$$V = \frac{-GM_E}{r_E} = \frac{6.7 \times 10^{-11}\,\text{N m}^2\,\text{kg}^{-2} \times 6.0 \times 10^{24}\,\text{kg}}{6.4 \times 10^6\,\text{m}}$$

$$= -63\,\text{MJ kg}^{-1}$$

The gravitational potential energy of an object of mass 500 kg is therefore

$$-500\,\text{kg} \times 63 \times 10^6\,\text{J kg}^{-1} = \mathbf{-3.1 \times 10^{10}\,J}$$

(This is the gravitational potential energy that 500 kg would lose in falling to the surface of the Earth from an infinite distance.)

b To get back to infinity by being shot into space, a 500 kg object will need initial kinetic energy equal to the $\mathbf{3.1 \times 10^{10}\,J}$ it lost falling to Earth.

c To escape from Earth,

$$\tfrac{1}{2}mv^2 = 3.1 \times 10^{10}\,\text{J}$$

which you can rearrange to give

$$v^2 = \frac{2 \times 3.1 \times 10^{10}\,\text{J}}{500\,\text{kg}} = 124 \times 10^6\,\text{m}^2\,\text{s}^{-2}$$

so $v = \mathbf{11.1\,km\,s^{-1}}$

d Both the kinetic and the potential energy are proportional to mass, so m will cancel in this problem and will not affect the velocity:

$$mV = \tfrac{1}{2}mv^2$$

$$\therefore \quad v = \sqrt{2V}$$

32 Electric fields

An electric field is a region in space where electric forces act on electric charge. It is represented by field lines.

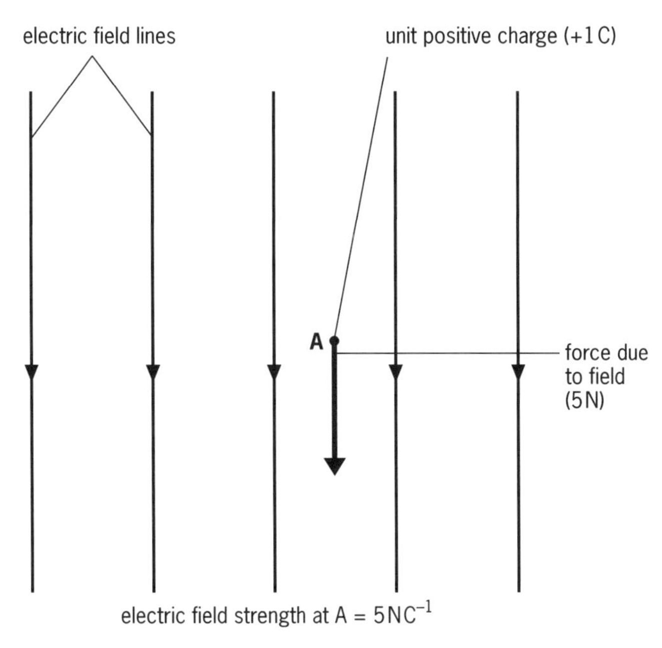

Figure 32.1

The strength of an electric field at some point A, Figure 32.1, is defined as being numerically equal to the force in newtons exerted on a unit positive charge (1 coulomb) placed at that point. **Electric field strength**, E, has the unit N C^{-1}. The direction of the field is the direction in which a positive charge placed at A would move.

If a charge q experiences a force F when placed in an electric field, the field strength, E, at this point is given by

$$E = \frac{F}{q}$$

If a larger charge is placed in the field it will experience a proportionally larger force, since $F = Eq$.

Example 1

Calculate the force exerted upon a particle carrying a charge of +3 μC placed in an electric field of strength 4 N C^{-1}.

$$F = Eq$$
$$= 4\,\text{N C}^{-1} \times 3 \times 10^{-6}\,\text{C}$$
$$= 1.2 \times 10^{-5}\,\text{N}$$

Electric field around a hollow charged conductor

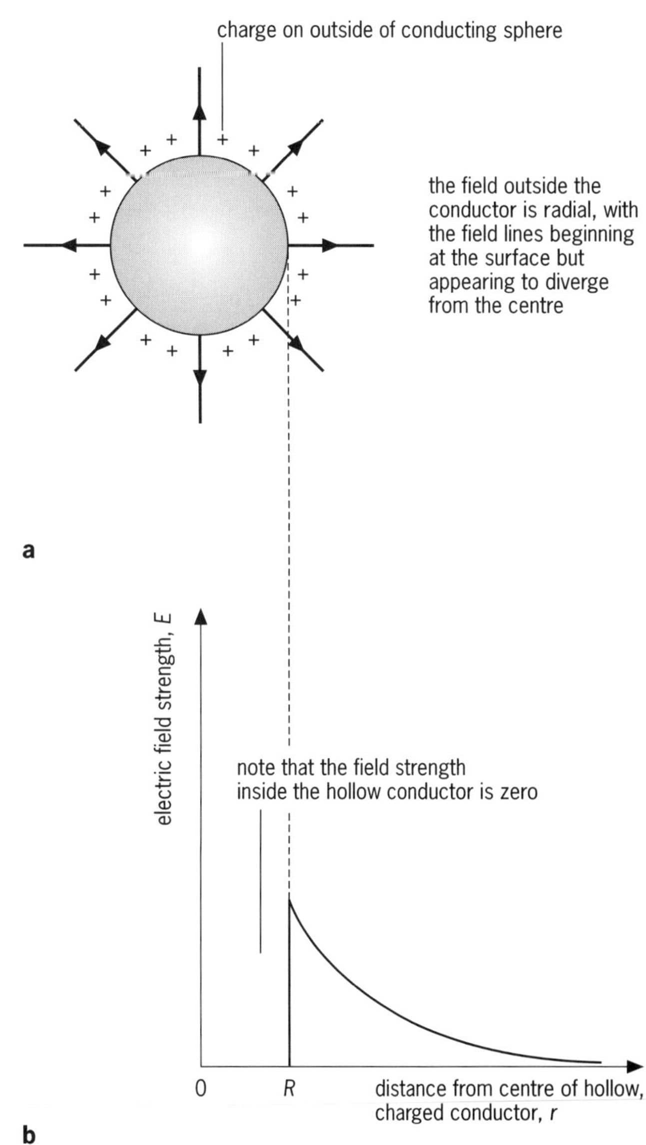

Figure 32.2 Electric field due to a hollow charged conducting sphere

For a charged isolated conducting sphere the external field strength and field shape is identical to that of a point charge at the centre of the sphere.

Forces between charged particles

Coulomb's Law describes the size of the force that exists between two small charged objects placed close together. It states that **the magnitude of the electric force that one particle exerts on another is directly proportional to the product of their charges and inversely proportional to the square of the distance between them**.

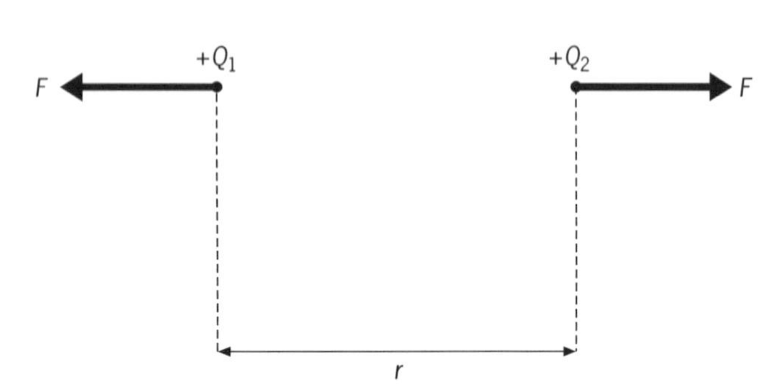

Figure 32.3

Using the annotation of Figure 32.3, the law can be written as

$$F \propto \frac{Q_1 Q_2}{r^2}$$

or $$F = k \times \frac{Q_1 Q_2}{r^2}$$

where k is the constant of proportionality. The value of this constant depends upon the medium between the two charges. For air and a vacuum, k has approximately the same value of $9 \times 10^9\,\mathrm{N\,m^2\,C^{-2}}$, or, more simply, $9 \times 10^9\,\mathrm{F^{-1}\,m}$. In practice this constant is often written as $1/4\pi\epsilon$, giving

$$\boxed{F = \frac{1}{4\pi\epsilon}\left(\frac{Q_1 Q_2}{r^2}\right)}$$

where ϵ is known as the **permittivity** of the medium. If the medium is a vacuum, $\epsilon = \epsilon_0$; this is known as the **permittivity of free space** and has the value $8.85 \times 10^{-12}\,\mathrm{F\,m^{-1}}$.

Example 2

Calculate the force between two small charged particles 50 cm apart in a vacuum. One of the particles carries a charge of $+5.0\,\mu\mathrm{C}$ and the other a charge of $-7.0\,\mu\mathrm{C}$. ($1/4\pi\epsilon_0 = 9 \times 10^9\,\mathrm{N\,m^2\,C^{-2}}$)

$$F = \frac{1}{4\pi\epsilon_0} \times \frac{Q_1 Q_2}{r^2}$$

$$= 9 \times 10^9\,\mathrm{N\,m^2\,C^{-2}} \times \frac{(+5.0 \times 10^{-6}\,\mathrm{C}) \times (-7.0 \times 10^{-6}\,\mathrm{C})}{(0.5\,\mathrm{m})^2}$$

$$= 9 \times 10^9\,\mathrm{N\,m^2\,C^{-2}} \times \frac{-35.0 \times 10^{-12}\,\mathrm{C^2}}{0.25\,\mathrm{m^2}}$$

$$= -1.3\,\mathrm{N}$$

The negative sign indicates that the force between the two particles is an attractive force.

Electric field strength due to a point charge

Consider a small positively charged particle, Q. If a second particle with a charge q is placed in the first particle's electric field it will experience a force, F:

$$F = \frac{1}{4\pi\epsilon}\left(\frac{Qq}{r^2}\right)$$

If the charge on the second particle $q = 1\,\mathrm{C}$ the force exerted upon it will, by the definition of E, be numerically equal to the field strength, E, at that point:

$$\boxed{E = \frac{1}{4\pi\epsilon}\left(\frac{Q}{r^2}\right)}$$

Example 3

Calculate the field strength 25 cm from a point charge of $40\,\mu\mathrm{C}$ in a vacuum. ($1/4\pi\epsilon_0 = 9 \times 10^9\,\mathrm{N\,m^2\,C^{-2}}$)

$$E = \frac{1}{4\pi\epsilon_0} \times \frac{Q}{r^2}$$

$$= 9 \times 10^9\,\mathrm{N\,m^2\,C^{-2}} \times \frac{40 \times 10^{-6}\,\mathrm{C}}{(0.25\,\mathrm{m})^2}$$

$$= 5.8 \times 10^6\,\mathrm{N\,C^{-1}}$$

Note that the formula also applies for the field strength outside a uniformly charged sphere.

Electric potential

The **electric potential**, V, at a point A, Figure 31.4, is defined as being the work done in moving *unit positive charge* from infinity to that point.

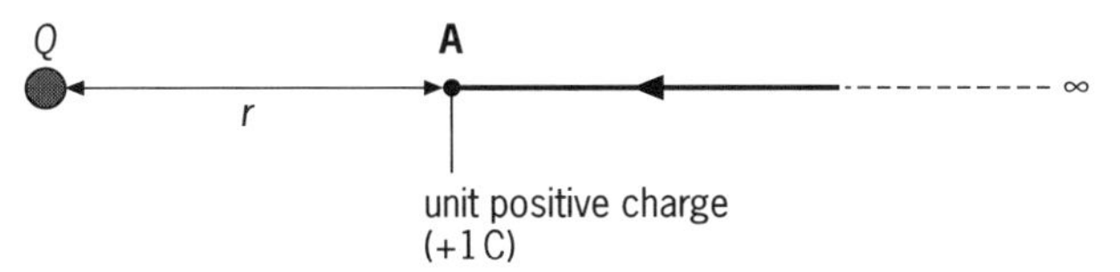

electrical potential at A = work done in moving unit positive charge from infinity

Figure 32.4

The electric potential at any point in a field around a point charge Q can be calculated using the equation

$$V = \frac{1}{4\pi\epsilon}\left(\frac{Q}{r}\right)$$

where r is the distance from the charge to the point being considered. Electric field strength, E, has the unit $\mathrm{N\,C^{-1}}$. Comparing the formulae for V and E you can see that the unit of V is $\mathrm{N\,C^{-1}\,m = J\,C^{-1}}$. This is the **volt**, V (see topic 3).

Again, the formula also applies for the potential outside a uniformly charged sphere.

Example 4

Calculate the electric potential at a point 50 cm from an isolated conducting sphere carrying a charge of 5 μC. (Assume $\epsilon = \epsilon_0$; $1/4\pi\epsilon_0 = 9 \times 10^9\,\mathrm{F^{-1}\,m}$.)

$$V = \frac{1}{4\pi\epsilon_0} \times \frac{Q}{r}$$

$$= 9 \times 10^9\,\mathrm{F^{-1}\,m} \times \frac{5 \times 10^{-6}\,\mathrm{C}}{0.5\,\mathrm{m}}$$

$$= 9 \times 10^4\,\mathrm{F^{-1}\,C} \text{ or } 9 \times 10^4\,\mathrm{V}$$

Electric potential difference

In order to move a charge from point to point in an electric field, work may have to be done on the charge. The work done, ΔW, in moving a charge of q between the two points with a potential difference of ΔV is given by

$$\Delta W = q\Delta V$$

(In topic 3 potential difference was defined as work done per unit charge, which can be seen from this equation rearranged to give $\Delta V = \Delta W/q$.)

Example 5

Calculate the work done when a charge of 15 C is moved through a potential difference of 4.0 V.

$$\Delta W = Q\Delta V$$

$$= 15\,\mathrm{C} \times 4.0\,\mathrm{V}$$

$$= 60\,\mathrm{J}$$

The electronvolt

If an electron is accelerated by an electric field, the energy it gains as it moves through a potential difference of one volt is

$$\Delta W = e\Delta V$$

where e is the charge on an electron (1.6×10^{-19} C).

$$\therefore \Delta W = 1.6 \times 10^{-19}\,\mathrm{C} \times 1\,\mathrm{V}$$

$$= 1.6 \times 10^{-19}\,\mathrm{J}$$

This quantity of energy is known as an **electronvolt**, eV.

$$1\,\mathrm{eV} = 1.6 \times 10^{-19}\,\mathrm{J}$$

The electronvolt is often used as a unit of energy when very small values are considered.

Lines of equipotential

It is often useful to draw on a field diagram not only the field lines but also lines of equipotential; these are lines joining places that have the same potential.

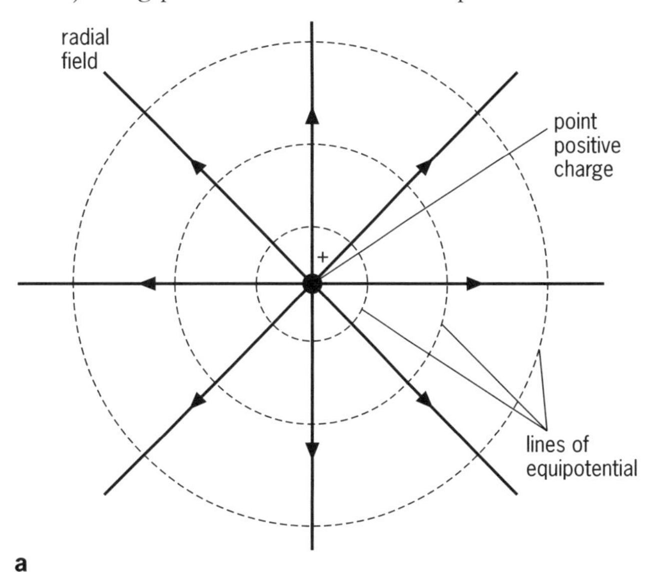

uniform field

lines of equipotential

+

−

b

Figure 32.5 Lines of equipotential for a radial field and a uniform field

If an object is moved along a line of equipotential there is no change in its potential, so there is no work done.

The relationship between electrical field strength and potential

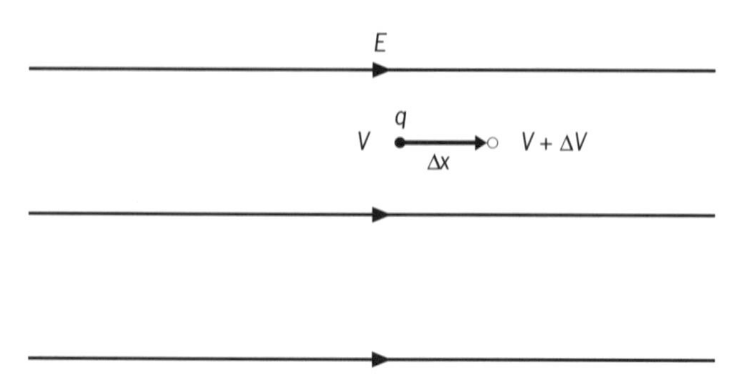

Figure 32.6

If a charge q is moved a small distance Δx between two points in an electric field with a potential difference of ΔV, Figure 32.6, then the work done is given by

$$\Delta W = q\Delta V$$

But, by definition, work done = force × distance moved in the direction of the force, so

$$\Delta W = F\Delta x$$

The force exerted on the charge in the field is given by

$$F = Eq$$

where E is the electric field strength.

It follows that

$$\Delta W = -Eq\Delta x$$

where the minus sign has been inserted to show that, when E and Δx are in the same direction, the work done to move the charge is negative (it loses potential energy).

Therefore

$$-Eq\Delta x = q\Delta V$$

and $$E = -\frac{\Delta V}{\Delta x}$$

or, as the changes become infinitesimally small,

$$\boxed{E = -\frac{\mathrm{d}V}{\mathrm{d}x}}$$

The quantity $\mathrm{d}V/\mathrm{d}x$ is called the **potential gradient** and is a measure of the rate of change of potential with distance in an electric field.

Note from this definition we can see that electric field strength can be measured in $\mathrm{V\,m^{-1}}$ as well as in $\mathrm{N\,C^{-1}}$.

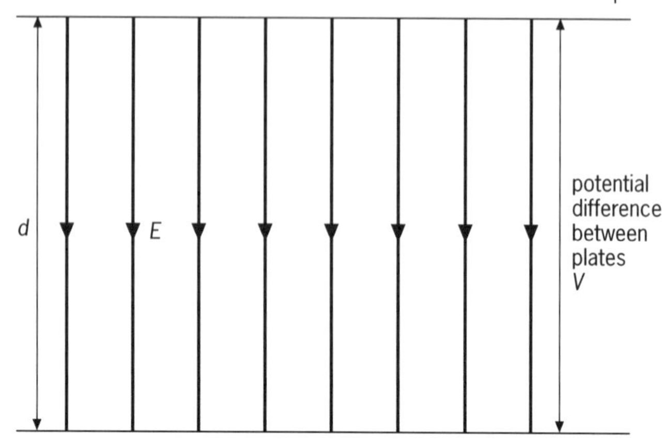

Figure 32.7

If the field being studied is uniform, like that between the plates of a capacitor, Figure 32.7, the above equation can be simplified to

$$\boxed{E = \frac{V}{d}}$$

where E is the *magnitude* of the electric field between the plates.

Example 6

Calculate the field strength between two parallel metal plates 1.0 mm apart with a potential difference across them of 10 V.

$$E = \frac{V}{d}$$

$$= \frac{10\,\mathrm{V}}{1.0 \times 10^{-3}\,\mathrm{m}}$$

$$= 10\,000\,\mathrm{V\,m^{-1}} \text{ or } 10\,\mathrm{kV\,m^{-1}}$$

Charge on electron, $e = -1.6 \times 10^{-19}$ C
Mass of electron $= 9.1 \times 10^{-31}$ kg
Permittivity of free space, $\epsilon_0 = 8.85 \times 10^{-12}\,\mathrm{F\,m^{-1}}$;
take $1/4\pi\epsilon_0 = 9.0 \times 10^9\,\mathrm{F^{-1}\,m}$

Newton per coulomb (NC^{-1}) is used in most of the questions as the unit for electric field strength, E; volt per metre (Vm^{-1}) is equally correct.

Level 1

1 a Is electric field strength, *E*, a vector or a scalar?
b A charge, *q*, of $+3.2 \times 10^{-19}$ C has a force, *F*, of 8.2×10^{-15} N acting on it because of the presence of an electrical field, *E*. What is the strength of this field?
c What is the force on a charge of 4.8×10^{-19} C in a field with $E = 2.6 \times 10^4\,\mathrm{N\,C^{-1}}$?

2 A particle with a charge of $+2.0\,\mu$C is in an electrical field of strength $980\,\mathrm{N\,C^{-1}}$, 3.0 m from a negatively charged plate, Figure 32.8.

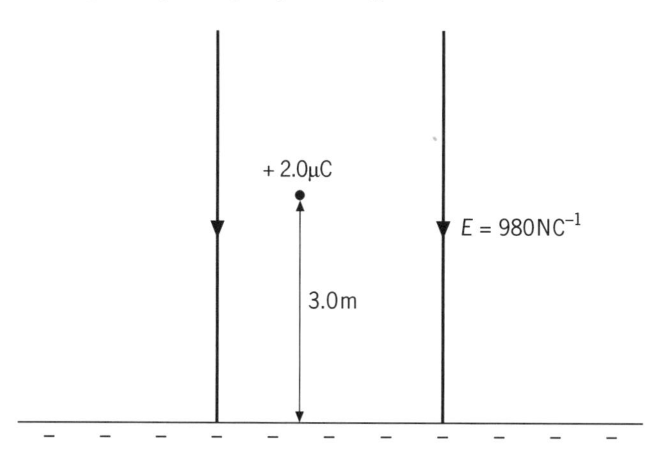

Figure 32.8

a What is the force on the particle?
b What is the work done *by the field* on the particle if it moves to the negatively charged plate?
c What type of energy is gained by the particle when it moves?
d What type of energy is lost by the particle when it moves?

Compare this question with question 2 of topic 31.

3 Explain why the electric field strengths at two points, A and B, the same distance from a positive point charge, Figure 32.9, are not *identical*.

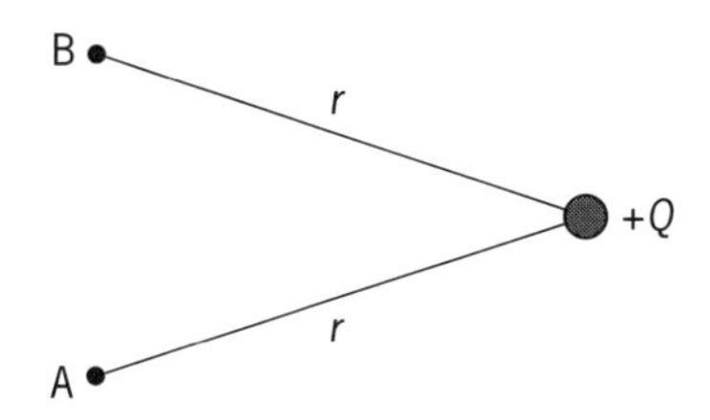

Figure 32.9

4 Use the concept of electric potential to calculate the energy change when a 12 C charge is moved from a point at potential $25\,\mathrm{J\,C^{-1}}$ (25 V) to one of potential $40\,\mathrm{J\,C^{-1}}$ (40 V).

5 Show that the alternative units of electrical field strength, $\mathrm{N\,C^{-1}}$ and $\mathrm{V\,m^{-1}}$, are equivalent.

6

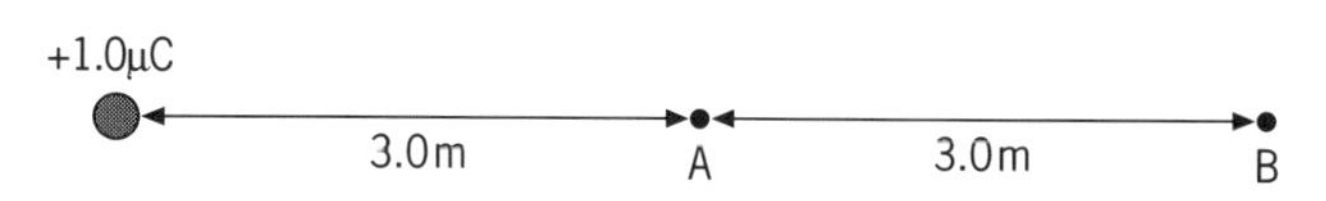

Figure 32.10

a Calculate the electric potential at A, 3.0 m away from a point charge of $+1.0\,\mu$C.
b Calculate the electric potential at B, 6.0 m away from the same charge of $+1.0\,\mu$C.
c How much work would be done moving a $+0.2\,\mu$C charge between A and B?
d Which way would you be moving this small charge if you were doing work *against* the field?

Level 2

7

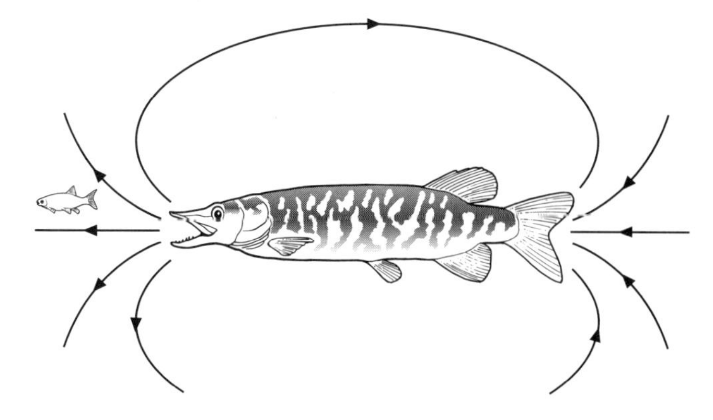

Figure 32.11

An electric fish surrounds itself with an electrical field as shown in Figure 32.11.
a How could it use this to detect objects?
b How does this illustrate the idea of a 'field' in physics?
c If a test mass or test charge is used to measure a field strength, why should it be small?

8 A charged, metal-coated polystyrene ball of mass 6.3 g, suspended on a fine thread, settles at an angle of 10° to the vertical when a charged rod is brought close to it, Figure 32.12.

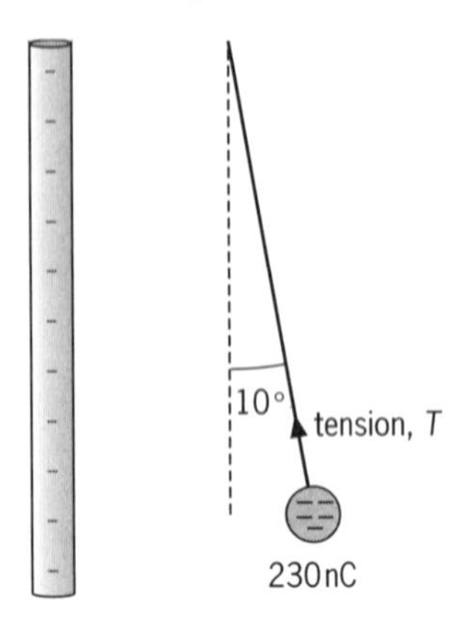

Figure 32.12

a What is the size of the component of the tension, *T*, that opposes the electric force?
b What is the size of the repulsive force between the rod and the ball?
c Assume the field is uniform close to the rod. If there is a charge of 230 nC on the ball, what is the electric field strength?

9 The force on a charged object 2.0 cm from the surface of a Van de Graaff dome of radius 10 cm is 36 mN. What will the force be when the object is at a distance of 14 cm from the surface of the dome?

10 On a dry day the leakage current, *I*, from a Van de Graaff machine through its support is 0.80 μA but it manages to maintain a constant dome potential, *V*.
a The resistance, *R*, of the support is measured as $2.6 \times 10^{11}\,\Omega$. What is the potential of the dome?
b The dome has radius, $r = 0.12$ m. The capacitance of a sphere in air is given by the equation $C = 4\pi\epsilon_0 r$. How much charge, *Q*, is on the dome when it is at this potential?
c What surface charge density, *Q*/*A*, is this? Take the dome as a complete sphere.
d Calculate the electric field strength close to the dome.

11 Electrophoresis is the movement of charged 'particles', suspended in a liquid, under the influence of an electric field. The process can be used to separate molecules with different 'mobilities' because they will move with different velocities.

In an electrophoresis experiment a potential difference of 25 V is put across a strip of paper 0.10 m long, Figure 32.13.

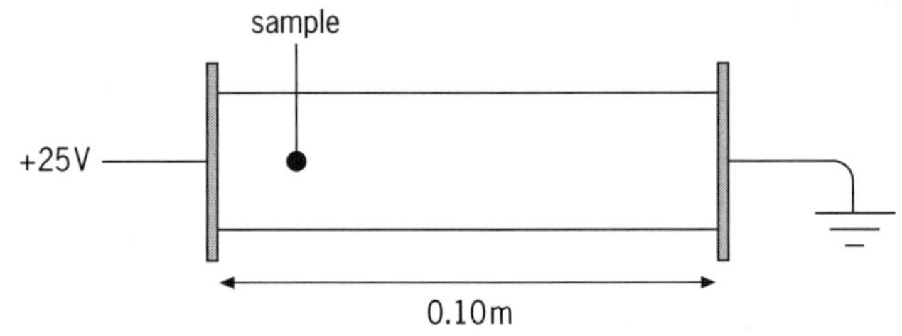

Figure 32.13

a What is the field intensity (strength)?
b What is the force on a singly ionised molecule?
c If the molecule is an anion, what sign charge does it have, and towards which electrode does it travel?

12 The gap in a spark plug is about 0.60 mm. A field of $3.0 \times 10^6\,\mathrm{V\,m^{-1}}$ is needed for a spark to occur and to ignite the petrol–air mixture. What potential difference must be produced by the coil to create this field in the gap?

13 The breakdown strength of an insulator is $2.5\,\mathrm{MV\,m^{-1}}$. It is to be used in a parallel plate capacitor designed to work at 50 V. How close may the conducting layers be?

14 a Using your result from question 6a, calculate the potential at the point X in Figure 32.14.
b Calculate the force experienced by an electron placed at X.

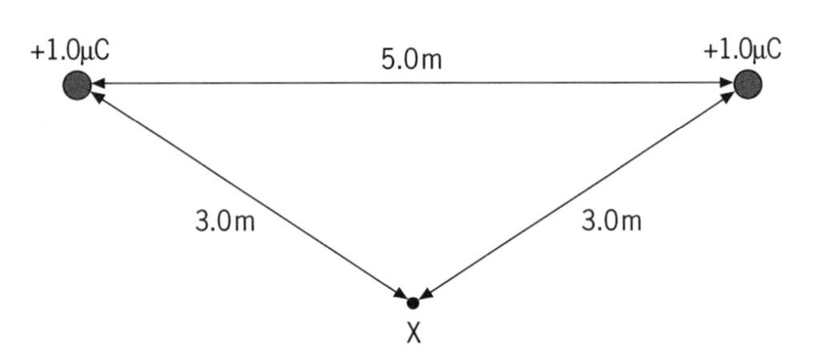

Figure 32.14

15 Shaped electrodes are used to produce an electric field with a field intensity that varies, with distance, as shown in Figure 32.15.

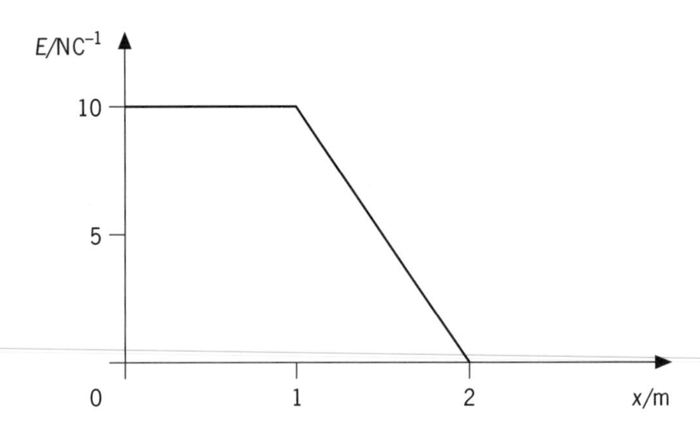

Figure 32.15

What is the potential difference between the origin and the point 2 m away?

Hint: Remember $E = -\dfrac{\Delta V}{\Delta x}$

16 A plant in the soil is effectively earthed. If fungicide is sprayed from a nozzle connected to a high voltage, explain how even the undersides of the leaves may receive spray.

Level 3

17 An electron is projected into the evacuated space between two charged plates of length 5.0 cm, as shown in Figure 32.16. The electric field strength, E, is $3.5 \times 10^4\,\text{N}\,\text{C}^{-1}$. The velocity of the electron, v, is $3.2. \times 10^7\,\text{m}\,\text{s}^{-1}$. Neglect any gravitational effects.

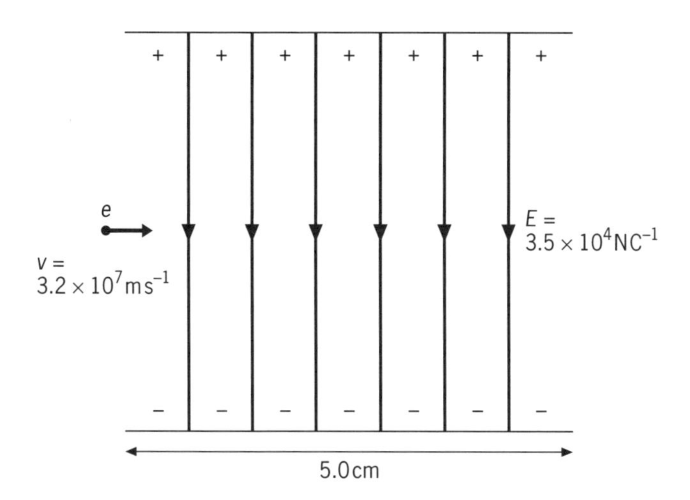

Figure 32.16

a What is the size and direction of the electric force on the electron?
b What is the horizontal velocity of the electron as it leaves the plates?
c How long does it take to pass between the plates?
d What is the vertical acceleration of the electron?
e What is the vertical velocity on leaving the plates?
f What is the size and direction of the final velocity?

18 What is the force on a charge of $+2.0\,\mu\text{C}$ placed at the centre of a square of side 10 cm with a $+1.0\,\mu\text{C}$ charge at each corner, as in Figure 32.17?

Think *before you do any calculations!*

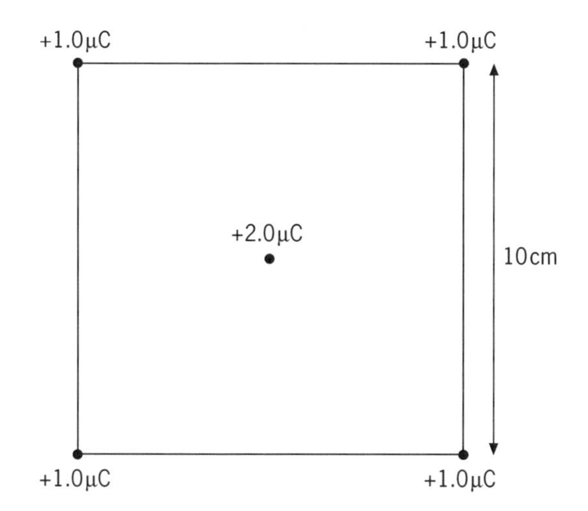

Figure 32.17

19 An alpha particle, charge $+2e$, is passing close to a nucleus with charge $+92e$, Figure 32.18.

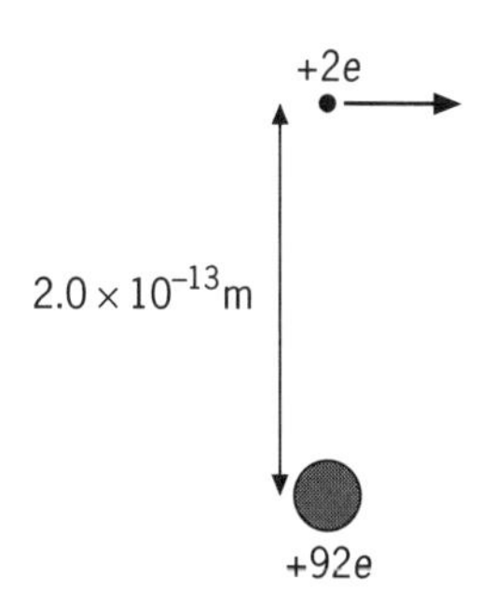

Figure 32.18

a What is the direction of the electric force on the alpha particle when it is at the position shown, and later on?
b What is the size of the force at the distance of closest approach, shown in the diagram?

20 A water molecule is shown in Figure 32.19.

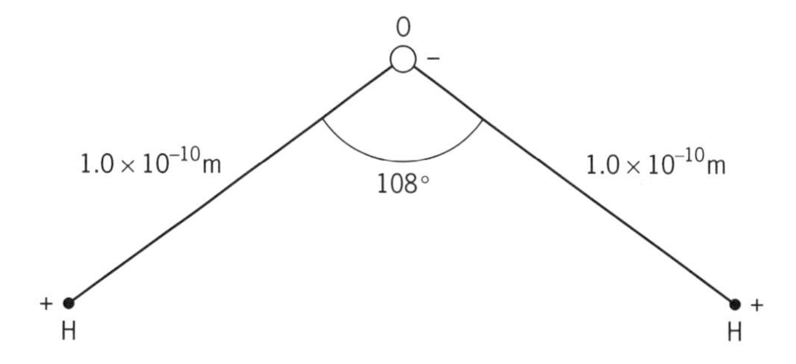

Figure 32.19

The oxygen atom has an effective charge of $-1.1 \times 10^{-19}\,\text{C}$ and the hydrogen atoms act as point charges of $+0.55 \times 10^{-19}\,\text{C}$.

Calculate the magnitude and direction of the resultant force on the oxygen atom due to the presence of the two hydrogen atoms.

21 In the hydrogen atom the electron and the proton are separated by $0.50 \times 10^{-10}\,\text{m}$ when the electron is in the lowest energy state.
a What is the electric potential at this distance from the proton?
b What is the potential energy of the electron in eV and in J?
c How can the ionisation energy of such an atom be measured?

22 An 8.0 MeV alpha particle heads straight for a gold nucleus with $Z = +79e$. How close does it get before it is stopped and repelled back along its original path?

Level 1

1 a Electric field strength is a vector; it is force per unit charge (F/q), and force is a vector.

b Electric field strength = force per unit charge

$$E = \frac{F}{q} = \frac{8.2 \times 10^{-15}\,\text{N}}{3.2 \times 10^{-19}\,\text{C}}$$

$$= \mathbf{2.6 \times 10^{4}\,N\,C^{-1}}$$

c Force = charge × electric field strength

$$F = qE = 4.8 \times 10^{-19}\,\text{C} \times 2.6 \times 10^{4}\,\text{N}\,\text{C}^{-1}$$
$$= \mathbf{1.2 \times 10^{-14}\,N}$$

2 a Force = charge × electric field strength

$$F = qE = 2.0 \times 10^{-6}\,\text{C} \times 980\,\text{N}\,\text{C}^{-1}$$
$$= \mathbf{1.96 \times 10^{-3}\,N}$$

b Work = force × distance moved in the direction of the force

$$= 1.96 \times 10^{-3}\,\text{N} \times 3.0\,\text{m} = \mathbf{5.88\,mJ}$$

c The particle gains **kinetic energy**.

d The particle loses **electric potential energy**.

3 Electric field strength is a vector, so although it will have the same *magnitude* at points A and B, the *direction* will be different. The field radiates out from the centre.

4 a Change in potential, $\Delta V = 40\,\text{J}\,\text{C}^{-1} - 25\,\text{J}\,\text{C}^{-1} = 15\,\text{J}\,\text{C}^{-1}$.

Total energy change $= q\Delta V = 12\,\text{C} \times 15\,\text{J}\,\text{C}^{-1} = \mathbf{180\,J}$.

5 By definition, $\text{V} = \text{J}\,\text{C}^{-1}$ and $\text{J} = \text{N}\,\text{m}$.

$\therefore \text{V} = \text{N}\,\text{m}\,\text{C}^{-1}$

So, dividing by m,

$\text{V}\,\text{m}^{-1} = \text{N}\,\text{C}^{-1}$

6 a The potential at a distance r from a point charge Q is given by the equation

$$V = \frac{1}{4\pi\epsilon_0} \times \frac{Q}{r}$$

So at 3.0 m from a +1.0 μC charge, with $1/4\pi\epsilon_0 = 9.0 \times 10^{9}\,\text{F}^{-1}\,\text{m}$,

$$V = 9.0 \times 10^{9}\,\text{F}^{-1}\,\text{m} \times \frac{1.0 \times 10^{-6}\,\text{C}}{3.0\,\text{m}} = +3000\,\text{F}^{-1}\,\text{C}$$
$$= \mathbf{+3000\,V}$$

b Potential is inversely proportional to distance, so at 6.0 m, when the distance has doubled, the potential will have halved to give $V = \mathbf{+1500\,V}$.

c The work done is given by

$$\Delta W = q\Delta V = 0.2 \times 10^{-6}\,\text{C} \times 1500\,\text{V} = \mathbf{3 \times 10^{-4}\,J}$$

d You would be pushing it closer, against the repulsive force, so from **B to A**.

Level 2

7 a An object in the field would change the shape and strength of the field. The fish could perhaps detect this change and know the direction and size of the object.

b A field affects objects in the space where it exists. The area close to the fish is filled with the field so for the fish the area is 'covered' at all times.

c If the test objects were large they would affect the field by their own field and the measurement would be incorrect.

8 a The force acting vertically downward is the weight of the ball,

$$mg = 6.3 \times 10^{-3}\,\text{kg} \times 9.8\,\text{N}\,\text{kg}^{-1} = 0.062\,\text{N}$$

This is balanced by the vertical component of the tension in the thread (see Figure 32.20):

$$T\cos 10° = mg$$

$$\therefore T = \frac{0.062\,\text{N}}{\cos 10°} = \frac{0.062\,\text{N}}{0.985} = 0.063\,\text{N}$$

The force opposing the electric force is the horizontal component of the tension in the thread:

$$T\sin 10° = 0.063\,\text{N} \times \sin 10°$$
$$= 0.063\,\text{N} \times 0.174$$
$$= \mathbf{0.011\,N}$$

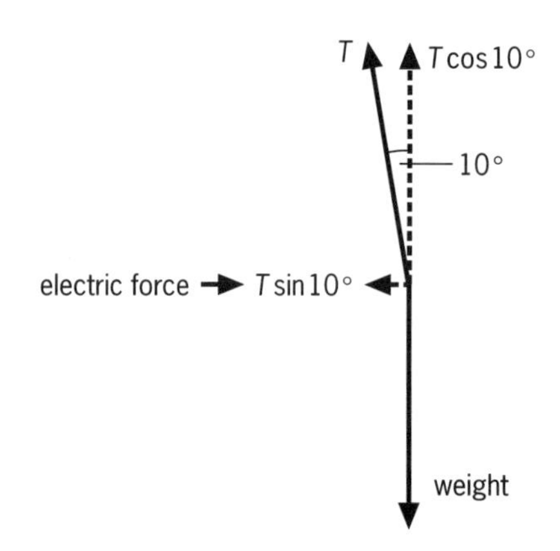

Figure 32.20

b The ball is in equilibrium so the force on the ball, because of the presence of the charged rod, is **0.011 N**.

c The electric field strength is given by $E = F/q$. In this example $F = 0.011\,\text{N}$ and $q = 230\,\text{nC}$,

$$\therefore E = \frac{0.011\,\text{N}}{230 \times 10^{-9}\,\text{C}} = \mathbf{4.8 \times 10^{4}\,N\,C^{-1}}$$

9 a Coulomb's Law applies for distances *from the centre* of a uniformly charged sphere. The first distance from the centre of the dome is 12 cm; the second distance is 24 cm. The force is inversely proportional to the distance squared, so if you double the distance you will reduce the force to one quarter. So

$$F = \frac{36\,\text{mN}}{4} = \mathbf{9\,mN}$$

10 a The potential difference between the dome and 'earth' produces the leakage current.

$$V = IR$$
$$= 0.80 \times 10^{-6}\,\text{A} \times 2.6 \times 10^{11}\,\Omega = \mathbf{2.1 \times 10^5\,V}$$

b Capacitance, $C = Q/V$

$$\therefore Q = C \times V = 4\pi\epsilon_0 r \times V$$
$$= 4\pi \times 8.85 \times 10^{-12}\,\text{F m}^{-1} \times 0.12\,\text{m} \times 2.1 \times 10^5\,\text{V}$$
$$= \mathbf{2.8\,\mu C}$$

c The charge density is

$$\frac{Q}{\text{surface area of dome}} = \frac{2.8 \times 10^{-6}\,\text{C}}{4\pi \times (0.12\,\text{m})^2}$$
$$= \mathbf{15.5\,\mu C\,m^{-2}}$$

d You can calculate the electric field strength using

$$E = \frac{1}{4\pi\epsilon_0} \times \frac{Q}{r^2}$$
$$= \frac{2.8 \times 10^{-6}\,\text{C}}{4\pi \times 8.85 \times 10^{-12}\,\text{F m}^{-1} \times (0.12\,\text{m})^2}$$
$$= \mathbf{1.8 \times 10^6\,N\,C^{-1}}$$

11 a The electric field strength, or intensity, is given by the potential gradient $E = V/d$. 25 V across 0.10 m gives

$$E = \frac{25\,\text{V}}{0.10\,\text{m}} = \mathbf{250\,V\,m^{-1}}$$

b The force is qE, where q here is the charge on an electron:

$$E = 1.6 \times 10^{-19}\,\text{C} \times 250\,\text{V m}^{-1}$$
$$= 4.0 \times 10^{-17}\,\text{J m}^{-1}$$
$$= \mathbf{4.0 \times 10^{-17}\,N}$$

c An anion has a negative charge. An anion will be attracted to the anode, which is the positive or higher voltage electrode.

12 The field strength is the potential gradient, $E = V/d$.

$$\therefore V = dE = 0.60 \times 10^{-3}\,\text{m} \times 3.0 \times 10^6\,\text{V m}^{-1} = \mathbf{1.8\,kV}$$

13 The field strength is given by $E = V/d$, which you can rearrange to give $d = V/E$. The minimum allowable separation for a given value of V is achieved when the field strength, E, is equal to the breakdown (maximum) field strength.

$$d_{min} = \frac{V}{E_{max}} = \frac{50\,\text{V}}{2.5 \times 10^6\,\text{V m}^{-1}} = \mathbf{20\,\mu m}$$

The separation must be greater than this.

14 a From question 6a, the potential at X is raised by +3000 V by each of the charges. As potential is a scalar, the total is **+6000 V**.

b Force, however, is a vector. The size of *each* force is given by

$$F = \frac{1}{4\pi\epsilon_0} \times \frac{Q_1 Q_2}{r^2}$$
$$F = \frac{9.0 \times 10^9\,\text{F}^{-1}\,\text{m} \times 1.6 \times 10^{-19}\,\text{C} \times 1.0 \times 10^{-6}\,\text{C}}{(3.0\,\text{m})^2}$$
$$= 1.6 \times 10^{-16}\,\text{N}$$

The directions are as shown in Figure 32.21.

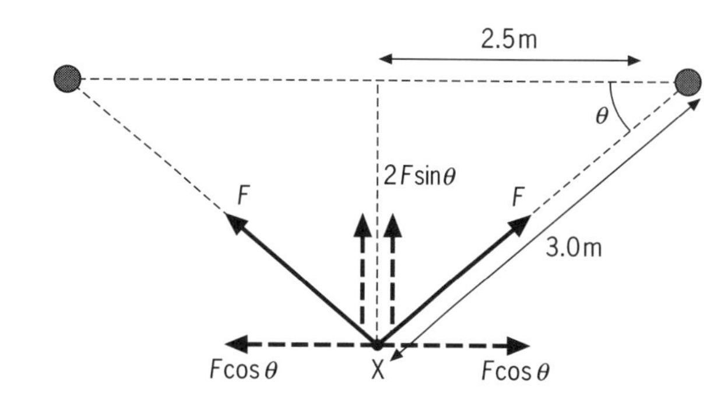

Figure 32.21

$$\cos\theta = \frac{2.5\,\text{m}}{3.0\,\text{m}} = 0.833$$
$$\therefore \theta = \cos^{-1} 0.833 = 33.6°$$

The components that are parallel to the line joining the two charges cancel. The two components at 90° to the line add to give

$$2F\sin\theta = 2 \times 1.6 \times 10^{-16}\,\text{N} \times 0.553$$
$$= \mathbf{1.77 \times 10^{-16}\,N}$$

This is the resultant force on the electron, directed towards the centre of the line joining the charges.

15 You can rearrange the equation $E = -\Delta V/\Delta x$ to give $\Delta V = -E\Delta x$, which means that the potential difference will be given by the area under the graph of E against x. In this case

$$\Delta V = -[(10\,\text{V m}^{-1} \times 1.0\,\text{m}) + (\tfrac{1}{2} \times 10\,\text{V m}^{-1} \times 1.0\,\text{m})\}$$
$$= \mathbf{-15\,V}$$

(The potential decreases as x increases.)

16 The field lines will start from the nozzle and will end on the earthed plant. The charged drops of spray will follow the field lines. See Figure 32.22.

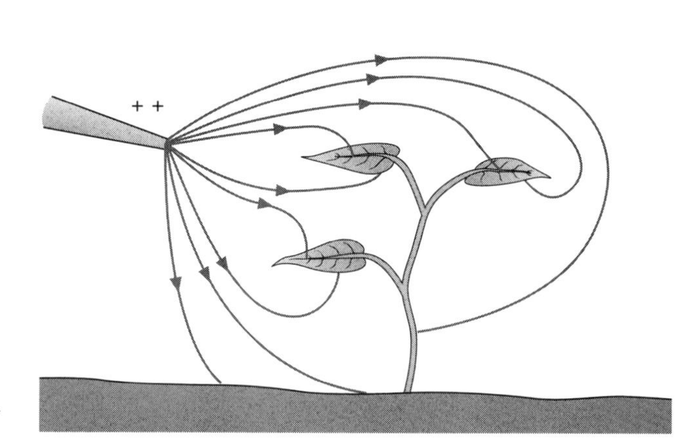

Figure 32.22

Level 3

17 a The size of the force on the electron is

$$F = qE = 1.6 \times 10^{-19}\,\text{C} \times 3.5 \times 10^{4}\,\text{N C}^{-1} = \mathbf{5.6 \times 10^{-15}\,N}$$

The force is **upwards** because the charge on the electron is negative.

b The horizontal velocity is unchanged because the force is always vertical.

c Any change in the vertical velocity does not affect the horizontal component, so

$$\text{time} = \frac{\text{horizontal distance}}{\text{horizontal speed}} = \frac{5.0 \times 10^{-2}\,\text{m}}{3.2 \times 10^{7}\,\text{m s}^{-1}}$$

$$= \mathbf{1.6\,ns}$$

d The force was found in part a.

$$\text{vertical acceleration, } a = \frac{\text{force}}{\text{mass}} = \frac{5.6 \times 10^{-15}\,\text{N}}{9.1 \times 10^{-31}\,\text{kg}} = \mathbf{6.1 \times 10^{15}\,m\,s^{-2}}$$

(How many g is this?)

e The final vertical velocity is given by $v = u + at$. The initial vertical velocity, $u = 0$.

$$\therefore v = 0\,\text{m s}^{-1} + 6.1 \times 10^{15}\,\text{m s}^{-2} \times 1.6 \times 10^{-9}\,\text{s} = \mathbf{9.8 \times 10^{6}\,m\,s^{-1}}\text{ upwards}$$

f The direction of the resultant velocity is given (see Figure 32.23) by

$$\tan\theta = \frac{9.8 \times 10^{6}\,\text{m s}^{-2}}{3.2 \times 10^{7}\,\text{m s}^{-2}} = 0.306$$

$$\therefore \theta = \tan^{-1} 0.306 = \mathbf{17°}$$

You can use Pythagoras' theorem to get the size of the final resultant velocity (see topic 6).

$$v_R = \sqrt{(9.8 \times 10^{6}\,\text{m s}^{-1})^2 + (3.2 \times 10^{7}\,\text{m s}^{-1})^2} = \mathbf{3.3 \times 10^{7}\,m\,s^{-1}}$$

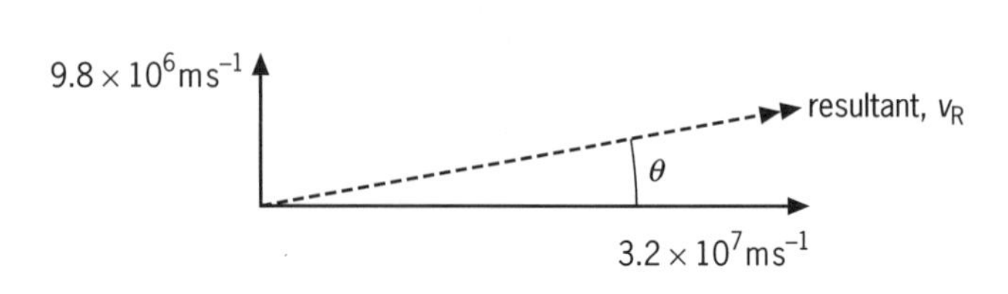

Figure 32.23

18 The forces from the diagonally opposite charges are equal and opposite so they all cancel out and there is zero force at the centre.

19 a The force is radial so it is at **90° to the velocity** at the point shown. The alpha particle will accelerate in the direction of the force but, as it moves, the direction of the force will change, being always radial.

b The size of the force at the closest distance is given by

$$F = \frac{Q_1 Q_2}{4\pi\epsilon_0 r^2}$$

Using the data supplied,

$$F = \frac{9.0 \times 10^{9}\,\text{F}^{-1}\,\text{m} \times 2e \times 92e}{(2.0 \times 10^{-13}\,\text{m})^2}$$

Substituting for $e = 1.6 \times 10^{-19}\,\text{C}$, this gives

$$F = \mathbf{1.1\,N}$$

This is a large force when you consider the mass of the alpha particle!

20 The two components of the attractive forces that are parallel to the line of symmetry will add up. The other components of each force (at right angles to the symmetry line) will cancel out.

The force between the oxygen atom and one of the hydrogen atoms is given by

$$F = \frac{1}{4\pi\epsilon_0} \times \frac{Q_1 Q_2}{r^2}$$

$$= 9.0 \times 10^{9}\,\text{F}^{-1}\,\text{m} \times \frac{1.1 \times 10^{-19}\,\text{C} \times 0.55 \times 10^{-19}\,\text{C}}{(1.0 \times 10^{-10}\,\text{m})^2}$$

$$= 5.4\,\text{nN}$$

Resolving and adding the two attractive forces gives

$$2F \times \cos 54° = 2 \times 5.4\,\text{nN} \times 0.588 = \mathbf{6.4\,nN}$$

towards the hydrogen atoms, along the line of symmetry. See Figure 32.24.

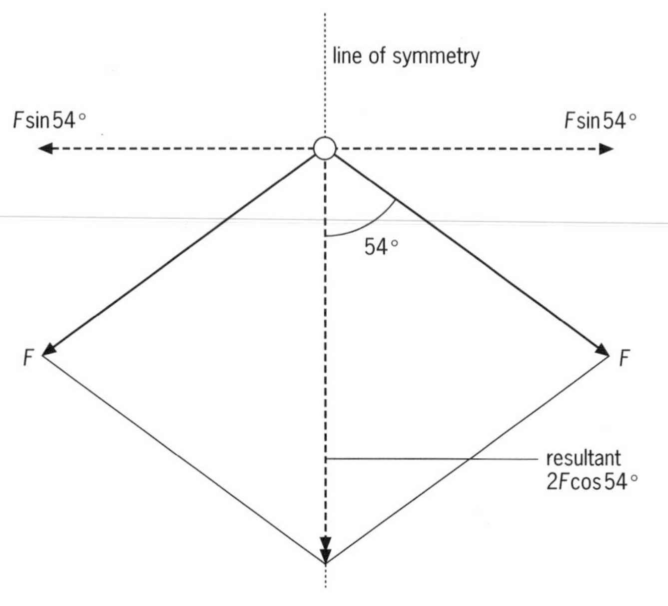

Figure 32.24

21 Each of the particles has a charge of size 1.6×10^{-19} C.

a The potential due to the proton is

$$V = \frac{1}{4\pi\epsilon_0} \times \frac{Q}{r}$$

$$= \frac{9.0 \times 10^9 \, \text{F}^{-1}\,\text{m} \times 1.6 \times 10^{-19}\,\text{C}}{0.50 \times 10^{-10}\,\text{m}} = \mathbf{28.8\,V}$$

b The potential at infinity is zero, so the work done to bring an electron in from infinity is negative because the electron is attracted by the proton. The potential energy of the electron, in its lowest energy state, is therefore

$$\mathbf{-28.8\,eV} \quad \text{or} \quad \mathbf{-4.6 \times 10^{-18}\,J}$$

c Find an account of the Franck–Hertz experiment, in which electrons are accelerated through a steadily increasing voltage until they gain enough energy to 'knock' the electrons of the surrounding atoms to higher energies. The collisions are inelastic and the bombarding electrons lose their energy. At the ionisation potential, the number of electrons passing through the gas suddenly drops so that you know the correct energy has been reached.

22 The original kinetic energy, E_k, of the alpha particle is lost as the particle slows down on meeting the repulsive force from the nucleus. The kinetic energy is changed into electric potential energy, E_p, as the alpha particle approaches the nucleus. When E_p is equal to the original value of E_k, the alpha particle has stopped and is at its distance of closest approach.

The initial kinetic energy of the alpha particle is 8.0 MeV. In joules this is

$$8.0 \times 10^6\,\text{eV} \times 1.6 \times 10^{-19}\,\text{J eV}^{-1} = 1.3 \times 10^{-12}\,\text{J}$$

This energy is converted into electric potential energy, E_p, as the alpha particle approaches the nucleus.

The electric potential at a distance r from the nucleus due to its charge is given by

$$V = \frac{1}{4\pi\epsilon_0} \times \frac{Q}{r}$$

where $Q = +79e$. At this distance the potential energy of the alpha particle, charge q, is

$$E_p = q \times V = \frac{qQ}{4\pi\epsilon_0 r}$$

You can rearrange this equation to make r the subject:

$$r = \frac{qQ}{4\pi\epsilon_0 E_p}$$

In this example the charge on the alpha particle is $q = 2e$ and the value of r at 'closest approach' will be when $E_p = E_k$, that is

$$r = \frac{9.0 \times 10^9\,\text{F}^{-1}\,\text{m} \times 2 \times 79 \times (1.6 \times 10^{-19}\,\text{C})^2}{1.3 \times 10^{-12}\,\text{J}}$$

$$= \mathbf{2.8 \times 10^{-14}\,m}$$

Check the units:
$F^{-1}\,m\,C^2\,J^{-1} = (C\,V^{-1})^{-1}\,m\,C^2\,(V\,C)^{-1} = C^{-1}\,V\,m\,C^2\,V^{-1}\,C^{-1} = m.$

33 Comparison of electric and gravitational fields

The inverse square laws

The force between two point charges a distance r apart is described by Coulomb's Law.

Coulomb's Law $F = \frac{1}{4\pi\epsilon}\left(\frac{Q_1Q_2}{r^2}\right)$

The force between two masses a distance r apart is described by Newton's Law of Universal Gravitation.

Newton's Law $F = G\left(\frac{M_1M_2}{r^2}\right)$

Note that both laws are inverse square laws; that is, if the distance between the two charges or the two masses is doubled the force between them will decrease to one quarter of its original value.

Gravitational forces are always attractive; forces between electrical objects can be either attractive or repulsive. Gravitational forces are extremely weak compared with electric forces unless very large masses are being considered.

Field strength

The field strength, E, at a point inside an electric field is defined as being the force that acts on unit charge placed at that point:

$$E = \frac{F}{q} \quad \text{or} \quad E = \frac{1}{4\pi\epsilon}\left(\frac{Q}{r^2}\right)$$

where Q is the point charge creating the field (or the charge on a uniformly charged sphere).

Similarly, the field strength, g, inside a gravitational field is defined as the force that acts on unit mass placed at that point.:

$$g = \frac{F}{m} \quad \text{or} \quad g = G\left(\frac{M}{r^2}\right)$$

where M is the spherical mass creating the field.

Potential difference and potential

If a charge q is moved within an electric field such that it gains or loses energy, ΔW, then there is a potential difference ΔV between the initial and final positions of the object. ΔV is defined as being the work done per unit charge:

$$\Delta V = \frac{\Delta W}{q}$$

Similarly, if a mass m is moved within a gravitational field and it gains or loses energy, the potential difference between the initial and final positions ΔV is defined as the work done per unit mass:

$$\Delta V = \frac{\Delta W}{m}$$

The potential, V, of a point in an electric field is described as being the work that must be done in moving the unit positive charge from infinity to that point.

For a radial field

$$V = \frac{1}{4\pi\epsilon}\left(\frac{Q}{r}\right)$$

where Q is the charge creating the field.

For a uniform field

$$V = Ed$$

where d is the distance from the surface of zero potential.

Similarly the potential, V, of a point in a gravitational field is described as being the work that must be done in moving the unit mass from infinity to that point.

For a radial field

$$V = -G\left(\frac{M}{r}\right)$$

where M is the spherical mass creating the field. The negative sign shows that the field is always attractive.

For a uniform field

$$V = gh$$

where h is the distance from the surface chosen to have zero potential, usually the surface of the Earth.

Field lines

We can represent both electric and gravitational fields using field lines and lines of equipotential.

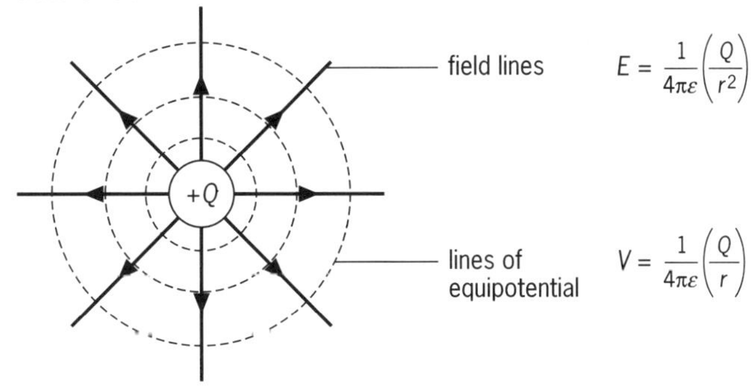

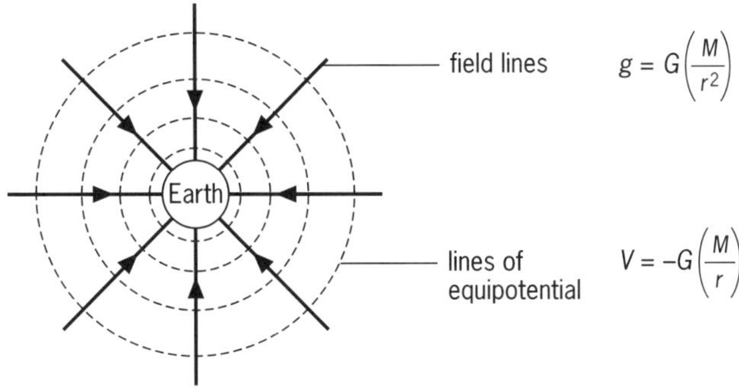

Figure 33.1 Representations of radial electric and gravitational fields

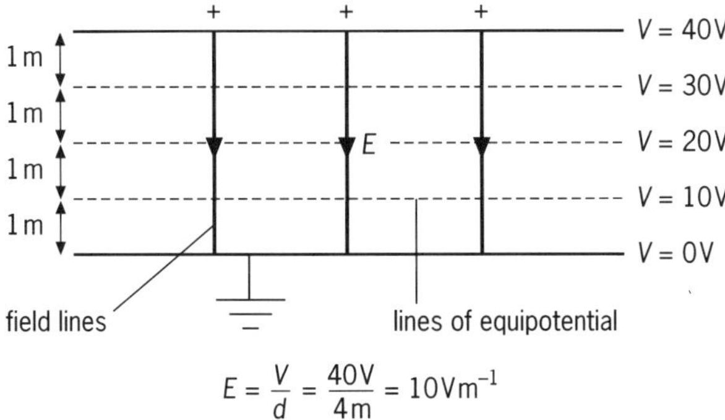

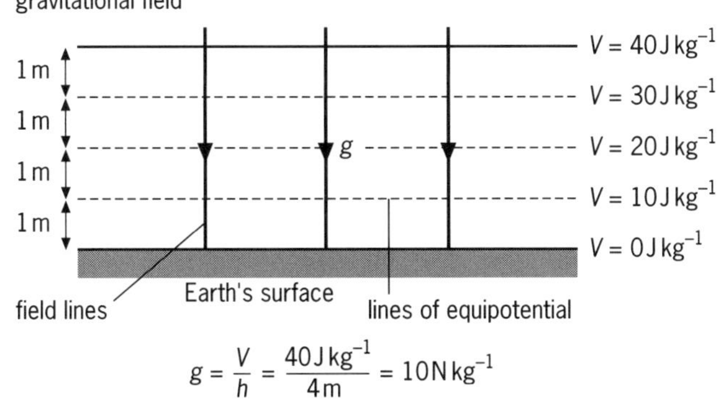

Figure 33.2 Representations of uniform electric and gravitational fields

Graphs

Figures 33.3 and 33.4 show graphically the variation in field strength and potential with distance from an electric charge or a mass.

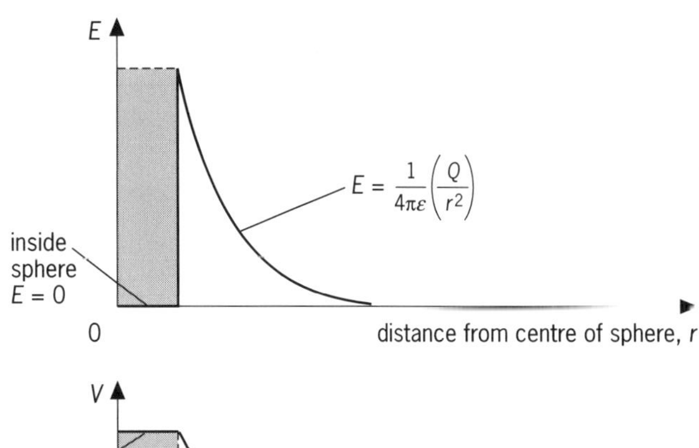

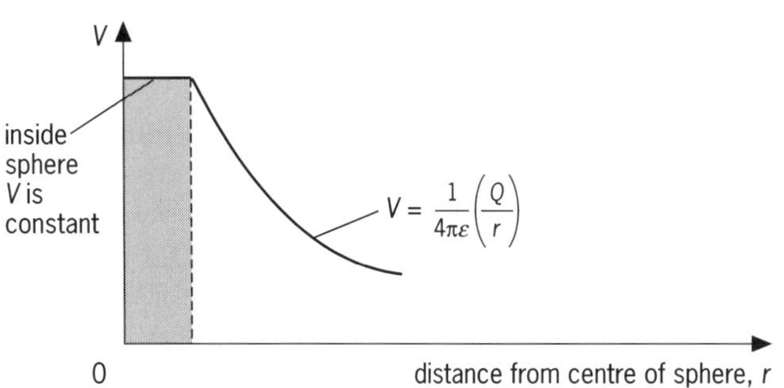

Figure 33.3 Electric field near a hollow, uniformly charged sphere

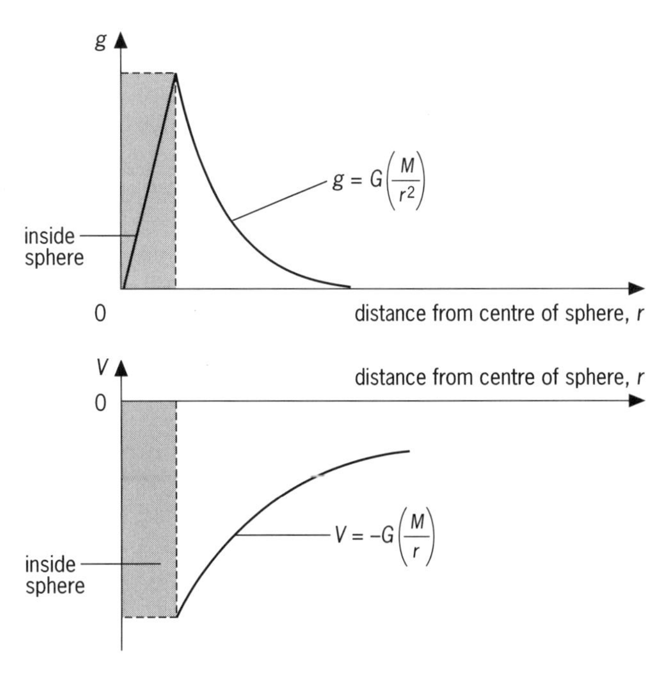

Figure 33.4 Gravitational field near a uniform spherical mass

Note that for both types of field:

- the area under a field strength graph between any two points is a measure of the potential difference between the points, that is the work done on unit charge or mass;
- the gradient at any point on a potential graph is a measure of the magnitude of the field strength at that point.

$G = 6.7 \times 10^{-11}\,\mathrm{N\,m^2\,kg^{-2}}$
$1/4\pi\epsilon_0 = 9.0 \times 10^9\,\mathrm{F^{-1}\,m}$

Level 1

1 a Write the units of both E and g in terms that involve newtons.

b What is a *major* difference between electric and gravitational fields?

c What is meant by a uniform field?

d Give an example and draw a diagram to show where you would use the equations that apply in a uniform electric field.

2 a Draw the gravitational field due to a point mass.

b Draw the electric field due to a positive point charge.

3 Compare the attractive force between two 1.0 kg masses that are 1.0 m apart with the force between a positive charge of 1.0 C and a negative charge of 1.0 C that are also 1.0 m apart in air. Comment on your answers.

4 Using the field strengths $g = 9.8\,\mathrm{N\,kg^{-1}}$ and $E = 9.8\,\mathrm{N\,C^{-1}}$, sketch and label, with values, the equipotentials every metre for distances of 0 to 3.0 m from

a the surface of the Earth

b an earthed plate with another plate positively charged and 5.0 m away.

5 a A Van de Graaff generator is working well on a dry clear day. The potential, measured with a flame probe, is 3000 V at 2.0 m away from the centre of the dome.

i What would you expect the potential to be at 4.0 m from the centre?

ii What would you expect the potential to be at 1.0 m from the centre?

iii Would the force on a charged object vary with distance in the same way?

b An asteroid, in space, is investigated by a space probe. The gravitational potential at an estimated 10 km from its centre is $-40\,\mathrm{kJ\,kg^{-1}}$, taking the potential of space as zero.

i What would you expect the potential to be at 20 km from the centre?

ii What would you expect the potential to be at 5 km from the centre?

iii Would the gravitational pull on the probe vary in the same way?

Level 2

6 a What charge is subject to a force of 4.2 N when in an electric field of strength $E = 8.6 \times 10^6\,\mathrm{N\,C^{-1}}$?

b What mass is subject to a force of 42 N when in a gravitational field of strength $g = 8.6\,\mathrm{N\,kg^{-1}}$?

7 a A small charged ball of mass 1.2 g hangs on a string close to the dome of a Van de Graaff machine as shown in Figure 33.5. The string is pulled to 30° from the vertical by the electrical attraction on the ball.

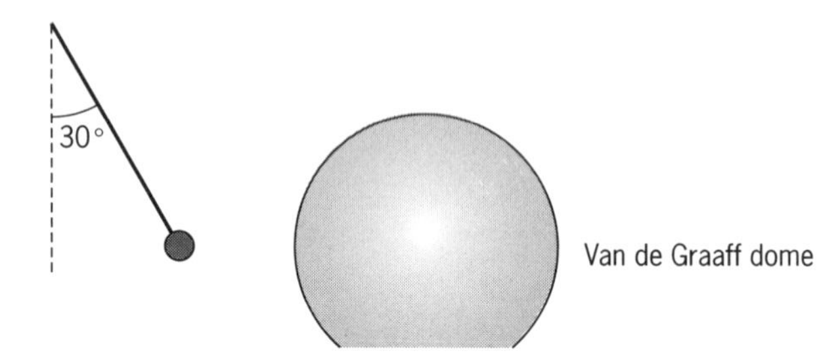

Figure 33.5

i Draw a free body diagram showing all the forces acting on the ball.

ii Draw a triangle of vectors showing that the forces are in equilibrium.

iii From your diagrams, work out the electric force on the ball.

iv The field strength in that position is $4.0 \times 10^5\,\mathrm{N\,C^{-1}}$. Calculate the charge on the ball.

b If the presence of a mountain produced a similar attraction to the Van de Graaff dome, but gravitational in origin,

i what would you use to detect it and what would you observe?

ii what factor complicates the gravitational case?

8 In this question do not worry about the *direction* of movement relative to the direction of the field. The potential energies will either increase or decrease and the work will be done either *on* the field or *by* the field.

a A mass, m, of 3.0 kg is raised a height, Δh, of 1.0 m from the surface of the Earth where the gravitational field strength, g, is $9.8\,\mathrm{N\,kg^{-1}}$. See Figure 33.6a.

i What is the change in gravitational potential energy caused by the work done on the mass?

ii Calculate the work done (defined as force × distance moved in the direction of the force) and compare with your answer to part i.

iii The work done (or energy change) per unit mass, $\Delta W/m$, is called the change in potential, ΔV. Calculate the change in potential for this mass.

iv Show that $\Delta V = g\Delta h$ and calculate $g\Delta h$ for this example.

v Write an expression for the energy change, ΔW, of a mass in terms of change in potential, ΔV.

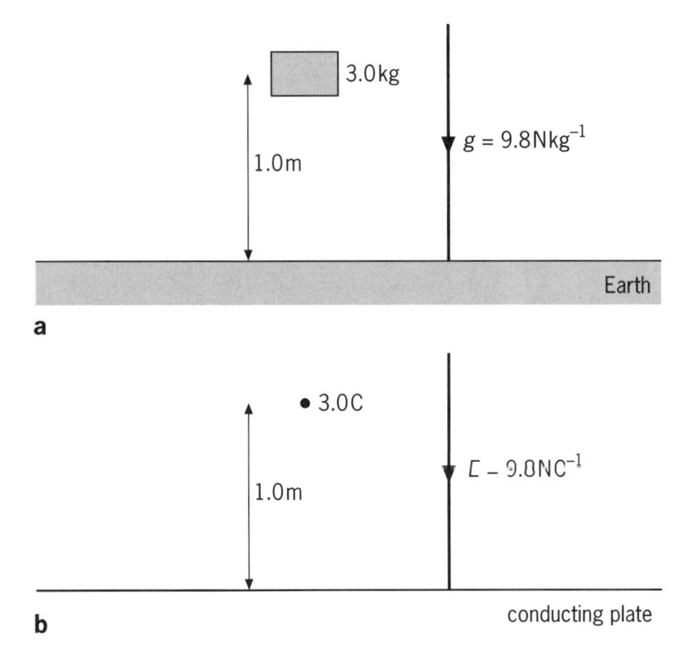

Figure 33.6

b A charge, q, of 3.0 C is moved a distance, Δx, of 1.0 m parallel to a uniform electric field of strength $E = 9.8 NC^{-1}$ (or Vm^{-1}). See Figure 33.6b.
i Calculate the change in electric potential energy, ΔW, where $\Delta W = qE\Delta x$.
ii Obtain this result using the definition of work in part a ii.
iii The work done (or energy change) per unit charge, $\Delta W/q$, is called the change in potential, ΔV. Calculate ΔV for the above charge.
iv Show that $\Delta V = E\Delta x$ and calculate $E\Delta x$ for the above example.
v Write an expression for the energy change of a charge in terms of change in potential, ΔV.

9 a The potential along a conductor is plotted in Figure 33.7a. What is the electric field strength?
b The gravitational potential in a region of space is plotted in Figure 33.7b. What is the gravitational field strength?

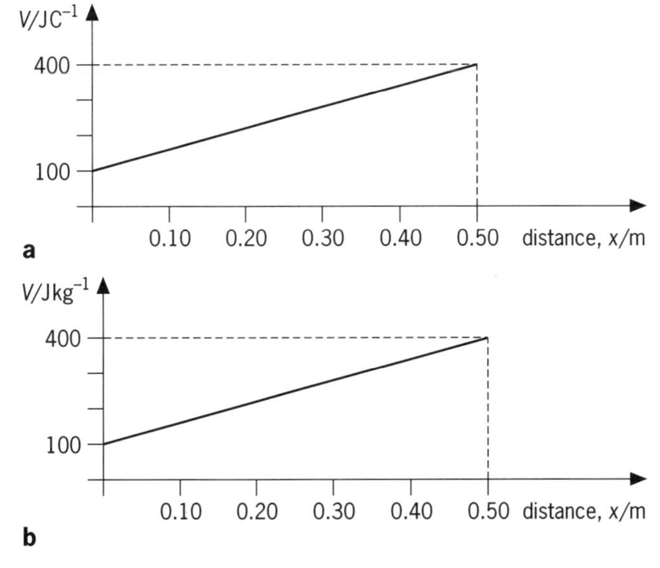

Figure 33.7

10 a Draw a graph to show how electric field intensity, E, varies with distance if the potential, V, varies as shown in the graph of Figure 33.8.

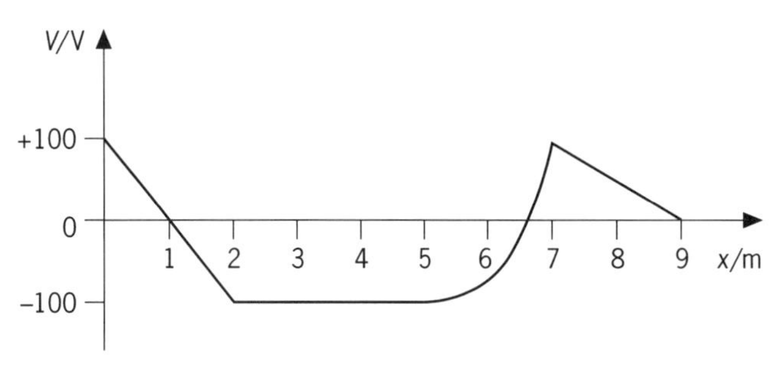

Figure 33.8

b Would your graph be different if the potential were gravitational rather than electric?

11 In Figure 33.9 graphs are plotted of **i** weight, **ii** electric field strength and **iii** gravitational potential, against distance moved along a field line in a region of uniform field.

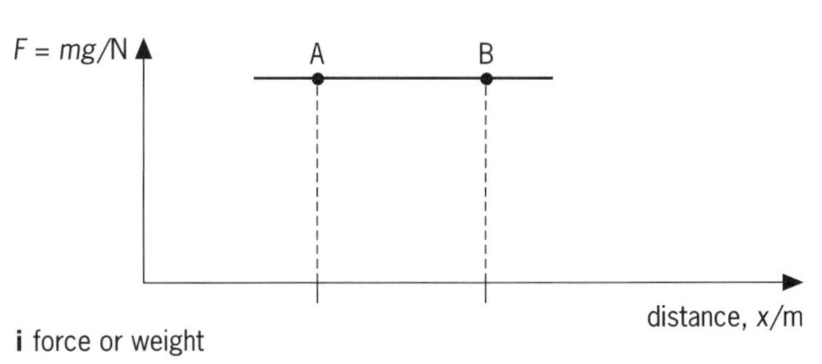

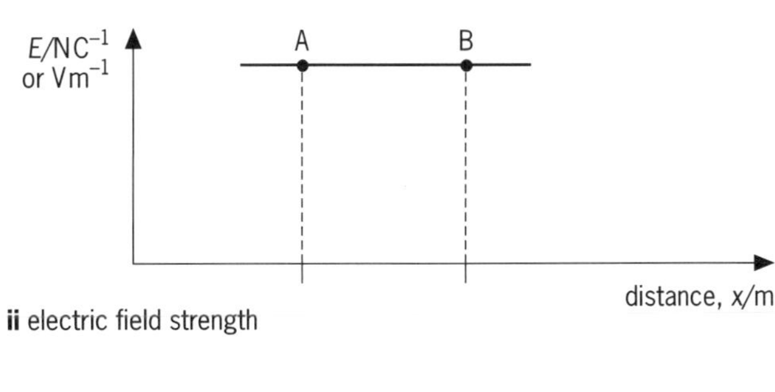

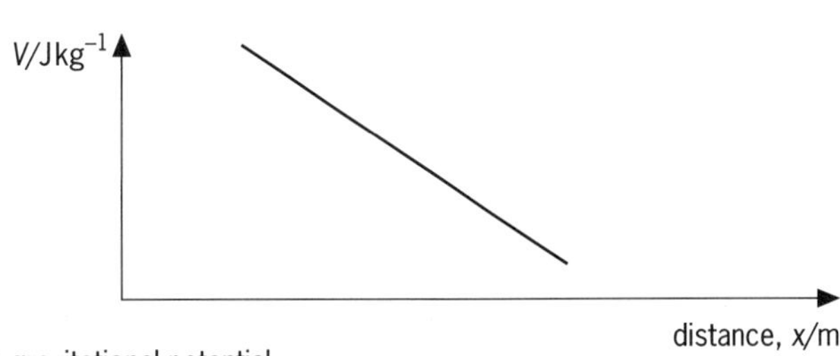

Figure 33.9

a What does the area under the line AB tell us in **i** and **ii**?
b What does the slope tell you in **iii**?

Level 3

12 A water drop is stationary at a height above the Earth where there is an electric field of strength 300 N C^{-1}, directed towards the surface of the Earth (Figure 33.10). The drop has 20 surplus electrons.

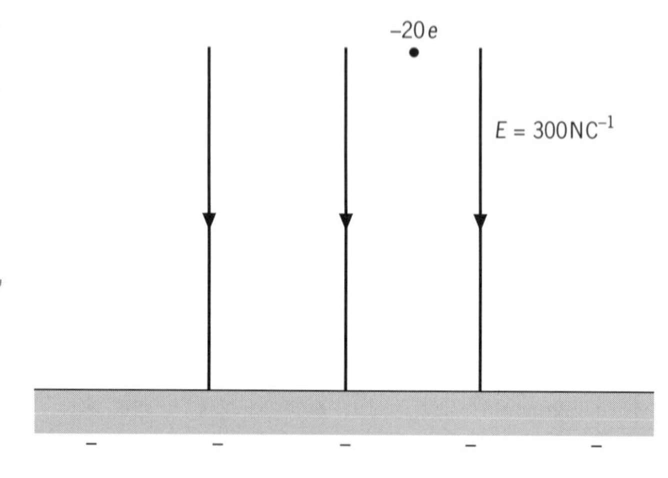

Figure 33.10

a Explain why the drop can be stationary.
b Find the mass of the drop.
c Find the diameter of the drop, given that the density of water is 1000 kg m^{-3}.

13 a A ball bearing is projected horizontally from the top of a table. An electron is projected horizontally into a uniform electric field acting vertically upwards. Qualitatively, compare the paths of the two projectiles.
b What value of electric field strength is needed to make the acceleration of the electron in the electric field 10^{12} times greater than its acceleration in the Earth's gravitational field?

14 a Figure 33.11 shows part of the gravitational field near the surface of a planet where $g = 10$ N kg^{-1}.

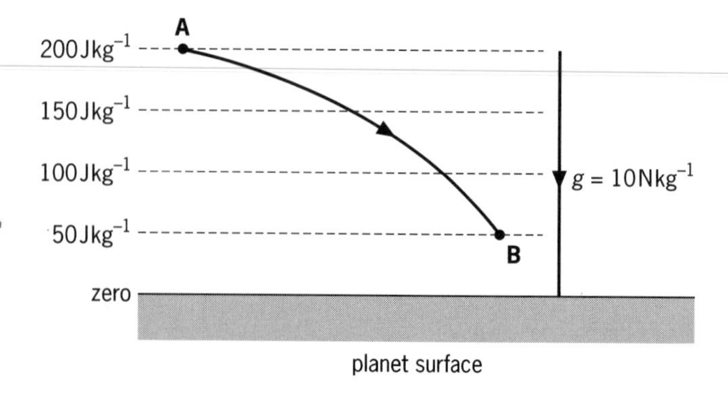

Figure 33.11

i How far apart are equipotential surfaces shown?
ii What is the gravitational force on an 8 kg cannon ball at A and at B?
iii What is the gravitational potential energy of this mass at A and at B?
iv How much work does the gravitational field do on the cannon ball, as it moves from A to B?
v At A, the ball's speed is 20 m s^{-1}. What is its speed at B?

b Figure 33.12 shows part of the electrical field near the surface of a conducting plate where $E = 10$ N C^{-1}.

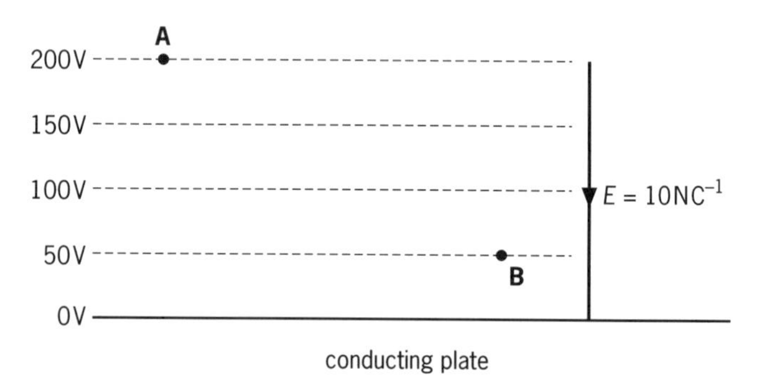

Figure 33.12

i How far apart are the equipotential surfaces shown?
ii What is the electric force on a +8 C charged object at A and at B?
iii What is the electric potential energy of this charge at A and at B?
iv How much work would the electrical field do on the charged object if it moved from A to B?

Level 1

1 a The unit of E is $\mathbf{NC^{-1}}$. The unit of g is $\mathbf{N\,kg^{-1}}$.

b The gravitational force seems to be always attractive, never repulsive.

c A uniform field is the same strength throughout (lines are equally spaced) and always in the same direction (lines are straight and parallel).

d The field in the centre of a pair of parallel oppositely charged plates is uniform.

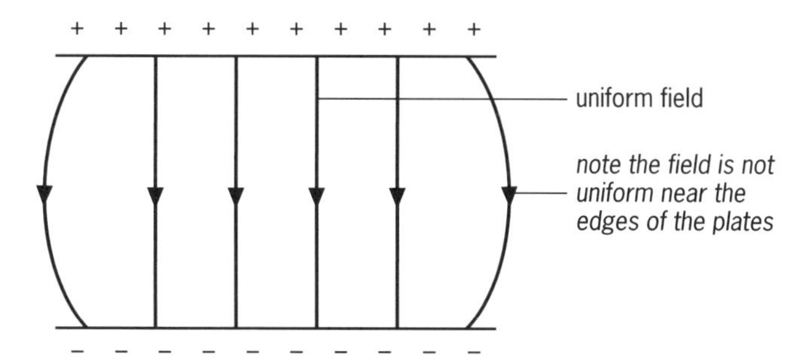

Figure 33.13

2

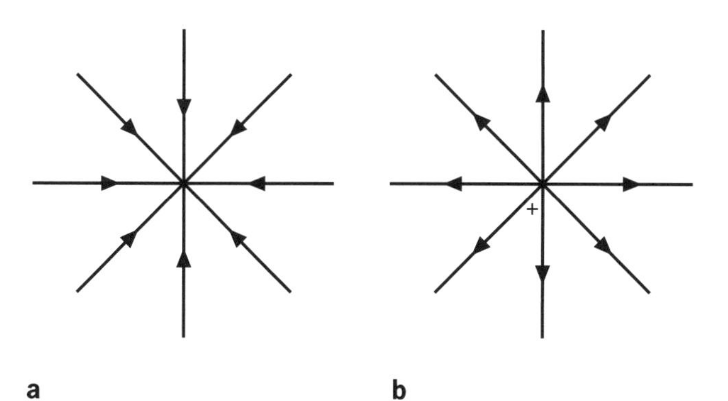

Figure 33.14

3 The force of gravitational attraction is given by

$$F = \frac{GM_1M_2}{r^2}$$

where $G = 6.7 \times 10^{-11}\,\mathrm{N\,m^2\,kg^{-2}}$. Because M_1, M_2 and r are all of value 1.0, $F = \mathbf{6.7 \times 10^{-11}\,N}$.

The force of electrostatic attraction is given by

$$F = \frac{1}{4\pi\epsilon_0}\,\frac{Q_1Q_2}{r^2}$$

Again, all the values Q_1, Q_2 and r to be substituted into the equation are 1.0, so

$$F = \frac{1}{4\pi\epsilon_0} = \mathbf{9.0 \times 10^9\,N}$$

It is worth remembering that $1/4\pi\epsilon_0 = 9.0 \times 10^9$ in SI units.

4

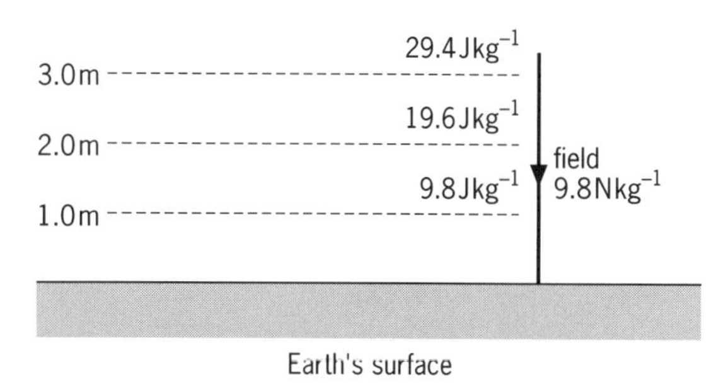

a

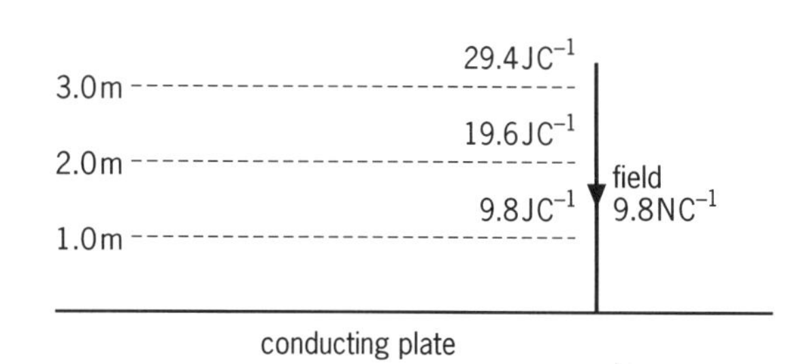

b

Figure 33.15

5 a Potential is inversely proportional to the distance from the centre.

i At double the distance the potential will be half, so it will be **1500 V**.

ii At half the distance it will be double, so it will be **6000 V**.

iii The force is proportional to the field intensity (strength), which is inversely proportional to the *square* of the distance from the centre. So the force **does not vary in the same way** as the electric potential. It would be $\frac{1}{4}$ in part i and 4 times in part ii.

b Potential is inversely proportional to the distance from the centre.

i At double the distance the potential will be half, so it will be $-\mathbf{20\,kJ\,kg^{-1}}$.

ii At half the distance it will be double, so it will be $-\mathbf{80\,kJ\,kg^{-1}}$.

iii The force is proportional to the field intensity (strength), which is inversely proportional to the *square* of the distance from the centre. So the force **does not vary in the same way** as the gravitational potential. It would be $\frac{1}{4}$ in part i and 4 times in part ii.

Level 2

6 a You can rearrange the equation for electric field strength, $E = F/q$ to give an equation with the charge as the subject: $q = F/E$. In this example,

$$q = \frac{4.2\,\text{N}}{8.6 \times 10^{6}\,\text{N C}^{-1}} = \mathbf{4.9 \times 10^{-7}\,C} \text{ or } \mathbf{490\,nC}$$

b Similarly, you can rearrange $g = F/m$ to give $m = F/g$. In this example,

$$m = \frac{42\,\text{N}}{8.6\,\text{N kg}^{-1}} = \mathbf{4.9\,kg}$$

7 a

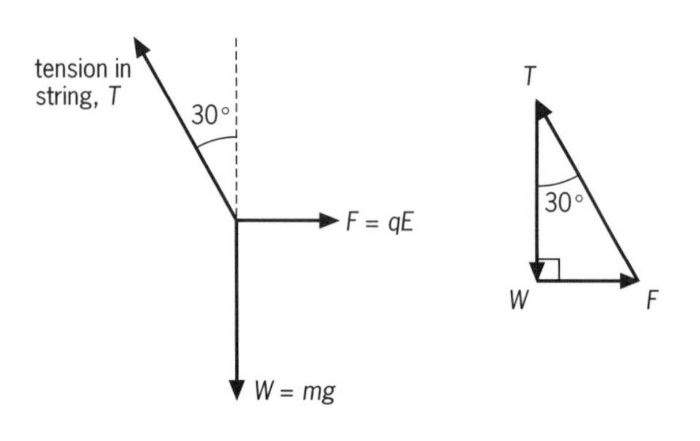

Figure 33.16

iii The weight of the ball is

$$W = mg = 1.2 \times 10^{-3}\,\text{kg} \times 9.8\,\text{m s}^{-2} = 12\,\text{mN}$$

From the triangle of forces,

$$\tan 30° = \frac{F}{W}$$

$$\therefore \quad F = W \tan 30° = \mathbf{6.8\,mN}$$

iv The electric force $= qE$, so

$$q = \frac{F}{E} = \frac{6.8 \times 10^{-3}\,\text{N}}{4.0 \times 10^{5}\,\text{N C}^{-1}} = \mathbf{17\,nC}$$

b i You would try to measure any deviation from the vertical using a pendulum bob; this would be very small.
ii You cannot isolate the mountain but always have the Earth's pull as well, which is much larger. See Figure 33.17.

Figure 33.17

8 a i The change in gravitational potential energy is given by

$$mg\Delta h = 3.0\,\text{kg} \times 9.8\,\text{m s}^{-2} \times 1.0\,\text{m} = \mathbf{29.4\,J}$$

ii The work done = force × distance moved in the direction of the force,

$$\Delta W = F\Delta x$$

The force here is the weight, $mg = 29.4\,\text{N}$.

$$\therefore \quad \Delta W = 29.4\,\text{N} \times 1.0\,\text{m} = \mathbf{29.4\,J}$$

This is the same as the potential energy gained.
iii The change in potential = work done per unit mass,

$$\Delta V = \frac{\Delta W}{m} = \frac{29.4\,\text{J}}{3.0\,\text{kg}} = \mathbf{9.8\,J\,kg^{-1}}$$

iv The work done $\Delta W = mg\Delta h$ so $\Delta W/m = g\Delta h = \Delta V$, the change in gravitational potential.

$$g\Delta h = 9.8\,\text{N kg}^{-1} \times 1.0\,\text{m} = \mathbf{9.8\,J\,kg^{-1}}$$

v $\Delta W/m = \Delta V$

$$\therefore \quad \Delta W = m\Delta V$$

The energy change, in terms of potential difference, is given by mass times change in potential.

The change in potential when you move 1.0 m parallel to the Earth's field (that is, towards or away from the surface) is always 9.8 J kg^{-1} (as long as you are fairly close to the surface). In fact it is only the start and finish points that matter – you can move between them in any fashion.

b i $\Delta W = qE\Delta x = 3.0\,\text{C} \times 9.8\,\text{N C}^{-1} \times 1.0\,\text{m} = \mathbf{29.4\,J}$

Notice the similarity with $mg\Delta h$ here.

ii The work done = force × distance moved in the direction of the force,

$$\Delta W = F\Delta x$$

the force here is

$$qE = 3.0 \times 9.8\,\text{N C}^{-1} = 29.4\,\text{N}$$

so

$$\Delta W = qE\Delta x = 29.4\,\text{N} \times 1.0\,\text{m} = \mathbf{29.4\,J}$$

This is the same as the potential energy gained.
iii The change in potential = work done per unit charge,

$$\Delta V = \frac{\Delta W}{q} = \frac{29.4\,\text{J}}{3.0\,\text{C}} = \mathbf{9.8\,J\,C^{-1}}$$

iv $\Delta W = qE\Delta x$, so the work done (energy gained) per unit charge, $\Delta W/q = E\Delta x = \Delta V$.

$$E\Delta x = 9.8\,\text{N C}^{-1} \times 1.0\,\text{m} = \mathbf{9.8\,J\,C^{-1}}$$

v $\Delta W/q = \Delta V$

$$\therefore \quad \Delta W = q\Delta V$$

The energy change is equal to the charge times the change in potential.

The change in potential when you move 1.0 m parallel to this field is always 9.8 J C^{-1}. In fact it is only the start and finish points that matter – you can move between them in any fashion.

9 a The electric field strength is the potential gradient (volts per m):

$$\frac{\Delta V}{\Delta x} = \frac{400\,V - 100\,V}{0.50\,m} = \frac{300\,V}{0.50\,m} = \mathbf{600\,V\,m^{-1}}$$

b The gravitational field strength is the potential gradient ($J\,kg^{-1}$ per m):

$$\frac{\Delta V}{\Delta x} = \frac{400\,J\,kg^{-1} - 100\,J\,kg^{-1}}{0.50\,m} = \frac{300\,J\,kg^{-1}}{0.50\,m}$$

$$= \mathbf{600\,J\,kg^{-1}\,m^{-1}}$$

($J = N\,m$ so $J\,kg^{-1}\,m^{-1} = N\,m\,kg^{-1}\,m^{-1} = N\,kg^{-1}$, which is the more usual unit.)

10 a E varies as the slope of the V graph; E is zero if V is constant.

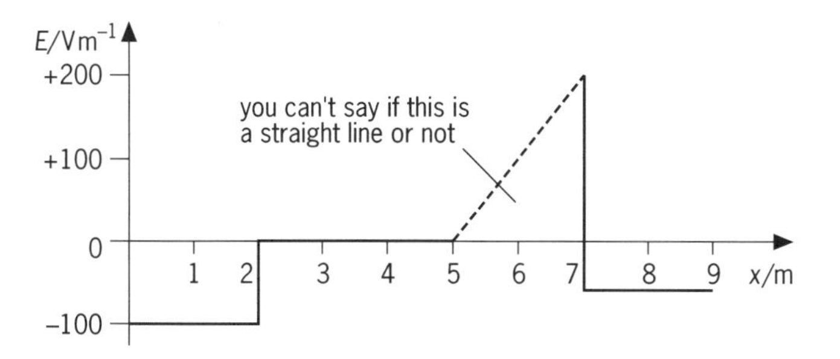

Figure 33.18

For the first 2 m the gradient is $-$**100 V m,$^{-1}$**, then the gradient is **zero** for the next 3 m. The slope **gradually increases** to a maximum, of about +**200 V m^{-1}**, over the next 2 m and is then constant at $-$**50 V m^{-1}** for the final 2 m.

b The shape of the field strength graph would be identical if it were for a gravitational field, but the vertical axis would be g in $N\,kg^{-1}$.

11 a i The weight, mg, is the gravitational force on an object. The area under the graph is force × distance moved in the direction of the force = **work done** moving the object from A to B.

ii When the electric field strength, which is equal to the force per unit charge, is plotted against distance, the area under the line is equal to $(F/q) \times \Delta x = \Delta W/q$, which is the **work done per unit charge** or the **change in potential**.

b The slope of the graph in **iii** is the potential gradient and is equal to the gravitational **field strength**, g.

Level 3

12 a If the field is towards the Earth, this is the way a positive charge would move, so the Earth is charged negatively and the negatively charged drop is repelled. If the electrostatic force upwards is equal to the weight downwards (Figure 33.19) the drop will remain suspended.

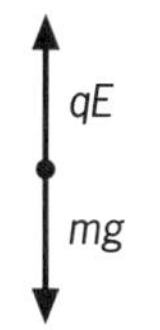

Figure 33.19

b The force on the drop due to the electric field is

$$qE = 20 \times 1.6 \times 10^{-19}\,C \times 300\,N\,C^{-1} = 9.6 \times 10^{-16}\,N$$

This must be equal to the weight of the drop, mg, because the drop is stationary. So the mass of the drop is

$$m = \frac{qE}{g} = \frac{9.6 \times 10^{-16}\,N}{9.8\,N\,kg^{-1}} = \mathbf{9.8 \times 10^{-17}\,kg}$$

c density = mass/volume, $\rho = m/V$, which you can rearrange to give $m = \rho V$. Assuming that the drop is spherical, $V = 4\pi r^3/3$. So $m = \rho \times 4\pi r^3/3$, which you can rearrange to give

$$r^3 = \frac{3m}{4\pi\rho}$$

Using the values in this example,

$$r^3 = \frac{3 \times 9.8 \times 10^{-17}\,kg}{4\pi \times 1000\,kg\,m^{-3}} = 2.3 \times 10^{-20}\,m^3$$

$$\therefore \quad r = 2.8 \times 10^{-7}\,m$$

$$d = 2 \times r = \mathbf{5.6 \times 10^{-7}\,m} \text{ or } \mathbf{0.56\,\mu m}$$

13 a Both fields produce a constant force downwards on the object. This produces a vertical acceleration, but the horizontal velocity is unchanged. Both objects follow parabolic paths.

b The acceleration due to gravity, g, is approximately $10\,m\,s^{-2}$ so the required acceleration of the electron is $10^{13}\,m\,s^{-2}$.

force = mass × acceleration = charge × electric field strength

or

$$ma = qE$$

which you can rearrange to give

$$E = \frac{ma}{q}$$

Using the values in this example, the required field strength is

$$\frac{ma}{q} = \frac{9.1 \times 10^{-31}\,kg \times 10^{13}\,m\,s^{-2}}{1.6 \times 10^{-19}\,C} = \mathbf{57\,N\,C^{-1}}$$

14 a The gravitational case:

i $g = 10\,\text{N kg}^{-1}$ so the potential gradient is also $10\,\text{N kg}^{-1}$. The newton, $\text{N} = \text{J m}^{-1}$, and if the potential gradient is expressed as $10\,\text{J kg}^{-1}\text{m}^{-1}$ it is clear that a difference of $50\,\text{J kg}^{-1}$ will occur every **5 m**.

ii The force on a mass, *m*, in a gravitational field is equal to *mg*. At point A,

$$mg = 8\,\text{kg} \times 10\,\text{N kg}^{-1} = \mathbf{80\,N}$$

This is the force throughout the field, so at B it will also be **80 N**.

iii At point A, the mass is on the $200\,\text{J kg}^{-1}$ equipotential so the potential energy is

$$mV = 8\,\text{kg} \times 200\,\text{J kg}^{-1} = \mathbf{1600\,J}$$

At B it is on the $50\,\text{J kg}^{-1}$ equipotential so the potential energy is

$$mV = 8\,\text{kg} \times 50\,\text{J kg}^{-1} = \mathbf{400\,J}$$

iv The work done on the mass, is

$$mg\Delta h = 80\,\text{N} \times 15\,\text{m} = \mathbf{1200\,J}$$

This is the change in potential energy (from iii):

$$1600\,\text{J} - 400\,\text{J} = 1200\,\text{J}$$

You can also see that change in potential energy = mass × change in potential.
In this example $\Delta V = 200\,Jkg^{-1} - 50\,Jkg^{-1} = 150\,Jkg^{-1}$

$$\therefore \Delta W = m\Delta V = 8\,kg \times 150\,Jkg^{-1} = 1200\,J$$

v At A, the initial kinetic energy is

$$\tfrac{1}{2}mu^2 = \tfrac{1}{2} \times 8\,\text{kg} \times (20\,\text{m s}^{-1})^2 = 1600\,\text{J}$$

At B, it has lost 1200 J of gravitational potential energy, which will have been converted to kinetic energy, so the kinetic energy at B is $1600\,\text{J} + 1200\,\text{J} = 2800\,\text{J}$.

$$\tfrac{1}{2}mv^2 = 2800\,\text{J}$$

which you can rearrange to give

$$v^2 = \frac{2800\,\text{J}}{\tfrac{1}{2} \times 8\,\text{kg}} = 700\,\text{m}^2\text{s}^{-2}$$

$$\therefore \quad v = \mathbf{26\,m\,s^{-1}}$$

The question asks for speed so you don't need to worry about components or directions.

b The electrical case:

i $E = 10\,\text{N C}^{-1}$ so the potential gradient is also $10\,\text{N C}^{-1}$. This can be written as $10\,\text{V m}^{-1}$ or $10\,\text{J C}^{-1}\text{m}^{-1}$. Expressed as $10\,\text{J C}^{-1}\text{m}^{-1}$ it is clear that a difference of $50\,\text{J C}^{-1}$ occurs every **5 m**.

ii The force on a charge, *q*, in an electrical field is $F = qE$. At point A,

$$F = 8\,\text{C} \times 10\,\text{N C}^{-1} = \mathbf{80\,N}$$

This is the force throughout the field; it doesn't matter whether the charge is at A or B.

iii At A, the charged object is on the $200\,\text{J C}^{-1}$ equipotential so the potential energy is

$$qV = 8\,\text{C} \times 200\,\text{J C}^{-1} = \mathbf{1600\,J}$$

At B it is on the $50\,\text{J C}^{-1}$ equipotential so the potential energy is

$$qV = 8\,\text{C} \times 50\,\text{J C}^{-1} = \mathbf{400\,J}$$

iv The work done on the charged object is

$$F\Delta x = qE\Delta x = 80\,\text{N} \times 15\,\text{m} = \mathbf{1200\,J}$$

This is the change in potential energy (from iii).

$$1600\,\text{J} - 400\,\text{J} = 1200\,\text{J}$$

You can also see that work done = charge × change in potential

$$\Delta V = 200\,JC^{-1} - 50\,JC^{-1} = 150\,JC^{-1}$$

$$\therefore \quad \Delta W = q\Delta V = 8\,C \times 150\,JC^{-1} = 1200\,J$$

34 Magnetic fields and forces

Around any magnetised object there is a volume of space where we can detect magnetic forces. This space is called a **magnetic field**. As with electric fields and gravitational fields the shape, strength and direction of a magnetic field can be described using field lines, Figure 34.1. These lines are called **lines of magnetic flux**.

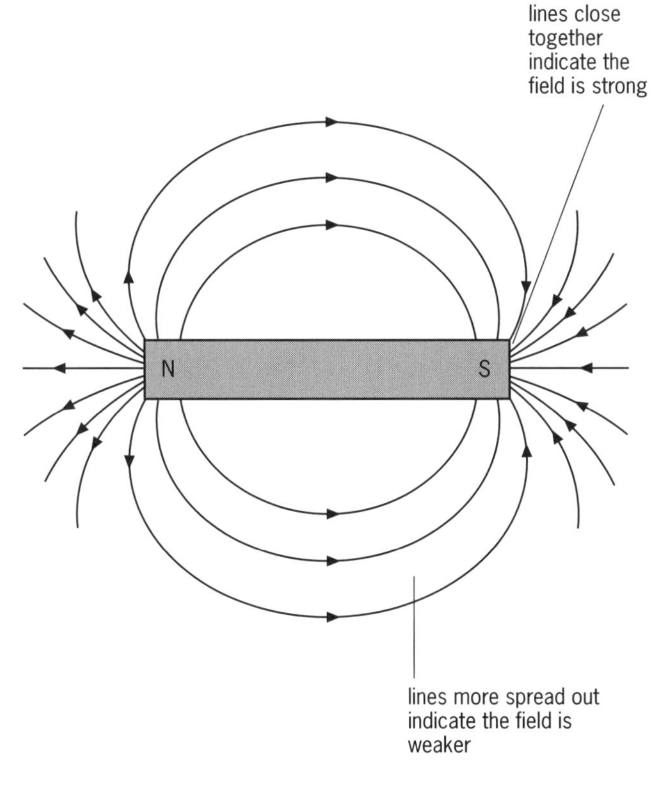

a The magnetic field around a bar magnet

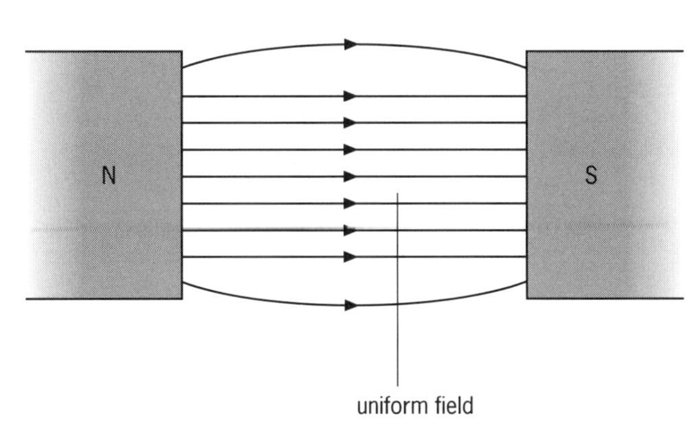

b A uniform magnetic field

Figure 34.1 Magnetic fields represented by lines of flux

Magnetic effect of an electric current

In 1819 a Danish scientist named Oersted discovered that an electric current passing through a wire created a magnetic field around it. The magnetic field is cylindrical in shape, centred on the wire, and is strongest closest to the wire. If the current is turned off the magnetism disappears. If the direction of the current is reversed, so too is the direction of the magnetic field; see Figure 34.2.

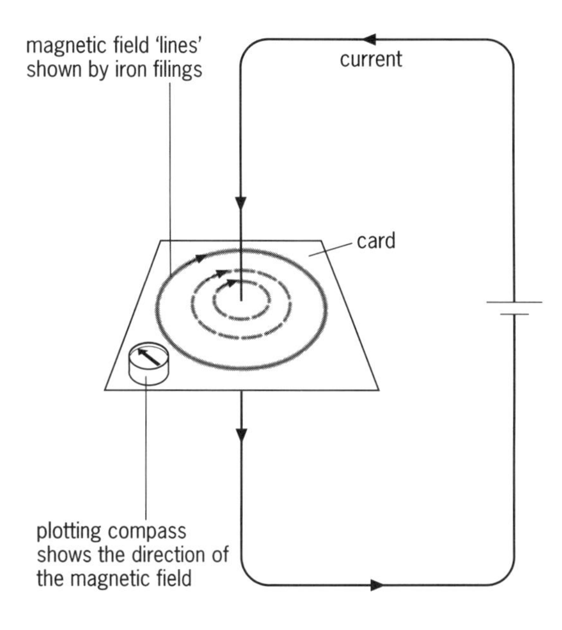

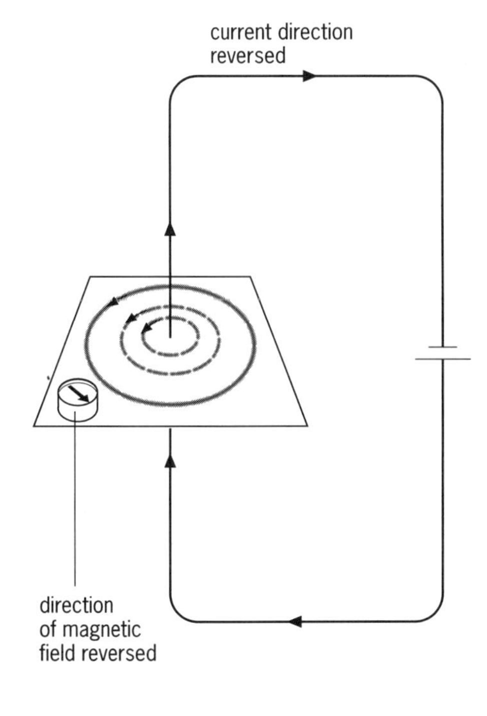

Figure 34.2 The magnetic field around a wire carrying a current

The direction of the magnetic field created by the current can be found using the **right hand grip rule**. If the right hand grips the wire such that the thumb points in the direction in which the current is flowing, the fingers point in the direction of the magnetic field; see Figure 34.3.

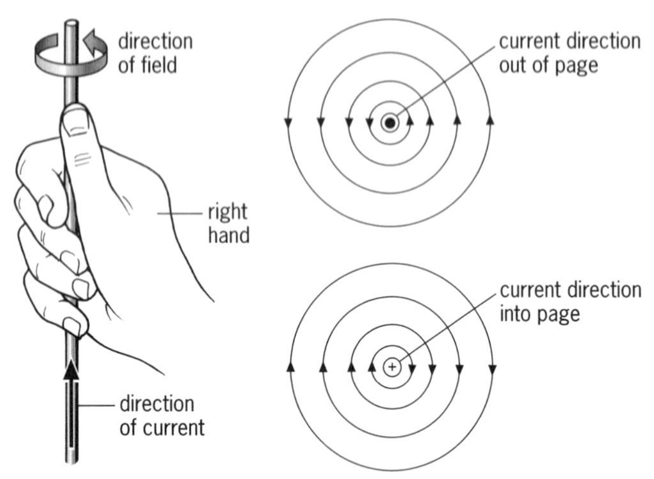

Figure 34.3 The right hand grip rule for a straight wire

If we wind a piece of wire round and round to make a coil and pass a current through it, the magnetic fields from the current in adjacent pieces of the wire overlap and reinforce each other, creating a much stronger magnetic field. A long coil is called a **solenoid**. Its magnetic field is similar in shape to that of a bar magnet; see Figure 34.4a. The polarity of the magnetic field surrounding a solenoid can be determined again using a right hand grip rule. Imagine gripping the solenoid with the right hand such that the fingers indicate the direction of the current in the coil, Figure 34.4b. The thumb will then point to the North-seeking pole of the solenoid, that is in the direction of the field lines.

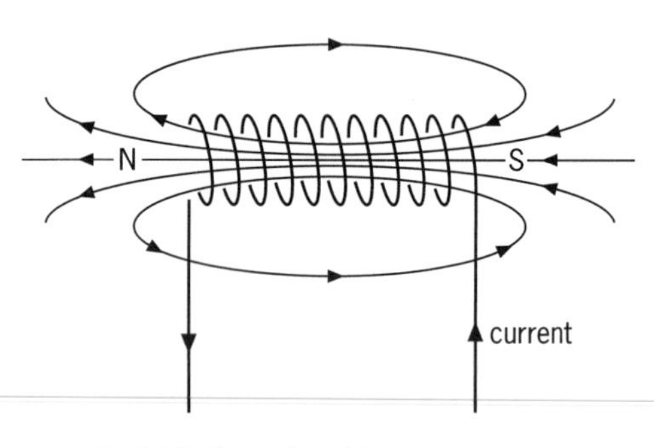

a The magnetic field of a solenoid

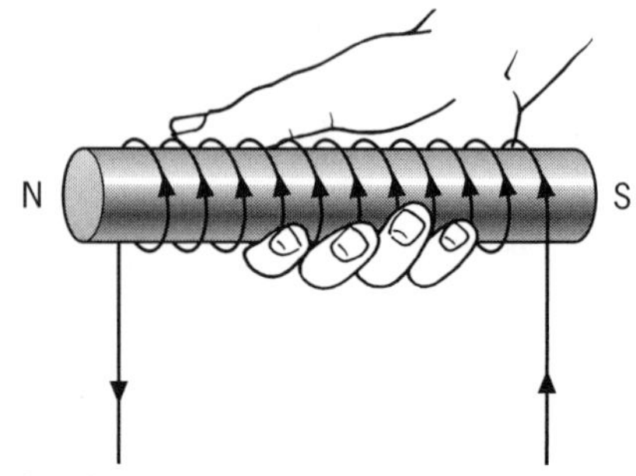

b The right hand grip rule to determine the polarity

Figure 34.4

Force on a current-carrying conductor in a magnetic field.

A straight length of wire carrying a current placed at right angles to a magnetic field experiences a force which is at right angles to both the field and the current, Figure 34.5a. The direction of the force on the wire can be predicted using **Fleming's left hand rule**; see Figure 34.5b.

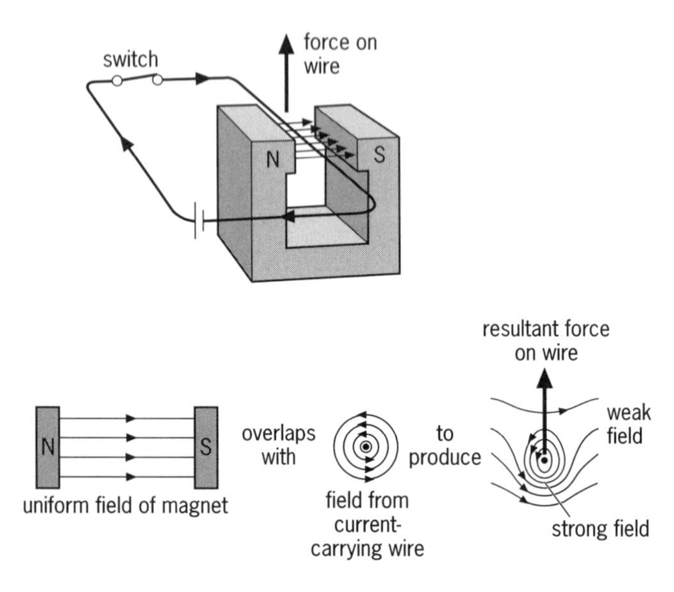

a The force on a straight wire in a magnetic field

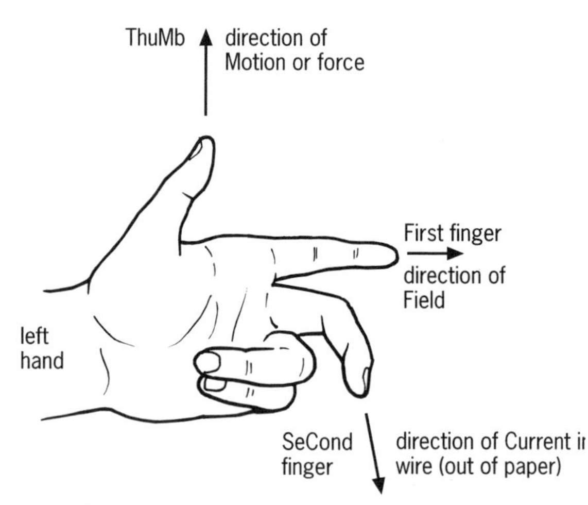

b Fleming's left hand rule to determine the direction of the force

Figure 34.5

This effect is called the **motor effect** and can be readily investigated using the simple current balance shown in Figure 34.6. A magnetic field is applied to one side of the wire balance by means of a U-shaped magnet. When a current is passed through the wire via the knife edges the balance is deflected up or down. The deflecting force can be measured by balancing it against the weight of a small rider on the end of the balance or by using a top pan balance.

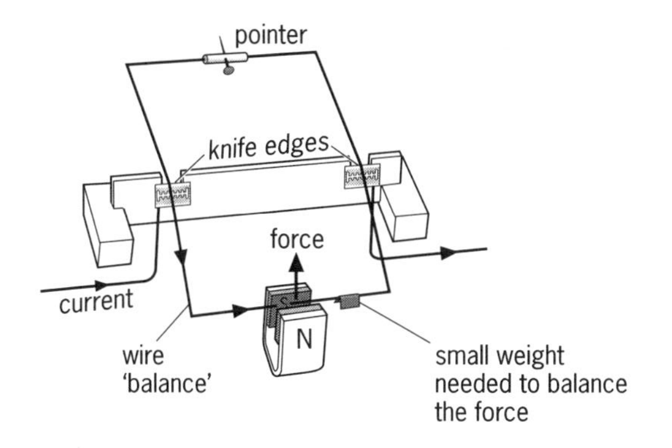

Figure 34.6 A simple current balance to investigate the motor effect

If the experiment is repeated several times, varying

1. the length of wire inside the magnetic field (using longer magnets);
2. the size of the current flowing through the wire;
3. the strength of the magnetic field (using stronger magnets);

it can be readily demonstrated that the size of the force F experienced by the wire is directly proportional to

- the length of the wire within the magnetic field (l);
- the current in the wire (I);
- the strength of the magnetic field.

This effect can be used to define the strength of a magnetic field using the quantity known as **magnetic flux density**, B. The results can be summarised as

$$F = BIl$$

where B is the magnitude of the magnetic flux density, which is a vector quantity. B is measured in tesla, T; $B = F/Il$, so $1\,\text{T} = 1\,\text{N}\,\text{A}^{-1}\,\text{m}^{-1}$. Magnetic flux density can be measured (using this effect) by an instrument as.

Example 1

Calculate the magnitude of the force exerted on a 1.5 m length of wire placed at right angles to a field of flux density 2.0 T and carrying a current of 3.0 A.

$$F = BIl$$
$$= 2.0\,\text{T} \times 3.0\,\text{A} \times 1.5\,\text{m}$$
$$= 9.0\,\text{N}$$

If the wire is at an angle θ to the field, Figure 34.7, the equation becomes

$$F = BIl \sin\theta$$

Figure 34.7

Force on moving charges

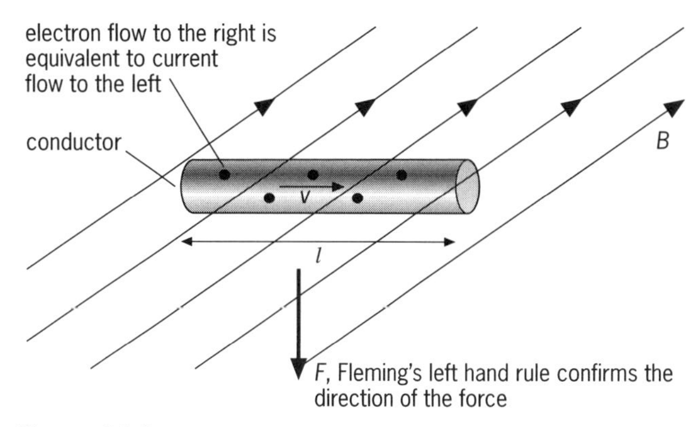

Figure 34.8

Consider n electrons travelling a distance, l, with a velocity, v, through a magnetic field, Figure 34.8.

$$\text{current} = \text{rate of flow of charge}, \; I = \frac{Q}{t}$$

$$\therefore I = \frac{ne}{t}$$

where e is the charge on an electron. But from velocity = displacement/time,

$$t = \frac{l}{v}$$

$$\therefore I = \frac{ne}{l/v} = \frac{nev}{l}$$

If the electrons are moving at right angles to the magnetic field of flux density B then the force exerted on the current is

$$F = BIl$$

From the above, substituting for I,

$$F = B \times \frac{nev}{l} \times l = Bnev$$

The force exerted on each electron is therefore

$$F = Bev$$

For a particle carrying a charge, q, and travelling at right angles to a magnetic field of flux density B, the magnitude of the force, F, exerted on the particle is given by

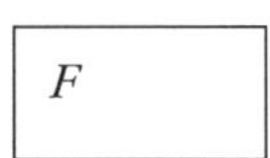

If the particle travels at an angle θ to the magnetic field the equation becomes

$$F = Bqv \sin\theta$$

Example 2

Calculate the magnitude of the force exerted upon an electron travelling through a uniform magnetic field of flux density 0.20T. The electron is travelling at right angles to the field lines with a velocity of $3.0 \times 10^{7}\,\mathrm{m\,s^{-1}}$. The charge carried by an electron is $1.6 \times 10^{-19}\,\mathrm{C}$.

$$F = Bqv$$
$$= 0.20\,\mathrm{T} \times 1.6 \times 10^{-19}\,\mathrm{C} \times 3.0 \times 10^{7}\,\mathrm{m\,s^{-1}}$$
$$= 9.6 \times 10^{-13}\,\mathrm{N}$$

Note that the force exerted on an electron travelling through a uniform magnetic field is always at right angles to the direction in which it is moving. The electron, while in the field, will therefore follow a circular path.

Flux density due to a current-carrying conductor

The magnitude of the magnetic flux density a distance, r, from a straight current-carrying conductor in air (Figure 34.9) is given by the equation

$$B = \frac{\mu_0 I}{2\pi r}$$

where μ_0 is called the **permeability of free space** and has the value $4\pi \times 10^{-7}\,\mathrm{N\,A^{-2}}$.

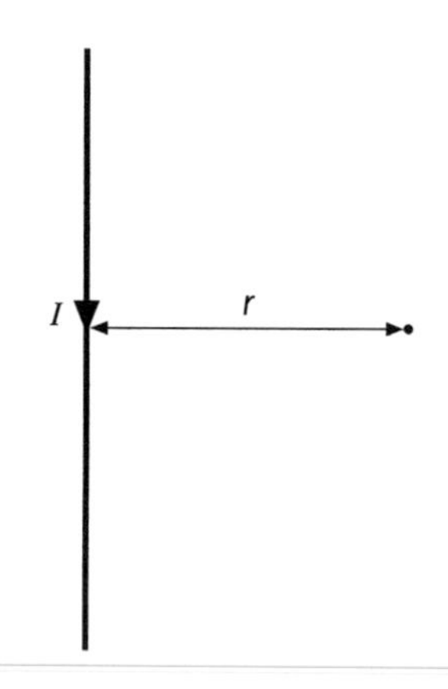

Figure 34.9

Example 3

Calculate the magnetic flux density at a point 5.0 cm from a straight piece of wire carrying a current of 10 A.

$$B = \frac{\mu_0 I}{2\pi r}$$

$$= \frac{4\pi \times 10^{-7}\,\mathrm{N\,A^{-2}} \times 10\,\mathrm{A}}{2\pi \times 0.050\,\mathrm{m}}$$

$$= 4.0 \times 10^{-5}\,\mathrm{T}$$

Force between two current-carrying conductors

If two current-carrying wires are placed close together the magnetic field around each wire acts on the other wire. If the currents are in the same direction the wires attract. If the currents are in opposite directions the wires repel. A clear demonstration uses strips of aluminium foil; see Figure 34.10.

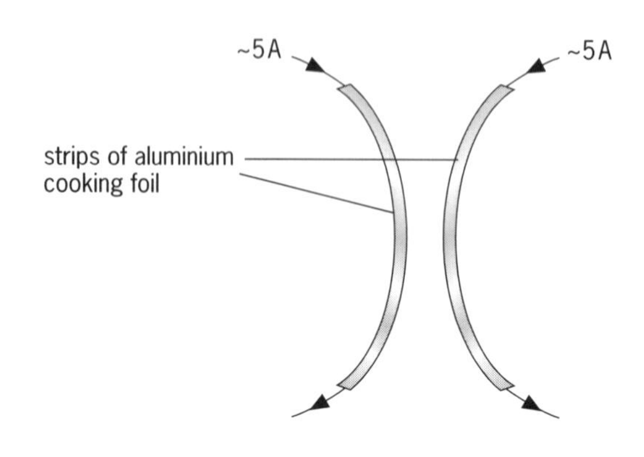

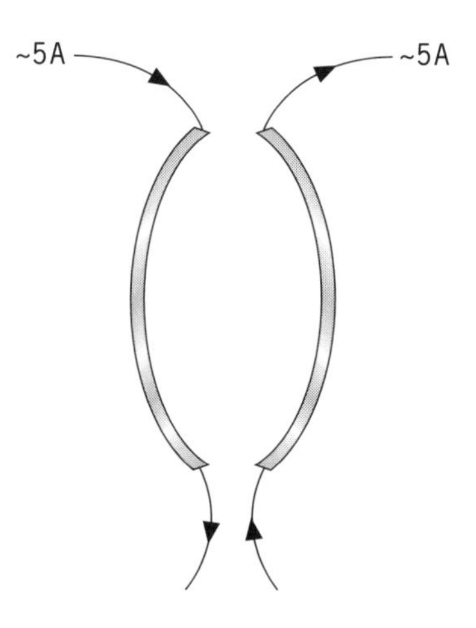

Figure 34.10 Demonstrating the force between two parallel conductors

The magnitude of the force between the two parallel wires is directly proportional to the current flowing in them and inversely proportional to the distance between them. This arrangement is used to formulate the definition of the ampere. The force per unit length of conductor is given by

$$\frac{F}{l} = \frac{\mu_0 I_1 I_2}{2\pi r}$$

where r is the distance between the conductors and $\mu_0/2\pi$ is the constant of proportionality.

The **ampere** is defined as **the current in two infinitely long wires 1 m apart in a vacuum which produces a force of 2×10^{-7} N per metre of their length**.

Forces on a coil in a magnetic field

In a uniform magnetic field, B, opposite sides of a rectangular current-carrying coil experience forces which are equal in size but opposite in direction. Thus a **couple** (see topic 7) acts on the coil, causing it to turn.

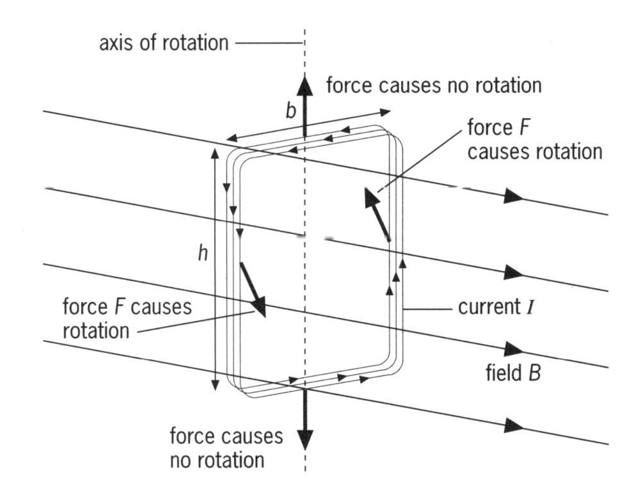

a Forces on a rectangular coil in a magnetic field

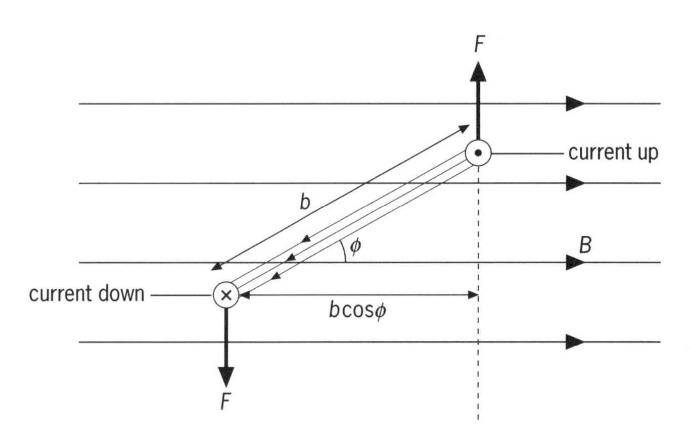

b View of coil looking down axis of rotation

Figure 34.11

Consider a rectangular coil of sides b and h, having N turns and carrying a current I. It is free to rotate about an axis which is at right angles to a field of flux density B, Figure 34.11a. The forces acting on the top and bottom of the coil are parallel to this axis and so produce no turning moment. The total force acting on each of the two sides is given by $F = BIl = BIhN$. These two sides remain at right angles to the field as the coil rotates so that the force has the same magnitude in all positions.

When the angle between the plane of the coil and the field is ϕ, Figure 34.11b, the perpendicular distance between the lines of action of the two forces is $b\cos\phi$. The moment of the couple, or **torque**, T, acting on the coil is given by

$$\text{torque} = \text{one force} \times \text{perpendicular distance between them}$$

$$\therefore \quad T = Fb\cos\phi \quad \text{or} \quad T = BIhNb\cos\phi$$

But $b \times h =$ area of coil, A. Therefore

$$\boxed{T = BAIN\cos\phi}$$

When the plane of the coil is parallel to the field, $\phi = 0°$ and $\cos\phi = 1$, so

$$T = BAIN$$

has its maximum value.

Example 4

Calculate the torque on a rectangular coil of sides 5.0 cm by 6.0 cm. The coil has 300 turns carrying a current of 0.25 A. The plane of the coil is at an angle of 60° to a uniform magnetic field of flux density 0.20 T.

torque,

$$T = BAIN\cos\phi$$
$$= 0.20\,\text{T} \times 5.0 \times 10^{-2}\,\text{m} \times 6.0 \times 10^{-2}\,\text{m} \times 0.25\,\text{A} \times 300 \times 0.5$$
$$= 2.3 \times 10^{-2}\,\text{N m}$$

Charge on an electron, $e = -1.6 \times 10^{-19}$ C
Speed of light, in a vacuum, $c = 3.0 \times 10^{8}\,\mathrm{m\,s^{-1}}$
Mass of an electron, $m_e = 9.1 \times 10^{-31}$ kg
Charge/mass ratio of an electron, $e/m_e = 1.8 \times 10^{11}\,\mathrm{C\,kg^{-1}}$
Permeability of free space, $\mu_0 = 4\pi \times 10^{-7}\,\mathrm{N\,A^{-2}}$

Level 1

Assume in these questions that all currents and magnetic fields are at right angles to each other.

1 a What flux density B will give a force of 2.0 N on a 50 cm length of wire carrying a current of 85 A?
b What current is flowing in a 3.0 m length of wire in the Earth's field of 55 μT if it experiences a force of 12 mN?
c What length of wire would experience a force of 0.10 N when carrying a current of 3.5 A in a magnetic field of flux density 14 mT?

2 a What is the size of the force on an electron travelling at $3.0 \times 10^{6}\,\mathrm{m\,s^{-1}}$ in a magnetic field of flux density $B = 200$ T?
b What flux density produces a force of magnitude 3.0×10^{-12} N on a doubly ionised copper atom travelling at $6.0 \times 10^{4}\,\mathrm{m\,s^{-1}}$?

3 A current balance has a 5.0 cm wide arm between the poles of a magnet, Figure 34.12. The flux density, B, between the poles is measured as 23 mT. When the current is 1.5 A, what is the force on the arm?

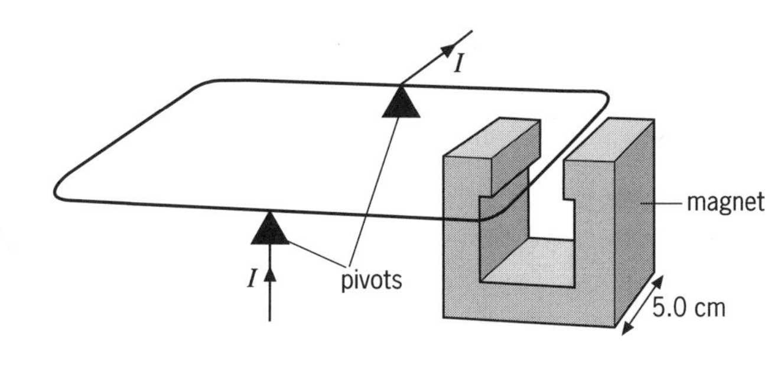

Figure 34.12

4 What experiments could you carry out to prove that the force $F = B \times I \times l$ when a current I flows in a length of wire l in a magnetic field with a flux density B?

5 Magnetic flux density, B, is measured by the force on a test object.
a What is the test object for a magnetic field?
b What is the unit of B, expressed in terms of force?
c There is one major difference in the measurement of magnetic flux density and the measurement of the strength of electric and gravitational fields. What is this?
d There is also an important difference in the direction of the magnetic force. Explain this.
e Write an equation for B in terms of force.
f What is μ_0 called?

6 Figure 34.13 shows the magnetic field lines resulting from currents flowing in each of two wires. Each set of field lines is due to one of the individual currents. *The effects have not been added.*

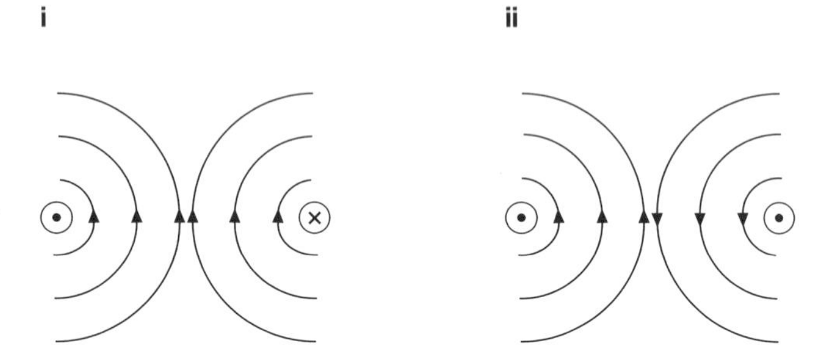

Figure 34.13

a What type of magnetic field is there between the wires shown in diagram **i** (strong or weak)?
b Is there a force on the wires in diagram **i**, and, if so, which way does it act?
c What type of magnetic field is there between the wires shown in diagram **ii** (strong or weak)?
d Is there a force on the wires in diagram **ii**, and, if so, which way does it act?
e Why can't magnetic field lines cross?
f How do you know what effect the shape of field in Figure 34.14 is going to have, just by looking at it?

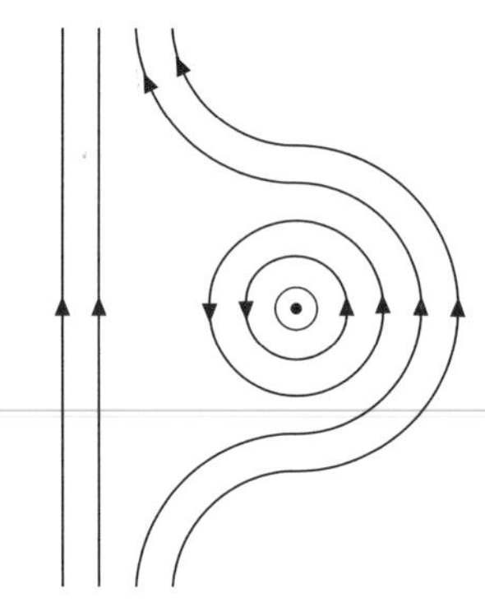

Figure 34.14

7 a Your 'thoughts' are just electrical signals in your brain. How can we 'look' at this mental activity from outside?
b A cable buried 1.0 m below the surface carries a.c. with a peak value of 80 A. What is the maximum flux density above ground?

Level 2

8 An electron travelling at 10% of the speed of light enters a region of a uniform magnetic field of flux density 2.2 mT. It is initially travelling at right angles to the field. What is the radius of the circle that the electron then follows?

9 A positively charged cosmic particle travelling at $6.0 \times 10^6\,\mathrm{m\,s^{-1}}$ enters the Earth's magnetic field along a radius and in the plane of the equator, Figure 34.15.

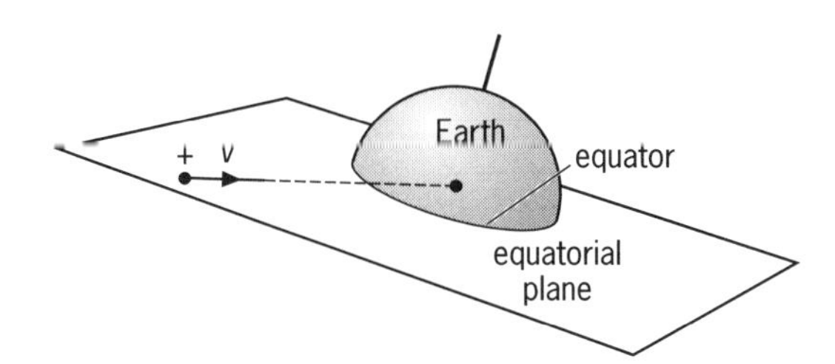

Figure 34.15

The flux density, B, at the equator is 55 μT.

a What is the direction of the Earth's field at the equator?

b What is the angle between the particle's path and the field?

c What is the force on the particle?

d What path will the particle follow?

10 Two copper 'bus-bars' (a term used for conductors of large currents) designed to distribute power are 200 mm apart, Figure 34.16.

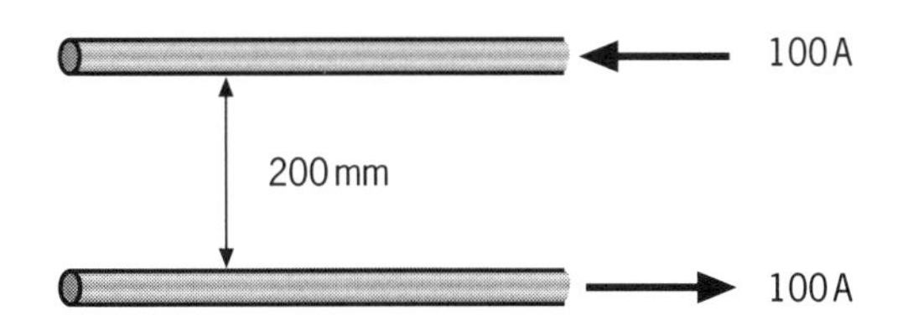

Figure 34.16

What is the force per metre on them when they carry 100 A each, in opposite directions?

11 Ions in a plasma inside a toroid (doughnut shape) carry currents in circles, Figure 34.17. This creates magnetic fields. In what way will the interaction between these fields affect the shape of the plasma?

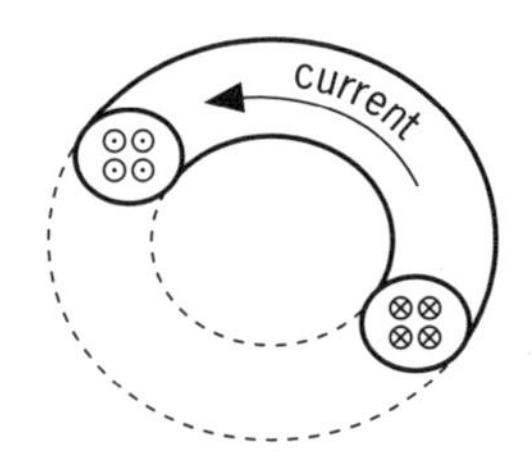

Figure 34.17

Level 3

12 Niobium can be a superconductor. A wire of niobium of diameter 0.50 mm could carry a large current. Niobium has a density of $8600\,\mathrm{kg\,m^{-3}}$.

a Calculate the weight of 1.0 m of this wire.

b Find what current would have to flow to levitate the 1.0 m length of wire where the Earth's magnetic field is 56 μT horizontally, in a northerly direction. Include a diagram to show the direction of the field and the direction of the current in the wire.

13 The plane rectangular coil in a d.c. motor is 0.040 m × 0.060 m and has 200 turns, Figure 34.18.

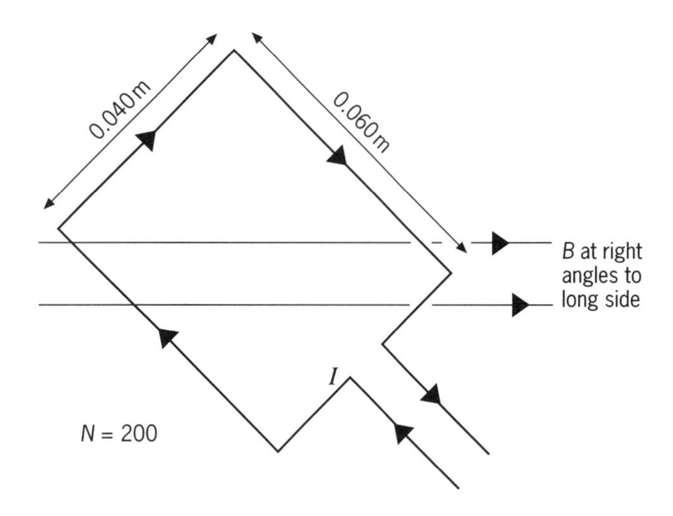

Figure 34.18

Find the maximum torque acting on it when the magnetic flux density B is 0.020 T and the current is 50 mA.

14 a A vertical wire carrying a current has a neutral point (zero field strength) 5.0 cm away to the East, at a place where the horizontal component of the Earth's field is 18 μT. Find the size and direction of the current in the wire.

b For the same horizontal component of the Earth's flux density, at what distance from a long vertical wire carrying a current of 3.5 A will there be a neutral point and in which direction would it be from the wire?

15 a What do you think is meant by *relative* permeability?

b Find out about magnetic shielding. What is its purpose?

c Find out what a 'flux guide' is. What is its purpose?

Level 1

1 a The force on a conductor, length l, carrying a current, I, at right angles to a magnetic field with flux density, B, is given by the equation $F = BIl$, which you can rearrange to give

$$B = \frac{F}{Il}$$

In this example $F = 2.0\,\text{N}$, $I = 85\,\text{A}$, $l = 50\,\text{cm} = 0.50\,\text{m}$.

$$\therefore B = \frac{2.0\,\text{N}}{85\,\text{A} \times 0.50\,\text{m}} = 0.047\,\text{NA}^{-1}\text{m}^{-1} = \mathbf{0.047\,T}$$

b To find the current, you can rearrange the equation to give

$$I = \frac{F}{Bl}$$

In this example $l = 3.0\,\text{m}$, $B = 55 \times 10^{-6}\,\text{T}$, $F = 12 \times 10^{-3}\,\text{N}$.

$$\therefore I = \frac{12 \times 10^{-3}\,\text{N}}{55 \times 10^{-6}\,\text{T} \times 3.0\,\text{m}} = 6.1\,\text{NT}^{-1}\text{m}^{-1}$$
$$= 6.1\,\text{N}(\text{N}^{-1}\text{Am})\text{m}^{-1}$$
$$= \mathbf{6.1\,A}$$

(All these units are to help you check; they would not normally need to be included.)

c To find the length, you can rearrange the equation to give

$$l = \frac{F}{BI}$$

In this example $F = 0.10\,\text{N}$, $I = 3.5\,\text{A}$, $B = 14 \times 10^{-3}\,\text{T}$.

$$\therefore l = \frac{0.10\,\text{N}}{14 \times 10^{-3}\,\text{T} \times 3.5\,\text{A}} = 2.0\,\text{N}(\text{N}^{-1}\text{Am})\text{A}^{-1}$$
$$= \mathbf{2.0\,m}$$

2 a The size of the force on a charged particle moving at right angles to a magnetic field is given by the equation

$$F = Bqv$$

In this example $v = 3.0 \times 10^{6}\,\text{m s}^{-1}$, $B = 200\,\text{T}$, $q = 1.6 \times 10^{-19}\,\text{C}$.

$$\therefore F = 200\,\text{T} \times 1.6 \times 10^{-19}\,\text{C} \times 3.0 \times 10^{6}\,\text{m s}^{-1}$$
$$= 9.6 \times 10^{-11}(\text{NA}^{-1}\text{m}^{-1})(\text{As})\text{m s}^{-1}$$
$$= \mathbf{9.6 \times 10^{-11}\,N} \text{ or } \mathbf{96\,pN}$$

b To find the flux density you can rearrange the equation for the force on a moving charged particle, to give

$$B = \frac{F}{qv}$$

In this example $F = 3.0 \times 10^{-12}\,\text{N}$, $q = 2 \times e = 2 \times 1.6 \times 10^{-19}\,\text{C}$, $v = 6.0 \times 10^{4}\,\text{m s}^{-1}$.

$$\therefore B = \frac{3.0 \times 10^{-12}\,\text{N}}{2 \times 1.6 \times 10^{-19}\,\text{C} \times 6.0 \times 10^{4}\,\text{m s}^{-1}}$$
$$= 1.6 \times 10^{2}\,\text{N}(\text{A}^{-1}\text{s}^{-1})\text{m}^{-1}\text{s}$$
$$= \mathbf{1.6 \times 10^{2}\,T} \text{ or } \mathbf{160\,T}$$

3 The force on a conductor, length l, carrying a current I, at right angles to a field with flux density B, is given by the equation $F = BIl$. In this example

$$F = 23 \times 10^{-3}\,\text{T} \times 1.5\,\text{A} \times 0.050\,\text{m}$$
$$= 1.7 \times 10^{-3}\,\text{TAm}$$
$$= 1.7 \times 10^{-3}(\text{NA}^{-1}\text{m}^{-1})\text{Am}$$
$$= \mathbf{1.7 \times 10^{-3}\,N}$$

4 Using a current balance (see Figure 34.6) or an arrangement on an electronic top-pan balance (Figure 34.19), you can measure the force exerted by the magnetic field on the current-carrying wire.

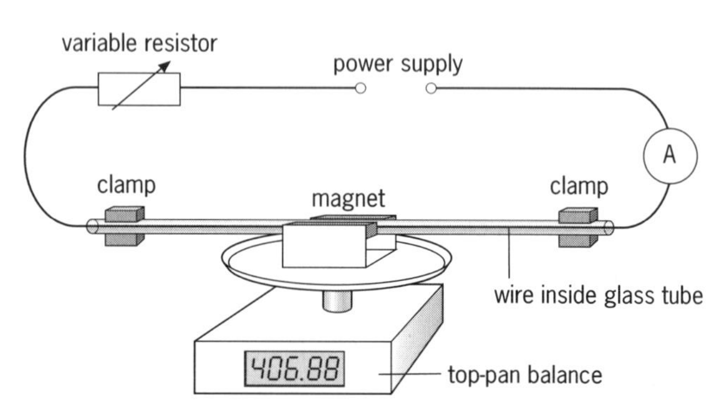

Figure 34.19 Using a top-pan balance to measure the force

You can measure the force, F, for a range of values of the current I, measured using an ammeter in the circuit, and **plot F against I**.

You can measure the force, F, for a range of values of the *length* of conductor perpendicular to the field. This may mean having a range of current balances with different widths, or putting several magnets side by side in a row (Figure 34.20a). You can then **plot F against l**. Three values of l would be enough.

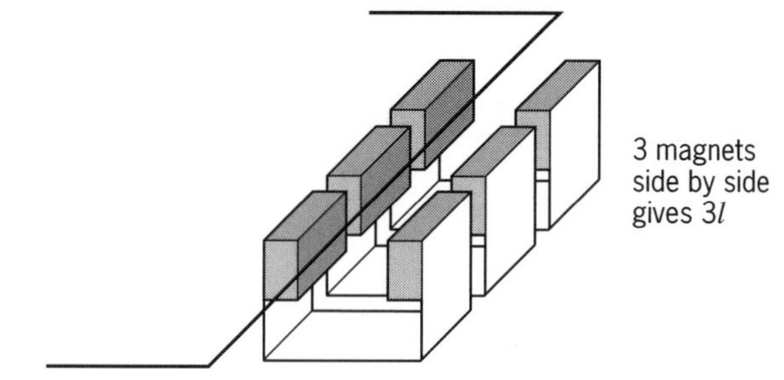

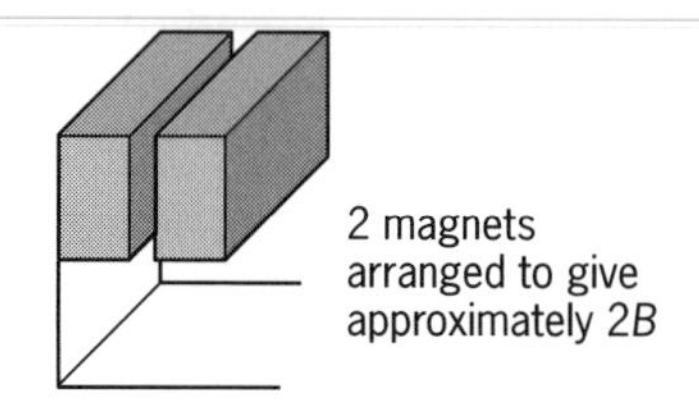

Figure 34.20

To vary the value of B, you can put the pole faces on top of one another (Figure 34.20b) or use different strength magnets, but the same length of wire must be in the field. In either case you will have to measure the field strength with a Hall probe and **plot F against B**.

5 a The test object is a wire of known length carrying a known current which is at right angles to the magnetic field.

b The unit of B is newton per amp per metre, $\mathbf{NA^{-1}m^{-1}}$. *(See part e.)*

c The current must be at right angles to the field.

d The magnetic force is at right angles to both the field and the current. (Electric and gravitational forces act in the plane of the field.)

e $B = \dfrac{F}{Il}$

f μ_0 is the **permeability of free space**.

Note this is different from the permittivity; that's ϵ_0.

6 a There is a **strong** magnetic field between the wires because the fields are in the same direction.

b There will be a force tending to **push them apart** (by Fleming's left hand rule); the wires will repel each other.

c There is a **weak** field with a neutral (zero) point between the wires because the magnetic fields go in opposite directions.

d The wires will be **pushed together** by the stronger field round the outside.

e If there were a point where two field lines crossed, the field would seem to be going in two directions at once.

f You can see the strong field to the right of the wire and the much weaker field, with a neutral (zero) point, on the left. The wire is pushed to the left by the imbalance of the fields.

This is called a catapult field. The field lines may be thought of as elastic, trying to become shorter.

7 a The magnetic fields caused by the electrical signals can be detected from outside the skull. Interesting work is being carried out to correlate activity in particular areas of the brain with certain types of mental process.

b The flux density at a distance r from a wire carrying a current I is given by the equation

$$B = \frac{\mu_0 I}{2\pi r}$$

Check your syllabus to see whether you are expected to learn this equation.

In this example $I = 80\,\text{A}$, $r = 1.0\,\text{m}$ and $\mu_0 = 4\pi \times 10^{-7}\,\text{NA}^{-2}$.

$$\therefore B = \frac{4\pi \times 10^{-7}\,\text{NA}^{-2} \times 80\,\text{A}}{2\pi \times 1.0\,\text{m}}$$

$$= 160 \times 10^{-7}\,(\text{NA}^{-1}\text{m}^{-1})$$

$$= \mathbf{16\,\mu T}$$

It's worth knowing that tesla, $T = NA^{-1}m^{-1}$.

Level 2

8 The force on a charged particle moving at right angles to a magnetic field is given by the equation

$$F_{mag} = Bqv$$

The force required to keep a particle of mass m moving in a circle of radius r is given by the equation

$$F_{cent} = \frac{mv^2}{r}$$

The force due to the magnetic field acts (always) at right angles to the motion of a charged particle and produces circular motion. So

$$Bqv = \frac{mv^2}{r}$$

which you can rearrange to give the radius,

$$r = \frac{mv}{Bq}$$

In this example

$v = 0.10 \times c = 0.10 \times 3.0 \times 10^8\,\text{m s}^{-1} = 3.0 \times 10^7\,\text{m s}^{-1}$
$B = 2.2 \times 10^{-3}\,\text{T}$
$q = e = (-)1.6 \times 10^{-19}\,\text{C}$
$m = m_e = 9.1 \times 10^{-31}\,\text{kg}$

$$\therefore r = \frac{9.1 \times 10^{-31}\,\text{kg} \times 3.0 \times 10^7\,\text{m s}^{-1}}{2.2 \times 10^{-3}\,\text{T} \times 1.6 \times 10^{-19}\,\text{C}}$$

$= \mathbf{0.078\,m}$ or $\mathbf{7.8\,cm}$

You could use the specific charge of the electron, e/m_e in this calculation, but you would have to invert it at some point:

$$r = \frac{v}{B \times e/m_e} = \frac{v}{B} \times \frac{m_e}{e}$$

9 a The magnetic field at the equator is parallel to the Earth's surface, Figure 34.21.

b This means the particle travels through it at right angles.

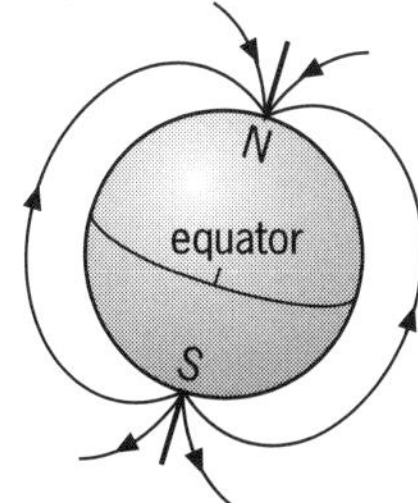

Figure 34.21 The Earth's magnetic field

c The field points North so the force on a positive charge heading for the centre of the Earth is to the East (by Fleming's left hand rule). The size of the force on the particle is given by the equation

$F = Bqv$
$= 55 \times 10^{-6}\,\text{T} \times 1.6 \times 10^{-19}\,\text{C} \times 6.0 \times 10^6\,\text{m s}^{-1}$
$= \mathbf{5.3 \times 10^{-17}\,N}$

(For a proton, this is 10^9 times its weight! Check the arithmetic.)

d Because the particle travels at right angles to the field, the force is always at right angles to the motion, and the particle will go in a circle.

(When there is a component of velocity parallel to the field, the path is a spiral around the field lines. The particles are led towards the North or South poles.)

10 The force per unit length acting on parallel wires carrying currents is given by

$$\frac{F}{l} = \frac{\mu_0 I_1 I_2}{2\pi r}$$

where $\mu_0 = 4\pi \times 10^{-7}\,\text{NA}^{-2}$. In this example $I_1 = I_2 = 100\,\text{A}$ and $l = 200\,\text{mm}$.

$$\therefore \frac{F}{l} = \frac{4\pi \times 10^{-7}\,\text{NA}^{-2} \times 100\,\text{A} \times 100\,\text{A}}{2\pi \times 200 \times 10^{-3}\,\text{m}} = \mathbf{0.010\,Nm^{-1}}$$

Don't forget that force is a vector so you should include its direction. This force will be a repulsion between the wires.

11 The positive ions all go one way round and the electrons go the other way round, so that the currents are all in the same direction. This means that there is an attractive force between the magnetic fields that are created, and the plasma is pulled in, away from the sides. (This is called the 'pinch' effect.)

Level 3

12 a The volume, V, of a cylinder of radius r and length l is given by $V = \pi r^2 l$.

$$\text{Density} = \frac{\text{mass}}{\text{volume}}, \quad \rho = \frac{m}{V}$$

$\therefore$ mass = density × volume, $m = \rho V$

So the weight of the cylinder $= mg = \rho\pi r^2 lg$.

Here density $= 8600\,\text{kg}\,\text{m}^{-3}$, radius $= \frac{1}{2} \times 0.5 \times 10^{-3}\,\text{m}$, $g = 9.8\,\text{m}\,\text{s}^{-2}$.

$\therefore$ weight of 1.0 m of wire
$= 8600\,\text{kg}\,\text{m}^{-3} \times \pi \times (0.25 \times 10^{-3}\,\text{m})^2 \times 1.0\,\text{m} \times 9.8\,\text{m}\,\text{s}^{-2}$
$= 0.017\,\text{kg}\,\text{m}\,\text{s}^{-2} = \mathbf{0.017\,N}$

b The maximum force on a conductor, length l, carrying a current I, in a field with flux density B, is given by the equation $F = BIl$. You can rearrange this equation to find the current needed to produce a specified force in a field of given flux density:

$$I = \frac{F}{Bl}$$

In this example the force required is equal to the weight of the length of wire, $F = 0.017\,\text{N}$. The flux density of the Earth's magnetic field is $56\,\mu\text{T}$, so

$$I = \frac{0.017\,\text{N}}{56 \times 10^{-6}\,\text{T} \times 1.0\,\text{m}} = 304\,\text{NT}^{-1}\text{m}^{-1}$$
$$= 304\,\text{N}(\text{N}^{-1}\,\text{Am})\,\text{m}^{-1} = \mathbf{304\,A}$$

The directions are given by the left hand rule.

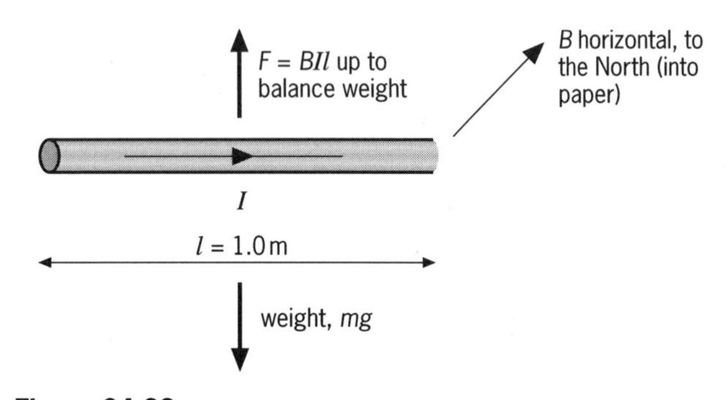

Figure 34.22

13 a The forces that tend to rotate the coil act on the longer sides.

(Check using the left hand rule.)

The forces on the short sides will cancel.

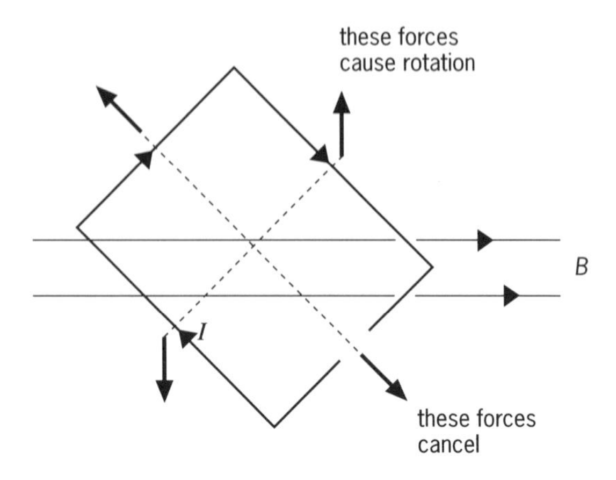

Figure 34.23

The force on each wire on the long side is given by the equation $F = BIl$. In this case

$$F = 0.020\,\text{T} \times 50 \times 10^{-3}\,\text{A} \times 0.060\,\text{m} = 6.0 \times 10^{-5}\,\text{N}$$

This stays the same as the coil rotates.

You need to multiply the force on each side by the number of turns on the coil:

$$BIlN = 6.0 \times 10^{-5}\,\text{N} \times 200 = 0.012\,\text{N}$$

The pair of forces on opposite sides of the coil form a couple which has a moment (torque), given by

one force × perpendicular distance between them

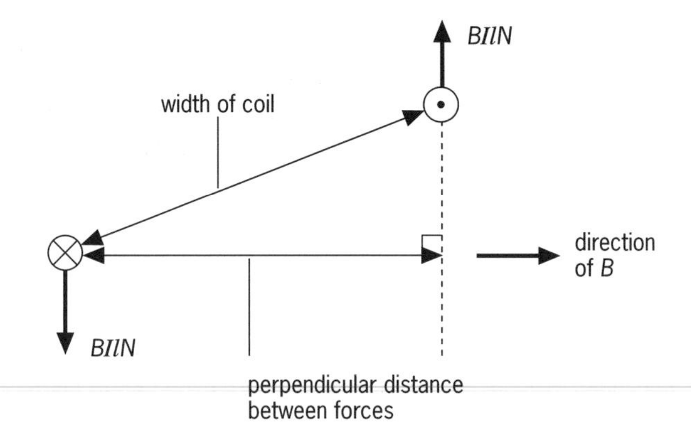

Figure 34.24

The perpendicular distance between the two sides of the coil varies as the coil rotates (Figure 34.24), and is a maximum when the plane of the coil is parallel to the field. The distance is then equal to the width of the coil, 0.040 m. So

$$\text{maximum torque} = 0.012\,\text{N} \times 0.040\,\text{m} = \mathbf{4.8 \times 10^{-4}\,Nm}$$

Think about what happens when the plane of the coil is perpendicular to the field.

14 a In order to give a neutral point, the flux density due to the current in the wire must be equal in magnitude but opposite in direction to the horizontal component of the flux density due to the Earth's magnetic field.

So, the flux density at $r = 5.0\,\text{cm}$ from the wire, which is given by the equation $B = \mu_0 I/2\pi r$, must be equal to $18 \times 10^{-6}\,\text{T}$. You can rearrange the equation to give

$$I = 18 \times 10^{-6}\,\text{T} \times \frac{2\pi r}{\mu_0}$$

$$= 18 \times 10^{-6}\,\text{T} \times \frac{2\pi \times 0.05\,\text{m}}{4\pi \times 10^{-7}\,\text{NA}^{-2}} = \mathbf{4.5\,A}$$

(Note $T = NA^{-1}m^{-1}$)

The fields cancel to the East of the wire, so the current must be **downwards** to give a Southward field (by the right hand grip rule). See Figure 34.25.

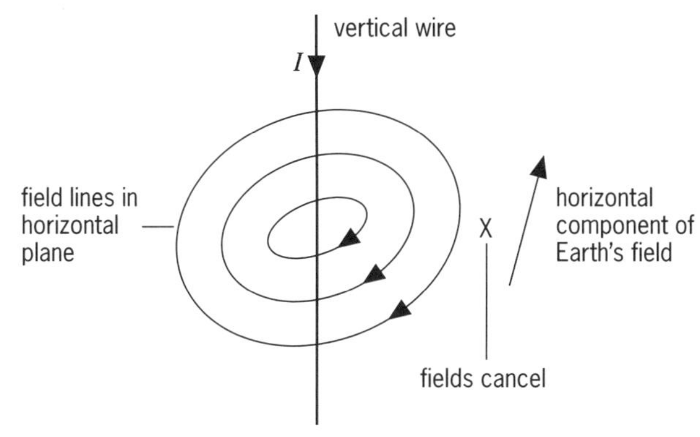

Figure 34.25

b You need to find the distance from the wire at which the flux density, B, due to the current flowing in the wire is $18\,\mu\text{T}$. You can rearrange the equation $B = \mu_0 I/2\pi r$ to give

$$r = \frac{\mu_0 I}{2\pi B}$$

$$= \frac{4\pi \times 10^{-7}\,\text{NA}^{-2} \times 3.5\,\text{A}}{2\pi \times 18 \times 10^{-6}\,\text{T}} = \mathbf{0.039\,m} \text{ or } \mathbf{3.9\,cm}$$

If the current is flowing **downwards** then the neutral point is to the **East** of the wire (see the answer to part a). If the current is flowing **upwards** then the neutral point is to the **West** of the wire.

15 a Relative permeability is the factor by which the magnetic flux density is changed by the medium, compared with the field in a vacuum. (The relative permeability of iron is about 1000 and alloys have been developed which increase the flux density by factors of up to 10000. Copper actually reduces the field strength!)

b If a non-magnetic object (or airspace) is surrounded by ferromagnetic material (one with high relative permeability) then magnetic flux lines will tend to go round the outside of the object, Figure 34.26. The purpose is to shield an object from magnetic influence.

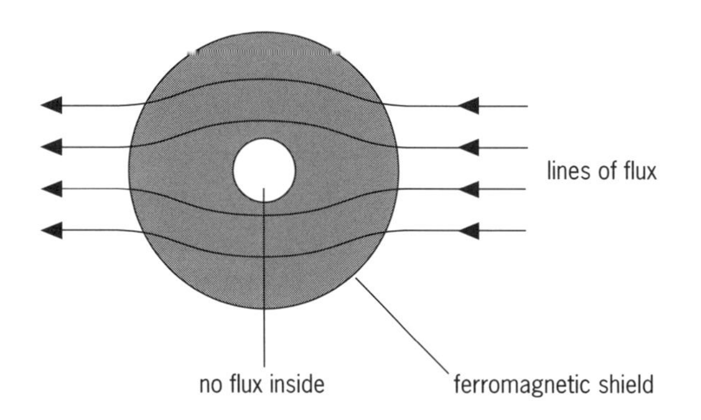

Figure 34.26 Magnetic shielding

c A flux guide is similar but is built around objects that create large magnetic effects. A box made of ferromagnetic material will trap the flux and not let it get out into the surroundings, Figure 34.27.

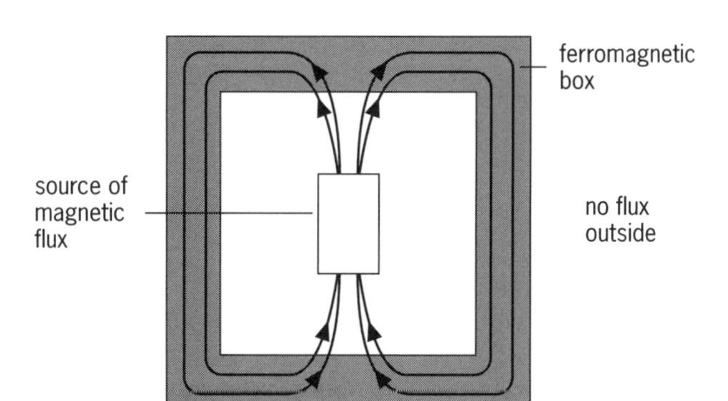

Figure 34.27 A flux guide

35 Flux and induced emf

Magnetic flux, Φ

The total number of magnetic field lines that pass perpendicularly through an area (Figure 35.1) indicates the **magnetic flux**; this is defined by the equation

flux = flux density × area

$$\Phi = BA$$

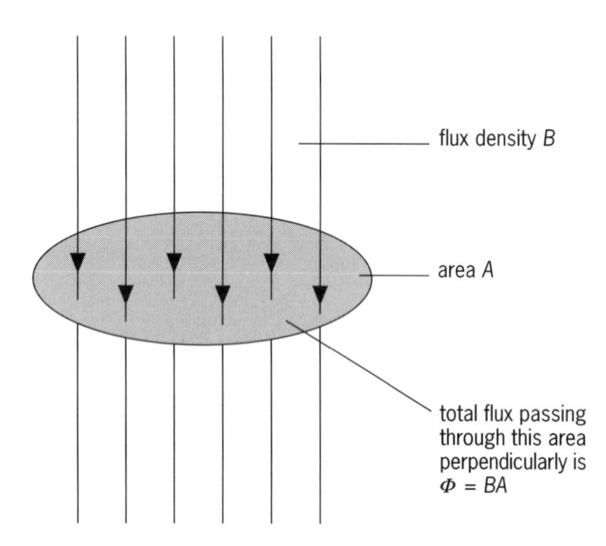

Figure 35.1

The SI unit of magnetic flux is the **weber** (Wb). The flux density is measured in tesla, T; $1\,\text{T} = 1\,\text{Wb}\,\text{m}^{-2}$.

If the field lines pass through the area, A, at an angle θ to the normal (Figure 35.2) the flux can be calculated using

$$\Phi = BA\cos\theta$$

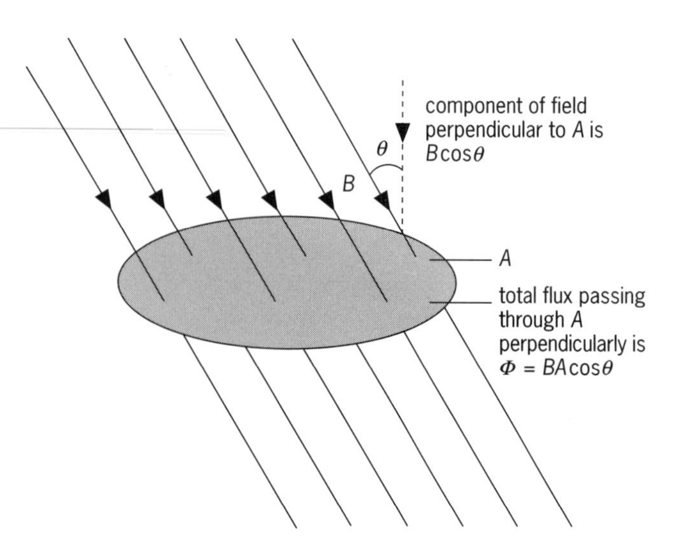

Figure 35.2

Electromagnetic induction

When a conductor is moved across a magnetic field it 'cuts through' the lines of magnetic flux and this creates a potential difference between its ends. If the conductor is part of a complete circuit the potential difference provides an emf and current will flow (Figure 35.3a). This effect is called **electromagnetic induction**.

If the conductor is held stationary within the field or is moved parallel to the lines of flux there is no cutting of the lines and there is no induced emf or current (Figure 35.3b).

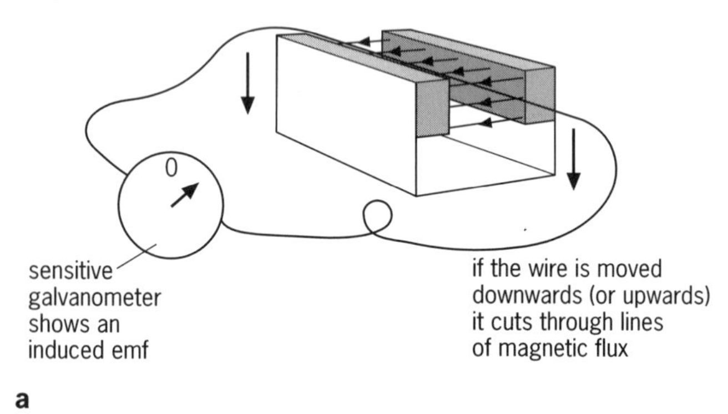

a

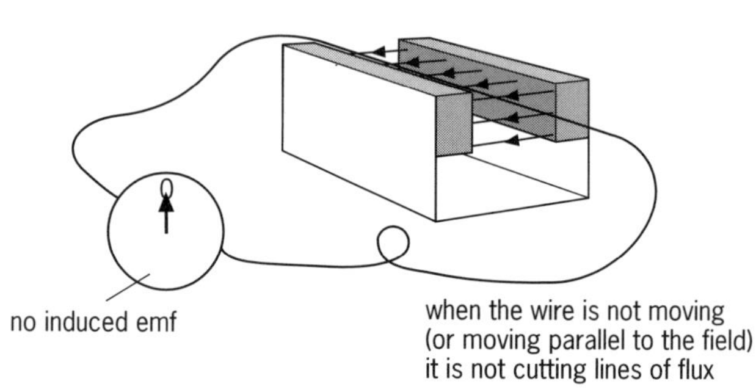

b

Figure 35.3 Demonstrating electromagnetic induction

Further experiments with this apparatus show that the size of the induced emf (and hence current) depends upon

- the speed of the movement of the conductor;
- the length of the conductor within the magnetic field;
- the strength of the magnetic field.

Similarly, if a conductor such as a coil of wire is held stationary and a magnetic field is moved, an emf and current can again be generated (Figure 35.4).

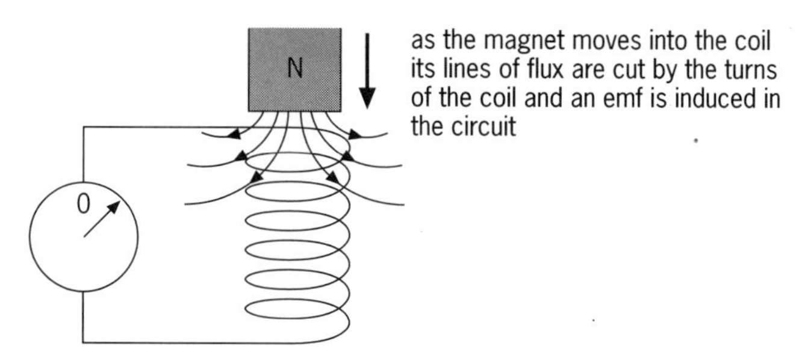

Figure 35.4 Alternative demonstration of electromagnetic induction

Experiments with this apparatus show that the size of the induced emf and current depends upon

- the speed of the movement of the bar magnet;
- the area of the coil;
- the number of turns on the coil;
- the strength of the magnetic field around the pole of the magnet.

The results from both these experiments are summarised by **Faraday's Law** of electromagnetic induction, which states that **the induced emf, *E*, across a conductor is proportional to the rate at which flux is cut**.

For a single length of wire this relationship may be described by the equation

$$E = \frac{-\mathrm{d}\Phi}{\mathrm{d}t}$$

where $\mathrm{d}\Phi/\mathrm{d}t$ is the rate of change in flux (indicating the number of lines of flux cut per second) and the minus sign indicates that the emf generated opposes the change creating it (see page 283). This shows that 1 volt = 1 weber per second.

If the wire is in the form of a coil of N turns, the equation becomes

$$E = \frac{-\mathrm{d}(N\Phi)}{\mathrm{d}t}$$

The quantity $N\Phi$ is known as the **flux linkage**, measured in weber (Wb) since N is a number and has no unit. $\mathrm{d}(N\Phi)/\mathrm{d}t$ is therefore the rate of change of flux linkage. See Figure 35.5.

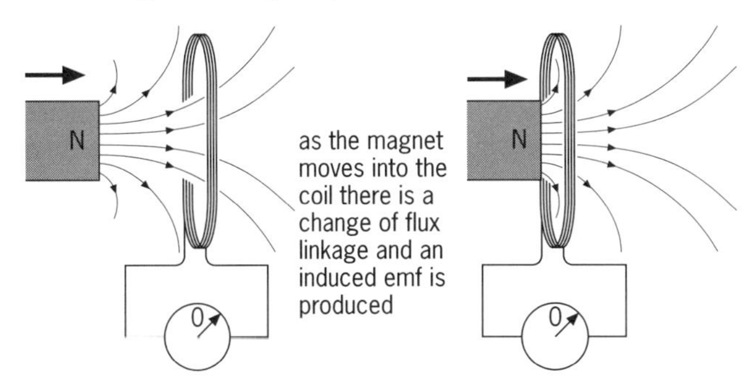

Figure 35.5 A change of flux linkage

Example 1

Calculate the induced emf in a coil of 500 turns placed in a magnetic field when the rate of change of flux is $0.02\,\mathrm{Wb\,s^{-1}}$.

$$E = \frac{-\mathrm{d}(N\Phi)}{\mathrm{d}t}$$

N is constant so this can be rewritten as

$$E = -N\left(\frac{\mathrm{d}\Phi}{\mathrm{d}t}\right) = -500 \times 0.02\,\mathrm{Wb\,s^{-1}}$$
$$= -10\,\mathrm{V}$$

(Note 1 $\mathrm{Wb\,s^{-1}} = 1\,\mathrm{V}$)

Induced emf in a straight conductor

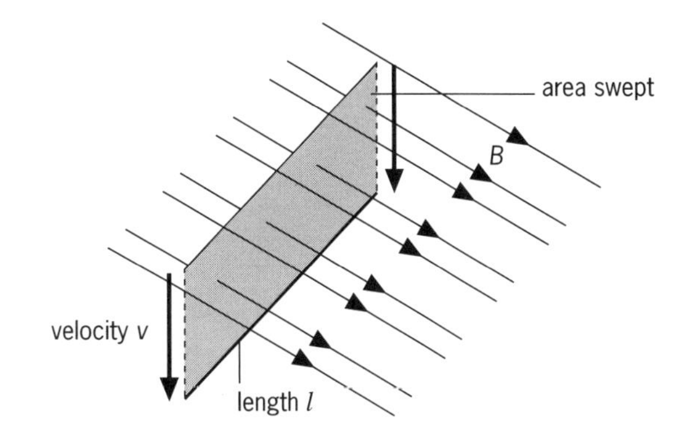

Figure 35.6

If a straight conductor of length l moves with a velocity v perpendicular to its length, it will sweep each second through an area, A, equal to $l \times v$, Figure 35.6.

$$\frac{\mathrm{d}A}{\mathrm{d}t} = lv$$

If the conductor cuts through a magnetic field of flux density B at 90° to the flux, the rate of cutting of the lines will be

$$\frac{\mathrm{d}\Phi}{\mathrm{d}t} = B\frac{\mathrm{d}A}{\mathrm{d}t} = Blv$$

From Faraday's Law it follows that the size of the induced emf, E, will be

$$E$$

If the conductor cuts through the field at an angle θ to the flux, Figure 35.7, the induced emf, E, will be

$$E = Blv\sin\theta$$

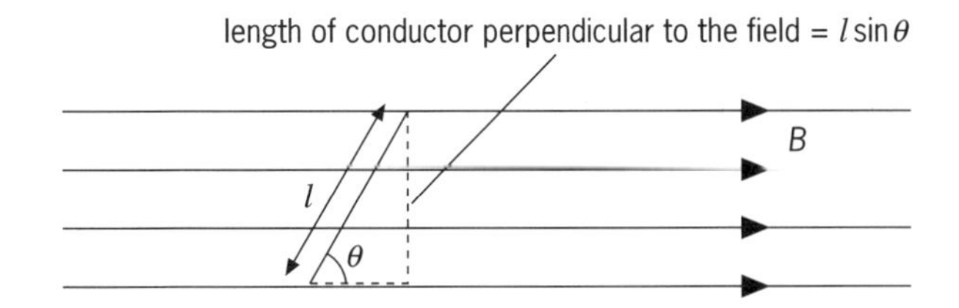

Figure 35.7

Example 2

Calculate the emf induced across a horizontal metal bar 0.50 m long moving horizontally, perpendicular to its length, with a speed of $0.40\,\mathrm{m\,s^{-1}}$, through a vertical magnetic field of flux density 2.5 T.

$$E = Blv$$
$$= 2.5\,\mathrm{T} \times 0.50\,\mathrm{m} \times 0.40\,\mathrm{m\,s^{-1}}$$
$$= 0.5\,\mathrm{V}$$

($\mathrm{Wb\,m^{-2}} \times \mathrm{m} \times \mathrm{m\,s^{-1}} = \mathrm{Wb\,s^{-1}} = \mathrm{V}$)

Induced emf in a rotating coil

If a coil is rotated between the poles of a magnet its wires will cut magnetic lines of force and an induced emf will be produced.

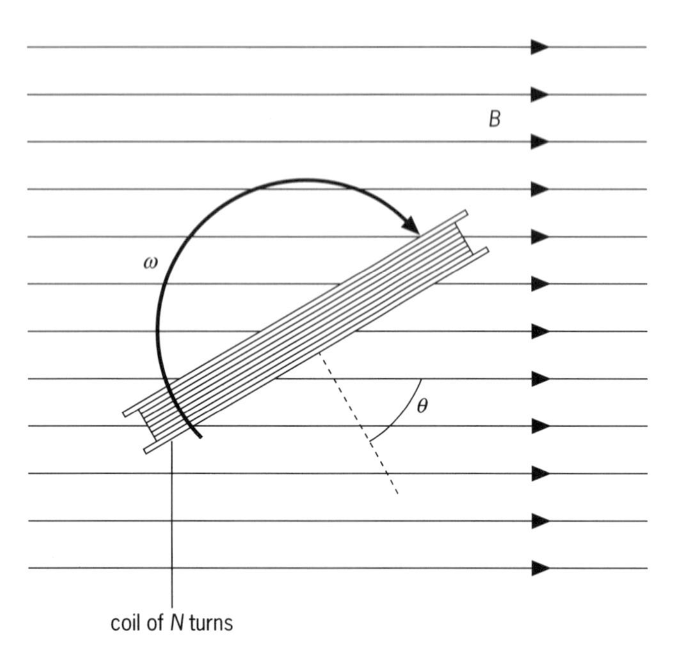

Figure 35.8

Consider a coil of N turns and area A being rotated with a constant angular velocity ω in a uniform magnetic field of flux density B, its axis being perpendicular to the field. When the normal to the coil is at an angle θ to the field, Figure 35.8, the flux through the coil is

$$\Phi = BAN\cos\theta$$

But $\theta = \omega t$,

$$\therefore \quad \Phi = BAN\cos\omega t$$

By definition the induced emf is

$$E = \frac{-\mathrm{d}\Phi}{\mathrm{d}t}$$

$$= \frac{-\mathrm{d}(BAN\cos\omega t)}{\mathrm{d}t}$$

Since BAN is constant this can be written as

$$E = -BAN\frac{\mathrm{d}(\cos\omega t)}{\mathrm{d}t}$$

But $\frac{\mathrm{d}(\cos\omega t)}{\mathrm{d}t} = -\omega\sin\omega t$

$$\therefore \quad E = BAN\omega\sin\omega t$$

From the previous equation we can see that the emf induced in the coil will vary sinusoidally with time, as shown in Figure 35.9.

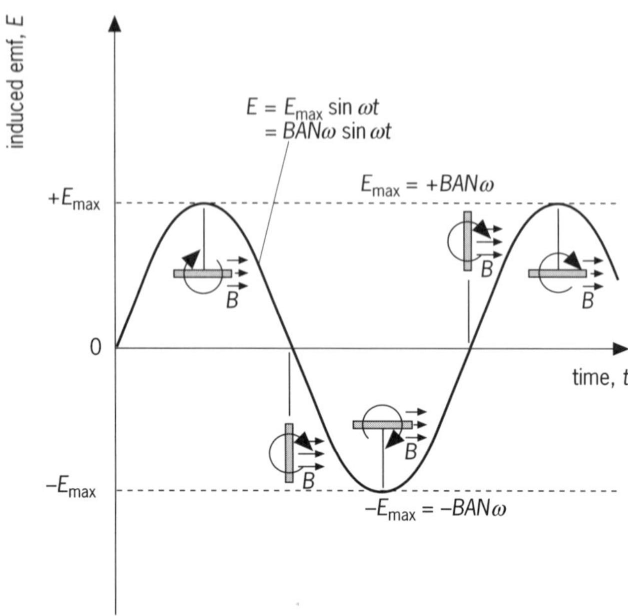

Figure 35.9 Induced emf in a rotating coil

This alternating emf has a peak or maximum value of $BAN\omega$ which occurs when the plane of the coil lies *parallel to the field*; then $\sin\omega t$ has the maximum value of ± 1. The output from the coil can be described by the equation

$$\boxed{E = E_{max}\sin\omega t}$$

where $E_{max} = BAN\omega$.

Example 3

Calculate the maximum emf induced in a coil with 500 turns, of area $1.0\times10^{-3}\,\mathrm{m}^2$, rotating at 20 radians per second in a field of flux density 0.20 T.

$$E_{max} = BAN\omega$$

$$= 0.20\,\mathrm{T}\times1.0\times10^{-3}\,\mathrm{m}^2\times500\times20\,\mathrm{s}^{-1}$$

(Note $\mathrm{Wb\,m^{-2}}\times\mathrm{m^2}\times\mathrm{s^{-1}} = \mathrm{V}$)

$$\therefore \quad E_{max} = 2.0\,\mathrm{V}$$

Direction of induced emfs and currents

In straight conductors

The direction of the induced current in a complete circuit can be predicted using **Fleming's right hand rule**; see Figure 35.10.

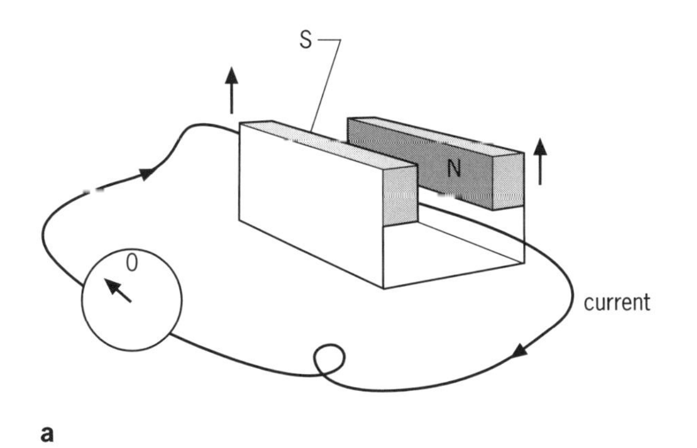

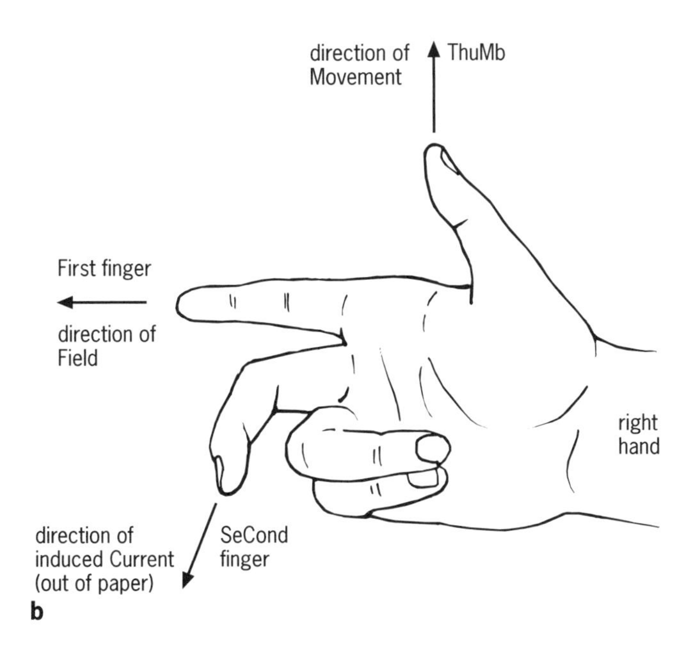

Figure 35.10 Fleming's right hand rule to determine the direction of the induced current in a straight conductor

In coils

The direction of induced emfs and currents in coils is described by **Lenz's Law**, which states that **the direction of the induced emf or current is such as to oppose the change that produces it**.

When the North-seeking pole of a bar magnet is pushed into a coil, for example, there is a change in flux linkage and current is generated. This current flows in a direction that produces a North-seeking pole at the top of the coil, which tries to repel (oppose) the North-seeking pole of the magnet. See Figure 35.11a; the direction of the current in the coil can be determined by the right hand grip rule (page 270).

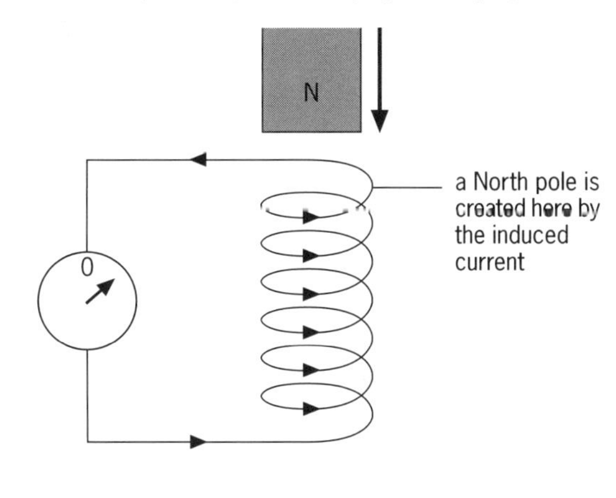

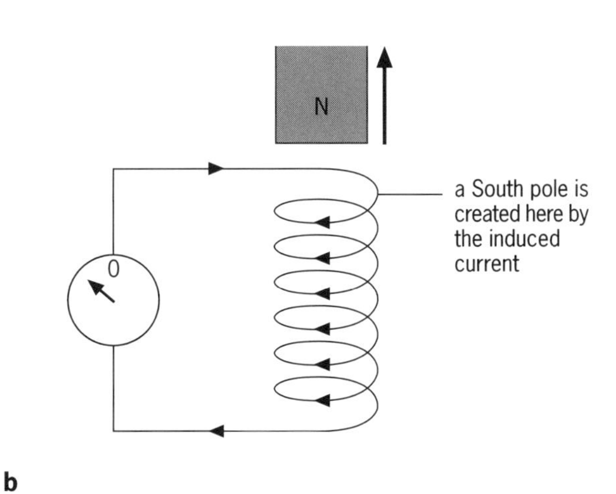

Figure 35.11 Using Lenz's Law to determine the polarity of the coil cause by the induced current

When the North-seeking pole is withdrawn, current is again generated but this time it flows in the opposite direction, Figure 35.11b. The induced current produces a South-seeking pole, which tries to prevent the North pole from being withdrawn.

Since the induced emf opposes the change which is producing it, it is often referred to as a **back emf**, particularly in electric motors where the induced emf opposes the rotation of the coil.

$\mu_0 = 4\pi \times 10^{-7}\,\mathrm{N\,A^{-2}}$

Level 1

1 A magnet with a pole face 0.020 m × 0.010 m has flux density $B = 1.8\,\mathrm{mT}$, measured with a probe held parallel to the face, Figure 35.12.

a What is the total flux leaving the pole face?

b The probe face is half the area of the pole face. How much flux does it detect?

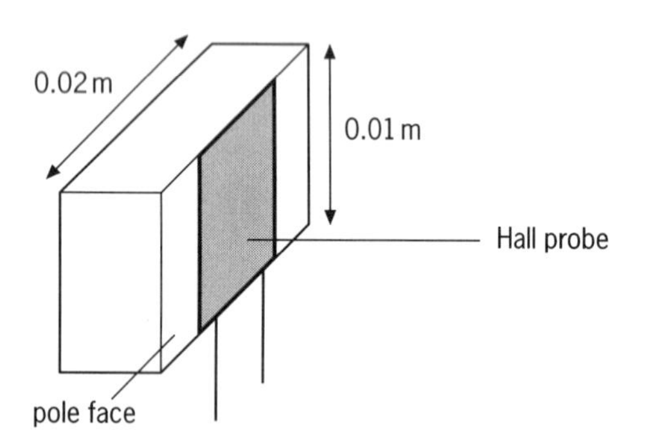

Figure 35.12

2 A 200-turn coil has 4.8 mWb of flux threaded through it. What is the flux linkage?

3 A coil of 10 turns has a flux linkage of 50 Wb. What is the flux entering it?

4 In 1830 the Swiss physicist Colladon was trying to detect induced currents. He connected a coil to a sensitive meter. To avoid any influence *on the meter* when he altered the magnetic field, he used long leads and put the meter in the next room. Every time he put a magnet into the coil and went to look at the meter it read zero. Explain how he missed his chance to make an important discovery!

5 Figure 35.13 shows a wire loop at right angles to a uniform magnetic field of flux density 0.30 T.

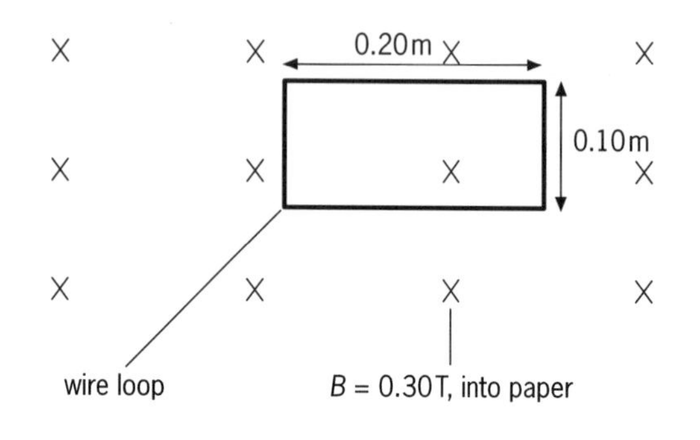

Figure 35.13

a How much flux threads the loop?

b By how much does the flux linkage change when the following movements are made?

i The loop slides 0.10 m to the left.

ii The loop is lowered (into the paper) by 0.10 m.

iii The loop is rotated 90° clockwise about its central axis in its own plane.

iv The loop is rotated about a short side to an upright position.

v The loop is first returned to its original position and then turned through 180°, about one of its long sides.

6 If a conductor is placed inside a changing magnetic field, currents which are continually changing direction are produced. These swirling induced currents are called 'eddy currents'. Their heating effect, known as induction heating, can be large enough to melt a metal.

a What factors will increase the heating effect?

b Why would the steel reinforcing rods in a concrete floor get hot when there is a coil carrying a large alternating current at mains frequency in the vicinity?

Level 2

7 In the UK, the Earth's magnetic field has a flux density of about 53 μT at an angle of 70° to the horizontal, Figure 35.14.

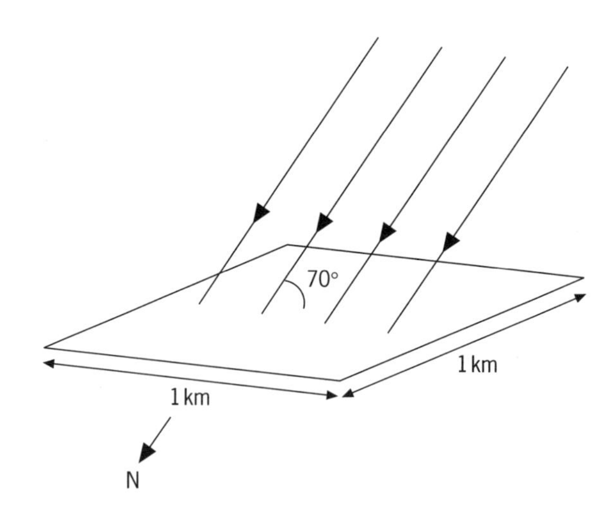

Figure 35.14

What is the magnetic flux through an area of land 1 km square ($10^6\,\mathrm{m}^2$)?

8 Flux density, B, inside a long solenoid of length l, with a total of N turns carrying a current I, is given by the equation

$$B = \frac{\mu_0 N I}{l}$$

This is often written as $B = \mu_0 n I$, where n is the number of turns per unit length, or turns per metre.

Calculate the flux density, B, and the flux, Φ, at the centre of the rectangular solenoids in Figure 35.15, when the current flowing is 1.3 A in each case.

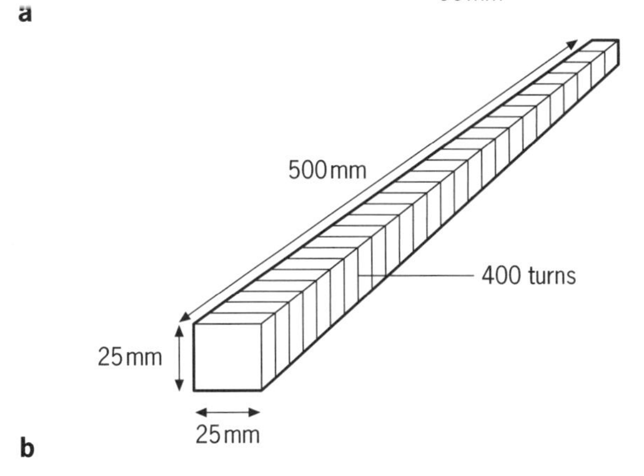

Figure 35.15

9 There is a uniform flux density of 2.6 mT in the centre of a solenoid. What would be the flux through a 10 mm × 5 mm probe

a when it is held with its plane at right angles to the flux

b when its plane makes an angle of 50° with the flux?

10 Several attempts have been made to trail a cable from a satellite so that it cuts the Earth's magnetic field and develops an emf.

a What would that emf be if the cable was 20 km long, the velocity of the satellite was 7 km s^{-1} and the flux density was 60 μT at right angles to the cable?

b If any power could be obtained by this means, what effect would it have on the motion of the satellite?

11 A 'ramp generator' is a device that increases current at a steady rate, as shown in the graph, Figure 35.16.

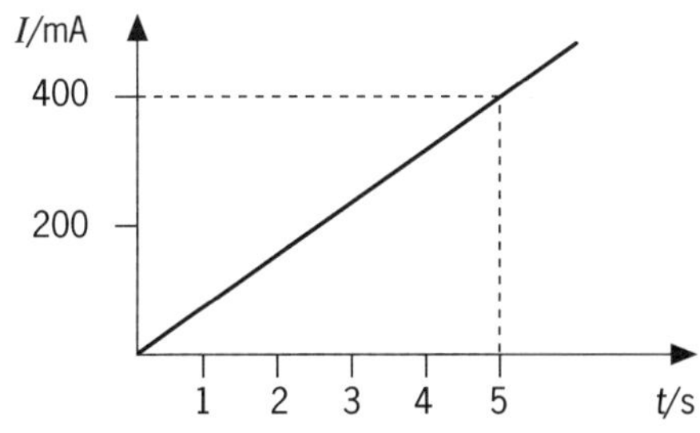

Figure 35.16

A ramp generator with a rate of current growth as shown in the graph is connected to a coil with many turns. Calculate the rate of change of flux in the coil, if it is 1.0 cm long, has 20 turns and an area of $2.5 \times 10^{-3}\,\text{m}^2$.

Hint: The flux density generated in the coil is given by the equation $B = \mu_0 nI$, and the total flux through the coil is given by the equation $\Phi = BA$. So

$$\Phi = \mu_0 nIA \quad \textit{and} \quad \frac{d\Phi}{dt} = \mu_0 nA \frac{dI}{dt}$$

where $\mu_0 = 4\pi \times 10^{-7}\,NA^{-2}$, n = number of turns per metre, I = current and A = area of coil.

12 a What emf is induced in a 500-turn coil when the flux through it is changing by 30 Wb s^{-1}?

b What is the rate of change of flux if 50 V is induced in a coil of 20 turns?

13 A car travels at 80 km h^{-1} (22 m s^{-1}). It will cut the vertical component of the Earth's field, which is about 50 μT downwards.

a *Estimate* the potential difference across the bonnet.

b Say which side will be positive, if the car is heading West.

14 A copper ring is suspended as shown in Figure 35.17.

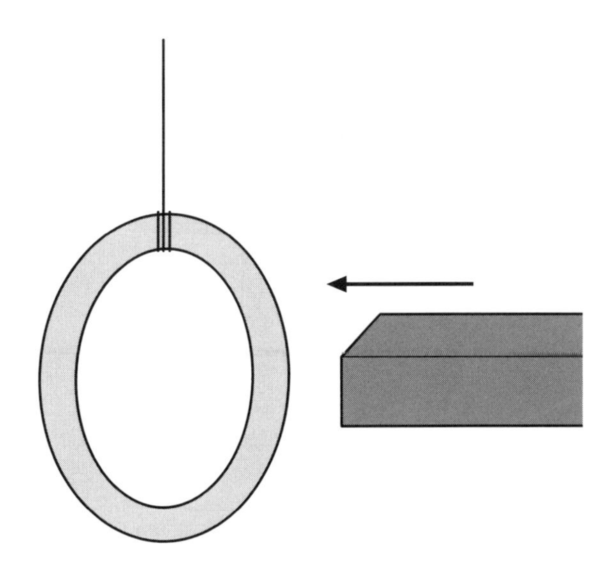

Figure 35.17

What would you expect to happen if

a an unmagnetised bar of steel is brought up to the ring

b a bar magnet is brought up to it, with the North-seeking pole towards the ring, and then withdrawn? Try the experiment!

15

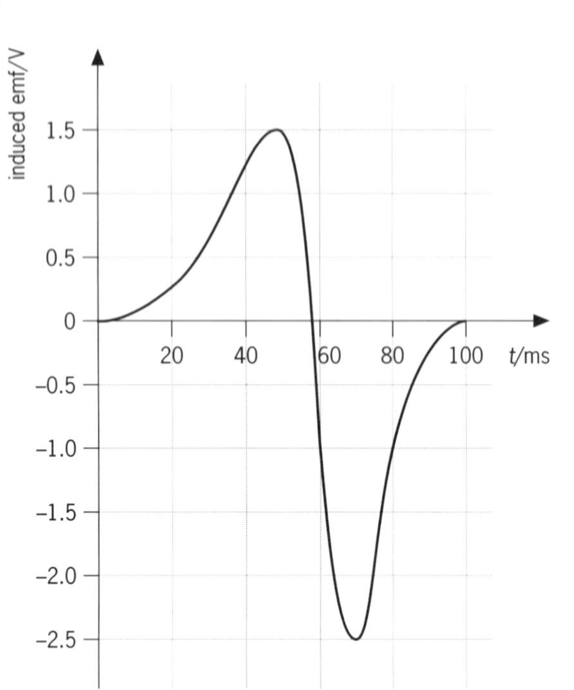

Figure 35.18

The graph in Figure 35.18 was printed out by a storage oscilloscope. It shows the emf induced in a coil when a magnet fell through it.

a At what time was the rate of change of flux a maximum?

b How large was it?

c How would the graph be different if the magnet had been dropped from higher up?

16 Show that $E = Blv$ when a conductor moves through a field B with velocity v, at right angles to its length and to the field (Figure 35.19). Use the basic rate-of-flux-cutting principle.

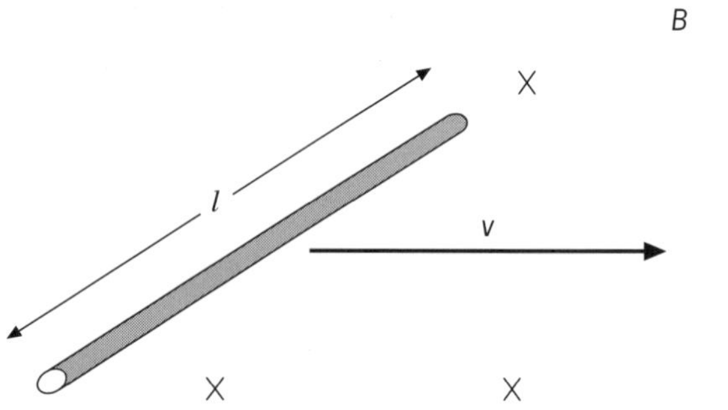

Figure 35.19

Level 3

17 The rate of rotation of a shaft may be measured by an instrument called a tacho-generator, which has a toothed wheel of ferromagnetic material attached to the rotating shaft. The arrangement is shown in Figure 35.20.

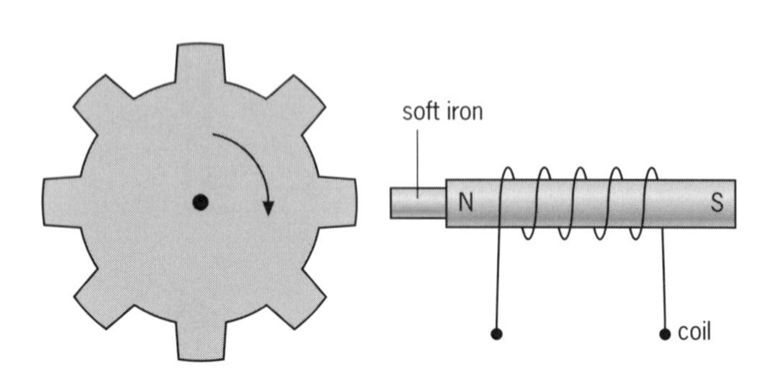

Figure 35.20

As each tooth comes close to the soft iron, the magnetic flux linked with the coil changes and an emf is induced.

Sketch a graph to show the variation of emf with time you might expect if the wheel has eight teeth and rotates at 1800 rpm.

18 A plane coil of area $2.0 \times 10^{-3}\,\text{m}^2$ is wound with 100 turns. It is rotated at a constant angular velocity ω in a magnetic flux density of 45 mT.

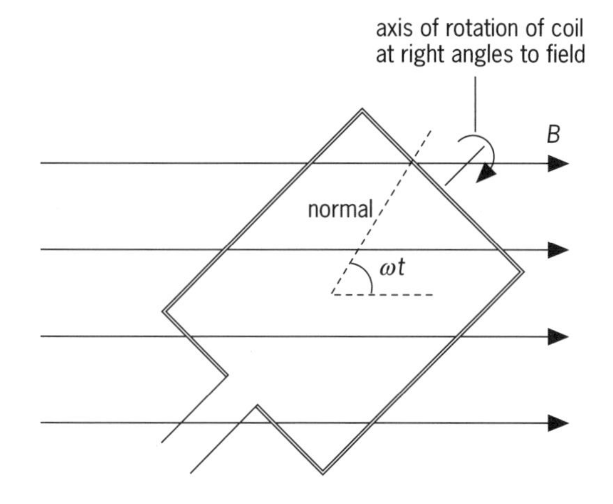

Figure 35.21

a Taking the angle ωt to be the angle that the normal to the coil makes with the field, write an expression for the flux linkage with the coil at this angle.

b At what rate is this flux linkage changing?

c What will be the emf at any instant if $\omega = 50\pi\,\text{rad}\,\text{s}^{-1}$?

d Find the peak emf generated at this frequency.

e What is the angle of the coil with the field when the emf is a maximum?

Level 1

1 You can rearrange the equation

$$\text{flux density} = \frac{\text{flux}}{\text{area}}, \quad B = \frac{\Phi}{A}$$

to give flux = flux density × area, $\Phi = B \times A$

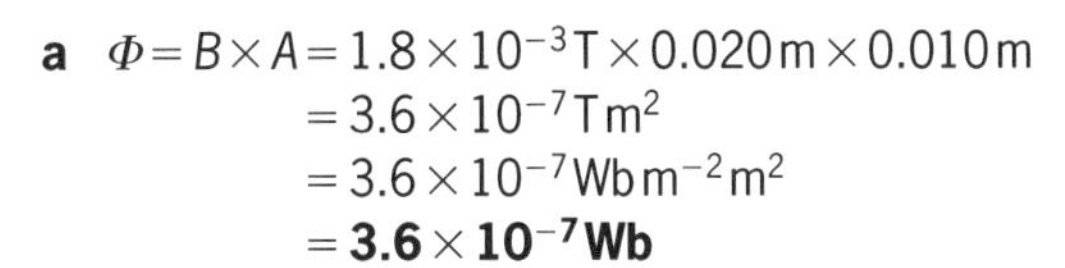

a $\Phi = B \times A = 1.8 \times 10^{-3}\,\text{T} \times 0.020\,\text{m} \times 0.010\,\text{m}$
$= 3.6 \times 10^{-7}\,\text{T}\,\text{m}^2$
$= 3.6 \times 10^{-7}\,\text{Wb}\,\text{m}^{-2}\,\text{m}^2$
$= \mathbf{3.6 \times 10^{-7}\,Wb}$

b The probe has only half the area of the pole face so, at best, only half the flux will go through the probe: $\mathbf{1.8 \times 10^{-7}\,Wb}$.

2 Flux linkage = $N \times \Phi$ where N is the number of turns in the coil and Φ is the flux through the coil. In this example,

flux linkage = 200 turns × 4.8×10^{-3} Wb = **0.96 Wb**

3 Flux linkage = $N \times \Phi$ where N is the number of turns in the coil and Φ is the flux through the coil. You can rearrange this equation to give

$$\Phi = \frac{\text{flux linkage}}{N}$$

$$= \frac{50\,\text{Wb}}{10} = \mathbf{5\,Wb}$$

4 The induced current can flow only when the magnetic field is *changing*. By the time Colladon got into the next room, the magnetic field would be steady so no current would be flowing. He needed to watch the meter *while* he was putting the magnet in.

5 a The flux through the loop is given by the equation

$\Phi = BA$

B, the flux density = 0.30 T = 0.30 Wb m^{-2}
A, the area of the loop = 0.20 m × 0.10 m = 0.020 m^2

$\therefore \Phi = 0.30\,\text{T} \times 0.020\,\text{m}^2 =$ **0.006 Wb** or **6 mWb**

b i The whole loop is still threaded by a field which is uniform, so the flux linkage does not change.
ii The same argument applies, so there is no change in the flux linkage.
iii Because the loop stays completely in the field, and at 90° to it, the flux linkage will again stay the same.
iv When the loop is 'upright', it is parallel to the field so no flux threads through it. The flux linkage changes from 6 mWb to zero. The change is **−6 mWb**.
v After the coil is turned over, the same amount of flux will thread the loop but in the opposite direction. Flux (like flux density) is a vector, so the change is from +6 mWb to −6 mWb = **−12mWb**.

6 a To produce large eddy currents within the metal there must be **large and rapid changes in the flux density**. This is achieved by using a coil carrying a large alternating current that changes rapidly (high frequency); see Figure 35.22.

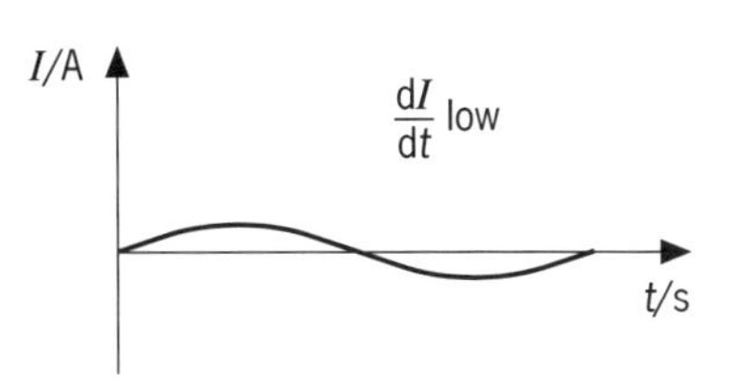

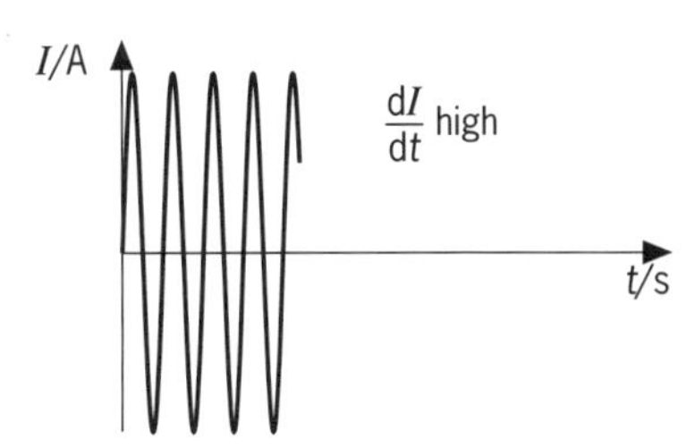

Figure 35.22

b Magnetic flux passes easily into ferromagnetic materials such as steel. Because the coil carries a.c. current, eddy currents will be induced in the rods and they will experience a heating effect.

Level 2

7

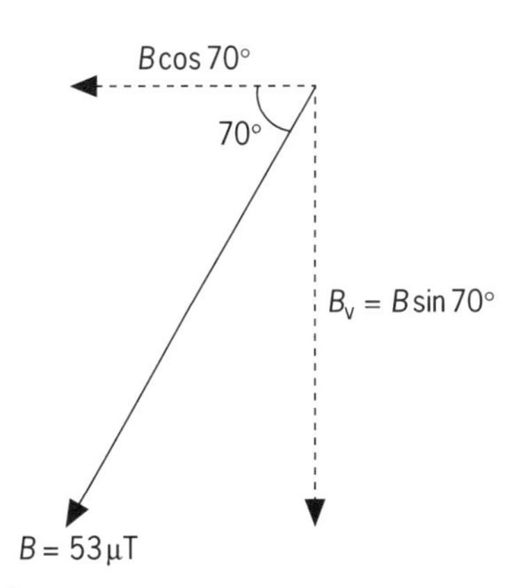

Figure 35.23

The vertical component of the flux density (Figure 35.23) is

$B_v = 53\,\mu\text{T} \times \sin 70°$
$= 53\,\mu\text{T} \times 0.94 = 50\,\mu\text{T}$

The area is 1 km square, so

flux, $\Phi = B_v \times \text{area} = 50 \times 10^{-6}\,\text{T} \times 10^6\,\text{m}^2 =$ **50 Wb**

($\text{T}\,\text{m}^2 = \text{Wb}\,\text{m}^{-2}\,\text{m}^2 = \text{Wb}$)

8 a $B = \dfrac{\mu_0 N I}{l}$

$$= \frac{4\pi \times 10^{-7}\,\text{N}\,\text{A}^{-2} \times 400 \times 1.3\,\text{A}}{0.25\,\text{m}}$$

$= \mathbf{2.6 \times 10^{-3}\,T}$ or **2.6 mT**

$\therefore \Phi = BA = 2.6 \times 10^{-3}\,\text{T} \times 0.050\,\text{m} \times 0.050\,\text{m}$
$= 6.5 \times 10^{-6}\,\text{T}\,\text{m}^2 = \mathbf{6.5\,\mu Wb}$

b $B = \dfrac{4\pi \times 10^{-7}\,\text{N}\,\text{A}^{-2} \times 400 \times 1.3\,\text{A}}{0.50\,\text{m}}$

$= \mathbf{1.3 \times 10^{-3}\,T}$ or **1.3 mT**

Did you spot that the flux density is half the value in part a because N/l is half?

The cross-sectional area of this solenoid is reduced by a factor of 4. So the flux will be $\frac{1}{8}$ of that in part b: **0.8 μWb**.

Look out for patterns like this. They can save time and let you check your answer.

9 a When the probe is at right angles to the flux, the whole area of the probe is threaded so

$$\Phi = BA = 2.6 \times 10^{-3}\,\text{T} \times 0.010\,\text{m} \times 0.005\,\text{m}$$
$$= \mathbf{1.3 \times 10^{-7}\,Wb}$$

b When the probe is at an angle you must find the projection of the area (like a shadow). The area at right angles to the flux is $A \times \sin 50°$; see Figure 35.24.

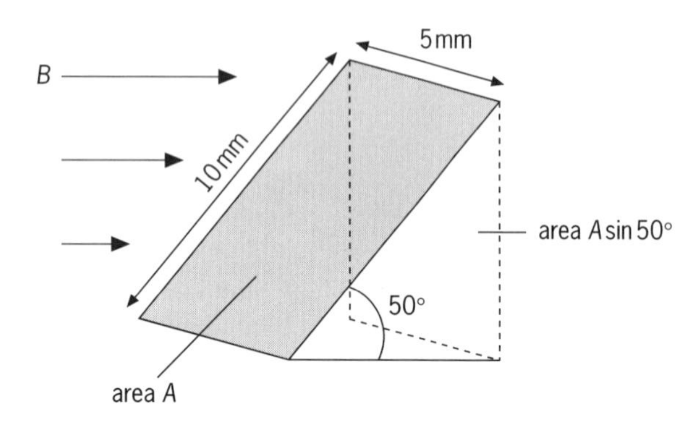

Figure 35.24

$$\therefore \Phi_{\text{probe}} = \Phi_{\text{solenoid}} \times \sin 50°$$
$$= 1.3 \times 10^{-7}\,\text{Wb} \times 0.766$$
$$= \mathbf{9.96 \times 10^{-8}\,Wb}$$

10 a The induced emf in a conductor, length l, moving at velocity v, at right angles to a magnetic field with flux density B, is given by the equation $E = Blv$.

In this example $B = 60\,\mu\text{T}$ at right angles to the conductor, $l = 20\,\text{km}$ and the velocity, $v = 7\,\text{km s}^{-1}$.

$$\therefore E = 60 \times 10^{-6}\,\text{T} \times 20 \times 10^{3}\,\text{m} \times 7 \times 10^{3}\,\text{m s}^{-1}$$
$$= 8400\,\text{T m}^2\,\text{s}^{-1}$$
$$= \mathbf{8.4\,kV}$$

You need to remember to convert the length and speed into SI units to get the correct numerical answer. And to get the units correct you need to remember that $T = Wb\,m^2$ and $Wb = V\,s$. If you put these in, everything except the V cancels out.

b From Lenz's Law (which states that the direction of the induced emf will tend to cause a current to flow in such a direction as to oppose the change causing it), there would obviously be a retarding force on the orbiting satellite and it would not maintain a constant-radius orbit unless there was some power input.

11 The gradient of the graph $= dI/dt = 400\,\text{mA}/5\,\text{s}$
$= 80\,\text{mA s}^{-1}$

Don't forget to check the units on the axes when you calculate a gradient.

The equations

$$\Phi = \mu_0 n I A \quad \text{and} \quad \frac{d\Phi}{dt} = \mu_0 n A \frac{dI}{dt}$$

are given in the question.
The number of turns per metre, n, is

$$n = \frac{N}{l} = \frac{20}{1.0 \times 10^{-2}\,\text{m}} = 2 \times 10^{3}\,\text{m}^{-1}$$

$$\therefore \frac{d\Phi}{dt} = 4\pi \times 10^{-7}\,\text{N A}^{-2} \times 2 \times 10^{3}\,\text{m}^{-1} \times 2.5 \times 10^{-3}\,\text{m}^2 \times 80 \times 10^{-3}\,\text{A s}^{-1}$$
$$= 5.0 \times 10^{-7}\,\text{N A}^{-2}\,\text{m}^{-1}\,\text{m}^2\,\text{A s}^{-1}$$
$$= 5.0 \times 10^{-7}\,\text{N m A}^{-1}\,\text{s}^{-1}$$
$$= \mathbf{5.0 \times 10^{-7}\,Wb\,s^{-1}}$$

12 a The emf induced in a coil by a changing magnetic flux, $N\Phi$, is given by the equation

$$E = -N\frac{d\Phi}{dt}$$

The minus sign shows that the induced emf opposes the change that is causing it.

$$\therefore E = -500 \times 30\,\text{Wb s}^{-1} = 15 \times 10^{3}\,\text{V} = \mathbf{-15\,kV}$$

Remember $Wb = V\,s$.

b You can rearrange the equation $E = -N d\Phi/dt$ to give

$$\frac{d\Phi}{dt} = \frac{-E}{N} = \frac{-50\,\text{V}}{20} = -2.5\,\text{V} = \mathbf{-2.5\,Wb\,s^{-1}}$$

(If the flux is decreasing, the emf is in a direction such that the flux will tend to be restored.)

13 a The induced emf is equal to the rate of flux cutting ($N = 1$).

Estimating the bonnet to be 2 m wide, then, travelling at $22\,\text{m s}^{-1}$, the car sweeps out an area $= 22\,\text{m s}^{-1} \times 2\,\text{m} = 44\,\text{m}^2\,\text{s}^{-1}$.

Total flux cut per second, $\dfrac{\Phi}{t} = \dfrac{BA}{t}$

$$= 50 \times 10^{-6}\,\text{T} \times 44\,\text{m}^2\,\text{s}^{-1}$$
$$= 2.2 \times 10^{-3}\,\text{T m}^2\,\text{s}^{-1}$$
$$= 2.2 \times 10^{-3}\,\text{Wb s}^{-1}$$

$\therefore$ induced emf $E = 2.2 \times 10^{-3}\,\text{Wb s}^{-1} = \mathbf{2.2\,mV}$

Figure 35.25

b You can use Fleming's right hand rule. The field is downwards (which is the case in the Northern hemisphere). If the car goes West, the positive charges (conventional current) will move towards the **South** side (the passenger side for the right hand drive vehicles). This makes that side positive by this small amount.

(You could also use the left hand rule, which would tell you that there is a force to the South when positive charges are carried towards the West.)

14 a An unmagnetised bar will not have any effect.

b The magnet will change the magnetic field around the ring. This will induce currents in the ring. They will travel in a direction to oppose the change causing them. So if a North-seeking pole approaches, the current will flow anticlockwise to create a field to oppose it, that is another North-seeking pole. See Figure 35.26a.

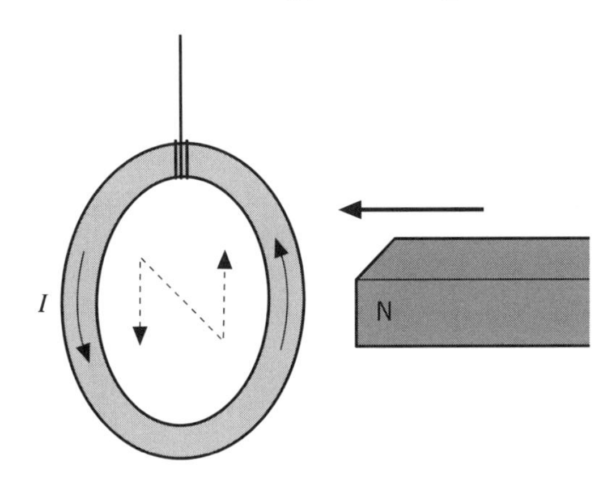

a

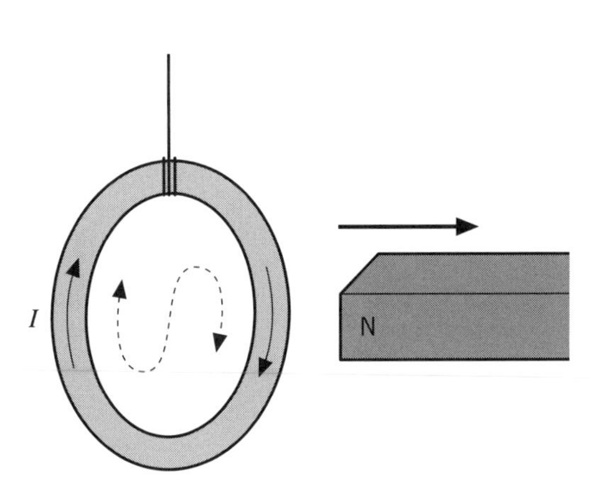

b

Figure 35.26

Because of the repulsive force, the ring will move away from the magnet. If the magnet is then withdrawn the current reverses to reinforce the field. The induced magnetic pole becomes a South-seeking pole, Figure 35.26b, and the ring will swing towards the magnet. When the magnet is stationary there will be no magnetic force on the ring.

Note the mnemonic used in Figure 35.26 for determining the polarity of a circulating current. Check with the right hand grip rule.

15 The induced emf is equal to the rate of change of flux.

a The rate of change of flux is a maximum when the emf is a maximum, that is at about 70 ms from the start of the graph. Although there is a peak in the emf before this, at about 50 ms, it isn't as big as the peak in the other direction at 70 ms. The magnet is accelerating as it falls through the coil, and leaves it at a higher velocity than it entered. So the flux changes more rapidly as the magnet leaves the coil.

b The peak emf is 2.5 V, so the flux is changing at 2.5 V or 2.5 Wb s^{-1}.

c If the magnet had fallen further it would be travelling faster so the rate of change of flux would be greater. The sizes of the peaks in emf would be greater and the time taken for the magnet to fall through the coil would be smaller.

16 The emf induced, $E = -N\mathrm{d}\Phi/\mathrm{d}t$ where N is the number of coils and $\mathrm{d}\Phi/\mathrm{d}t$ is the rate of change or rate of cutting of flux. In this case $N = 1$.

$$\frac{\mathrm{d}\Phi}{\mathrm{d}t} = B\frac{\mathrm{d}A}{\mathrm{d}t}$$

where $\mathrm{d}A/\mathrm{d}t$ is the rate at which the conductor, length l, sweeps over an area of the magnetic field.

$$\therefore \frac{\mathrm{d}A}{\mathrm{d}t} = lv$$

where v is the velocity at right angles to l and to B, Figure 35.27.

$$E = -\frac{\mathrm{d}\Phi}{\mathrm{d}t} = -B\frac{\mathrm{d}A}{\mathrm{d}t}$$

so the magnitude of E is $\boldsymbol{Blv}$.

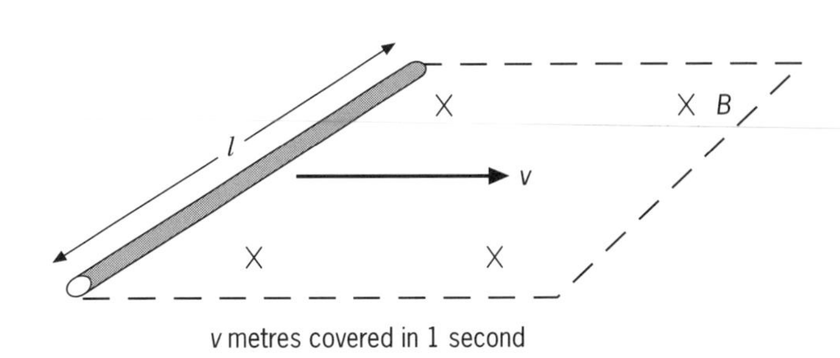

Figure 35.27

Level 3

17 You would expect some sort of regularly varying output, which might be a series of pulses or something like a sine wave. You need to find the frequency. You can't calculate the amplitude of the signal because you haven't been told anything about the flux density or the size of any of the components.

The shaft rotates at 1800 rpm which is 30 revs per second, so the time for one rotation, $T = 1/f = 0.033\,\text{s} = 33\,\text{ms}$. There are eight teeth on the wheel so the time between one tooth and the next passing a fixed point is $33\,\text{ms}/8 = 4.1\,\text{ms}$. If, by some happy chance, the flux variation is proportional to $\cos \omega t$, then the emf will be proportional to $\sin \omega t$.

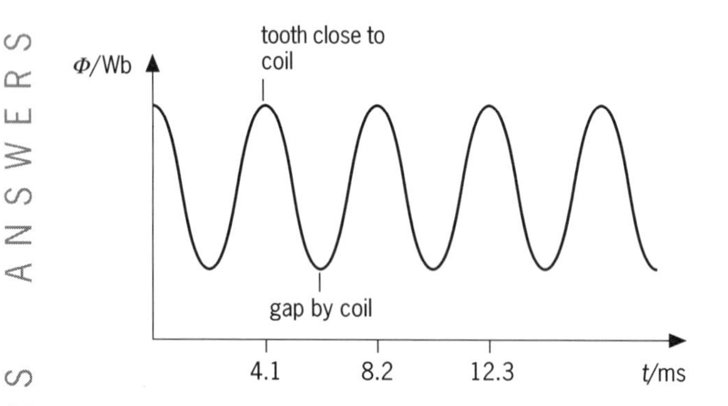

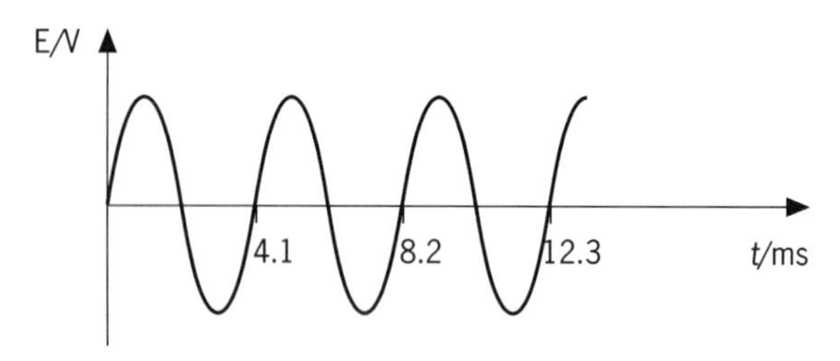

Figure 35.28

(If $\Phi = k \times \cos \omega t$ then $E = -N\,d\Phi/dt = -Nkd(\cos \omega t)/dt = Nk\omega \sin \omega t$. The output will not depend on which way round the wheel goes but it will depend on which way round you connect a meter to the coil.)

18 a With the angle defined like this then the flux linkage is given by the equation

$$N\Phi = BAN \cos \omega t$$

(Use the component of the field at right angles to the coil or the projected area of the coil at right angles to the field.)

b The rate of change of flux linkage is

$$N\frac{d\Phi}{dt} = -BAN\omega \sin \omega t$$

c $E = -N\dfrac{d\Phi}{dt} = BAN\omega \sin \omega t$

$$= 45 \times 10^{-3}\,\text{Wb m}^{-2} \times 2.0 \times 10^{-3}\,\text{m}^2 \times 100 \times 50 \times \pi\,\text{s}^{-1} \times \sin(50\pi t)$$

$$= 1.4 \sin(50\pi t)\,\text{Wb s}^{-1} = \mathbf{1.4 \sin(50\pi t)\,V}$$

The radian (rad) is the dimensionless SI unit of plane angle, defined as the ratio of two lengths; so, although it appears in the units of ω, it can be ignored elsewhere.

d The peak values of E occur when $\sin \omega t = \pm 1$, which gives $E = \pm\mathbf{1.4\,V}$

e $\sin \omega t = \pm 1$ when ωt is 90°, that is when the plane of the coil is parallel with the field. This is when there is the greatest change in the amount of flux threaded through the coil, for a small change in the angle.

Synoptic questions

Examples of three different types of synoptic questions are included in this section:

1. those that require knowledge from across the specifications;
2. data analysis questions;
3. comprehension questions.

Using information from across the specifications

In the Module Tests the examiners can only ask questions that are based on the topics studied in that module. They are not allowed to ask questions that link aspects from two or more modules. But they *can* do this in the final papers. They can ask questions in which knowledge, understanding and skills learnt in different parts of the course are drawn together. Topics covered at GCSE or even earlier can be included.

If you think about complicated equipment or systems, such as aeroplanes, hot-air balloons, nuclear power stations, hi-fi systems, Geiger counters, X-ray tubes, cathode ray tubes, cars, boats, swimming pools, or an entire leisure centre or home, you will see that topics such as heat, motion, electricity, waves, and so on, can be included in just one problem. The examiners will expect you to point out similarities and differences between different branches of physics, sometimes without prompting. You must always be aware of analogies such as the similarity between electric and gravitational fields. (More are listed below.) Topics not specifically covered by the syllabus may be introduced because they are similar to something that *has* been studied.

Some papers may be laboratory-based and you may be asked to describe experiments you have done or to design a new experiment. Others will just concentrate on the physics of the system concerned.

Be prepared in particular for:

- alternating current linked with oscillation theory (see question 2 of this section);
- energy changes – examiners see all sorts of opportunities here;
- circular motion – satellites, roller-coasters, spin-dryers, atoms, cyclotrons – think about what type of physical force provides the necessary centripetal force in each case; it could be a mechanical, gravitational, electrostatic or magnetic force;
- electrical effects – heat, light, sound, magnetic fields, forces;
- parallels between flow of charge, which you will have studied, and heat flow or fluid flow, which may not have been covered (see question 2 of this section);
- materials – you must give reasons for a choice of a material;
- photoelectric effect and ionisation producing electric currents;
- motors, generators and transducers that change one type of signal into another;
- the inverse square law – this applies to *all* waves that are not absorbed in the medium;
- exponential decrease – charge on a capacitor, or number of undecayed nuclei.

Sample questions

$g = 9.8\,\mathrm{m\,s^{-2}}$
$c = 3.0 \times 10^{8}\,\mathrm{m\,s^{-1}}$
$e = -1.6 \times 10^{-19}\,\mathrm{C}$
$m_e = 9.1 \times 10^{-31}\,\mathrm{kg}$

1 A hydraulic system with fluid flowing through pipes, Figure 36.1, behaves like another system you have studied.

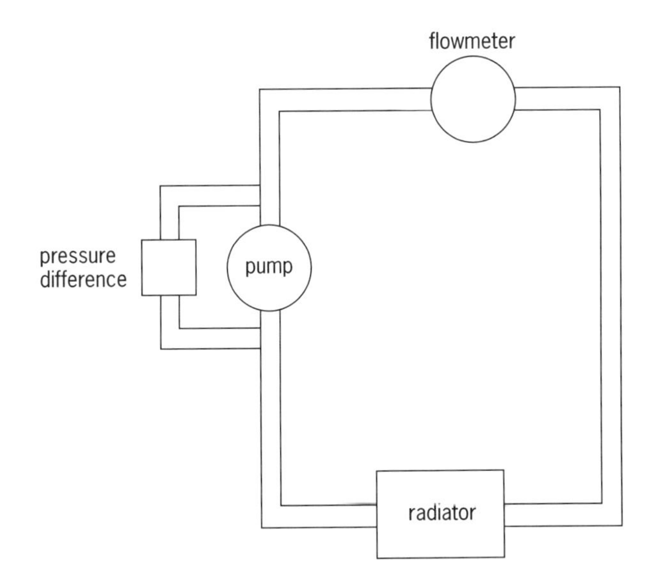

Figure 36.1

a What would be the equivalent, in this other system, of the volume rate of flow of fluid in a hydraulic system?

b What would be the equivalent, in this other system, of the pressure difference created by the pump in a hydraulic system?

c How do the 'cross-sectional area of the pipe' and the 'velocity of the fluid' in a hydraulic system relate to the 'volume flow per unit time'?

d Using your answers so far, or otherwise, obtain an expression for power delivered by a hydraulic system.

2 The movement of an electron beam in a cathode ray oscilloscope lets us make accurate measurements of the input signal. The oscilloscope shown in Figure 36.2 has the voltage sensitivity set on 200 mV/div.

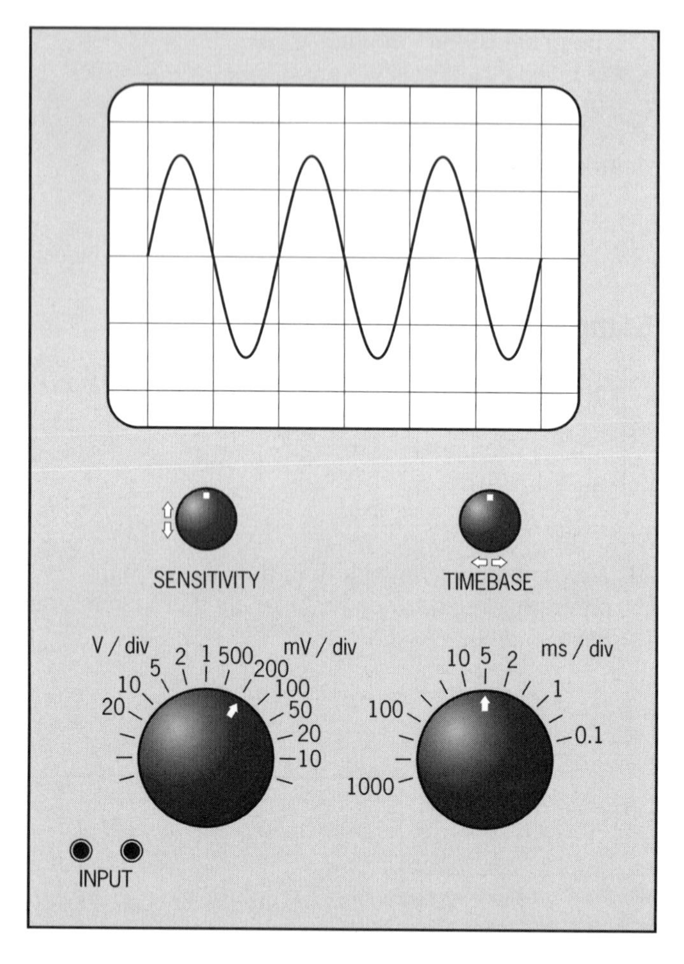

Figure 36.2

a What is the peak-to-peak voltage of the signal?
b What is the amplitude of the beam motion at the screen?

The time base is set on 5 ms/div.

c How long does one complete cycle take? *(This is the time period, T.)*
d Calculate the frequency of the signal in Hz. *(Don't forget the time is in ms.)*
e Write an equation describing the motion of the beam at the screen.
f Write an equation describing the variation of the voltage of the electrical signal.

Or you might simply be asked to describe the motion quantitatively *without being led through these steps.* ***Quantitative*** *means including numerical values.* ***Qualitative*** *means describe in words, which is difficult to do well – so practise!*

3 Figure 36.3 shows a trace from a cathode ray oscilloscope used to measure the speed of an electrical pulse travelling in a cable.

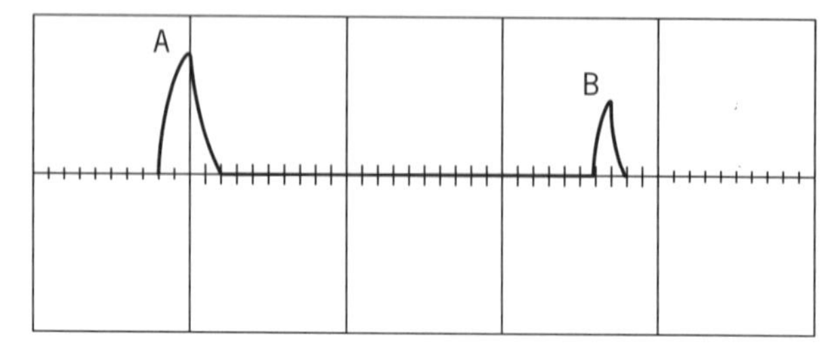

Figure 36.3

A is the original pulse and B is the pulse received after travelling down a 300 m length of the cable and returning after reflection from the far end.

a The time base of the oscilloscope is set to 1 μs/div. Calculate the speed of the pulse.
b Give an example of the use of a pulse of known speed that is used to find distances to, or to create a picture of, an object which reflects the waves.

4 Study the diagram of the device in Figure 36.4 and try to decide:

a what it is measuring
b what type of output signal is produced
c how the change in output is produced.

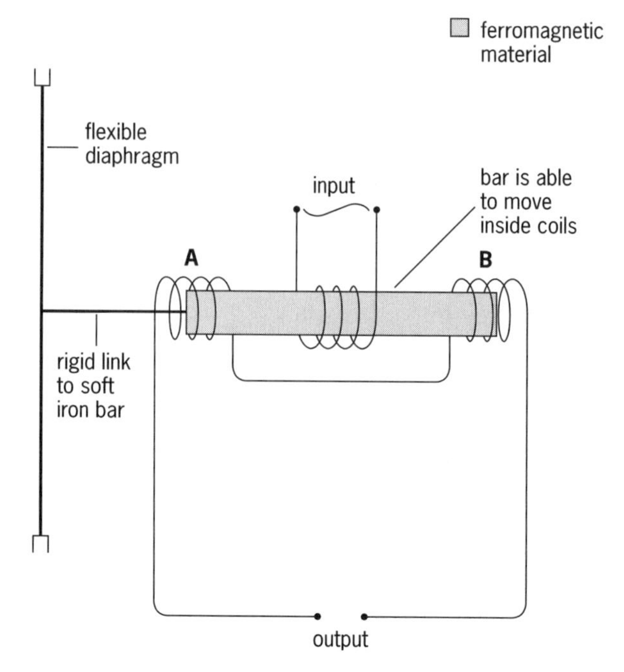

Note: the direction of emf induced in **A** is opposite to that of **B**.

Figure 36.4

5 A helium balloon of mass 1.0 g (when empty) contains 0.36 g of gas with a total volume of $2.0 \times 10^{-3}\,m^3$. Archimedes' Principle tells us that the upthrust on an object immersed in a fluid (gas or liquid) is equal to the weight of the fluid displaced.

The density of air is $1.3\,kg\,m^{-3}$.

a Calculate the upthrust on the balloon.
b Calculate the resultant force on the balloon.
c Calculate the acceleration of the balloon that this will produce.

6 An electron beam enters the space between a pair of horizontal parallel plates, Figure 36.5. A potential difference between the plates produces a vertical deflecting electric field.

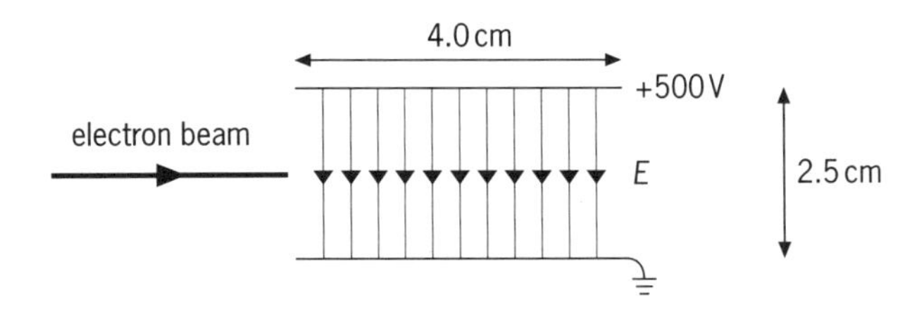

Figure 36.5

The speed of the electrons is $1.0 \times 10^8\,m\,s^{-1}$ and the plates are 4.0 cm long.

a What is the time spent by an electron between the plates?

The potential difference across the plates is 500 V and they are 2.5 cm apart.

b Calculate the electric field strength, E.
c Find the force on an electron in this field and justify ignoring the weight of the electron.
d Find the vertical acceleration of the electron.
e Find the distance travelled in the vertical direction, during the time the electron is between the plates.

Remember that vertical and horizontal motions are independent.

f Give an example of *similar* motion of an object but in a *different* type of field.

7 Study the diagram of the device in Figure 36.6 and try to decide:

a what it is measuring
b what type of output signal is produced
c how the change in output is produced.

• electrical connections

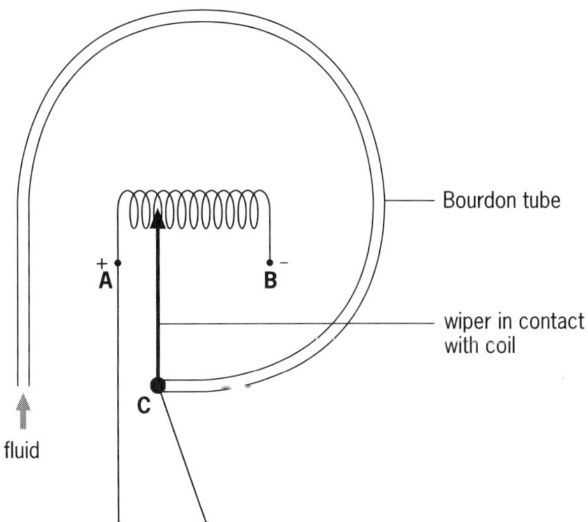

Figure 36.6

8 The electron beam current in a cathode ray oscilloscope is 0.12 mA and the electrons are accelerated through a potential difference of 1.5 kV. Calculate:

a the number of electrons striking the screen per second
b the momentum of each electron
c the average force experienced by the screen from the electrons
d the energy transferred to the screen each second.

Describe a similar situation in a different branch of physics.

9 The material PZT is a piezoelectric crystal. This means that when it is squashed or stretched a charge imbalance is produced and a potential difference is created across the crystal.

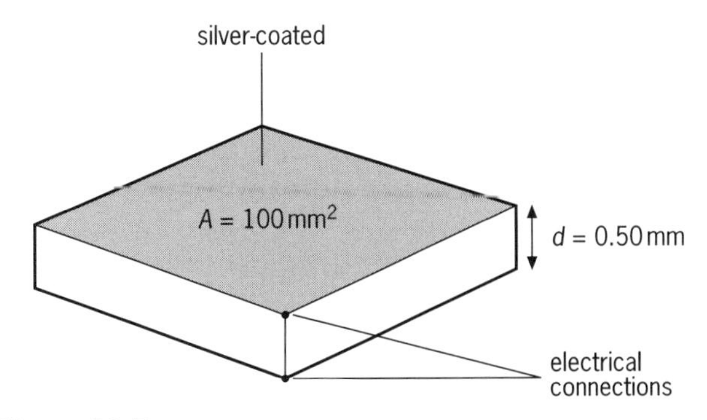

Figure 36.7

A PZT crystal produces $265\,pC\,N^{-1}$ when subjected to a force and has a value of relative permittivity, $\epsilon_r = 1500$ ($\epsilon_0 = 8.8 \times 10^{-12}\,F\,m^{-1}$). The end faces of the crystal each have an area of $100\,mm^2$. The faces are coated with silver and separated by 0.50 mm, Figure 36.7.

Calculate the change in potential difference between the end faces when the pressure on the end faces changes by 1 kPa.

10 The power of a motor may be measured by making it do work against friction. A motor turns a copper cylinder of mass 0.45 kg and radius 0.040 m against a friction band linked to two forcemeters. At 4.0 Hz the readings on the forcemeters are as shown in Figure 36.8.

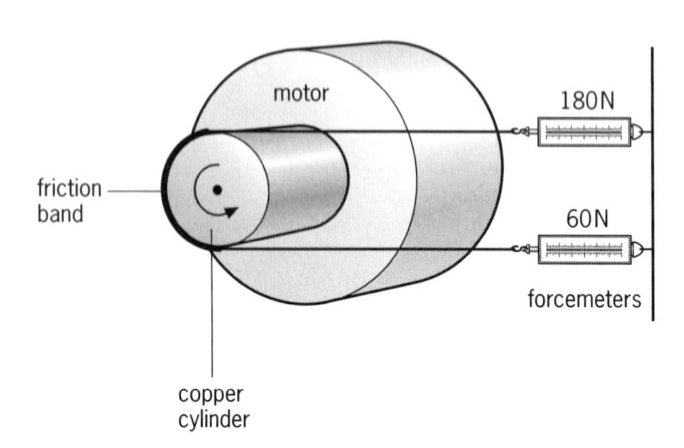

Figure 36.8

a Calculate the work done per revolution.
b Calculate the power of the motor.
c What will happen to the copper cylinder? The specific heat capacity of copper = 390 J kg^{-1} K^{-1}.

11 Figure 36.9 shows the trail of a beta particle in a cloud chamber as it passes through a lead sheet.

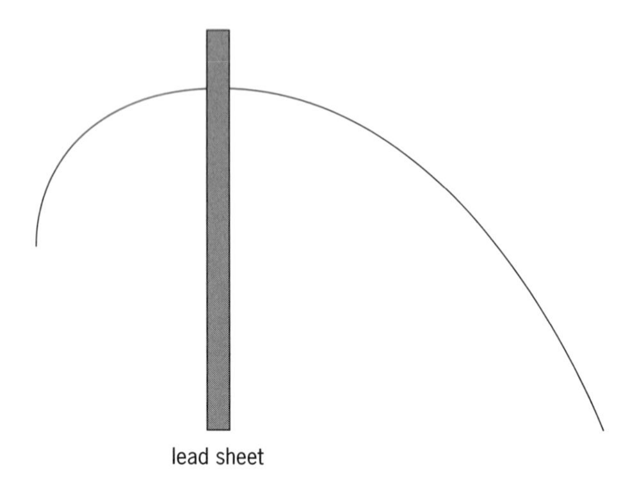

Figure 36.9

a Why is there a visible trail?
b There is a magnetic field directed into the paper. Is the particle travelling from left to right or right to left? Explain your answer.

Hint: Where is it travelling faster?

c Does this beta particle carry a positive or a negative charge? Explain your answer.

Synoptic answers

1 a The volume rate of flow, the volume of liquid flowing past a point in the system per second, or V/t, is equivalent to an electric current which is rate of flow of charge or Q/t.

b The pressure difference set up by the pump, ΔP, between the ends of a tube, makes the fluid flow. This is equivalent to the potential difference between the ends of an electrical conductor, which makes charge flow.

c The volume per unit time flowing through a tube is equal to the area of cross-section × the velocity of the fluid:

$$\frac{V}{t} = A \times v$$

See Figure 36.10.

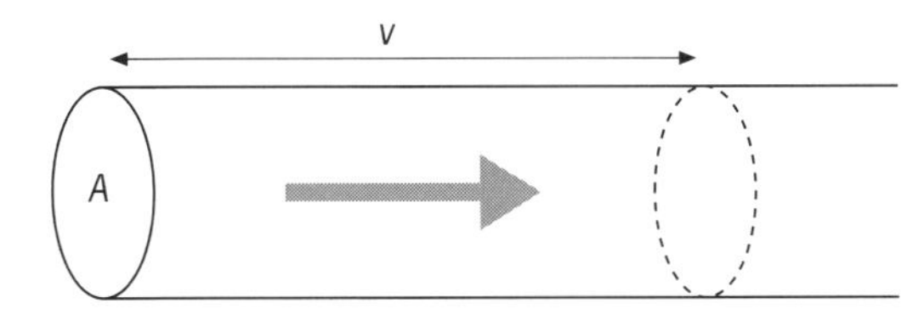

Figure 36.10

d Electrical power = potential difference × current = VI. So, by analogy, for the hydraulic case,

$$\text{power} = \text{pressure difference} \times \text{volume flow rate} = \Delta P \times \frac{V}{t}$$

But pressure $= \dfrac{\text{force}}{\text{area}}$, and $V/t = A \times v$, so

$$\text{power} = \frac{F}{A} \times A \times v = F \times v$$

This equation, power = force × velocity, should be familiar to you from mechanics.

2 a A 'division' is one square on the screen. It usually has sides 10 mm long. The trace is 3 squares from top to bottom. At 200 mV per square, this means that the peak-to-peak voltage is 3 div × 200 mV/div = **600 mV**.

b The beam moves with an amplitude ($\frac{1}{2}$ peak-to-peak) of **15 mm = 0.015 m** or **$1\frac{1}{2}$ divisions** on the screen.

c One complete cycle of the motion shown in the trace takes 2 squares (divisions) of the time (horizontal) axis so, at 5 ms/div, the time taken is

$$2\text{ div} \times 5\text{ ms/div} = \mathbf{10\,ms} = 10 \times 10^{-3}\text{ s}$$
$= T$, the time period

d Frequency, $f = \dfrac{1}{T}$

$$= \frac{1}{10 \times 10^{-3}\text{ s}} = 100\text{ s}^{-1} = \mathbf{100\,Hz}$$

If you did not get 100 as the answer, check your calculator key presses. The sequence should be

[1] [÷] [10] [EXP (or EE)] [+/−] [3] [=]

or [10] [EXP] [+/−] [3] [1/x]

or, as $10 \times 10^{-3} = 1 \times 10^{-2}$, you could press

[1] [EXP] [+/−] [2] [1/x]

(1/x may be a 2nd function)

e The motion is described by the equation for simple harmonic motion,

$$x = x_0 \sin(2\pi f t)$$

where x_0 is the amplitude and f is the frequency. In this example

$$x = 0.015 \times \sin(200\pi t)$$

f The electrical signal has the same frequency and the amplitude is half the peak-to-peak voltage, so

$$V = V_0 \sin(2\pi f t) = 300 \sin(200\pi t)\text{ mV}$$

3 a The pulse will have travelled a total distance of 2 × 300 m = 600 m down the cable and back. The total time shown on screen between the peaks of the pulses is

$$2.7\text{ div} = 2.7\,\mu\text{s} = 2.7 \times 10^{-6}\text{ s}$$

$$\therefore \text{pulse speed} = \frac{\text{distance}}{\text{time}} = \frac{600\text{ m}}{2.7 \times 10^{-6}\text{ s}}$$
$$= \mathbf{2.2 \times 10^{8}\,m\,s^{-1}}$$

(This result is used to demonstrate that electrical signals in wires travel at speeds close to the speed of light.)

b A description is required of the use of *one* example out of radar, sonar echo-sounding, ultrasound scans of babies, or a bat's echo-location technique.

4 a This device responds to movements of the diaphragm, which are related to changes in **pressure**.

b The output coils, A and B, will have induced alternating currents in them because there is fluctuating magnetic flux linkage along the iron bar from the alternating input. The output signal will be a **change in the induced a.c. voltage**.

c If the ferromagnetic core moves, one coil will have more material inside and the other will have less. This will change the size of the induced emfs. (The coils are actually arranged so that the emfs oppose one another but give a small reading for 'zero' so that movement one way increases V and the other way decreases it.)

5 a To calculate the upthrust we need to know the weight of the air displaced by the balloon. The volume of air displaced will be the volume of the balloon.

$$\text{density} = \frac{\text{mass}}{\text{volume}}$$

which you can rearrange to give

mass = density × volume

In this example, mass of air $= 1.3\,\text{kg}\,\text{m}^{-3} \times 2.0 \times 10^{-3}\,\text{m}^3$
$= 2.6 \times 10^{-3}\,\text{kg}$

The lift (upthrust) is the weight of this displaced air

$mg = 2.6 \times 10^{-3}\,\text{kg} \times 9.8\,\text{m}\,\text{s}^{-2}$
$= \mathbf{25.5 \times 10^{-3}\,N = 25.5\,mN}$

b The weight of balloon plus gas
$= (1.0 \times 10^{-3} + 0.36 \times 10^{-3})\,\text{kg} \times 9.8\,\text{m}\,\text{s}^{-2}$
$= 13.3 \times 10^{-3}\,\text{N} = 13.3\,\text{mN}$

The resultant force is 25.5 mN (upwards) – 13.3 mN (downwards)
= **12.2 mN (upwards)**

c This unbalanced force acts on the mass of the balloon and gas. You can rearrange the equation force = mass × acceleration to give

$$\text{acceleration} = \frac{\text{force}}{\text{mass}}$$

$$= \frac{12.2 \times 10^{-3}\,\text{N}}{1.36 \times 10^{-3}\,\text{kg}}$$

$= \mathbf{9.0\,m\,s^{-2}}$ **(upwards)**

6 a The horizontal speed of the electrons, v_H = distance/time, which you can rearrange to give time = distance/v_H. In this example, horizontal distance = 4.0 cm = 0.040 m and $v_H = 1.0 \times 10^8\,\text{m}\,\text{s}^{-1}$. So the electrons are between the plates for a time given by

$$\text{time} = \frac{0.040\,\text{m}}{1.0 \times 10^8\,\text{m}\,\text{s}^{-2}} = \mathbf{4.0 \times 10^{-10}\,s}$$

b Electric field strength = potential gradient between the plates

The plates are separated by $d = 2.5\,\text{cm} = 0.025\,\text{m}$, and $V = 500\,\text{V}$. So

$$E = \frac{V}{d} = \frac{500\,\text{V}}{0.025\,\text{m}} = \mathbf{2.0 \times 10^4\,V\,m^{-1}}\ \textbf{(downwards)}$$

c The force on a charge in an electric field is given by $F = qE$, so for an electron in the field between the plates

$F = -1.6 \times 10^{-19}\,\text{C} \times 2.0 \times 10^4\,\text{V}\,\text{m}^{-1}$
$= \mathbf{-3.2 \times 10^{-15}\,N}$

The minus sign indicates that the force will be upwards.

The weight of an electron is approximately

$9 \times 10^{-31}\,\text{kg} \times 10\,\text{N}\,\text{kg}^{-1} = 9 \times 10^{-30}\,\text{N}$

which is 10^{15} times smaller than the electrical force and so can be ignored.

d You can rearrange the equation force = mass × acceleration to give

$$\text{acceleration} = \frac{\text{force}}{\text{mass}}$$

$$= \frac{3.2 \times 10^{-15}\,\text{N}}{9.0 \times 10^{-31}\,\text{kg}}$$

$= \mathbf{3.6 \times 10^{15}\,m\,s^{-2}}$

This is a very large acceleration!

e The electrons have *zero* vertical velocity when they first enter the space between the plates. While they are between the plates they are accelerated vertically upwards. The vertical displacement after time t is given by the standard equation of motion

$s = ut + \frac{1}{2}at^2$

where $u = 0$ and $t = 4.0 \times 10^{-10}\,\text{s}$ (from part a). So

$s = \frac{1}{2} \times 3.6 \times 10^{15}\,\text{m}\,\text{s}^{-2} \times (4.0 \times 10^{-10})^2\,\text{s}^2$
$= \mathbf{2.8 \times 10^{-4}\,m = 0.28\,mm}$

f You could give an example of a projectile travelling initially horizontally and acted upon by gravitational force. A ball rolling off the edge of a table is one example.

7 a This is a device that measures the **pressure of a fluid** (gas or liquid).

b The output is a change in potential difference between say, A and C, as the wiper moves.

c With increased pressure the Bourdon tube uncurls and the wiper moves along the coil towards A, altering the potential difference between A and C. If the pressure falls, the tube curls up more and the wiper moves towards B. As connected in Figure 36.6, the potential difference between A and C would increase as the pressure decreased.

8 a The electron beam current (0.12 mA) is equivalent to $0.12\,\text{mC}\,\text{s}^{-1}$. This current is produced by a flow of electrons, say N per second, each with a charge of $e = 1.6 \times 10^{-19}\,\text{C}$.

In this question there is no need to worry about the fact that electrons are negatively charged.

So the rate at which charge passes down the beam is

$\Delta q/\Delta t = N \times e = 0.12\,\text{mC}\,\text{s}^{-1}$

which you can rearrange to give

$$N = \frac{0.12 \times 10^{-3}\,\text{C}\,\text{s}^{-1}}{1.6 \times 10^{-19}\,\text{C}} = \mathbf{7.5 \times 10^{14}\,(electrons)\,s^{-1}}$$

b Momentum, $p = m_e \times v$. You have been given a value for m_e (at the start of the questions) so you need to calculate a value for v. Consider the energy of the electrons when they hit the screen. Their kinetic energy will be equal to the potential energy lost moving through the potential difference, so

$$\tfrac{1}{2}m_e v^2 = e\Delta V$$

which you can rearrange to give

$$v = \sqrt{\frac{2e\Delta V}{m_e}} = \sqrt{\frac{2 \times 1.6 \times 10^{-19}\,\text{C} \times 1.5 \times 10^3\,\text{V}}{9.1 \times 10^{-31}\,\text{kg}}}$$

$$= 2.3 \times 10^7\,\text{m s}^{-1}$$

So the momentum is

$$p = m_e v = 9.1 \times 10^{-31}\,\text{kg} \times 2.3 \times 10^7\,\text{m s}^{-1}$$
$$= \mathbf{2.1 \times 10^{-23}\,N\,s}$$

Can you show that $\sqrt{\frac{C \times V}{kg}} = m\,s^{-1}$*?*

c Average force on the screen = average rate of change of momentum

Assuming that the electrons lose all their forward velocity on hitting the screen, and using the answer to parts a and b,

change in momentum per second = electrons per second × momentum change of each

$$= N \times \Delta p$$
$$= 7.5 \times 10^{14}\,\text{s}^{-1} \times 2.1 \times 10^{-23}\,\text{N s}$$
$$= \mathbf{1.6 \times 10^{-8}\,N}$$

d Kinetic energy gained by each electron $= e\Delta V$

$$= 1.6 \times 10^{-19}\,\text{C} \times 1.5 \times 10^3\,\text{V}$$
$$= 2.4 \times 10^{-16}\,\text{J}$$

N electrons give up this energy each second, so the total rate of energy transfer is

$$7.5 \times 10^{14}\,\text{s}^{-1} \times 2.4 \times 10^{-16}\,\text{J} = \mathbf{0.18\,W}$$

Examples of similar situations could include: a hose pipe playing water onto a wall; water falling onto a water wheel; a stream of ping-pong balls from a gun; the kinetic theory of the pressure of a gas.

9 The crystal behaves as a parallel plate capacitor. The capacity of the crystal is therefore given by

$$C = \frac{\epsilon_0 \epsilon_r A}{d}$$

Using the figures given,

$$C = \frac{8.8 \times 10^{-12}\,\text{F m}^{-1} \times 1500 \times 100 \times 10^{-6}\,\text{m}^2}{0.50 \times 10^{-3}\,\text{m}}$$

$$= 2.6 \times 10^{-9}\,\text{F}$$
$$= 2.6\,\text{nF}$$

When the pressure changes there is a change in the force on the faces. $P = F/A$ can be rearranged to give $F = PA$, from which $\Delta F = \Delta P \times A$. For $\Delta P = 1$ kPa,

$$\Delta F = \Delta P \times A = 1.0 \times 10^3\,\text{Pa} \times 100 \times 10^{-6}\,\text{m}^2 = 0.10\,\text{N}$$

The crystal produces 265 pC N^{-1}, so for $\Delta P = 1$ kPa, the charge imbalance is

$$\Delta Q = 265\,\text{pC N}^{-1} \times 0.10\,\text{N} = 26.5 \times 10^{-12}\,\text{C}$$

The crystal has a capacitance of 2.6×10^{-9} F (from above) and you can rearrange the equation $C = Q/V$ to give $V = Q/C$, from which $\Delta V = \Delta Q/C$. So the change in potential, for a pressure change of 1 kPa, is

$$\Delta V = \frac{26.5 \times 10^{-12}\,\text{C}}{2.6 \times 10^{-9}\,\text{F}} = \mathbf{10\,mV}$$

This is the change in output of the crystal, which is a measure of the pressure change.

10 a In one revolution, a point on the surface of the cylinder moves a distance equal to its circumference, $2\pi r = 2\pi \times 0.040\,\text{m} = 0.25\,\text{m}$.

The frictional resisting force is the difference in the two readings of the forcemeters,

$$F_{\text{friction}} = 180\,\text{N} - 60\,\text{N} = 120\,\text{N}$$

The work done per revolution = force × distance $= 120\,\text{N} \times 0.25\,\text{m} = \mathbf{30\,J}$.

b The motor is running at 4.0 Hz, so each revolution takes 0.25 s and

$$\text{power} = \frac{\text{work}}{\text{time}} = \frac{30\,\text{J}}{0.25\,\text{s}} = \mathbf{120\,W}$$

You could get the same answer by saying there are 4 revs in 1 s, so the work done in 1 second = 4 revs × 30 J rev^{-1} *=* ***120 W****.*

c The energy transferred will raise the temperature of the copper cylinder. Assuming that all the energy goes into the copper block (which is unlikely) the temperature rise $\Delta\theta$ will be given by the equation $mc_c\Delta\theta = \Delta Q$, where m is the mass of the copper cylinder and c_c its specific heat capacity.

The *rate* of rise in temperature will be proportional to the rate of supply of energy or the power,

$$mc_c \frac{\Delta\theta}{t} = \text{power}, \Delta Q/t$$

which you can rearrange to give

$$\frac{\Delta\theta}{t} = \frac{\text{power}}{mc_c}$$

$$= \frac{120\,\text{W}}{0.45\,\text{kg} \times 390\,\text{J kg}^{-1}\,\text{K}^{-1}} = \mathbf{0.68\,K\,s^{-1}}$$

The temperature of the copper block will rise at a rate of 0.68 kelvin per second.

You may prefer to say ***in one second*** *120 J is supplied, so*

$$\Delta\theta = \frac{120\,\text{W}}{0.45\,\text{kg} \times 390\,\text{J kg}^{-1}\,\text{K}^{-1}} = 0.68\,\text{K}$$

11 a There is a visible trail because the fast-moving charged particle knocks electrons from the atoms of air in the chamber, forming ions. Water vapour then condenses on the ions, revealing the path followed. The path of an uncharged particle is not visible.

b Notice that the radius of the arc on the left of the lead sheet (Figure 36.9) is smaller than that of the arc on the right. The force of a magnetic field B on a charged particle is given by the equation $F = Bqv$. This force acts at right angles to the direction of motion of the particle and will force it to move in a circle. The centripetal force required to maintain a particle, of mass m and velocity v, in motion in a circle of radius r, is equal to mv^2/r. So

$$Bqv = \frac{mv^2}{r}$$

which you can simplify to give

$$Bqr = mv$$

From this equation you can see that the radius of the circular path, r, is proportional to the velocity, v, all other factors being constant. A smaller radius means the particle must have a smaller velocity, as it needs less force to change the direction of motion. It is harder to deflect a faster particle from its straight-line path. Your conclusion should be that the particle is going faster on the right of the lead sheet than it is on the left. The lead must have slowed the particle down so it must have been travelling from **right to left**. (The lead can't have made the particle speed up.)

c Using the **left hand rule**, you should be able to see that the particle must be charged **positively**.

These tracks were the first evidence for the existence of the positron.

Data analysis

Data analysis questions should not be a problem if you have processed experimental results in your course.

You should be well practised in obtaining straight line graphs. There is often a section explaining this at the start of practical manuals (for example *Advanced Practical Physics*, ed. L. Beckett, John Murray) or at some point in standard textbooks. If you want to show proportionality, don't forget that the straight line *must go through the origin*, so your graph must include the origin!

You should also be experienced at:

- measuring gradients (the rate of change of y with x) – remember to use *large* triangles and show the points that you used on the graph;
- estimating the area between a curve and the x-axis
- recognising turning points (maxima and minima).

Don't be put off by unfamiliar or complex equations. Decide which factors are constants and which are variables. Then look at the general shape or form of the equation and rearrange it, if possible, to give the form $y = mx + c$. Remember that y or x may represent a complicated function, and m or c may be an assemblage of several constants. For example,

$$Z^2 = \frac{1}{4\pi^2 f^2 C^2} + R^2$$

where only Z and C vary, will give a straight line if you plot Z^2 against $1/C^2$.

If you always plot the 'y' (left hand side) on the ordinate (upwards) and the variable from the right hand side on the abscissa (along), then the gradient will automatically equal the constant in front of the right hand variable. In the above example, the m, or gradient, is $1/4\pi^2 f^2$. The intercept on the y-axis is R^2. Often, the more complex the subject *seems*, the easier the actual questions are likely to be!

Some questions require you to take readings from one graph, perform some calculations and then plot a new graph. Care is needed at each stage.

Keep an eye on units, particularly of the values plotted on graphs, because the gradient will have these units. Don't worry if the units seem complicated, just quote them. Remember a ratio of two similar quantities does not have a unit. Neither does a logarithm.

When performing practical work, it will stand you in good stead if you think about what you are doing, and don't just follow instructions. A favourite question, for examiners, is for you to design further investigations. You need to have a realistic appreciation of what it is possible to measure *directly* and what will need more subtle approaches, such as the projection of a shadow. Take note if your teacher says you can't use such-and-such a piece of equipment. Imagine yourself ordering the equipment, chosen from what you know is available in your laboratory. Then think through setting up the apparatus and completing the measurements. Don't be afraid of stating the obvious, such as stopping a stopwatch when the event ends, or pulling the blinds down if the experiment needs a dark room.

Sample questions

1 The resistance, R, of a coil of wire at temperature t can be expressed by the equation $R_t = R_0(1 + \alpha t)$ where R_0 is the resistance at 0 °C and α is the change in resistance per ohm per degree temperature change.

This equation crops up in a number of other places; for example in thermal expansion, it appears as $l_t = l_0(1 + \alpha t)$ where α is the change in length per unit length per degree temperature change.

Use the data in Table 36.1 to test the validity of this equation by drawing a graph, and then obtain a value for α from the graph you have drawn.

Table 36.1

R_t/Ω	t/°C
23.7	14
25.2	29
26.4	44
27.8	58
29.8	81
31.7	100

2 The power, P, in watts, dissipated by a wire filament of resistance R is supposed to be given by a power law of the form $P = kR^n$ where k and n are constants. (In fact, n should be an integer.) The behaviour is shown in the curve of power plotted against resistance, Figure 36.11.

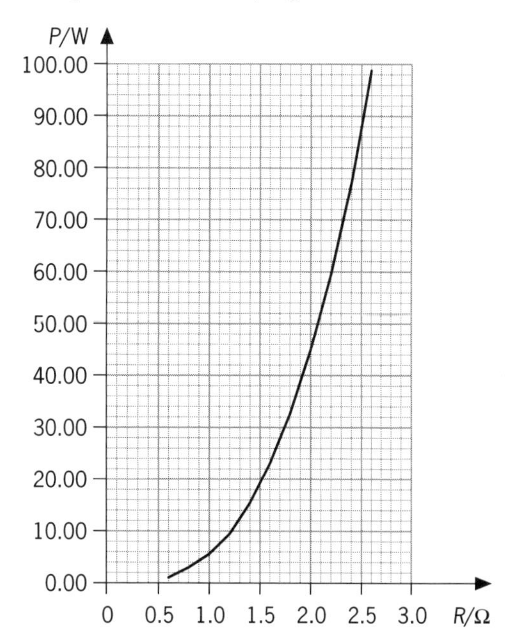

Figure 36.11

Prepare a table of five values of P and R taken from this graph. Then prepare a new table which will allow you to demonstrate that the expression above is followed, by obtaining a straight line graph. Use this graph to find the values of k and n.

1 You know that R_0 and α are constants and R_t and t are variables. If you expand the expression to give $R_t = R_0 + R_0\alpha t$ and then rearrange to

$$R_t = R_0\alpha t + R_0$$

you should see that this has the form $y = mx + c$ with $m = R_0\alpha$ and $c = R_0$. So a graph of R_t against t, Figure 36.11, should be a straight line with intercept on the y-axis R_0.

Remember to include the origin on the x-axis in order to be able to find the intercept on the y-axis.

You need to plot the values given, then measure the intercept to give R_0. Find the gradient using a *large triangle*. To find α, divide the gradient by the intercept, $R_0\alpha/R_0 = \alpha$.

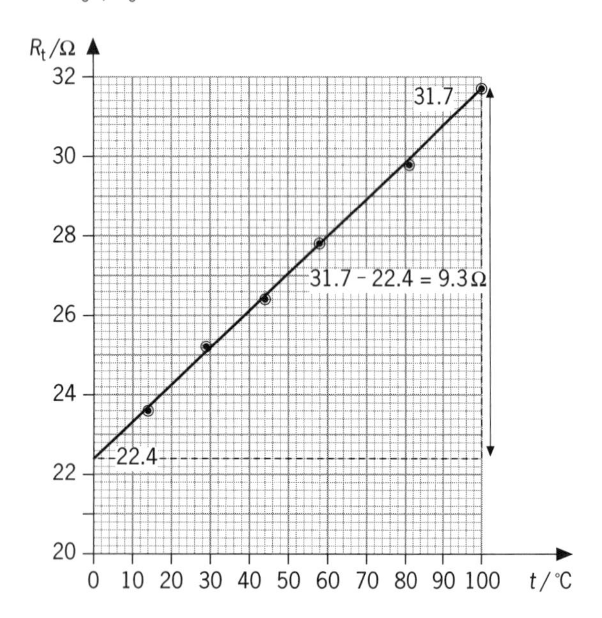

Figure 36.12

The intercept is at $22.4\,\Omega = R_0$.

The gradient is

$$\frac{9.3\,\Omega}{100\,°\text{C}} = 9.3 \times 10^{-2}\,\Omega\,°\text{C}^{-1} = R_0\alpha$$

So $$\alpha = \frac{9.3 \times 10^{-2}\,\Omega\,°\text{C}^{-1}}{22.4\,\Omega} = \mathbf{4.2 \times 10^{-3}\,°C^{-1}}$$

Note the units! Check by looking back at the equation. Each term in $R_t = R_0 + R_0\alpha t$ must be a resistance so the unit of α must be reciprocal temperature to cancel with the temperature t in the term $R_0\alpha t$. The 'change in resistance per ohm per degree' quoted previously was to clarify the significance of α.

2 To demonstrate that the rule is followed and to find the complete equation, a straight-line graph must be plotted. To obtain a straight-line graph from a power law you have to use logarithms. When logs (to base 10) are taken on both sides, the equation becomes

$$\log P = \log k + n \log R$$

which rearranges to give

$$\log P = n \log R + \log k$$

Because $\log k$ and n are constants, this has the form $y = mx + c$.

When $\log P$ is plotted against $\log R$, a straight line should be obtained; $\log k$ can be found from the intercept on the y-axis and n from the gradient.

Table 36.2 shows the format your table should have, with some sample values.

Table 36.2

P/W	log P	R/Ω	log R
9.7	0.99	1.2	0.079
22.9	1.36	1.6	0.200

Choose values of P that range from below 10, spread fairly evenly up to the maximum possible, but choose points which look clear on the grid. It doesn't matter which type of log is taken but $\log_{10}$ (log to the base 10) is more usual.

Figure 36.13 shows the resulting graph.

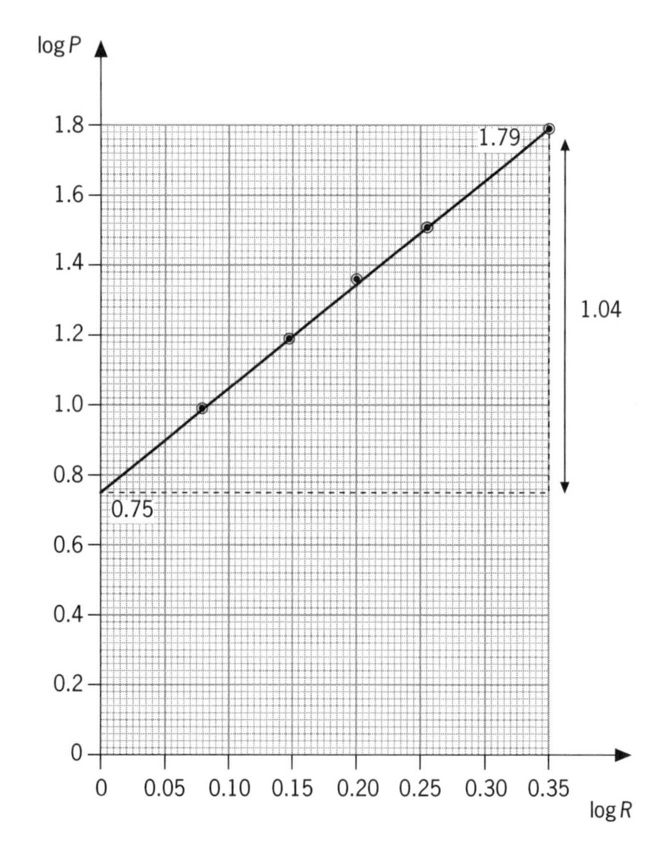

Figure 36.13

The intercept on the y-axis is at $\log P = 0.75 = \log k$

$\therefore k = \text{inverse} \log_{10} 0.75 = \mathbf{5.6}$

The gradient $= \dfrac{1.04}{0.35} = 2.97$, so $\boldsymbol{n} = \mathbf{3}$.

(It was supposed to be an integer.)

So the law has been shown to be $\boldsymbol{P = 5.6R^3}$.

Comprehension passages

The best practice for this type of question is to read widely as you study. You should have several textbooks and should read one of the popular scientific magazines. When reading, check your understanding. Have a dictionary handy and build up your scientific vocabulary. You might like to keep a special vocabulary notebook.

If there are calculations or algebraic manipulations within a piece of text, check them yourself. (It is surprising how many errors you find!)

Where there are graphs, look closely at the scales and axes to be sure you know what they are showing and test yourself by estimating the co-ordinates of points. This is tricky with logarithmic scales. If you have to take readings from a graph with a logarithmic scale, remember that the distance between numbers gets smaller as the number gets larger; see Figure 36.14.

Figure 36.14 A logarithmic scale

If you have to find somewhere between 1 and 10, 10 and 100, or 100 and 1000, it is helpful to remember that

$\log_{10} 2$ is close to 0.3
$\log_{10} 4$ is close to 0.6
$\log_{10} 5$ is close to 0.7
$\log_{10} 8$ is close to 0.9

These will give you a good indication where intermediate points on a log scale lie. Don't be put off by unfamiliar topics; you should have confidence that you can grasp the main ideas.

In a comprehension passage there will not be any questions on prior knowledge except topics covered by the specification or assumed background knowledge. It is best to read the whole passage first, as sometimes there may be help with a question set on the first paragraph further on in the article. You may be asked for a definition of any words covered by the specification.

Sample questions

1

Rubber

The two most striking mechanical properties of rubber are (*a*) its range of elasticity is great – some rubbers can be stretched to more than ten times their original length (i.e. 1000% strain) before the elastic limit is reached, and (*b*) its value of the Young modulus is about 10^4 times smaller than most solids and *increases* as the temperature rises, an effect not shown by any other material ... [This rise] can be attributed to the greater disorder among the chains when the material is heated; their resistance to alignment by a stretching force therefore increases ... if a sample of stretched rubber is 'photographed' by a beam of high-energy electrons, sharp spots are obtained similar to those produced by X-rays and a crystal.

(Sample passage from *Advanced Physics* Fifth Edition, Tom Duncan, John Murray, page 36)

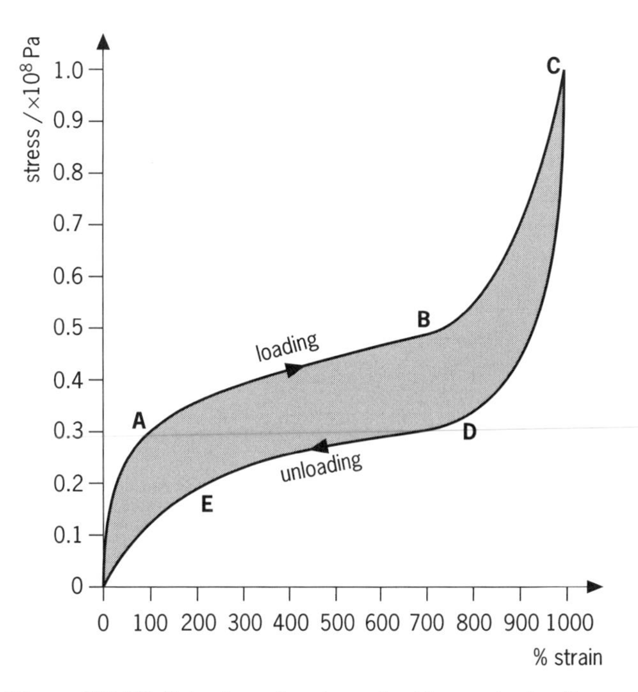

Figure 36.15 Behaviour of a piece of rubber under loading and unloading

The phrases underlined in the passage are technical words covered by most specifications, so definitions and explanations may be tested.

a Explain why the Young modulus does *not* increase with temperature for other substances, for example metals.

b If the Young modulus of steel is 210 GPa, estimate the Young modulus of rubber.

The figure '10^4 times smaller' invites a quick calculation here.

c The word 'photographed' is in inverted commas in the passage because it is not strictly the correct term. What would be a better expression?

d At stress 0.3×10^8 Pa, how many times greater is the strain upon contracting than upon initial stretching? (See Figure 36.15.)

e Comment on the *different gradients* shown in the graph.

f What does the *shaded area* of the graph show?

g A *small* area enclosed on the stress–strain graph means the rubber is 'resilient'. What are the implications of a large area enclosed?

Here you are expected to establish a link between two pieces of information in the text.

2

X-rays are high-energy photons. They are used in medicine for diagnosis and therapy. In an X-ray tube (Figure 36.16) electrons are emitted thermionically and are accelerated through a high potential difference to hit a target, usually of tungsten, placed in their path. X-rays are emitted as the electrons hit the target.

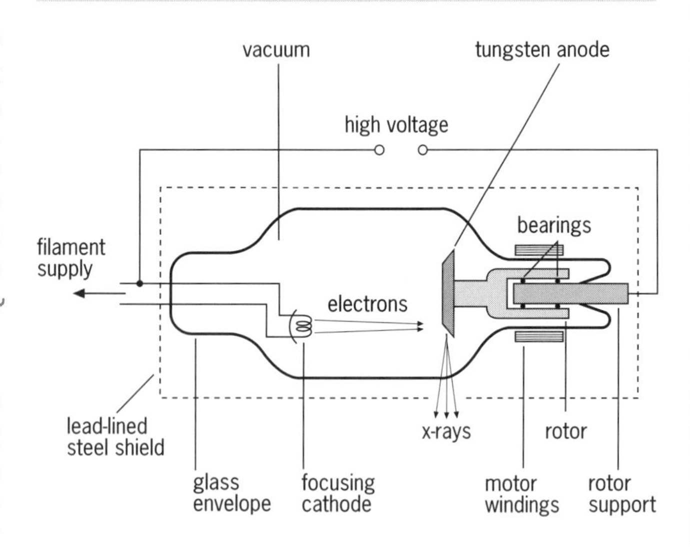

Figure 36.16 X-ray tube

Increasing the voltage across an X-ray tube increases the maximum photon energy. The quality or penetrating power is described by the Half Value Thickness, HVT or $x_{1/2}$ – the thickness of material needed to cut the intensity of the beam to one half of its original value. $x_{1/2} = 0.693/$, where is called the attenuation coefficient.

You are supposed to recognise the similarity with radioactive equations.

The variation of *m* with X-ray photon energy is shown in Figure 36.17. *Study the axes.*

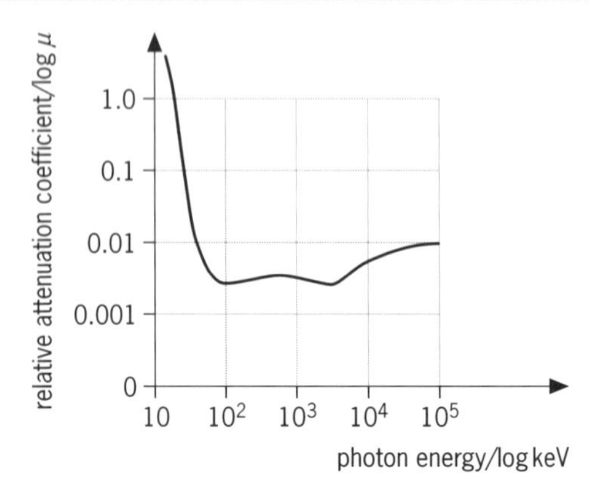

Figure 36.17 Variation of attenuation coefficient with X-ray photon energy

Quality may be altered by using filters, low energy photons are readily absorbed so the average energy of the beam increases and hence the HVT.

Be suspicious of 'hence'; you may be asked to explain or justify it.

Superficial radiotherapy does not need the X-rays to penetrate the skin. Radiotherapy requires the beam to get deep inside the body and for radiography it must pass right through soft tissue. The tube current affects the intensity but not the penetrating power. The overall intensity is proportional to the current but to the *square* of the tube voltage.

It was found that there were two parts to the spectrum of X-rays, a continuous part known as *Bremsstrahlung* (braking radiation) and intense emission at certain frequencies characteristic of the target material.

You are expected to recognise this description of spectral lines.

An electron in an inner shell of the target atom is ejected. Where there is a vacancy at a lower energy level, a higher energy electron may drop down with the emission of an X-ray photon. Below a certain tube potential no lines are emitted because the incident electrons do not have enough energy.

a Does a high energy photon have a longer or a shorter wavelength than a low energy photon?

This tests your knowledge of the equation for the energy of a photon and also the wave equation showing an inverse relationship between frequency and wavelength.

b Describe how an X-ray photograph (strictly a shadowgram) may help in diagnosis of a broken bone.

You are expected to know about everyday applications.

c Explain the meaning of the underlined words.

Look at the makeup of the words and make an intelligent guess if you are not sure.

d Explain why increasing the accelerating voltage increases the maximum energy of the X-rays.

This tests knowledge of basic energy changes.

e Why does the tube current affect the intensity of the beam but not the penetrating power?

This tests your understanding, probably based on the photoelectric effect if you have not covered a medical physics option.

f Write an equation linking the energy of a line spectrum photon with the energy levels in the target atoms.

This tests basic understanding of line spectra.

g From the description of 'Half Value Thickness' you should be able to suggest an equation relating the intensity I after a beam has travelled through x metres of absorber.

You are expected to work it out from the hints and parallels with the radioactive equations.

h From the equation in part g give details of a graph which could be drawn to enable μ to be found with some accuracy.

This is a standard technique so make sure you are familiar with it.

i The most energetic X-rays produced in a certain tube have a wavelength of 2.0×10^{-11} m. What is the operating pd of the tube? ($h = 6.6 \times 10^{-34}$ J s)

This tests your use of the energy equations from electrical energy to mechanical to photon energy.

j A tube, which is 1% efficient, produces X-ray energy at 25 J s^{-1}. Calculate the tube current if the pd is 50 kV.

Efficiency and power equations are needed but you also have to recognise that the tube current and voltage determine the power available.

k Explain why the Half Value Thickness increases with filtration of the beam (third paragraph).

Justify the 'hence'.

l Using Figure 36.17:
i By appproximately what factor does the attenuation coefficient fall as the photon energy increases from 10 keV to 100 KeV?
ii There are two points at which the attenuation falls to a minimum. At roughly what photon energy is the higher energy minimum?

Practise reading log scales – refer to Figure 36.14.

1 **a** In other substances such as metals, the increased vibrations and energy of the atoms will weaken the interatomic bonding and make it easier to extend a regular array of atoms.

When rubber is stretched, at first the tangled chains of molecules straighten; the change in length of the specimen does not involve interatomic forces.

b For steel the Young modulus is 210×10^9 Pa; in standard form this is 2.1×10^{11} Pa. For rubber it will be about 10^4 times smaller, of the order of 10^{11} Pa/10^4 = **10^7 Pa**.

c You would be expected to know that patterns of spots are obtained when X-rays are diffracted by the atoms of a crystal or other ordered substance, so the photograph records the **diffraction pattern** of the electrons.

d Comparing about 700% with 100% (from the graph) gives **7 times greater**.

e Clearly the rubber is harder to stretch at first and at the end, but easier in the central portion where a large strain is produced for little change in stress.

f It is the difference between the energy supplied on stretching (area between OABC and the strain axis) and the energy released by the rubber upon contraction (area below OEDC). So it is the **energy retained** within the rubber.

g A large area means that a large amount of energy remains in the rubber; the rubber gets hot. Thus if a rubber that is not resilient is used for a tyre, the tyre could overheat.

2 **a** $E = hf = hc/\lambda$
The energy is inversely proportional to the wavelength so if the energy is large the wavelength must be small.

b A shadow photograph is taken. Bone absorbs the X-rays but tissue does not. The shape of the bone is seen as light on a dark background as the X-rays do not get through to blacken the film.

c Photon – a quantum of electromagnetic radiation.
Thermionically – (electrons are) emitted from a heated filament.

You could guess 'therm' indicates heat; 'ion' indicates a charged particle.

Quality – the penetrating power.

(X-rays with high penetrating power are called 'hard'; with lower penetrating power they are called 'soft'.)

Superficial – on the surface.

(In the treatment of skin cancer.)

Radiotherapy – treatment (therapy, *but don't repeat the word from the text*) using X-rays (or gamma rays).
Radiography – taking pictures using X-rays (or gamma rays).
Characteristic – unique to that element.

(As are all line spectra.)

d The electrons will gain more kinetic energy and be travelling faster when they hit the target. If this increased energy is transferred to an electron in the target a higher energy photon can be emitted.

e The tube current is proportional to the number of electrons travelling across the tube. If there are more electrons there will be more X-rays emitted but the energy (hence penetrating power) of each will be the same unless the tube voltage increases.

f $E_1 - E_2 = hf$
The difference between energy levels in the target atom equals the photon energy.

A simple diagram would help (see Figure 36.18).

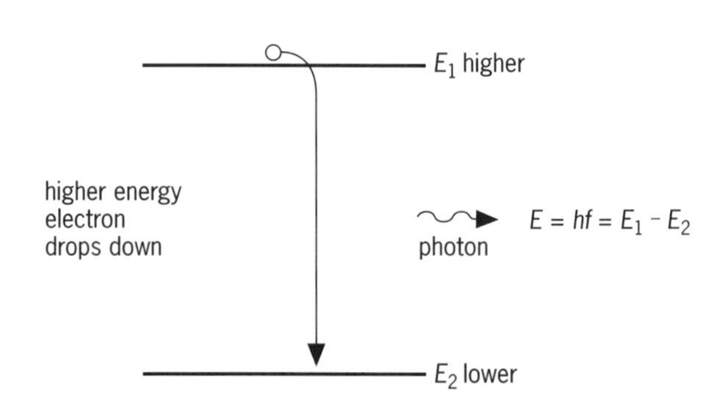

Figure 36.18

g Half-values are used with exponential relationships. Clearly the intensity will decrease with distance so try

$$I = I_0 e^{-\mu x}$$

h If a graph of natural log intensity is plotted against thickness of absorber then μ is found from the gradient (see Figure 36.19), since

$$\log_e I = -\mu x + \log_e I_0$$

Compare with $y = mx + c$.

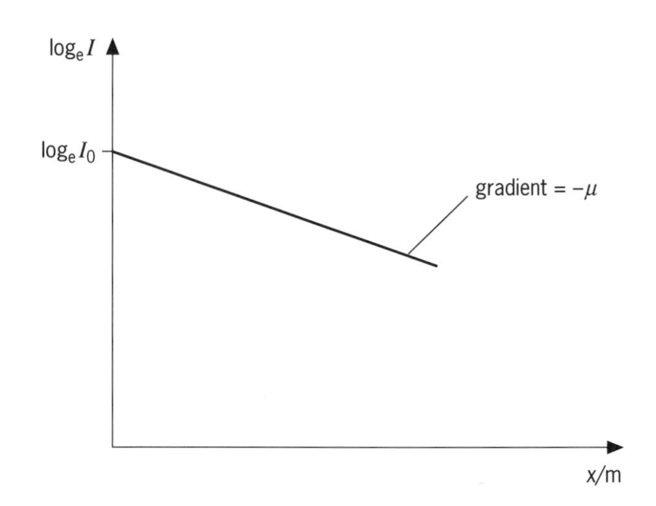

Figure 36.19

Note that $\log_e$ is sometimes written ln.

i Energy of photon = maximum electron energy gained by passing through the potential difference:

$$\frac{hc}{\lambda} = eV$$

$$V = \frac{hc}{e\lambda} = \frac{6.6 \times 10^{-34}\,\text{J s} \times 3.0 \times 10^{8}\,\text{m s}^{-1}}{1.6 \times 10^{-19}\,\text{C} \times 2.0 \times 10^{-11}\,\text{m}}$$

$$= \mathbf{62\,kV}$$

j Efficiency of 1% gives 25 W, so 100% = 2500 W.

$$V_{\text{tube}} \times I_{\text{tube}} = \text{power} = 2500\,\text{W}$$

$$\therefore I_{\text{tube}} = \frac{2500\,\text{W}}{50\,000\,\text{V}} = \mathbf{0.05\,A}$$

k *(You are expected to see that)* the more energetic X-rays travel further within the absorber. If low energy photons are removed and the average energy increases then the penetrating power of that beam should increase too.

l **i** Falling from 1.0 to 0.01 would be a fall of a factor of 100; in fact the curve drops from almost 10 to well below 0.01, so **by a factor of 1000** up to a factor of about 3000.

ii The X-ray photon energy is about halfway between 10^3 and 10^4 keV *(and using the scale in Figure 36.14 the number nearest to halfway is 3)* so the photon energy is roughly 3000 keV, up to 4000 keV.

Note that no matter what the subject matter, the same type of questions are likely to be asked.

Index

Italic entries indicate references to a question. **Bold** entries indicate references to a figure